7

Springer Series on Fluorescence

Methods and Applications

Series Editor: O.S. Wolfbeis

For further volumes:
http://www.springer.com/series/4243

Springer Series on Fluorescence

Series Editor: O.S. Wolfbeis

Recently Published and Forthcoming Volumes

Advanced Fluorescence Reporters in Chemistry and Biology III
Applications in Sensing and Imaging
Volume Editor: A.P. Demchenko
Vol. 10, 2011

Advanced Fluorescence Reporters in Chemistry and Biology II
Molecular Constructions, Polymers and Nanoparticles
Volume Editor: A.P. Demchenko
Vol. 9, 2010

Advanced Fluorescence Reporters in Chemistry and Biology I
Fundamentals and Molecular Design
Volume Editor: A.P. Demchenko
Vol. 8, 2010

Lanthanide Luminescence
Photophysical, Analytical and Biological Aspects
Volume Editors: P. Hänninen and H. Härmä
Vol. 7, 2011

Standardization and Quality Assurance in Fluorescence Measurements II
Bioanalytical and Biomedical Applications
Volume Editor: Resch-Genger, U.
Vol. 6, 2008

Standardization and Quality Assurance in Fluorescence Measurements I
Techniques
Volume Editor: U. Resch-Genger
Vol. 5, 2008

Fluorescence of Supermolecules, Polymeres, and Nanosystems
Volume Editor: M.N. Berberan-Santos
Vol. 4, 2007

Fluorescence Spectroscopy in Biology
Volume Editor: M. Hof
Vol. 3, 2004

Fluorescence Spectroscopy, Imaging and Probes
Volume Editor: R. Kraayenhof
Vol. 2, 2002

New Trends in Fluorescence Spectroscopy
Volume Editor: B. Valeur
Vol. 1, 2001

Lanthanide Luminescence

Photophysical, Analytical and Biological Aspects

Volume Editors:
Pekka Hänninen · Harri Härmä

With contributions by

T. Ala-Kleme · J.-C.G. Bünzli · H. Bazin · S.V. Eliseeva · S. Faulkner · P. Hänninen · H. Härmä · I. Hemmilä · I. Hyppänen · J. Kankare · S. Kulmala · A. Kuusisto · V. Laitala · G. Mathis · T. Nann · M. Schäferling · T. Soukka · C. Spangler · U.-H. Stenman · J. Suomi · D. Sykes · H.J. Tanke · P.A. Tanner · H.-Q. Wang · M.H.V. Werts

Springer

Volume Editors
Prof. Dr. Pekka Hänninen
Dr. Harri Härmä
University of Turku
Department of Cell Biology and Anatomy
Tykistökatu 6A
20520 Turku, Finland
pekka.hanninen@utu.fi
harri.harma@utu.fi

ISSN 1617-1306 e-ISSN 1865-1313
ISBN 978-3-642-21022-8 e-ISBN 978-3-642-21023-5
DOI 10.1007/978-3-642-21023-5
Springer Heidelberg Dordrecht London New York

Library of Congress Control Number: 2011930252

Cover design: WMXDesign GmbH, Heidelberg, Germany

Printed on acid-free paper

Springer is part of Springer Science+Business Media (www.springer.com)

Series Editor

Prof. Dr. Otto S. Wolfbeis

Institute of Analytical Chemistry
Chemo- and Biosensors
University of Regensburg
93040 Regensburg
Germany
otto.wolfbeis@chemie.uni-regensburg.de

Aims and Scope

Fluorescence spectroscopy, fluorescence imaging and fluorescent probes are indispensible tools in numerous fields of modern medicine and science, including molecular biology, biophysics, biochemistry, clinical diagnosis and analytical and environmental chemistry. Applications stretch from spectroscopy and sensor technology to microscopy and imaging, to single molecule detection, to the development of novel fluorescent probes, and to proteomics and genomics. The *Springer Series on Fluorescence* aims at publishing state-of-the-art articles that can serve as invaluable tools for both practitioners and researchers being active in this highly interdisciplinary field. The carefully edited collection of papers in each volume will give continuous inspiration for new research and will point to exciting new trends.

Preface

Lanthanides have fascinated scientists for more than two centuries now – first starting at the end of eighteenth century because they presented previously unknown metals from "missing" block of the periodic table and further because of their properties. This book is about luminescence of these elements, maybe the most intriguing feature of lanthanides; luminescence that has so many uses – industrial and research.

These unique luminescence properties have led to broad industrial exploitation of these f-block elements of the periodic table: bioanalytics, lasers, illumination, and displays – the role of luminescent lanthanides is extensive and rapidly growing. Lanthanides plus yttrium and scandium are also known as "rare earths" – a historical misinterpretation that stems from the way that these metals were first discovered. For those who have been working with sensitive lanthanide-based assays and especially for those who have been building instrumentation for time-resolved measurements, the luminescent lanthanides are far from rare – especially glass components of almost any glass will show unwanted background at lanthanide emission wavelengths. The word lanthanide has its origins in Greek and its meaning "lying hidden" is maybe the best explanation for why these naturally abundant elements were considered rare: separation of them from other minerals was not trivial for early day metallurgists since the lanthanides were "hiding" from other minerals. Once efficient separation techniques were established 50 years ago, the use of the lanthanides found their way into industrial exploitation and our everyday life.

- For once so scarce they lie no longer hidden but illuminate our lives and serve us well -

Luminescence of lanthanides in crystalline form is maybe the best-known application for the common people. The lighting and display industry has utilized the unique long lifetime luminescence as phosphors and laser industry has exploited lanthanide crystals since population inversion is easy to achieve with long lifetime laser materials. Beyond the luminescence use, lanthanides are used in magnets, glass production, as colorants, contrast agents, in computer memories – the application field is broad, but as said, this book is about luminescence of these compounds with special emphasis on bioanalytics.

Whereas the applications of solid luminescent lanthanides can well be paired with the progress of the electronics industry, the use of the lanthanides in biomedical field started from the need to change the way the researchers and industry were operating, and it certainly took its own pace. Already in 1942, Weissman had discovered the sensitizing of lanthanide luminescence by organic ligands, but the first bioapplications were not proposed until in the 1970s when Vallarino and Leif suggested the use of lanthanide labels in cytometry – it took an industrialist to generate the "breakthrough" in the field of bioanalytics! Radioimmunoassays (RIAs) were dominating the bioassaying during the 1960s and 70s. Although sensitive, the problems related to the use of radioactive tracers were eminent, and already in the late 1960s, ideas of assays based on the use of enzymatic reactions or the use of fluorescent tracers were presented. However, the new methods could not compete with RIA in all of its flavors, the field was open for new ideas.

The use of rare-earths in immunoassays as labels was first discussed by two physicists when Erkki Soini, the director of research at company Wallac visited researcher Reino Laiho at the Wihuri Physical Laboratory, University of Turku. Research of time-resolved luminescence had already started at Wallac earlier but Wallac did not have a suitable label. The trials with conventional fluorochromes had led to serious problems on developing instrumentation that could be routinely used. The spectroscopic properties of lanthanides were well known from solid state physics, and after the discussion, the research at Wallac got a new course. Wallac, a small Turku-based company had grown in 1960s and 1970s building beta- and gamma counters for various purposes – in the 70s the fastest growing field was RIA but the company research was set a new course by these new ideas.

The use of lanthanides in bioassays evolved rapidly from Wallac's DELFIA principle to first directly luminescent chelates and to TR-FRET, first proposed by Morrison and then pioneered by Mathis – developments that took the use of lanthanides from clinical routines to research and discovery. With applications in microscopy pioneered by Tanke, Seveus, and others, the lanthanides are now among the most important tools in bioanalysis – listing of all applications would certainly be appropriate to honor everybody involved, but is well beyond this preface. The purpose of this book is not to give a complete overview on lanthanide luminescence but rather highlight the special aspects of their luminescence with emphasis toward biological research without forgetting their basic luminescent properties – both in liquid and in solid materials. If readers intend to step into the use of lanthanides or refine their ways of using them, we hope that this book can serve as a comprehensive handbook and help them toward their goals.

Although the history of utilizing lanthanide luminescence by now is long, they are still today's news since new ways of putting lanthanide luminescence into use appear constantly. We ourselves have been intrigued by the special properties of the lanthanides that can so well be utilized in bioanalytics – having worked with several different luminescent molecules and materials we still turn back to lanthanide labels whenever there is a problem with sensitivity. We also hope that this book can be seen as a tribute to all the people who have contributed to the field in our hometown. Turku has played a central role with discovery of "rare earths" and putting them into use. From professor Johann Gadolin who at the Academy of Turku discovered yttrium more than 200 years ago to Erkki Soini, Ilkka Hemmilä, and many others who at their turn have put lanthanide luminescence to practical work.

Turku, March 2011 Pekka Hänninen, Harri Härmä

Contents

Basics of Lanthanide Photophysics

Jean-Claude G. Bünzli and Svetlana V. Eliseeva

Abstract The fascination for lanthanide optical spectroscopy dates back to the 1880s when renowned scientists such as Sir William Crookes, LeCoq de Boisbaudran, Eugène Demarçay or, later, Georges Urbain were using luminescence as an analytical tool to test the purity of their crystallizations and to identify potential new elements. The richness and complexity of lanthanide optical spectra are reflected in an article published in 1937 by J.H. van Vleck: *The Puzzle of Rare Earth Spectra in Solids*. After this analytical and exploratory period, lanthanide unique optical properties were taken advantage of in optical glasses, filters, and lasers. In the mid-1970s, E. Soini and I. Hemmilä proposed lanthanide luminescent probes for time-resolved immunoassays (Soini and Hemmilä in Clin Chem 25:353–361, 1979) and this has been the starting point of the present numerous bio-applications based on optical properties of lanthanides. In this chapter, we first briefly outline the principles underlying the simplest models used for describing the electronic structure and spectroscopic properties of trivalent lanthanide ions Ln^{III} ($4f^n$) with special emphasis on luminescence. Since the book is intended for a broad readership within the sciences, we start from scratch defining all quantities used, but we stay at a descriptive level, leaving out detailed mathematical developments. For the latter, the reader is referred to references Liu and Jacquier, Spectroscopic properties of rare earths in optical materials. Tsinghua University Press & Springer, Beijing & Heidelberg, 2005 and Görller-Walrand and Binnemans, Rationalization of crystal field parameters. In: Gschneidner, Eyring (eds) Handbook on the physics and chemistry of rare earths, vol 23. Elsevier BV, Amsterdam, Ch 155, 1996. The second part of the chapter is devoted to practical aspects of lanthanide luminescent probes, both from the point of view of their design and of their potential utility.

J.-C.G. Bünzli (✉) and S.V. Eliseeva
Laboratory of Lanthanide Supramolecular Chemistry, École Polytechnique Fédérale de Lausanne, BCH 1402, 1015 Lausanne, Switzerland
e-mail: Jean-Claude.Bunzli@epfl.ch

P. Hänninen and H. Härmä (eds.), *Lanthanide Luminescence: Photophysical, Analytical and Biological Aspects*, Springer Ser Fluoresc (2011) 7: 1–46, DOI 10.1007/4243_2010_3,

Keywords Crystal-field analysis · Energy transfer · f–f Transition · Intrinsic quantum yield · Lanthanide bioprobe · Lanthanide luminescence · Lanthanide spectroscopy · Lifetime · Luminescence sensitization · Population analysis · Quantum yield · Radiative lifetime · Selection rule · Site symmetry · Stern–Volmer quenching

Contents

Abbreviations

AO	Acridine orange
CF	Crystal field
CT	Charge transfer
DMF	Dimethylformamide
DNA	Deoxyribonucleic acid
dpa	Dipicolinate (2,6-pyridine dicarboxylate)
dtpa	Diethylenetrinitrilopentaacetate
EB	Ethidium bromide
ED	Electric dipole

EQ	Electric quadrupole
FRET	Förster resonant energy transfer
hfa	Hexafluoroacetylacetonate
ILCT	Intraligand charge transfer
ISC	Intersystem crossing
JO	Judd–Ofelt
LLB	Lanthanide luminescent bioprobe
LMCT	Ligand-to-metal charge transfer
MD	Magnetic dipole
MLCT	Metal-to-ligand charge transfer
NIR	Near-infrared
PCR	Polymerase chain reaction
SO	Spin–orbit
tta	Thenoyltrifluoroacetylacetonate
YAG	Yttrium aluminum garnet

1 Electronic Structure of Trivalent Lanthanide Ions

1.1 Atomic Orbitals

In quantum mechanics, three variables depict the movement of the electrons around the positively-charged nucleus, these electrons being considered as waves with wavelength $\lambda = h/mv$ where h is Planck's constant (6.626×10^{-34} J s^{-1}), m and v the mass (9.109×10^{-31} kg) and velocity of the electron, respectively:

- The time-dependent Hamiltonian operator $\mathcal{H}$ describing the sum of kinetic and potential energies in the system; it is a function of the coordinates of the electrons and nucleus.
- The wavefunction, Ψ_n, also depending on the coordinates and time, related to the movement of the particles, and not directly observable; its square $(\Psi_n)^2$ though gives the probability that the particle it describes will be found at the position given by the coordinates; the set of all probabilities for a given electronic Ψ_n, is called an *orbital*.
- The quantified energy E_n associated with a specific wavefunction, and independent of the coordinates.

These quantities are related by the dramatically simple Schrödinger equation, which replaces the fundamental equations of classical mechanics for atomic systems:

$$\mathcal{H}\Psi_n = E_n\Psi_n. \tag{1}$$

Energies E_n are eigenvalues of Ψ_n, themselves called eigenfunctions. In view of the complexity brought by the multidimensional aspect of this equation

(3 coordinates for each electron and nucleus, in addition to time) several simplifications are made. Firstly, the energy is assumed to be constant with time, which removes one coordinate. Secondly, nuclei being much heavier than electrons, they are considered as being fixed (Born–Oppenheimer approximation). Thirdly, since the equation can only be solved precisely for the hydrogen atom, the resulting hydrogenoid or one-electron wavefunction is used for the other elements, with a scaling taking into account the apparent nucleus charge, i.e., including screening effects from the other electrons. Finally, to ease solving the equation for non-H atoms, the various interactions occurring in the electron-nucleus system are treated separately, in order of decreasing importance (*perturbation method*).

For hydrogen, the Hamiltonian simply reflects Coulomb's attraction between the nucleus and the electron, separated by a distance r_i, and the kinetic energy of the latter:[1]

$$\mathcal{H}_0 = -\frac{1}{r_i} - \frac{1}{2}\Delta_i \quad \left(\Delta = \frac{\partial^2}{\partial x^2} + \frac{\partial^2}{\partial y^2} + \frac{\partial^2}{\partial z^2}\right). \tag{2}$$

Each wavefunction (or orbital: the two terms are very often, but wrongly, taken as synonyms) resulting from solving (1) is defined by four quantum numbers reflecting the quantified energy of the two motions of the electrons: the orbital motion, defined by the angular momentum $\vec{\ell}$, and the spin, characterized by the angular momentum $\vec{s}$. If polar coordinates (r, ϑ, φ) are used, wavefunctions are expressed as the product of a normalizing factor N, of a radial function $\Re_{n,\ell}$, of an angular function Φ_{ℓ,m_ℓ}, and of a spin function S_{m_s}:

$$\Psi_{n,\ell,m_\ell,m_s} = N \cdot \Re_{n,\ell}(r) \cdot \Phi_{\ell,m_\ell}(\vartheta,\varphi) \cdot S_{m_s}. \tag{3}$$

The principal quantum number n is an integer (1, 2, 3, ...) and represents the radial expansion of the orbital. The angular quantum number ℓ varies from 0 to $(n - 1)$ and characterizes the shape of the orbital (designed by letters: $s, p, d, f, g, \ldots$ for $\ell = 0, 1, 2, 3, 4, \ldots$). The magnetic quantum number m_ℓ is the projection of the vector $\vec{\ell}$ onto the z axis and is linked to the orientation of the orbital in space; it varies between $-\ell$ and $+\ell$. Finally, m_s is the projection of the vector $\vec{s}$ and takes values of $\pm$ ½. Pauli's principle requires that two electrons of the same atom must at least differ by the value of one quantum number; this implies that only two electrons of opposite spin can be associated with a given orbital. An electronic shell consists in all electrons having the same quantum number n. A sub-shell regroups electrons with same n and ℓ numbers, has therefore $(2\ell + 1)$ orbitals, and may contain a maximum of $(4\ell + 2)$ electrons. The shapes of the seven 4f orbitals $(n = 4, \ell = 3)$ are represented on top of Fig. 1.

[1]We use the atomic system units (a.u.) in order to simplify the equations as much as possible.

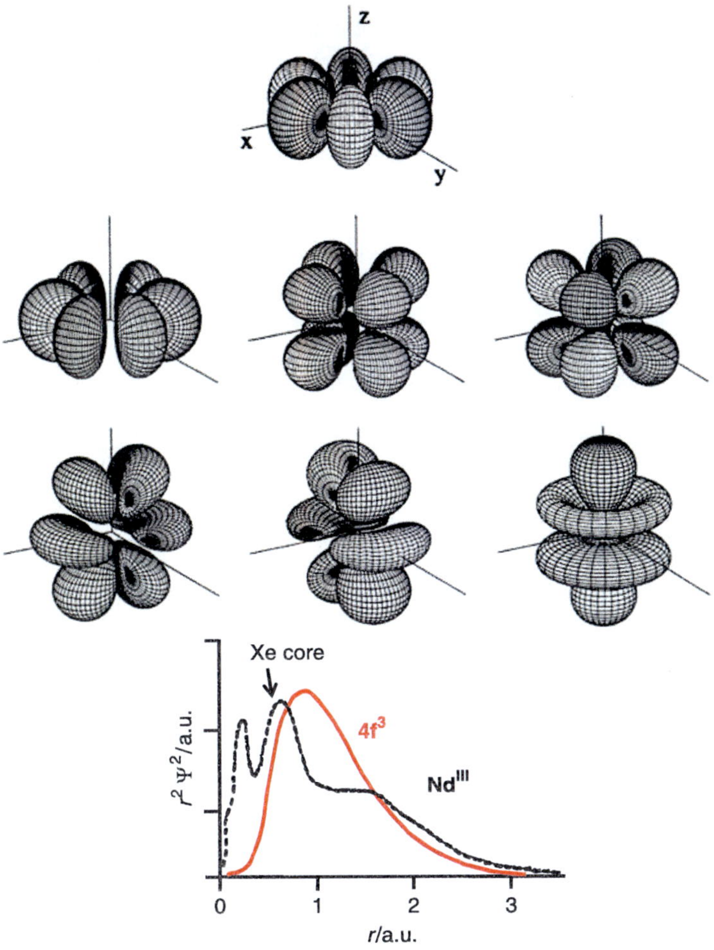

Fig. 1 *Top*: Shape of the one-electron (hydrogenoid) 4f orbitals in a Cartesian space. From *top* to *bottom* and *left* to *right*: $4f_{x(x^2-3y^2)}$, $4f_{y(3y^2-x^2)}$, $4f_{xyz}$, $4f_{z(x^2-y^2)}$, $4f_{xz^2}$, $4f_{yz^2}$, and $4f_{z^2}$ (combinations of Cartesian coordinates represent the angular functions). *Bottom*: Radial wavefunction of the three 4f electrons of Nd^{III} compared with the radial wavefunction of the xenon core (a.u. = atomic units); redrawn after [1]

1.2 Electronic Configuration

The ground state electronic configuration of Ln^{III} ions is $[Xe]4f^n$ ($n = 0$–14). It is energetically well separated from the $[Xe]4f^{n-1}5d^1$ configuration ($\Delta E > 32{,}000$ cm^{-1}). A far reaching fact is the shielding of the 4f orbitals by the xenon core (54 electrons), particularly the larger radial expansion of the $5s^25p^6$ subshells, making the valence 4f orbitals "inner orbitals" (bottom of Fig. 1). This is the key to the chemical and spectroscopic properties of these metal ions.

Each of the n electrons of the $4f^n$ configuration is associated with one of the seven 4f wavefunctions and may have a spin of ± ½. There are a number of ways of associating the n electrons with the 4f orbitals, taking the spin into consideration; this number corresponds to the multiplicity (or degeneracy) of the configuration and is given by the following combinatorial formula:

$$\frac{(4\ell+2)!}{n!(4\ell+2-n)!} = \frac{14!}{n!(14-n)!} \quad \text{if } \ell = 3. \tag{4}$$

Since there are more than one electron in the configuration, the Hamiltonian in (2) has to be adjusted to take into account the number of electrons, the apparent (screened) nucleus charge Z', and the repulsion between electrons located at a distance r_{ij}:

$$\mathcal{H} = \sum_{i=1}^{n}\left(-\frac{Z'}{r_i} - \frac{1}{2}\Delta_i\right) + \sum_{i\neq j}^{n}\frac{1}{r_{ij}}. \tag{5}$$

A given association of the electrons with the 4f wavefunctions $\{(m_\ell, m_s)_1, (m_\ell, m_s)_2, \ldots (m_\ell, m_s)_n\}$ is called a *micro state*. The latter is characterized with "overall" quantum numbers M_L and M_S derived from the projections of the sum of the angular momenta:

$$\vec{L} = \sum_{i=1}^{n}\vec{\ell}_i, \quad \vec{S} = \sum_{i=1}^{n}\vec{s}_i, \quad M_L = \sum_{i=1}^{n}(m_\ell)_i, \quad M_S = \sum_{i=1}^{n}(m_s)_i. \tag{6}$$

A set of micro states such that all the M_L and M_S quantum numbers correspond to the projections of one value of L and S, respectively, is called a *spectroscopic term*. It is written as $^{(2S+1)}\Gamma$, where Γ is a capital letter (S, P, D, F, G. . .) corresponding to the values of L (0, 1, 2, 3, 4. . .). The multiplicity of a term, i.e., the number of micro states it regroups, is given by $(2S + 1) \times (2L + 1)$. An electronic configuration contains several terms and the sum of their multiplicities is equal to the degeneracy of the configuration. The procedure for finding them out is rather tedious, except for the ground term, for which Hund's rules make its determination easy. According to these rules (to be applied in the given order), the ground term has:

Rule 1 The largest spin multiplicity
Rule 2 The largest orbital multiplicity

For instance, for Eu^{III}, $4f^6$, the largest multiplicity is obtained when each electron is associated with a unique 4f wavefunction: $S = 6 \times ½ = 3$; therefore, $(2S + 1) = 7$. To obtain the largest orbital multiplicity, these electrons have to be related with wavefunctions having the largest m_ℓ values, i.e., +3, +2, +1, 0, −1, and −2; the sum is 3, henceforth $L = 3$ and Γ = F: the ground term is a spin septet, ^7F with overall multiplicity $7 \times 7 = 49$.

The two movements of the electrons are in fact not independent, they couple and the strength of the interaction usually increases with the atomic number. To simplify the treatment of this interaction, Russel and Saunders have proposed to consider this coupling at the level of the overall angular momenta and not for each individual electron. This model is valid for lighter elements and not quite adequate for lanthanides for which an intermediate coupling scheme should be applied; however, it is in large use in view of its simplicity and we shall keep with it in this chapter. The Hamiltonian becomes:

$$\mathcal{H} = \sum_{i=1}^{n}\left(-\frac{Z'}{r_i} - \frac{1}{2}\Delta_i\right) + \sum_{i \neq j}^{n}\frac{1}{r_{ij}} + \lambda \cdot \vec{L} \cdot \vec{S} \quad \text{with} \quad \lambda = \pm\frac{\xi}{2S}. \tag{7}$$

The spin–orbit coupling constant λ is positive if the 4f subshell is less than half filled and negative if it is more than half filled. A new quantum number J, associated with the total angular momentum $\vec{J} = \vec{L} + \vec{S}$, has to be introduced with values ranging from $(L + S)$ to $(L - S)$. As a consequence, each term is further split into a number of spectroscopic levels $^{(2S+1)}\Gamma_J$ each with a $(2J + 1)$ multiplicity. Again, the sum of these multiplicities must be equal to the multiplicity of the term. For instance, the ground term of Eu^{III} is split into $^7F_0$, $^7F_1$, $^7F_2$, $^7F_3$, $^7F_4$, $^7F_5$, and $^7F_6$ with multiplicities $1 + 3 + 5 + 7 + 9 + 11 + 13 = 49$. The ground level can be found with third Hund's rule:

Rule 3 $\quad$ if $n < (2\ell + 1)$, $J = J_{\min}$; $\quad$ if $n > (2\ell + 1), J = J_{\max}$.

Note that if the sub-shell is half filled, then $L = 0$ and $J = S$. Additionally, J may take half-integer values if S is half-integer. The set of levels is referred to as a multiplet and this multiplet is a regular one if $n < (2\ell + 1)$, the energy of the levels increasing with increasing values of J, while it is inverted if $n > (2\ell + 1)$. This is illustrated with Eu^{III} ($4f^6$) for which the ground level is $^7F_0$ while it is $^7F_6$ for Tb^{III} ($4f^8$). Finally, the energy difference between two consecutive spin–orbit levels with quantum numbers J and $J' = J + 1$ is directly proportional to J':

$$\Delta E = \lambda \cdot J'. \tag{8}$$

The electronic properties of the trivalent 4f free ions are summarized in Table 1.

1.3 The Ions in a Ligand Field

The above developments are valid for free ions. When a Ln^{III} ion is inserted into a chemical environment, the spherical symmetry of its electronic structure is destroyed and the remaining $(2J + 1)$ degeneracy of its spectroscopic levels is partly lifted, depending on the exact symmetry of the metal–ion site. In view of the

Table 1 Electronic properties of Ln^{III} free ions

f^n		Multiplicity	No. of terms	No. of levels	Ground level		ζ/cm$^{-1,a}$		λ/cm$^{-1,a}$	
f^0	f^{14}	1	1	1	$^1S_0$	$^1S_0$	–	–	–	–
f^1	f^{13}	14	1	2	$^2F_{5/2}$	$^2F_{7/2}$	625	2,870	625	−2,870
f^2	f^{12}	91	7	13	$^3H_4$	$^3H_6$	740	2,628	370	−1,314
f^3	f^{11}	364	17	41	$^4I_{9/2}$	$^4I_{15/2}$	884	2,380	295	−793
f^4	f^{10}	1,001	47	107	$^5I_4$	$^5I_8$	1,000	2,141	250	−535
f^5	f^9	2,002	73	198	$^6H_{5/2}$	$^6H_{15/2}$	1,157	1,932	231	−386
f^6	f^8	3,003	119	295	$^7F_0$	$^7F_6$	1,326	1,709	221	−285
f^7		3,432	119	327	$^8S_{7/2}$		1,450		0	

[a]For aqua ions, except for Ce^{III} (Ce:$LaCl_3$) and Yb^{III} ($Yb_3Ga_5O_{12}$), from [2]. The first column refers to f^{1-7} and the second to f^{8-14}

inner character of the 4f wavefunctions their mixing with the surrounding orbitals remains small and so is the resulting level splitting (a few hundreds of cm^{-1}), so that this perturbation can be treated last. Nevertheless, the resulting Hamiltonian gets very complex, so that a simplifying concept has been put forward by H. Bethe in 1929: the ligands are replaced by (negative) point charges generating a crystal (or ligand) electrostatic field which, in turn, interacts with the moving 4f electrons, generating a ligand-field (or crystal-field, or Stark) splitting of the spectroscopic levels.

With the parameterization introduced by B. G. Wybourne in 1965, the final Hamiltonian becomes:

$$\mathcal{H} = \sum_{i=1}^{n}\left(-\frac{Z'}{r_i} - \frac{1}{2}\Delta_i\right) + \sum_{i \neq j}^{n}\frac{1}{r_{ij}} + \lambda \cdot \vec{L} \cdot \vec{S} + \sum_{k,q,i} B_q^k C_q^{(k)}(i), \tag{9}$$

where the summation involving i is on all the 4f electrons, B_q^k are ligand-field parameters, commonly treated as phenomenological parameters, and $C_q^{(k)}$ are components of tensor operators $\boldsymbol{C}^{(k)}$ which transform like the spherical harmonics used for the analytical form of the 4f wavefunctions. The running number k must be even and smaller than 2ℓ; for 4f electrons it can, therefore, take values of 0, 2, 4, and 6. The values for q are restricted by the point group of symmetry into which the Ln^{III} ion is embedded, but in any case, $|q| \leq k$.

The B_q^k parameters may be complex numbers but they have to be real for any symmetry group with a 180° rotation about the y axis or with the xy plane being a mirror plane. The relationship between the 32 crystallographic symmetry groups and the B_q^k parameters is given in Table 2. In order to compare the ligand field strengths in different compounds, F. Auzel has proposed the following expression for an "average" total ligand-field effect:

$$N_v = \left[\frac{1}{4\pi}\sum_{k,q}\frac{(B_q^k)^2}{(2k+1)}\right]^{1/2}. \tag{10}$$

Table 2 Non-zero crystal-field parameters for f^n electronic configurations and examples of corresponding crystal hosts [1]

Symmetry	Site symmetry	Crystal field parameters	Example
Monoclinic	$C_1, C_S, C_2, C_{2h},$	$B_0^2,\ B_0^4,\ B_0^6,\ \Re(B_2^2),\ B_2^4, B_2^6,\ B_4^4,\ B_4^6,\ B_6^6$	LaF_3
Rhombic	C_{2v}, D_2, D_{2h}	$B_0^2,\ B_0^4,\ B_0^6,\ \Re(B_2^2,\ B_2^4, B_2^6,\ B_4^4,\ B_4^6,\ B_6^6)$	$Y_3Al_5O_{12}$
Trigonal	C_3, S_6	$B_0^2,\ B_0^4,\ B_0^6,\ \Re(B_3^4),\ B_3^6,\ B_6^6$	$LiNbO_3$
	C_{3v}, D_3, D_{3d}	$B_0^2,\ B_0^4,\ B_0^6,\ \Re(B_3^4,\ B_3^6,\ B_6^6)$	Y_2O_2S
Tetragonal	C_4, S_4, C_{4h}	$B_0^2,\ B_0^4,\ B_0^6,\ \Re(B_4^4),\ B_4^6$	$LiYF_4$
	$C_{4v}, D_4, D_{2d}, D_{4h}$	$B_0^2,\ B_0^4,\ B_0^6,\ \Re(B_4^4,\ B_4^6)$	YPO_4
Hexagonal	$C_{3h}, D_{3h}, C_6, C_{6h}, C_{6v}, D_6, D_{6h}$	$B_0^2,\ B_0^4,\ B_0^6,\ \Re(B_6^6)$	$LaCl_3$
Cubic	T, T_d, T_h, O, O_h	$B_0^4,\ B_0^6,\ \Re(B_4^4,\ B_4^6)$ [a]	CeO_2

[a] $B_4^4 = \frac{5}{\sqrt{70}} B_0^4,\quad B_4^6 = -\sqrt{7/2}\, B_0^6$

Finally, it is worth noting that electrostatic ligand-field effects do not completely lift the degeneracy of the J levels of odd-numbered electronic configurations; in the latter case, all ligand-field sublevels are at least doubly degenerate (Kramer's doublets) and this degeneracy can only be removed by a magnetic field. A partial energy diagram, including crystal-field splittings is given in Fig. 2. Due to their large number, energy levels may extend up to 190,000 cm^{-1} for n = 6, 7, 8, and are not yet fully explored, although an extension of Carnall's diagram up to this energy has been recently published [4].

The maximum numbers of ligand-field (or Stark) sublevels depend on the point group of symmetry: they are given in Table 3 versus values of J. This can be exploited for the determination of the symmetry point group from f–f absorption or emission spectra, at least when J is integer.

2 Absorption Spectra

Description of the interaction between photons (massless elemental particles of light) and matter considers the former behaving as waves comprised of two perpendicular fields, electric and magnetic, oscillating in time (henceforth the denomination of electromagnetic wave or radiation). When a photon is absorbed, its energy is transferred to an electron which then may be "pushed" into an orbital with higher energy. The absorption is promoted by "operators" linked to the nature of light: the odd-parity electric dipole (ED) operator $\vec{P}$, the even-parity magnetic dipole (MD) $\vec{M}$ and electric quadrupole (EQ) $\vec{Q}$ operators:

$$\vec{P} = -e\sum_{i=1}^{n}\vec{r}_i \qquad \vec{M} = -\frac{eh}{4\pi mc}\sum_{i=1}^{n}(\vec{\ell}_i + 2\vec{s}_i) \qquad \vec{Q} = \frac{1}{2}\sum_{i=1}^{n}(\vec{k}\cdot\vec{r}_i)\cdot\vec{r}_i. \tag{11}$$

There are three types of electronic transitions involving lanthanide ions: sharp intraconfigurational 4f–4f transitions, broader 4f–5d transitions, and broad

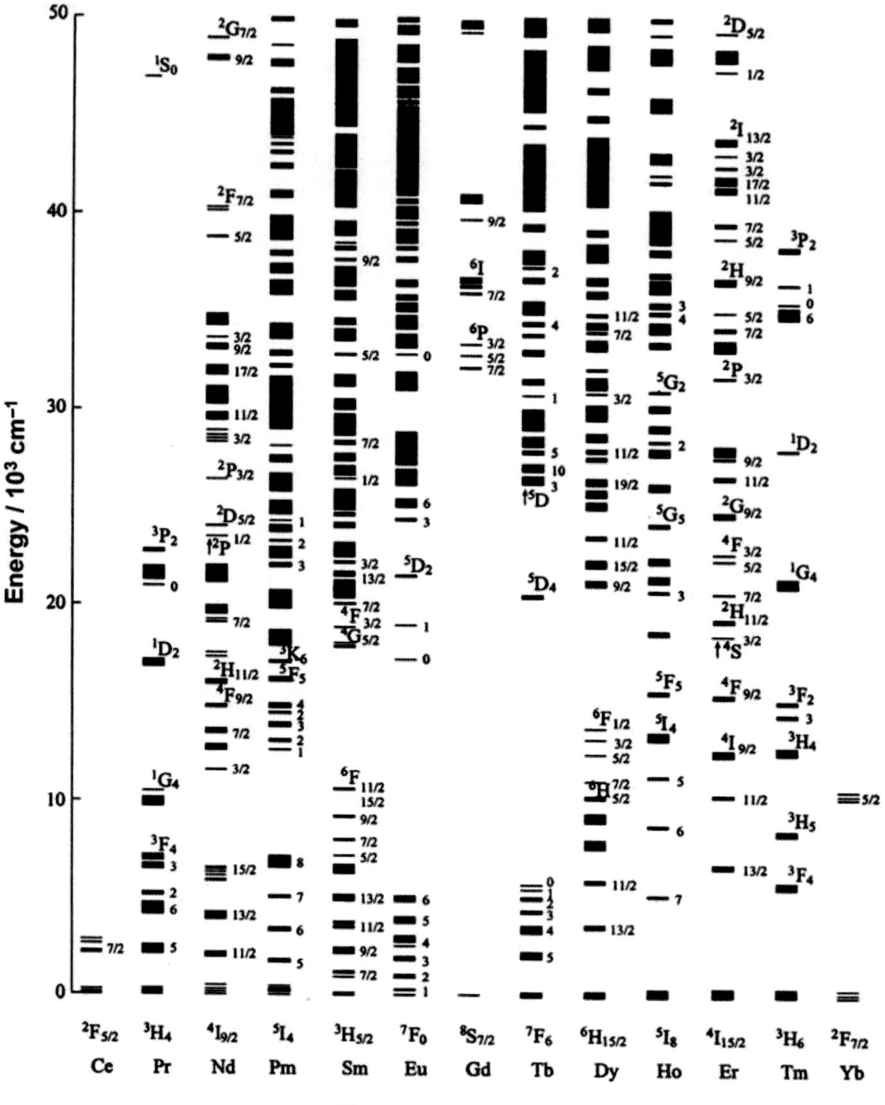

Fig. 2 Energy level diagram for Ln^{III} ions doped in a low-symmetry crystal, LaF_3. Redrawn from [3]

charge-transfer transitions (metal-to-ligand, MLCT or ligand-to-metal, LMCT). Not all transitions are permitted and the allowed ones are described by selection rules. *Laporte's parity selection rule* implies that states with the same parity cannot be connected by electric dipole transitions; as a consequence f–f transitions are forbidden by the ED mechanism. However, when the lanthanide ion is under the influence of a ligand-field, non-centrosymmetric interactions allow the mixing of

Table 3 Number of Stark levels versus the value of quantum number *J*

Symmetry	Site symmetry	Integer *J*								
		0	1	2	3	4	5	6	7	8
Cubic	T, T_d, T_h, O, O_h	1	1	2	3	4	4	6	6	7
Hexagonal	$C_{3h}, D_{3h}, C_6, C_{6h}, C_{6v}, D_6, D_{6h}$	1	2	3	5	6	7	9	10	11
Trigonal	$C_3, S_6\ C_{3v}, D_3, D_{3d}$									
Tetragonal	$C_4, S_4, C_{4h}, C_{4v}, D_4, D_{2d}, D_{4h}$	1	2	4	5	7	8	10	11	13
Low	$C_1, C_S, C_2, C_{2h}, C_{2v}, D_2, D_{2h}$	1	3	5	7	9	11	13	15	17
Symmetry	Site symmetry	Half-integer *J*								
		1/2	3/2	5/2	7/2	9/2	11/2	13/2	15/2	17/2
Cubic	T, T_d, T_h, O, O_h	1	1	2	3	3	4	5	6	6
All others[a]	See above	1	2	3	4	5	6	7	8	9

[a]All Stark sublevels are doubly degenerate (Kramer's doublets)

electronic states of opposite parity into the 4f wavefunctions, which somewhat relaxes the selection rules and the transition becomes partially allowed; it is called an *induced* (or forced) *electric dipole transition.* Magnetic dipole transitions are allowed, but their intensity is weak; in 4f–4f spectra, however, they often have intensity of the same order of magnitude as induced electric dipole transitions. Quadrupolar transitions are also parity allowed, but they are much weaker than MD transitions so that they are usually not observed. Some induced ED transitions are highly sensitive to minute changes in the Ln^{III} environment and are called *hypersensitive* or sometimes *pseudo-quadrupolar transitions* because they apparently follow the selection rules of EQ transitions. A listing of experimentally identified such transitions is presented in Table 4; note, that these transitions are not necessarily the most intense ones in the optical spectra.

In addition to the parity selection rule, other rules are operative, for instance, on ΔS (spin selection rule, requiring no change of spin for all three mechanisms, $\Delta S = 0$), ΔL, and ΔJ; they will be detailed below. The selection rules are derived under several hypotheses which are not always completely fulfilled in reality (in particular 4f wavefunctions are not completely pure), so that the terms "forbidden" and "allowed" transitions are not accurate. Let's say that a forbidden transition has a low probability and an allowed transition a high probability of occurring.

2.1 *Induced ED f–f Transitions: Judd–Ofelt Theory [5, 6]*

Judd–Ofelt (JO) theory has been established within the frame of the crystal-field concept and it provides a simple model for reproducing the intensities of f–f transitions both in solids and solutions. It only takes into account the $4f^n$ electronic configuration, that is inter-configurational $4f^n$–$4f^{n-1}5d^1$ interactions are neglected. On the other hand, spin–orbit coupling is treated within the frame of the intermediate

Table 4 Experimentally observed hypersensitive transitions for Ln^{III} ions in optical spectra [5]. Energies/wavelengths are approximate

Ln	Transition	$\tilde{\nu}$/cm^{-1}	λ/nm
Pr	$^3F_2 \leftarrow {}^3H_4$	5,200	1,920
Nd	$^4G_{5/2} \leftarrow {}^4I_{9/2}$ [a]	17,300	578
	$^2H_{9/2}, {}^4F_{5/2} \leftarrow {}^4I_{9/2}$	12,400	806
	$^4G_{7/2}, {}^3K_{13/2} \leftarrow {}^4I_{9/2}$	19,200	521
Sm	$^4F_{1/2}, {}^4F_{3/2} \leftarrow {}^6H_{5/2}$	6,400	1,560
Eu	$^5D_2 \leftarrow {}^7F_0$	21,500	465
	$^5D_1 \leftarrow {}^7F_1$	18,700	535
	$^5D_0 \rightarrow {}^7F_2$	16,300	613
Gd	$^6P_{5/2}, {}^6P_{7/2} \leftarrow {}^8S_{7/2}$	32,500	308
Tb	– [b]	–	–
Dy	$^6F_{11/2} \leftarrow {}^6H_{15/2}$	7,700	1,300
	$^4G_{11/2}, {}^4I_{15/2} \leftarrow {}^6H_{15/2}$	23,400	427
Ho	$^3H_6 \leftarrow {}^5I_8$	27,700	361
	$^5G_6 \leftarrow {}^5I_8$	22,100	452
Er	$^4G_{11/2} \leftarrow {}^4I_{15/2}$	26,400	379
	$^2H_{11/2} \leftarrow {}^4I_{15/2}$	19,200	521
Tm	$^1G_4 \leftarrow {}^3H_6$	21,300	469
	$^3H_4 \leftarrow {}^3H_6$	12,700	787
	$^3F_4 \leftarrow {}^3H_6$	5,900	1,695

[a] The transition $^4G_{5/2} \leftarrow {}^4I_{9/2}$ overlaps with $^2G_{7/2} \leftarrow {}^4I_{9/2}$

[b] None identified positively, but the $^5D_4 \rightarrow {}^7F_5$ transition shows sometimes ligand-induced pseudo-hypersensitivity

coupling scheme. The dipole strength in esu^{2}cm^2 (=10^{36} debye^2) of an induced ED f–f transition between states Ψ and Ψ' is given by:

$$D_{\mathrm{ED}} = e^2 \sum_{\lambda=2,4,6} \Omega_\lambda \left| \langle \Psi \| U^\lambda \| \Psi' \rangle \right|^2, \tag{12}$$

in which e is the electric charge of the electron, wavefunctions Ψ and Ψ' are full intermediate-coupled functions f^n[SL]J, U^λ are the irreducible tensor forms of the ED operator, and Ω_λ are the phenomenological JO parameters, expressed in cm^2. The bracketed expressions in (12) are dimensionless doubly-reduced matrix elements which are tabulated (and insensitive to the metal–ion environment). Mathematical treatment of the parity mixing by the crystal-field perturbation leads to the selection rules for f–f transitions reproduced in Table 5.

JO parameters are adjustable parameters and they are calculated from the absorption spectrum $\varepsilon(\tilde{\nu})$. For an isotropic crystal or a solution, the experimental dipole strength is defined as:

$$D(\exp) = \frac{10^{36}}{108.9 \cdot \tilde{\nu}_{\mathrm{mean}} \cdot X_{\mathrm{A}}} \left((2J+1) \cdot \frac{9n}{(n^2+2)^2} \right) \int \varepsilon(\tilde{\nu}) \mathrm{d}\tilde{\nu}, \tag{13a}$$

Table 5 Selection rules for intra-configurational f–f transitions

Operator	Parity	ΔS	ΔL	ΔJ[a]
ED	Opposite	0	≤ 6	≤ 6 (2,4,6 if J or $J' = 0$)
MD	Same	0	0	$0, \pm 1$
EQ	Same	0	$0, \pm 1, \pm 2$	$0, \pm 1, \pm 2$

[a] $J = 0$ to $J' = 0$ transitions are always forbidden

with X_A being the fractional population of the initial state while $\tilde{\nu}_{mean}$ is given by:

$$\tilde{\nu}_{mean} = \frac{\int \tilde{\nu} \cdot \varepsilon(\tilde{\nu}) d\tilde{\nu}}{\int \varepsilon(\tilde{\nu}) d\tilde{\nu}}. \tag{13b}$$

The above equations assume that the absorption bands are symmetrical, i.e., either Gaussian or Lorentzian. If not, (13a) has to be replaced with:

$$D(\exp) = \frac{10^{36}}{108.9 \cdot X_A} \left((2J + 1) \frac{9n}{(n^2 + 2)^2} \right) \int \frac{\varepsilon(\tilde{\nu})}{\tilde{\nu}} d\tilde{\nu}. \tag{14}$$

Finally, $(2J + 1)$ is the degeneracy of the initial state and the expression involving the refractive index n is known as Lorentz's local-field correction. Calculations of transition probabilities within the frame of JO theory are usually made assuming that all Stark sublevels within the ground level are equally populated and that the material under investigation is optically isotropic. The former hypothesis is only reasonable in some cases, e.g., when transitions initiate from non-degenerate states such as Eu($^7F_0$). Otherwise, there is a Boltzmann distribution of the population among the crystal-field sublevels. The second assumption is not valid for uniaxial or biaxial crystals, but, of course, holds for solutions.

The phenomenological JO parameters are determined from a fit of (12) to the experimental values defined by (13a), using adequate matrix elements. The exact procedure is described in details in reference [6]. In the case of Eu^{III} the procedure is quite simple since Ω_2, Ω_4, and Ω_6 can be directly extracted from the dipole strength of the $^5D_2 \leftarrow {}^7F_0$, $^5D_4 \leftarrow {}^7F_0$, and $^5L_6 \leftarrow {}^7F_0$ transitions, respectively. An example is shown on Fig. 3 for europium tris(dipicolinate). Extensive tabulations of JO parameters can be found in reference [5] while spectra for all Ln^{III} ions are presented in reference [8]; note that molar absorption coefficients are, with a few exceptions, smaller than 10 M^{-1} cm^{-1} and very often smaller than 1 or even 0.1 M^{-1} cm^{-1}.

2.2 *4f–5d and CT Transitions*

The promotion of a 4f electron into the 5d sub-shell is parity allowed; the corresponding transitions are broader than f–f transitions and their energy depends

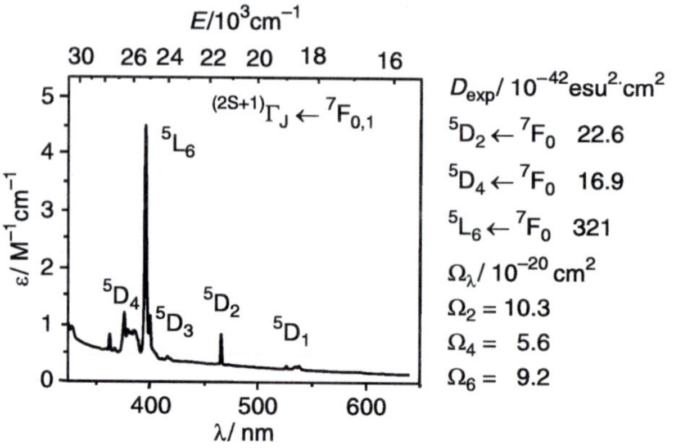

Fig. 3 Absorption spectrum of $Na_3[Eu(dpa)_3]$ 1.8×10^{-2} M in Tris–HCl 0.1 M and associated dipole strengths and JO parameters. Redrawn from [7]

Fig. 4 Energy of the 4f–5d transitions in Ln^{III}:CaF_2 (*squares*, [9]) and of the 2p (O)-4f LMCT transitions (*triangles*, [10])

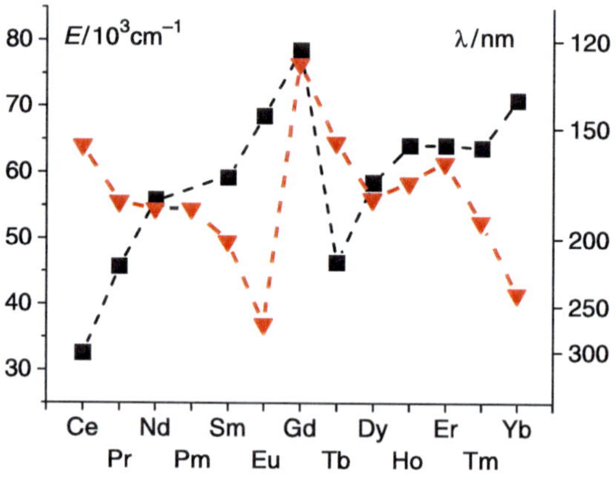

largely on the metal environment since the 5d orbitals are external and interact directly with the ligand orbitals. The 4f–5d transitions have high energies (Fig. 4) and only those of Ce^{III}, Pr^{III}, and Tb^{III} are commonly observed. Figure 5 shows the crystal-field splitting of both the $4f^1(^2F_{5/2}, {}^2F_{7/2})$ and $5d^1(^2D_{3/2}, {}^2D_{5/2})$ electronic configurations of Ce^{III} in D_{3h} symmetry. In the spectrum displayed, the third transition to $^2D_{5/2}$ is not observed because it lies at too high energy. Conversely, the Ce^{III} luminescence can be tuned from about 290 to 450 nm, depending on the matrix into which the metal ion is inserted, because of large crystal-field effects on the $5d^1$ excited configuration.

Charge-transfer transitions, both LMCT and MLCT, are allowed and have also high energies (Fig. 4), so that only the LMCT of Eu^{III} and Yb^{III} (possibly Sm^{III} and Tm^{III}) are commonly observed in ordinary solvents, contrary to d-transition metal

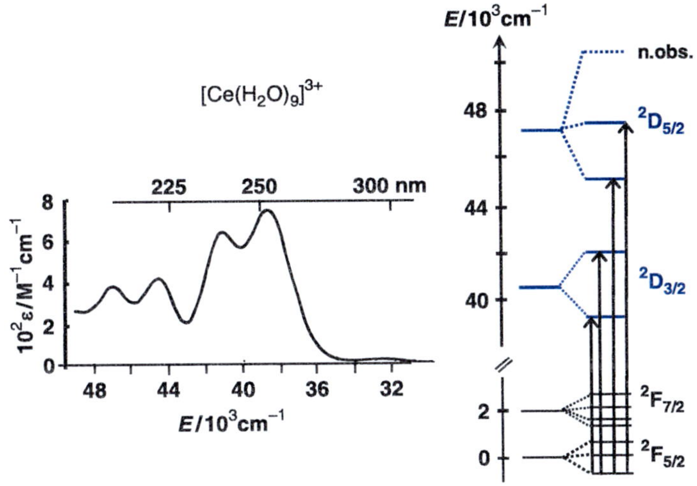

Fig. 5 *Left*: Absorption spectrum of $[Ce(H_2O)_9]^{3+}$ and *right*: its assignment (D_{3h} symmetry)

ions for which this type of transition is widespread. This is sometimes not well understood and the literature features many wrong assignments to MLCT transitions made by analogy to d-metal complexes.

3 Emission Spectra

With the exception of La^{III} and Lu^{III}, all Ln^{III} ions are luminescent and their f–f emission lines cover the entire spectrum, from UV (Gd^{III}) to visible (e.g., Pr^{III}, Sm^{III}, Eu^{III}, Tb^{III}, Dy^{III}, Tm^{III}) and near-infrared (NIR, e.g., Pr^{III}, Nd^{III}, Ho^{III}, Er^{III}, Yb^{III} [11]) ranges. Some ions are fluorescent ($\Delta S = 0$), others are phosphorescent ($\Delta S \neq 0$), and some are both. The f–f emission lines are sharp because the rearrangement consecutive to the promotion of an electron into a 4f orbital of higher energy does not perturb much the binding pattern in the molecules since 4f orbitals do not participate much in this binding (the covalency of a Ln^{III}–ligand bond is at most 5–7%). Therefore, the internuclear distances remain almost the same in the excited state, which generates narrow bands and very small Stokes' shifts. A different situation is met in organic molecules for which excitation leads frequently to a lengthening of the chemical bonds, resulting in large Stokes' shifts and since the coupling with vibrations is strong, in broad emission bands (Fig. 6). The main emission lines observed in Ln^{III} luminescence spectra are listed in Table 6, together with other key photophysical parameters.

As for absorption, emission of light through f–f transitions is achieved by either electric dipole or magnetic dipole mechanisms, and the selection rules detailed in

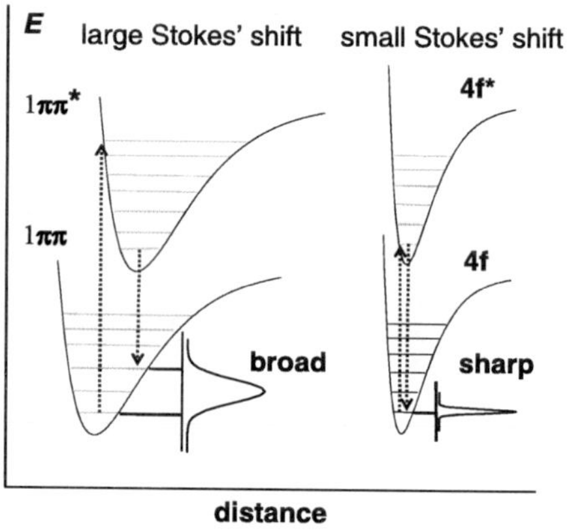

Fig. 6 Configurational coordinate diagram for emission from (*left*) an organic chromophore and (*right*) a lanthanide ion

Table 5 apply. Important parameters characterizing the emission of light from a Ln^{III} ion are the lifetime of the excited state $\tau_{obs} = 1/k_{obs}$ and the quantum yield Q. A general expression for the latter is simply

$$Q = \frac{\text{number of emitted photons}}{\text{number of absorbed photons}}. \tag{15}$$

The quantum yield is related to the rate at which the excited level is depopulated k_{obs} and to the radiative rate constant k^{rad}:

$$Q_{Ln}^{Ln} = \frac{k^{rad}}{k_{obs}} = \frac{\tau_{obs}}{\tau^{rad}}. \tag{16}$$

Subscript and superscript "Ln" have been added to avoid confusion with the other definition of quantum yield discussed below. The quantity defined in (16) is called the *intrinsic quantum yield*, that is, the quantum yield of the metal-centered luminescence upon direct excitation into the 4f levels. Its value reflects the extent of nonradiative deactivation processes occurring both in the inner- and outer-coordination spheres of the metal ion. The rate constant k_{obs} is the sum of the rates of the various deactivation processes:

$$k_{obs} = k^{rad} + \sum_n k_n^{nr} = k^{rad} + \sum_i k_i^{vibr}(T) + \sum_j k_j^{pet}(T) + \sum_k k'_k \,\text{nr}, \tag{17}$$

where k^{rad} and k^{nr} are the radiative and nonradiative rate constants, respectively; the superscript *vibr* points to vibration-induced processes while *pet* refers to photo-induced electron transfer processes such as those generated by LMCT states, for instance; the rate constants k' are associated with the remaining deactivation paths.

In absence of nonradiative deactivation processes, $k_{obs} = k^{rad}$ and the quantum yield would be equal to 1, which is very rare. Examples are, in solid state and under

Table 6 Ground (G), main emissive (I) and final (F) states for the most important f–f emission bands, approximate corresponding wavelengths (λ), energy gap between the emissive state and the highest SO level of the receiving state, and radiative lifetime of Ln^{III} ions. More NIR lines are listed in [11]

Ln	G	I	F	λ/μm or nm[a]	Gap/cm^{-1},[a]	τ^{rad}/ms[a]
Ce	$^2F_{5/2}$	5d	$^2F_{5/2}$	Tunable, 300–450	–	–
Pr	$^3H_4$	$^1D_2$	$^3F_4$, $^1G_4$, $^3H_4$, $^3H_5$	1.0, 1.44, 600, 690	6,940	(0.05[b]–0.35)
		$^3P_0$	^3H_J (J = 4–6)	490, 545, 615, 640,	3,910	(0.003[b]–0.02)
		$^3P_0$	^3F_J (J = 2–4)	700, 725		
Nd	$^4I_{9/2}$	$^4F_{3/2}$	^4I_J (J = 9/2–13/2)	900, 1.06, 1.35	5,400	0.42 (0.2–0.5)
Sm	$^6H_{5/2}$	$^4G_{5/2}$	^6H_J (J = 5/2–13/2)	560, 595, 640, 700, 775	7,400	6.26 (4.3–6.3)
		$^4G_{5/2}$	^6F_J (J = 1/2–9/2)	870, 887, 926, 1.01, 1.15		
		$^4G_{5/2}$	$^6H_{13/2}$	877		
Eu[c]	$^7F_0$	$^5D_0$	^7F_J (J = 0–6)	580, 590, 615, 650, 720, 750, 820	12,300	9.7 (1–11)
Gd	$^8S_{7/2}$	$^6P_{7/2}$	$^8S_{7/2}$	315	32,100	10.9
Tb	$^7F_6$	$^5D_4$	^7F_J (J = 6–0)	490, 540, 580, 620, 650, 660, 675	14,800	9.0 (1–9)
Dy	$^6H_{15/2}$	$^4F_{9/2}$	^6H_J (J = 15/2–9/2)	475, 570, 660, 750	7,850	1.85 (0.15–1.9)
		$^4I_{15/2}$	^6H_J (J = 15/2–9/2)	455, 540, 615, 695	1,000	3.22[b]
Ho[d]	$^5I_8$	$^5S_2$	^5I_J (J = 8,7)	545, 750	3,000	0.37 (0.51[b])
		$^5F_5$	$^5I_8$	650	2,200	0.8[b]
		$^5F_5$	$^5I_7$	965		
Er[e]	$^4I_{15/2}$	$^4S_{3/2}$	^4I_J (J = 15/2, 13/2)	545, 850	3,100	0.7[b]
		$^4F_{9/2}$	$^4I_{15/2}$	660	2,850	0.6[b]
		$^4I_{9/2}$	$^4I_{15/2}$	810	2,150	4.5[b]
		$^4I_{13/2}$	$^4I_{15/2}$	1.54	6,500	0.66 (0.7–12)
Tm	$^3H_6$	$^1D_2$	$^3F_4$, $^3H_4$, $^3F_3$, $^3F_2$	450, 650, 740, 775	6,650	0.09
		$^1G_4$	$^3H_6$, $^3F_4$, $^3H_5$	470, 650, 770	6,250	1.29
		$^3H_4$	$^3H_6$	800	4,300	3.6[b]
Yb	$^2F_{7/2}$	$^2F_{5/2}$	$^2F_{7/2}$	980	10,250	1.3 or 2.0[f]

[a]Values for the aqua ions [8], otherwise stated, and ranges of observed lifetimes in all media, if available, between parentheses
[b]Doped in Y_2O_3 or in $YLiF_4$ (Ho), or in $YAl_3(BO_3)_4$ (Dy)
[c]Luminescence from $^5D_1$, $^5D_2$, and $^5D_3$ is sometimes observed as well
[d]The laser transition $^5I_7 \rightarrow {}^5I_8$ (2.1–2.2 μm) is used in medical surgery of the eyes
[e]Luminescence from four other states has also been observed: $^4D_{5/2}$, $^2P_{3/2}$, $^4G_{11/2}$, $^2H_{9/2}$
[f]Complexes in solution: 1.2–1.3 ms; solid-state inorganic compounds: ≈2 ms

excitation at 254 nm, Y_2O_3:Eu (5%) with $Q = 0.99$ and terbium benzoate with $Q = 1$ [12]; in solution, a terbium complex with a dipyrazoylpyridine bearing aminocarboxylate coordinating groups was reported having $Q = 0.95$ [13].

Temperature-dependent vibrational deactivation processes can often be fitted to an Arrhenius-type of equation [14]:

$$\ln(k_{\mathrm{obs}} - k_0) = \ln A - \frac{E_{\mathrm{A}}}{RT}, \tag{18}$$

where k_0 is the rate constant at 0 K (practically: at 4 K, or even at 77 K) which allows one to decipher which vibration is responsible for it; examples are presented in references [15] and [16].

The intrinsic quantum yield essentially depends on the energy gap ΔE between the emissive state of the metal ion and the highest sublevel of its ground, or receiving, multiplet. The smaller this gap, the easier is its closing by nonradiative deactivation processes, for instance, through vibrations of bound ligands, particularly those with high energy such as O–H, N–H, or C–H. With the assumption that the deactivating phonons involved have all the same energy $\hbar\omega$, the temperature-dependent rate constant $k^{\text{vibr}}(T)$ for the quenching of a single excited level is described by the following expression [2]:

$$k^{\text{vibr}}(T) = k^{\text{vibr}}(0)\left(1 - e^{-\hbar\omega/k_BT}\right)^{-i} \quad \text{with} \quad i = \frac{\Delta E}{\hbar\omega}, \tag{19}$$

where k_B is Boltzmann's constant (1.38×10^{-23} J K^{-1} ≈ 0.695 cm^{-1}), i the number of phonons required to bridge the gap, and $k^{\text{vibr}}(0)$ the spontaneous rate at 0 K. The latter heavily depends on the order n of the process. In practice, the excited level possesses several crystal-field sublevels, the population of which is in thermal equilibrium. This equilibrium is reached in times short compared to the multi-phonon decay time, but since phonon-induced decay rates are significantly slower for the upper levels in view of the larger energy gaps, depopulation of the lower crystal field sublevel is the major contribution to the deactivation process. A rule of thumb is that radiative de-excitation will compete efficiently with multi-phonon processes if the energy gap is more than 6 quanta of the most energetic vibration present in the molecule. This type of nonradiative deactivation is especially detrimental to NIR luminescence [11]: for Er^{III}, for instance, a C–H vibrator located outside the inner coordination sphere at a distance between 20 and 30 Å from the emitting center induces a radiationless rate equal to the radiative one.

Determination of the intrinsic quantum yield with (16) requires evaluation of the radiative lifetime which is related to Einstein's rates of spontaneous emission A from an initial state $|\Psi_J\rangle$, characterized by a quantum number J, to a final state $|\Psi'_{J'}\rangle$:

$$A(\Psi_J, \Psi'_{J'}) = k^{\text{rad}} = \frac{1}{\tau^{\text{rad}}} = \frac{64\pi^4\tilde{\nu}^3}{3h(2J+1)}\left[\frac{n(n^2+2)^2}{9}D_{\text{ED}} + n^3D_{\text{MD}}\right], \tag{20}$$

where $\tilde{\nu}$ is the mean energy of the transition defined in (13b), h is Planck's constant, n is the refractive index; D_{ED} is given by (12) and D_{MD} by (21):

$$D_{\text{MD}} = \left(\frac{eh}{4\pi m_e c}\right)^2 |\langle\Psi||\text{L}+2\text{S}||\Psi'\rangle|^2. \tag{21}$$

The bracketed matrix elements are tabulated and the radiative lifetime can, therefore, be extracted from the spectral intensity, that is from (12), (20), and (21). Except in few cases, this calculation is not trivial and large errors may occur, including those pertaining to the hypotheses made within JO theory.

In particular, it has been assumed that the emitting and receiving levels are $(2J + 1)$-fold degenerate or, if split by crystal field effects, that all the sublevels are equally populated. This is obviously not true and in the case of Er^{III} this may lead to up to 20% errors. Even larger errors have been found for Ln^{III} tris(dipicolinate) [7].

On the other hand, if the absorption spectrum corresponding to an emission spectrum is known, which may be the case when the luminescence transitions terminate onto the ground level, the radiative lifetime can be simply calculated from the following equation where N_A is Avogadro's number (6.023×10^{23}):

$$\frac{1}{\tau^{\mathrm{rad}}} = 2303 \times \frac{8\pi c n^2 \tilde{\nu}^2 (2J+1)}{N_A(2J'+1)} \int \varepsilon(\tilde{\nu}) d\tilde{\nu}. \quad (22)$$

In the special case of Eu^{III} for which one transition ($^5D_0 \rightarrow {}^7F_1$) has pure magnetic origin, a convenient simplified equation can be derived [17]:

$$A(\Psi_J, \Psi'_{J'}) = \frac{1}{\tau^{\mathrm{rad}}} = A_{\mathrm{MD},0} \cdot n^3 \left(\frac{I_{\mathrm{tot}}}{I_{\mathrm{MD}}}\right), \quad (23)$$

with $A_{\mathrm{MD},0}$ being a constant equal to 14.65 s^{-1} and ($I_{\mathrm{tot}}/I_{\mathrm{MD}}$) the ratio of the total integrated emission from the Eu($^5D_0$) level to the ^7F_J manifold (J = 0–6) to the integrated intensity of the MD transition $^5D_0 \rightarrow {}^7F_1$.

An example of radiative lifetime calculations based on the various procedures evoked above can be found in reference [7], the results of which are collected in Table 7. The limits of JO approach are clearly seen both for Eu (error $\approx -20\%$) and Tb (error $> -100\%$). On the other hand, (22) and (23) yield trustworthy results: a radiative lifetime of 1.2 ms has been reported for $[Yb(dtpa)]^{2-}$ [17].

Finally, there are two important points to be stressed here in order to correct many errors reported in the literature. Firstly, the radiative lifetime is characteristic of one emitting state. If several excited states of an Ln^{III} ion emit light, then each of

Table 7 Experimental intrinsic quantum yields, observed and radiative lifetimes of $M_3[Ln(dpa)_3]$ (M = Na or Cs) samples in solution (0.1 M Tris–HCl, pH 7.4) and solid state at 295 K; 2σ are given between parentheses [7]

Sample	Q_{Ln}^{Ln} / %	τ_{obs}/ms	τ^{rad}/ms[a]		
			(i)	(ii)	(iii)
Eu, 1.8–3.7 × 10^{-2} M	41(2)	1.7(0.1)	4.1(3)	3.15	4.0
Eu, solid state[b]	68(4)	1.8(0.1)	2.6(2)	–	2.7
Tb, 2.0 × 10^{-2} M	–[c]	1.74(1)	–	1.0	–
Tb, solid state[b]	72(5)	1.36(2)	1.9(1)	–	–
Yb, 4.04 × 10^{-2} M	–[c]	2.23(1)[d]	–	–	1.31(2)

[a]Successively: (i) experimental, i.e., from (16), (ii) from JO theory (20), (iii) from (23) for Eu and (22) for Yb
[b]Refractive index = 1.517
[c]Determination not feasible
[d]In μs

them will have a characteristic radiative lifetime. Moreover, the radiative lifetime is not a constant for a given ion and a given electronic level. Indeed, there is a dependence on the refractive index, as clearly shown in (20), so that transposition of a literature value to a specific compound cannot be made directly, which explains the wide range of τ^{rad} values reported for an individual Ln^{III} ion (Table 6).

4 Sensitization of Lanthanide Luminescence

Since the dipole strengths of f–f transitions are very small, direct excitation into the 4f excited levels rarely yields highly luminescent materials, even if the intrinsic quantum yield is large, unless considerable excitation power is used (laser excitation, given the sharpness of the absorption bands). Therefore, an alternative path has been worked out which is called *luminescence sensitization* or *antenna effect*. The luminescent ion is imbedded into a matrix or an organic environment such that the latter is a good light harvester. Energy is then transferred from the excited surroundings onto the metal ion which eventually gives off its characteristic light. Note, that several of the Ln^{III} excited states may be implied in this process. The use of charge transfer or 4f5d states for collecting and transferring energy has long been well established in inorganic phosphors for lighting applications. On the other hand, the tuning of the electronic properties of organic ligands to achieve the same goal starts only to be understood since the process is more involved in view of the numerous electronic levels and mechanisms which may be implied.

Here, we focus on the latter case for which efficient light-harvesting is mainly performed by the aromatic ($\pi \rightarrow \pi^*$) and/or ($n \rightarrow \pi^*$) transitions of unsaturated ligands displaying large cross sections for one-photon absorption. Alternatively, singlet states, intra-ligand charge transfer states (ILCT), ligand-to-metal charge transfer states (LMCT), or ^{3}MLCT states localized on a transition-metal containing ligand may also play this role [18]. As a result of the poor expansion of the 4f orbitals, the Ln–ligand bonds are mainly electrostatic and only some minute mixing of metal and ligand electronic wavefunctions contributes to covalency. It, therefore, appears justified to consider separately ligand-centered and metal-centered excited states in lanthanide complexes, and a Jablonsky diagram is adequate for representing energy migration paths (Fig. 7). In this diagram, grey arrows representing energy transfer to the metal ion do not point to a specific excited state since several of them may intervene.

One of the main energy migration path implies Laporte- and spin-allowed ligand-centered absorptions followed by intersystem crossing ($^1S^* \rightarrow {}^3T^*$, k_{ISC}) reaching the long-lived ligand-centered triplet state, from which $^3T^* \rightarrow Ln^*(k^{et})$ energy transfer occurs. Spontaneous metal-centered radiative emission completes the light-conversion process. It is to be stressed that although important, this energy transfer path is by far not the only operative one. Kleinerman who studied over 600 lanthanide chelates pointed out as early as 1969 that excited singlet states may

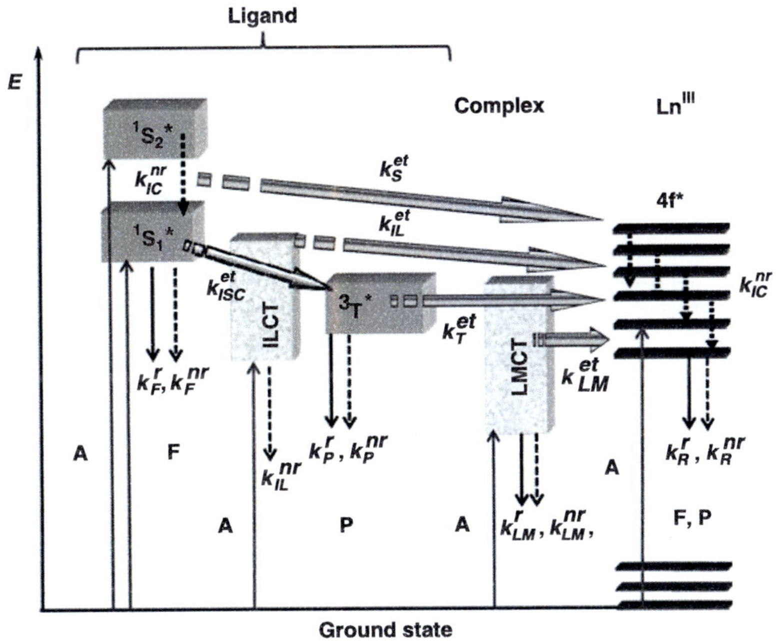

Fig. 7 Schematic representation of energy absorption, migration, emission (*plain arrows*) and dissipation (*dotted arrows*) processes in a lanthanide complex. ^{1}S* or S = singlet state, ^{3}T* or T = triplet state, A = absorption, F = fluorescence, P = phosphorescence, k = rate constant, r = radiative, nr = nonradiative, IC = internal conversion, ISC = intersystem crossing, ILCT (or IL) = intra-ligand charge transfer, LMCT (or LM) = ligand-to-metal charge transfer. Back transfer processes are not drawn for the sake of clarity

contribute to the transfer and may even be the privileged donor states, depending on the relative values of the rate constants for the various intervening processes [19]. Since triplet states are quenched by oxygen, an easy way of establishing their involvement in the energy transfer process is to bubble oxygen into the solution: a reduction in Ln^{III} luminescence proves their role but no effect may simply means $k^{et} > k^{O2}$. In fact a workable model of the entire energy-converting mechanism has shown that considering as many as 20–30 rate constants (including those describing back transfers) may be necessary [20]. All these potential energy funnels, particularly if ILCT, LMCT, as well as ^{3}MLCT states are included, render difficult the a priori precise design of highly luminescent lanthanide-containing edifices.

Once the ligand is excited, subsequent intramolecular energy migrations obey Fermi's golden rule governing resonant energy transfer (24), whereby W_{DA} is the probability of energy transfer, Ω_{DA} is the spectral overlap integral between the absorption spectrum of the acceptor A and the emission spectrum of the donor D, while $\mathcal{H}'$ is the perturbation operator in the matrix element $< D^*A|\mathcal{H}'|DA^* >$.

$$W_{DA} = (4\pi^2/h) \cdot | < D^*A|\mathcal{H}'|DA^* > |^2 \cdot \Omega_{DA}. \tag{24}$$

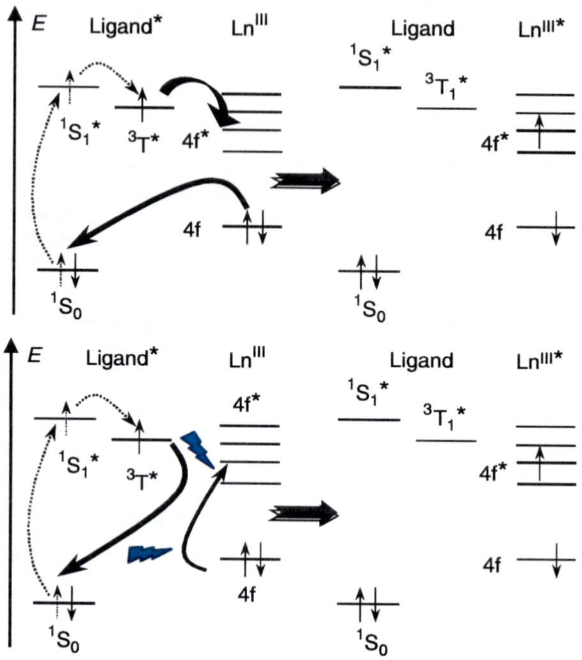

Fig. 8 Dexter (*top*) and Förster (*bottom*) energy transfer mechanisms

Depending on the electromagnetic nature of $\mathcal{H}'$, a double-electron exchange (Dexter) mechanism or an electrostatic multipolar (Förster) mechanism have been proposed and theoretically modeled. They are sketched on Fig. 8 for the simple ^{1}S*-^{3}T*-Ln* path. Their specific dependences on the distance d separating the donor D from the acceptor A, i.e., $e^{-\beta d}$ for double-electron exchange and d^{-6} for dipole–dipolar processes, respectively, often limit Dexter mechanism to operate at short distance (typically 30–50 pm) at which orbital overlap is significant, while Förster mechanism may extend over much longer distances (up to 1,000 pm).

In addition, a d^{-8}-dependent dipole–quadrupolar mechanism may also be quite effective at short to medium-range distances; in fact, depending on Ω_{DA}, it may be as efficient as the dipole–dipole mechanism up to distances as long as 300 pm.

For lanthanides possessing low-lying charge-transfer excited states (e.g., Eu^{III}, Sm^{III}) or for complexes having low-lying ILCT states, the energy transfer process is further affected by additional nonradiative quenching arising from back energy transfer onto the ligand (not shown on Fig. 7). Since in this case the accepting states are quite broad, minute differences in their energy may lead to large differences in the spectral overlap, and therefore, in the overall quantum yield. Yb^{III} represents a special case since it has only one, low-lying, excited state ($^2F_{5/2}$) and several excitation mechanisms have been proposed [11].

For the molecular edifices discussed here, another definition of quantum yield ought to be made: the *overall quantum yield*, that is the quantum yield of the metal-centered luminescence upon ligand excitation. It is related to the intrinsic quantum yield by the following equation:

$$Q_{\mathrm{L}}^{\mathrm{Ln}} = \eta_{\mathrm{pop}}^{\mathrm{D}}\, \eta_{\mathrm{et}}\, Q_{\mathrm{Ln}}^{\mathrm{Ln}} = \eta_{\mathrm{sens}}\, Q_{\mathrm{Ln}}^{\mathrm{Ln}}. \tag{25}$$

The three components in the middle term represent (1) the efficiency $\eta_{\mathrm{pop}}^{\mathrm{D}}$ with which the feeding level (^{3}T, ILCT, LMCT, ^{3}MLCT, possibly a 4f5d state) is populated by the initially excited state (the corresponding rate constant is $k_{\mathrm{ISC}}^{\mathrm{et}}$ if ^{1}S* is excited and ^{3}T* is the donor level, see Fig. 7), (2) the efficiency of the energy transfer (η_{et}) from the donor state to the accepting $\mathrm{Ln^{III}}$ ion, and (3) the intrinsic quantum yield. The overall sensitization efficiency, η_{sens} can be accessed experimentally if both the overall and intrinsic quantum yields are known or, alternatively, the overall quantum yield and the observed and radiative lifetimes:

$$\eta_{\mathrm{sens}} = \frac{Q_{\mathrm{L}}^{\mathrm{Ln}}}{Q_{\mathrm{Ln}}^{\mathrm{Ln}}} = Q_{\mathrm{L}}^{\mathrm{Ln}}\, \frac{\tau^{\mathrm{rad}}}{\tau_{\mathrm{obs}}}. \tag{26}$$

The lifetime method is especially easy to implement for $\mathrm{Eu^{III}}$ compounds since the radiative lifetime is readily determined from the emission spectrum via (23). Some data are reported in Table 8.

An important remark at this stage is that the intrinsic quantum yield is directly proportional to τ_{obs}, but not necessarily the overall quantum yield since a change in

Table 8 Quantum yields, observed and radiative lifetimes, as well as sensitization efficiency for $\mathrm{Eu^{III}}$ tris(dipicolinate) and bimetallic $\mathrm{Eu^{III}}$ helicates; all data are at room temperature, for solutions in Tris–HCl 0.1 M (pH = 7.4); 2σ are given between parentheses [7, 21]. See Fig. 9 for formulae

Sample	Q_L^{Eu}	Q_{Eu}^{Eu}	τ_{obs}/ms	τ^{rad}/ms	η_{sens}
$[\mathrm{Eu(dpa)_3}]^{3-}$	0.29(2)	0.41(2)	1.7(1)	4.1(3)	0.76(6)
$[\mathrm{Eu_2(L^{C1})_3}]$	0.24(2)	0.37(4)	2.4(1)	6.8(3)	0.67(10)
$[\mathrm{Eu_2(L^{C2})_3}]$	0.21(2)	0.37(4)	2.4(1)	6.9(3)	0.58(8)
$[\mathrm{Eu_2(L^{C3})_3}]$	0.11(2)	0.36(4)	2.2(1)	6.2(3)	0.30(5)

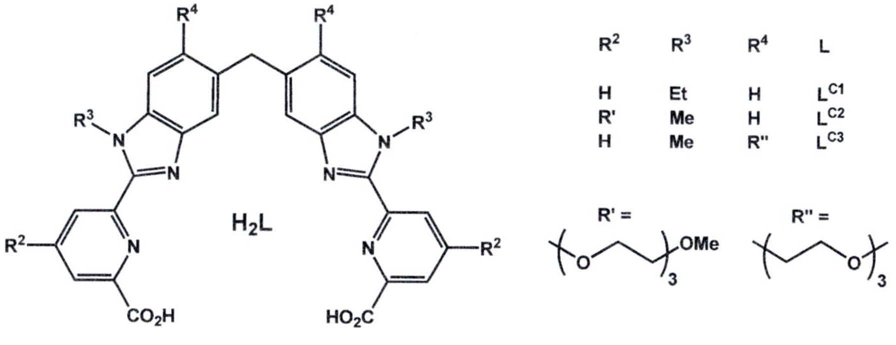

Fig. 9 Chemical structures of the ligands mentioned in Table 8

the inner coordination sphere inducing, for instance, a smaller nonradiative deactivation rate may also influence η_{sens} (in one way or the other) through the resulting electronic changes in the molecular edifice. Another point is that the distinction between intrinsic and overall quantum yields is often unclear in the literature, particularly for NIR-emitting ions for which direct experimental determination of the quantum yield is rarely performed. Authors commonly rely on (16) to evaluate the intrinsic quantum yield from lifetime determination and from a "literature" value of τ^{rad}, so that extreme caution must be exercised in discussing these estimates [11].

4.1 Design of Efficient Lanthanide Luminescent Bioprobes

The ligand design for building efficient lanthanide luminescent bioprobes (LLBs) must meet several requirements, both chemical, photophysical, and biochemical: (1) efficient sensitization of the metal luminescence, (2) embedding of the emitting ion into a rigid and protective cavity minimizing nonradiative deactivation, (3) long excited state lifetime, (4) water solubility, (5) large thermodynamic stability, (6) kinetic inertness, (7) intense absorption above 330 nm, and (8) whenever relevant, ability to couple to bioactive molecules while retaining their photophysical properties and not altering the bio-affinity of the host.

From the chemical point of view, it is best when the coordination sphere is saturated, i.e., when 8–10 donor atoms are bound to the metal ion, and when the coordinating groups are strong since in vivo experiments require large pLns (defined as $-\log[Ln^{III}]_{free}$ in water, at pH 7.4, $[Ln^{III}]_t = 1$ μM, and $[Ligand]_t = 10$ μM; ideally pLn should be >20). Carboxylates, aminocarboxylates, phosphonates, hydroxyquinolinates, and hydroxypyridinones are good candidates, while β-diketonates which have excellent photophysical properties have the tendency to be less stable. In aqueous solutions, the enthalpy and entropy changes upon complex formation between Ln^{III} cations and many ionic ligands are predominantly influenced by changes in hydration of both the cation and the ligand(s). Complexation results in a decrease in hydration, yielding positive entropy changes favorable to the complexation process. On the other hand, dehydration is endothermic and contribution from bond formation between the cation and the ligand(s) often does not compensate this unfavorable energy contribution to the variation in Gibbs free energy, so that the overall complexation process is generally entropy driven. Therefore, it is advantageous to resort to polydentate ligands for building a coordination environment around Ln^{III} ions. Macrocyclic complexes based on the cyclen framework [22] or on cryptands [23] are also proved to be quite adequate, as well as self-assembled mono- and bi-nuclear triple helical edifices [21].

The photophysical requirements are related to the two aspects described in (25): energy transfer (η_{sens}) and minimization of nonradiative processes (Q_{Ln}^{Ln}). The first one is difficult to master in view of the intricate processes going on (Fig. 7, [20]). Some authors have nevertheless tried to establish phenomenological rules. One has, however, to be cautious in applying them since these rules rely on a rather simple and naive picture: the $^1S^*$–$^3T^*$–Ln* energy transfer path on one hand, and the consideration that the only parameter of importance is the energy gap between $^3T^*$ and the emitting Ln^{III} level. Some of the relationships are exemplified on Fig. 10 and examination of these data clearly point to the difficulty in establishing a dependable relationship. The following lessons can be drawn from these data:

- The maximum values of the quantum yields usually occur when the triplet state energy is close to the energy of one of the higher excited states of the metal ion, consistent with the fact that the emissive level is usually not directly fed by the ligand excited states (except maybe in the case of the Schiff base complexes depicted at the bottom of Fig. 10). When the energy of the feeding state becomes closer to the energy of the emitting state, back-energy transfer operates and the quantum yield goes down: this is true for both Eu^{III} and Tb^{III} and a "safe" energy difference minimizing this process is around 2,500–3,000 cm^{-1}.
- Inspection of the Eu^{III} quantum yields clearly demonstrates that the energy of the triplet state corresponding to the larger values depends on the type of ligand: it is close to the $^5D_0$ level for Schiff base complexes, to the $^5D_1$ level for β-diketonates, and to the $^5D_2$ level for polyaminocarboxylates.
- For the two series of Eu^{III} and Tb^{III} complexes with the same polyaminocarboxylate ligands, the maximum values reached by the quantum yield of Tb^{III} are larger than those of Eu^{III}: this reflects the smaller Eu($^5D_0$–$^7F_6$) energy gap compared to the Tb($^5D_4$–$^7F_0$) gap (Table 6).

It has been shown for calixarenes that more efficient ISC transfers take place when the energy difference between the singlet and triplet states is around 5,000 cm^{-1}; therefore, ligand designers try to keep to the following phenomenological rules: $\Delta E(^1S^* - ^3T^*) \approx 5{,}000\ cm^{-1}$ and $\Delta E(^3T^* - Ln^*$ emissive level) in the range 2,500–3,500 cm^{-1}. These are, however, golden rules only and sometimes minute energy differences in the ligand states lead to large differences in overlap between the emission spectrum of the donor and the absorption spectrum of the acceptor, resulting in large differences in quantum yield [27].

The second aspect, namely minimization of nonradiative deactivation has two facets: avoiding low-lying LMCT states, essentially for Sm^{III}, Eu^{III}, and Yb^{III}, and avoiding high energy vibrations in the first and second coordination spheres; the latter aspect is dealt with in Sect. 5.4.

4.2 *Practical Measurements of Absolute Quantum Yields*

Quantum yield measurements are simple in their principle, but very difficult to carry out experimentally, particularly when it comes to the luminescence of lanthanide ions and to intrinsic quantum yield, f–f absorptions being faint. There are two main methods: the comparative method in which the sample under examination is compared to a standard with known quantum yield, and the absolute method which

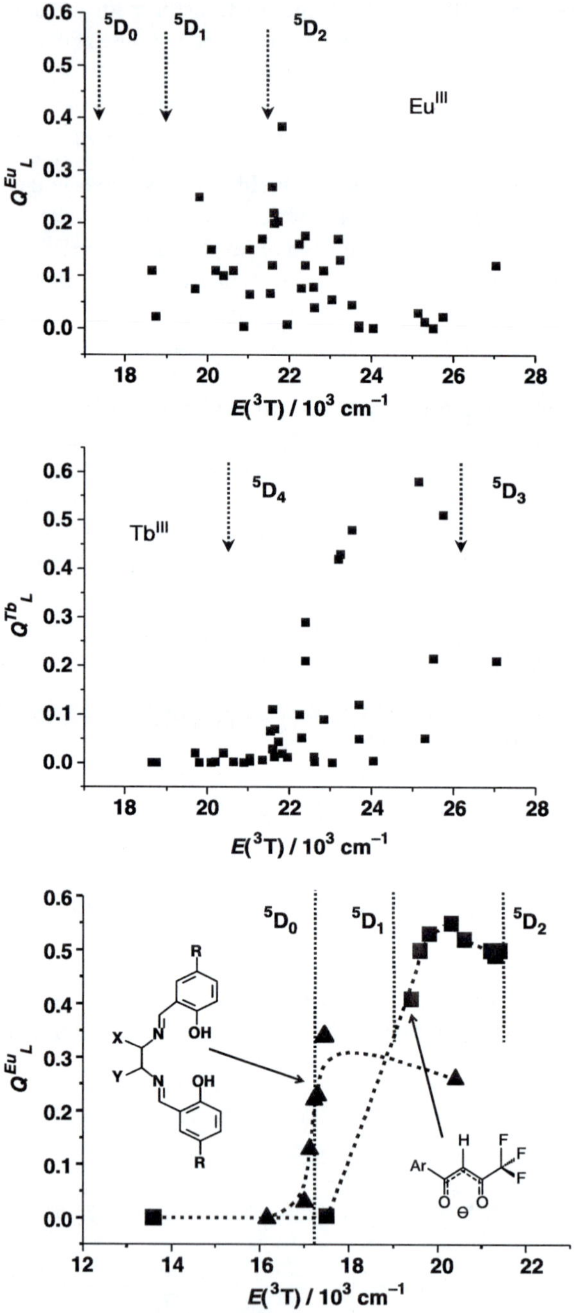

Fig. 10 Relationships between the triplet state energy of the feeding ligand and the quantum yields of 39 Eu^{III} (*top*) and Tb^{III} (*middle*) polyaminocarboxylates [24] and of Eu^{III} Schiff bases (*triangles*, [25]) and β-diketonates (*squares*, [26])

determines the amount of absorbed and emitted light with an integrating sphere since luminescence is emitted in all directions.

In the comparative method, the quantum yield of the unknown sample (indices x) is given relative to the quantum yield of the standard (indices S) by:

$$Q_{\mathrm{L}}^{\mathrm{Ln}} = Q_{\mathrm{S}} \times \frac{E_x}{E_{\mathrm{S}}} \times \frac{A_{\mathrm{S}}(\lambda_{\mathrm{S}})}{A_x(\lambda_x)} \times \frac{I_{\mathrm{S}}(\lambda_{\mathrm{S}})}{I_x(\lambda_x)} \times \frac{n_x{}^2}{n_{\mathrm{S}}{}^2}, \tag{27a}$$

with E being the integrated and corrected emission spectrum, A the absorbance at the excitation wavelength λ, I the intensity of the excitation source at the excitation wavelength, and n the refractive index. In principle, linearity between the intensity of the emitted light and the concentration of the sample is only achieved if $A < 0.05$, so that samples should be diluted to reach this value of absorbance, while being cautious not to dissociate the complex. In practice, if both the standard and the sample are excited at the same wavelength λ_{exc} (a highly desirable, although not always achievable situation), measurements can be safely carried out up to $A = 0.5$ since the inner-filter effect would be the same for both samples. In this case, (27a) simplifies to:

$$Q_{\mathrm{L}}^{\mathrm{Ln}} = Q_{\mathrm{S}} \times \frac{E_x}{E_{\mathrm{S}}} \times \frac{A_{\mathrm{S}}(\lambda_{\mathrm{exc}})}{A_x(\lambda_{\mathrm{exc}})} \times \frac{n_x{}^2}{n_{\mathrm{S}}{}^2}, \tag{27b}$$

or to (27c) if the standard and the unknown sample are in the same solvent:

$$Q_{\mathrm{L}}^{\mathrm{Ln}} = Q_{\mathrm{S}} \times \frac{E_x}{E_{\mathrm{S}}} \times \frac{A_{\mathrm{S}}(\lambda_{\mathrm{exc}})}{A_x(\lambda_{\mathrm{exc}})}. \tag{27c}$$

In case $A_x \neq A_S$ and at least one of them is >0.05, then the absorbances in (27a–c) should be replaced by:

$$A \rightarrow (1 - 10^{-A}), \tag{28}$$

to take into account the different inner-filter effects. The correction is small for small differences in absorbances but can become very important: e.g., if $A_S = 0.1$ and $A_x = 0.15$, 0.2, 0.4, and 0.6, respectively, then the corrected A_S/A_x ratios would be 0.42, 1.79, 2.93, and 3.64 instead of 0.5, 2, 4 and 6, respectively, corresponding to corrections of −5, −10, −19, −27, and −39%, respectively.

It is also essential that emission spectra are corrected for the instrumental function established with a standard calibrated lamp. It is wise not to use the calibration curve given by the manufacturer of the spectrometer and to re-measure this instrumental function at regular intervals because many items influence it, particularly the emission intensity of the excitation lamp and the quantum efficiency of the detector (which both decrease with time). In case (27a) is used, the excitation instrumental function has to be known as well. Regarding the standard, it is best when its emission spectrum overlaps the emission spectrum of the unknown

Table 9 Selected useful standards for quantum yield determinations at room temperature. More extensive listings can be found in [28, 29]

Compound[a]	conc./solvent[b]	Range (nm)	Q
Quinine sulfate	aq. H_2SO_4 (0.5 M)	400–600	0.546
Cresyl violet	Methanol	600–650	0.54(3)
$Cs_3[Tb(dpa)_3]$	6.5×10^{-5} M, Tris–HCl 0.1 M	480–670	0.22(2)
$[Ru(bpy)_3](ClO_4)_3$	10^{-5} M, aerated water	550–800	0.028(2)
	10^{-5} M, de-aerated water		0.043(2)
$Cs_3[Eu(dpa)_3]$	7.5×10^{-5} M, Tris–HCl 0.1 M	580–690	0.24(2)
$[Yb(tta)_3(H_2O)_2]$	10^{-3} M, toluene	950–1,080	0.0035

[a]bpy = bipyridine; tta = thenoyltrifluoroacetylacetonate
[b]When not given, the concentration should be such that $A < 0.1$ (usually $c < 10^{-5}$ M)

sample; a safe way to proceed is to use two different standards and to measure them against each other as well. Each measurement should also be repeated several times. Selected standards useful in lanthanide photophysics are listed in Table 9.

The absolute method using an integration sphere has been in use for a long time, particularly by physicists or scientists involved in the design of phosphor materials [30]. A specially designed integration sphere has been produced in our laboratory which fairly well fulfills the needs of chemists and biochemists. It has a small 2-inch diameter to ensure maximum sensitivity and is manufactured in Zenith® Teflon [7]. Samples are put in 2.4-mm I.D. quartz capillaries, themselves inserted into a protective quartz tube. The modified de Mello et al. method [30] requires the measurement of (1) L_a, the integrated intensity of light exiting the sphere when the empty capillary is illuminated at the excitation wavelength (Rayleigh scattering band); (2) L_c, the same integrated intensity at the excitation wavelength when the sample is introduced into the sphere; these two measurements often necessitate the use of attenuators (transmission 0.01–10%); (3) E_c the integrated intensity of the entire emission spectrum. The absolute quantum yield is then given by:

$$Q_{\text{abs}} = \frac{E_c}{[L_a(\lambda_{\text{exc}}) - L_c(\lambda_{\text{exc}})]F_{\text{att}}(\lambda_{\text{exc}})}, \tag{29}$$

whereby $F_{\text{att}}(\lambda_{\text{exc}})$ is the correction for the attenuators used. Reproducible and accurate data can be obtained when the fraction of absorbed light $\alpha = (L_a - L_c)/L_a$ is in the range 0.10–0.90 [7]; both solid state samples and solutions (minimum volume: 60 μL) can be measured. This method also requires carefully established instrumental functions; it is illustrated in Fig. 11.

5 Information Extracted from Lanthanide Luminescent Probes

The most important applications of lanthanide luminescent stains used as structural or analytical probes are summarized below. Any luminescent Ln^{III} may act as a luminescent probe, but some ions either bear more information or are more

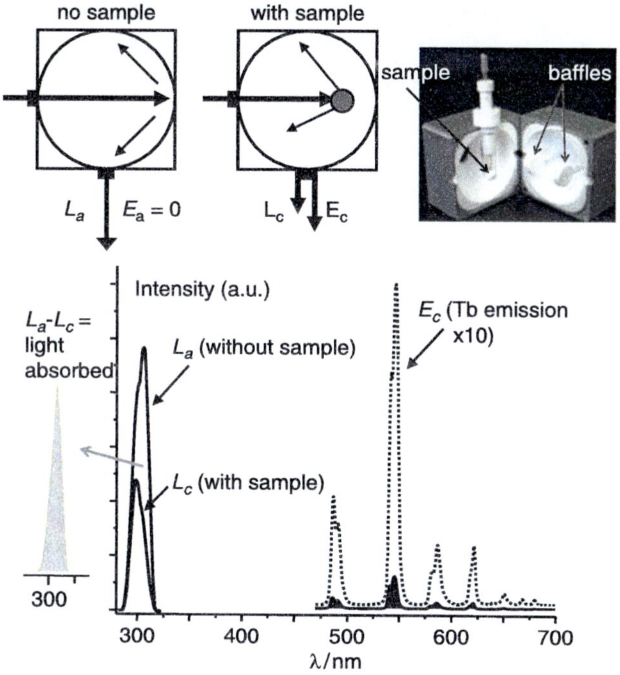

Fig. 11 *Top*: Integration sphere and *bottom*: example of quantum yield determination on a Tb^{III} sample (this work)

luminescent than others, so that they are preferentially used. This is the case of Eu^{III}, which will be extensively referred to in the following description [31].

5.1 Metal Ion Sites: Number, Composition, and Population Analysis

Lanthanide ions have been used as substitutes for Ca^{II} and Zn^{II} in proteins to obtain information on the number of metallic sites (by simple titration) and on their composition. This may of course be extended to any molecule or materials. One very useful transition in this respect is the highly forbidden and faint Eu ($^5D_0 \rightarrow {}^7F_0$) transition which is best detected in excitation mode by analyzing the emission of the hypersensitive transition $^5D_0 \rightarrow {}^7F_2$; since both the emitting and end states are non-degenerate, its number of components indicates the number of different metal–ion sites. Moreover, the energy of this transition depends on the nephelauxetic effect δ_i generated by coordinated atoms and ions; at 298 K:

$$\tilde{\nu}_{\text{calc}} = 17{,}374 + C_{\text{CN}} \sum_{i=1}^{CN} n_i\, \delta_i, \tag{30}$$

with C_{CN} being a constant depending on the coordination number and n_i the number of coordinating groups with nephelauxetic effect δ_i. The latter are tabulated for the most common ligands [32] and some predictions are rather accurate [21], which allows one to check the composition of the inner coordination sphere.

When several metal ion sites are present in a compound, population analysis can be carried out in two main ways. For Eu^{III}, since the intensity of the MD transition $^5D_0 \rightarrow {}^7F_1$ is independent of the metal–ion environment, a spectral decomposition of the transition recorded under broad band excitation into its components measured under selective laser excitation, followed by integration yields the population P_i of each site [31]. More generally, one can rely on lifetime measurement, since the luminescence decay will be a multi-exponential function which may be analyzed, for instance, with Origin®, using the following equations:

$$I(t) = A + I_0 \sum_{i=1}^{n} B_i e^{-k_i t}, \tag{31a}$$

$$P_i = \frac{B_i/k_i}{\sum\limits_{i=1}^{n} (B_i/k_i)}. \tag{31b}$$

In recording the decay, one has to make sure that (1) there is no artifact at the beginning of the decay (remaining light from the light pulse), (2) the decay is recorded during at least 5–6 lifetimes, (3) the signal at the end goes back to the background value, and (4) the decay is defined by a sufficient number of data points. Even if experimental data are of high quality, it is difficult to determine populations smaller than 5 % and to decompose decays with more than 2 or 3 exponential functions or when the two lifetimes are either very different or quite similar. The example given on Fig. 12 illustrates a bi-exponential analysis of an Eu($^5D_0$) decay.

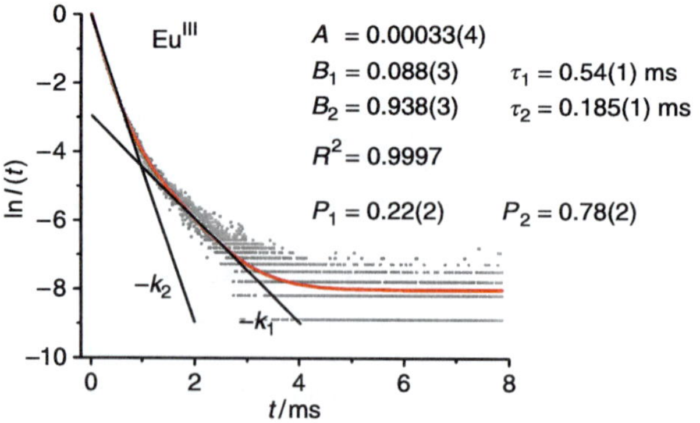

Fig. 12 Luminescence decay for an Eu^{III} sample with its bi-exponential analysis; straight lines correspond to the two decay rates and the red line is the calculated fit (this work)

5.2 Site Symmetry Through Crystal-Field Analysis

This aspect is related to the Stark splitting of the levels described in Table 3. Again here, Eu^{III} luminescence is the easiest to analyze given the non-degeneracy of the emissive $^5D_0$ level. When allied to high-resolution selective laser excitation of components of the $^5D_0 \leftarrow {}^7F_0$ transition, symmetry of multi-site molecules and materials can be worked out easily, based on group-theoretical considerations [31, 33]. Such detailed analyses are not discussed here.

On the other hand, the $Eu(^5D_0 \rightarrow {}^7F_1)$ transition represents an interesting case. A simple examination of its splitting tells immediately at which crystal system the compound belongs: cubic if only 1 component is seen, axial (i.e., hexagonal, tetragonal, or trigonal, labeled A and E in group-theoretical notation) if there are two components, and low symmetry if the maximum splitting of three appears. For truly low-symmetry species, the three components are equally spaced and tend to have the same intensity. However, when the coordination sphere is close to an idealized higher symmetry, the splitting is unsymmetrical. In this case, three important pieces of information can be extracted for symmetries close to axial symmetry: (1) the sign of the B^0_2 crystal-field parameter which depends on the relative energetic position of the A and E sublevels of $^7F_1$, (2) its value thanks to a phenomenological relationship between ΔE(A–E) and this parameter [34], and (3) the extent of the deviation from the idealized symmetry given by the splitting of the E sublevel. In the example depicted on Fig. 13, the crystal field parameter has a value of ca $-600\ cm^{-1}$ and the coordination polyhedron EuN_6O_3 appears to be only slightly distorted from the idealized D_{3h} symmetry with ΔE(E–E) equal to 31 cm^{-1}.

5.3 Strength of Metal–Ligand Bonds: Vibronic Satellite Analysis

The analysis depicted above requires high-resolution spectra. It is sometimes complicated by the occurrence of vibronic satellites which may artificially increase the number of components of a given transition, so again care has to be exercised. Vibronic transitions have the tendency to be strongest when associated with

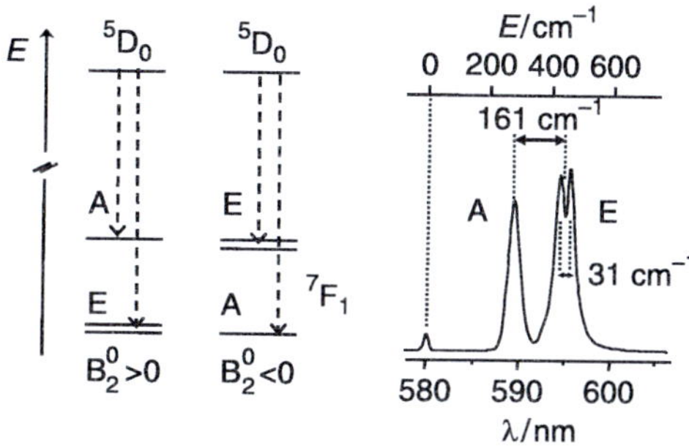

Fig. 13 *Left*: Ligand-field splitting of the $Eu(^7F_1)$ sublevel and sign of the B^0_2 parameter. *Right*: $^5D_0 \rightarrow {}^7F_1$ transition in $[Eu_2(L^{C2})_3]$ (redrawn from ref. [21], see Fig. 9 for formula)

hypersensitive transitions, for instance, Eu($^5D_0 \rightarrow {}^7F_2$); in addition, the $^7F_2$ level corresponds to an energy range (1,000–1,500 cm^{-1}) in which the density of phonon states is usually large. On the other hand, these satellites provide useful information on the strength of the Ln–ligand bond: the larger this strength, the more intense the satellites [35].

5.4 Solvation State of the Metal Ion

Quenching of the lanthanide luminescence by high-energy vibrations is a major concern in the design of highly luminescent probes. Multi-phonon deactivation is very sensitive to the metal–ligand distance and the phenomenon can be reasonably modeled by Förster's dipole–dipole mechanism [2]. The more phonons needed to bridge the energy gap, the less likely the quenching phenomenon to occur. Table 10 illustrates this phenomenon for solutions of complexes in water ($\tilde{\nu}$(O–H) = 3,600 cm^{-1}) and deuterated water ($\tilde{\nu}$(O–D) = 2,200 cm^{-1}) by listing the lifetimes of the excited level in these two solvents.

Although detrimental to the emission intensity, vibrational quenching allows one to assess the number of water molecules q interacting in the inner-coordination sphere. Several phenomenological equations have been proposed, based on the assumptions that O–D oscillators contribute little to deactivation and that all the other deactivation paths are the same in water and in deuterated water and can henceforth be determined by measuring the lifetime in the deuterated solvent. An important point for their application is to make sure that quenching by solvent vibrations is by far the most important deactivation process in the molecule. If other temperature-dependent phenomena (e.g., phonon-assisted back transfer) are operating, these relationships become unreliable. This has often been observed with Tb^{III} [27].

Table 10 Illustration of the energy gap law with respect to quenching of the Ln^{III} luminescence by high-energy vibrations. Samples are dilute solutions of perchlorates or triflates at room temperature [11]

Ln	ΔE/cm^{-1}	No. of phonons		Lifetime/μs	
		OH	OD	H_2O	D_2O
Gd	32,100	9	15	2,300	n.a.
Tb	14,800	4	7	467	3,800
Eu	12,300	3–4	5–6	108	4,100
Yb	10,250	3	4.5	0.17[a]	3.95
Dy	7,850	2–3	3–4	2.6	42
Sm	7,400	2	3	2.7	60
Er	6,600	2	3	n.a.	0.37
Nd	5,400	1–2	2–3	0.031	0.14

[a]Estimated from quantum yields in water and deuterated water and from $\tau(D_2O)$

Altogether, such relationships, which exist for Nd^{III}, Sm^{III}, Eu^{III}, Tb^{III}, Dy^{III}, and Yb^{III}, are to be used with care and bearing in mind their peculiar calibration. The general form of these relationships is:

$$q = A \times (\Delta k_{obs} - B) - C, \tag{32a}$$

$$\Delta k_{obs} = k_{H2O} - k_{D2O} = 1/\tau(H_2O) - 1/\tau(D_2O), \tag{32b}$$

where A, B, and C are phenomenological Ln-depending (and sometimes ligand-depending) parameters determined using series of compounds with known hydration numbers. Parameter A describes the inner-sphere contribution to the quenching, parameter C the outer-sphere contribution of closely diffusing solvent molecules, while the corrective factor B, which has the same units as k, accounts for the presence of other deactivating vibrations (e.g., N–H or C–H oscillators). Here are the most reliable relationships for H_2O (and MeOH):

$$q_{Eu}(H_2O) = 1.11 \times (\Delta k_{obs} - 0.31) \qquad \Delta k_{obs} \text{ in ms}^{-1}, \ [36] \tag{33}$$

$$q_{Eu}(H_2O) = 1.2 \times (\Delta k_{obs} - 0.25 - 1.20q^{NH} - 0.075q^{CONHR}) \quad \Delta k_{obs} \text{ in ms}^{-1}, \ [37] \tag{34}$$

with q^{NH} being the number of N–H oscillators in the first coordination sphere and q^{CONHR} the number of coordinated amide groups.

$$q_{Tb}(H_2O) = 5.0 \times (\Delta k_{obs} - 0.06) \qquad \Delta k_{obs} \text{ in ms}^{-1}, \ [37] \tag{35}$$

$$q_{Yb}(H_2O) = 1.0 \times (\Delta k_{obs} - 0.20) \qquad \Delta k_{obs} \text{ in } \mu s^{-1}, \ [37] \tag{36}$$

$$q_{Yb}(MeOH) = 2.0 \times (k_{MeOH} - k_{CD3OD} - 0.1) \qquad \Delta k_{obs} \text{ in } \mu s^{-1}. \ [37] \tag{37}$$

A flaw to avoid is to use (33)–(37) with $\tau_{obs}(D_2O)$ set equal to the observed lifetime measured on the hydrated sample at 77 K because it is not granted that all vibrational quenching is switched off at this temperature.

Equations calibrated with polyaminocarboxylates and relying on the sole determination of $\tau(H_2O)$ have also been suggested, but they are less reliable:

$$q_{Nd}(H_2O) = 0.36 \times k_{obs} - 2.0 \qquad k_{obs} \text{ in } \mu s^{-1}, \ [38] \tag{38}$$

$$q_{Sm}(H_2O) = 25.4 \times k_{obs} - 0.37 \qquad k_{obs} \text{ in } \mu s^{-1}, \ [39] \tag{39}$$

$$q_{Dy}(H_2O) = 21.1 \times k_{obs} - 0.6 \qquad k_{obs} \text{ in } \mu s^{-1}. \ [39] \tag{40}$$

Establishing q for Nd^{III} is really problematic and other relationships have been put forward, which do not yield very satisfying results, owing to too limited calibration range, as, by the way, for (38)–(40).

The best way to minimize vibration-induced deactivation processes is to design a rigid metal–ion environment, devoid of high-energy vibrations and protecting the Ln^{III} ion from solvent interactions. Such an environment also contributes to reduce collision-induced deactivation in solution. Further protection may be gained by inserting the luminescent edifice into micelles, a strategy used in bioanalyses [23]. Recent reports have also demonstrated a considerable weakening of the quenching ability of O–H vibrations if the coordinated water molecules are involved in strong intra- or inter-molecular H-bonding. Combining this effect with encapsulation into a rigid receptor turns the weakly emitting aqua ions into entities with sizeable luminescence.

5.5 *Energy Transfers: Donor–Acceptor Distances and Control of the Photophysical Properties of the Acceptor by the Donor*

Distances between a chromophore and a metal–ion site, as well as between metal ions, may be inferred from the determination of energy transfer efficiency within the frame of Förster's dipole–dipole mechanism. In this case, the following simplified equations hold to estimate the efficiency of transfer between the donor D and the acceptor A:

$$\eta_{et} = 1 - \frac{\tau_{obs}}{\tau_0} = \frac{k_0}{k_{obs}} = \frac{1}{1 + (R_{DA}/R_0)^6}, \quad (41)$$

in which τ_{obs} and τ_0 are the lifetimes of the donor in presence and in absence of the acceptor, respectively, R_{DA} is the distance between the donor and the acceptor, and R_0 is the critical distance for 50 % transfer, which depends on (1) an orientation factor κ having an isotropic limit of 2/3, (2) the quantum yield Q_D of the donor (in absence of the acceptor), (3) the refractive index n of the medium, and (4) the overlap integral J between the emission spectrum $E(\tilde{\nu})$of the donor and the absorption spectrum $\varepsilon(\tilde{\nu})$ of the acceptor:

$$R_0^6 = 8.75 \times 10^{-25}(\kappa^2 \cdot Q_D \cdot n^{-4} \cdot J), \quad (42a)$$

$$J = \frac{\int \varepsilon(\tilde{\nu}) \cdot E(\tilde{\nu}) \cdot (\tilde{\nu})^{-4} d\tilde{\nu}}{E(\tilde{\nu}) d\tilde{\nu}}. \quad (42b)$$

Estimation of R_0 is, therefore, accessible from the experimental optical and structural properties of the system. If a crystal structure is at hand, the problem simplifies in that R_{DA} is known and if the lifetimes of (41) can be determined, then calculation of R_0 is straightforward. Energy transfer between lanthanide probes has allowed determining the distance between Ca^{II} and Zn^{II} ions in proteins (substituted by Ln^{III} ions), as well as the distance between tryptophan chromophores and these metal–ion sites [40].

Another interesting application of directional energy transfer is the control of the photophysical properties of a metal ion (nd or 4f) by another one. For instance, Cr^{III} can be used to populate the excited state of Nd^{III} or Yb^{III}. If the rate constant of the energy transfer is fast enough and if $k_{obs}(Cr) << k_{obs}(Ln)$, then the excited Ln^{III} ions will decay with an apparent lifetime equal to the (long) lifetime of the 3d partner. In this case, the lifetime on Nd^{III} and Yb^{III} can be "shifted" in the millisecond range, which is an advantage for time-resolved detection [41, 42].

5.6 FRET Analysis

In fact, FRET (Förster resonant energy transfer) analysis has the same basis as the energy transfer described in the above section. It is used either in simple bio-analyses or to detect protein interactions and DNA hybridization. Its principle is shown on Fig. 14 in the case of a homogeneous immunoassay [43].

In homogeneous immunoassays, the analyte is biochemically coupled to two specific antibodies labeled one with a LLB and the other by an organic acceptor. Emission from the organic acceptor is detected in time-resolved mode because the population of its excited state by intramolecular transfer from the LLB shifts its lifetime in the millisecond range. In this way, it is easy to discriminate between the luminescence emitted by uncoupled and coupled antibody molecules labeled with A; similarly, since the luminescence of A is spectrally different from that of the LLB, interference from Ln^{III} luminescence emitted by the uncoupled antibody labeled with the Ln^{III} chelate is also discriminated. There is, therefore, no need to wash out unused reactants. A method using FRET for the

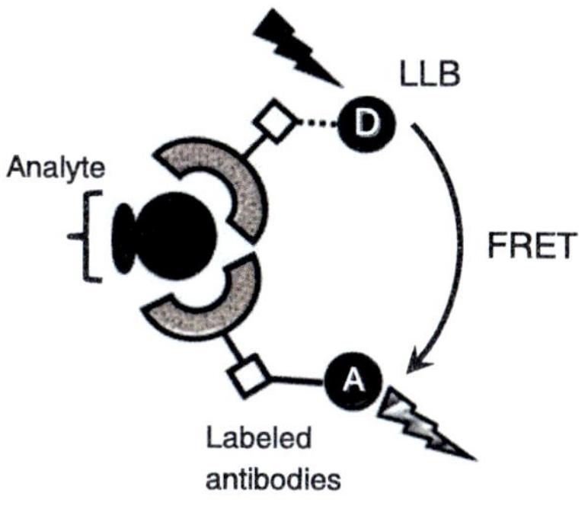

Fig. 14 Principle of a homogeneous immunoassay based of FRET technology (redrawn from Ref. [43])

determination of quantum yields of lanthanide chelates and organic dyes has also been described [44].

5.7 Ligand Exchange Kinetics

As a starting point, assume that two Ln^{III} complexes are present simultaneously in solution at equilibrium and are related by a ligand exchange process. The time-dependence of the luminescence emission following an excitation pulse will depend on the rate of chemical exchange relative to the photophysical deactivation rates. If the exchange is slow or fast, no information can be gathered from lifetime measurements. However, if the chemical exchange process occurs at a rate comparable to the de-excitation rate, the time dependence of the luminescence decay is a function of both the excited state lifetimes and the interconversion rate. Equations have been worked out and the exchange kinetics of several Eu^{III} and Tb^{III} complexes has been elucidated. The Eu^{III} ion lends itself more easily to such experiments because selective excitation of one species through the $^5D_0 \leftarrow {}^7F_0$ transition can be easily achieved [45, 46].

5.8 Analytical Probes

In analytical applications, the Ln^{III} absorption or emission properties are either simply detected or modulated by a process depending on the concentration of the analyte, itself reversibly binding to the lanthanide tag. If absorption is used, hypersensitive transitions (Table 4) are good reporters in view of their sensitivity to minute changes in the Ln^{III} environment. When it comes to luminescence, a much more sensitive technique (especially if time-resolved detection is used), there are several ways of modulating the emission (Fig. 15). One obvious way is to modulate the solvation in the first coordination sphere (a); alternatively, interaction of the analyte with the ligand molecules may modify the energy transfer ability of the latter (b), and the analyte itself may transfer energy onto the reporter ion (c). Note that situations may be reverse, in that either sensitization or quenching may be induced. Cations, anions, pH, pO_2, aromatic molecule sensors have been designed along these lines, while time-resolved immunoassays often take advantage of FRET technology [23].

Molecular interaction between the luminescent tag and other molecules present in solution can also result in luminescence quenching and quantitative investigation of the phenomenon provides both analytical and photophysical information. In collisional (dynamic) quenching, the quencher molecule diffuses to the luminescent probe during the lifetime of the excited state; upon collision, the latter returns to the ground state without emission of light. The average distance that a molecule having

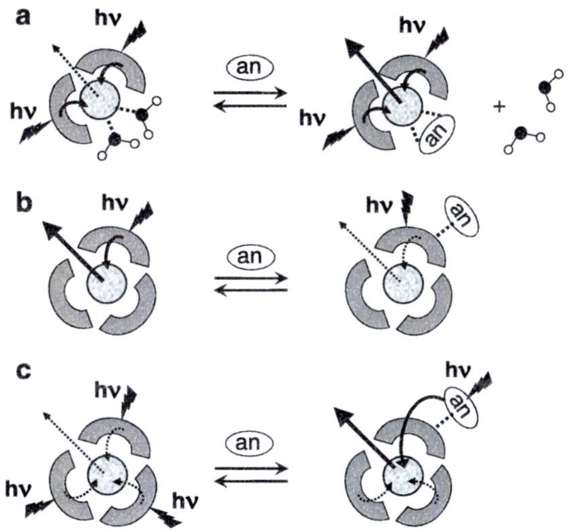

Fig. 15 Modulation of lanthanide luminescence by an analyte through reversible binding: (**a**) removal of solvent quenching, (**b**) modulation of the ligand ability to transfer energy onto the Ln^{III} ion, and (**c**) binding of a sensitizing analyte to the ligand(s). Redrawn from [42]

a diffusion coefficient D can travel in solution during the lifetime of the excited state is given by:

$$\bar{x} = \sqrt{2D\tau_{\text{obs}}}. \tag{43}$$

A typical collisional quencher is molecular oxygen which has a diffusion coefficient of 2.5×10^{-5} $cm^2{\cdot}s^{-1}$ in water at 298 K. During the lifetime of the $Eu(^5D_0)$ level, typically 1 ms, it can, therefore, diffuse over 2.2 μm, that is a distance comparable to the size of a biological cell. In some instances, a non-luminescent complex may result from the collision (static quenching), similar to the cases shown on Fig. 15. Stern and Volmer have worked out the equation for dynamic quenching:

$$\frac{E_0}{E} = 1 + K_{\text{D}}[Q] = 1 + k_q\tau_0[Q], \tag{44}$$

in which K_{D} is the dynamic quenching constant, k_{q} the bimolecular rate constant, E_0 and E the emission intensities in absence and in presence of quencher, respectively, and τ_0 the observed lifetime in absence of quencher. When collisional quenching occurs, the lifetime decreases in parallel to the luminescence intensity:

$$\frac{E_0}{E} = \frac{\tau_0}{\tau}. \tag{45}$$

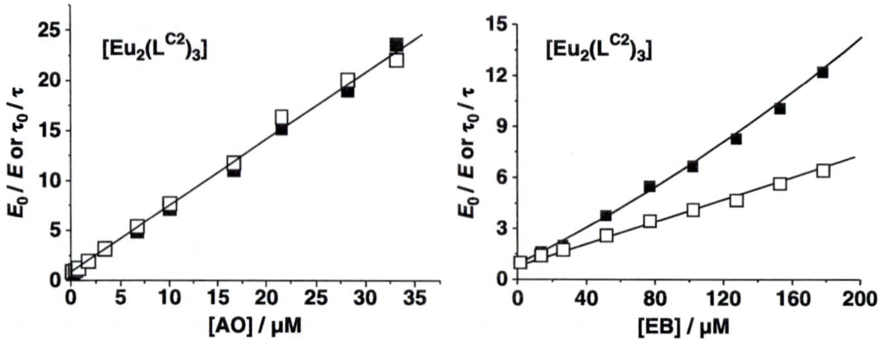

Fig. 16 Stern–Volmer plots for the quenching of the luminescence of $[Eu_2(L^{C2})_3]$ by acridine orange (*left*) and ethidium bromide (*right*) in Tris–HCl buffer (pH 7.4). *Open squares*: lifetimes, *solid squares*: intensities. Redrawn after [47]

A similar equation can be derived for static quenching, with K_S being the static quenching constant:

$$\frac{E_0}{E} = 1 + K_S[Q]. \tag{46}$$

When both dynamic and static quenching occurs, the equations combine into:

$$\frac{E_0}{E} = (1 + K_D[Q])\,(1 + K_S[Q]\,), \tag{47a}$$

$$\frac{E_0}{E} - 1 = (K_D + K_S)[Q] + K_D K_S[Q]^2. \tag{47b}$$

Here again, (45) holds and luminescence intensities may be substituted with lifetimes. It turns out from these equations that if the Stern–Volmer plot is linear, it reflects the sole presence of dynamic quenching. This is, for instance, the case for the quenching of the bimetallic $[Eu_2(L^{C2})_3]$ helicate with acridine orange (AO), as shown on Fig. 16. The corresponding constants are $K_D = 6.7(1) \times 10^5$ M^{-1} and $k_q = 2.7(1) \times 10^8$ M^{-1} s^{-1}. On the other hand, quenching of the same chelate with ethidium bromide (EB) is typical of both dynamic and static quenching with $K_D = 3.0(1) \times 10^4$ M^{-1}, $K_S = 2.0(1) \times 10^3$ $M^{-1}$and $k_q = 1.23(4) \times 10^7$ M^{-1} s^{-1}. The bimolecular rate constants are relatively small compared to diffusion ($\approx 10^{10}$ M^{-1} s^{-1}) because of the shielding of the Eu^{III} ion embedded inside the helical edifice. This quenching has been taken advantage of to develop a versatile and robust method for the detection of various types of DNA and of PCR products [47].

Acknowledgments This work is supported through grants from the Swiss National Science Foundation. The authors are grateful to Frédéric Gumy for his help in recording luminescence data and to Dr Jonathan Dumke for pertinent comments.

6 Appendices

6.1 *Site Symmetry Determination from Eu^III^ Luminescence Spectra*

The exact site symmetry can be determined usually only if polarization measurements are made, that is, on single crystals. A light with σ polarization has its electric vector perpendicular to the crystallographic *c* axis and its magnetic vector parallel to it; the reverse holds for π polarization. The following scheme sketches how the site symmetries other than cubic (T_x, O_x, see Table 3) may be found from Eu($^5D_0$) emission spectra. This is the simplest procedure since $^5D_0$ is non-degenerate. However, site symmetry can be worked out with other luminescent ions as well and, also, from absorption spectra.

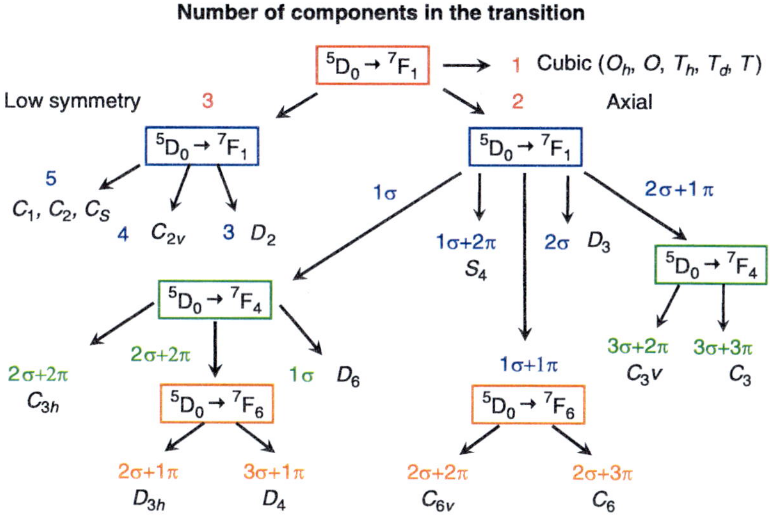

6.2 *Examples of Judd–Ofelt Parameters*

Table 11 Judd–Ofelt parameters for Ln^{III} aqua ions in dilute acidic solution [8]

Ln	10^{20} Ω_2/cm^2	10^{20} Ω_4/cm^2	10^{20} Ω_6/cm^2
Pr	32.6	5.7	32.0
Nd	0.93	5.00	7.91
Sm	0.91	4.13	2.70
Eu	1.46	6.66	5.40
Gd	2.56	4.70	4.73
Tb	0.004	7.19	3.45
Dy	1.50	3.44	3.46
Ho	0.36	3.14	3.07
Er	1.59	1.95	1.90
Tm	0.80	2.08	1.86

Table 12 Judd–Ofelt parameters for Nd^{III} in selected environments [11, 48]

Sample	$10^{20}\ \Omega_2/cm^2$	$10^{20}\ \Omega_4/cm^2$	$10^{20}\ \Omega_6/cm^2$
Nd:YAG	0.37	2.29	5.97
$Cs_3[Nd(dpa)_3]/H_2O$	7.13	3.78	13.21
$Nd(ClO_4)_3$/MeCN	1.2	7.7	7.8
$Nd(ClO_4)_3$/DMF	1.2	8.9	8.4
$Nd(NO_3)_3$/MeCN	11.8	2.1	6.6
$Nd(NO_3)_3$/DMF	6.7	5.0	7.6

6.3 Examples of Reduced Matrix Elements

Table 13 Doubly reduced matrix elements used in the calculations of the dipole strengths for absorption and emission of $Cs_3[Eu(dpa)_3]$, from [7]

Transition	Element	Value	Transition	Element	Value
$^5D_2 \leftarrow {}^7F_0$	$\lvert\langle\Psi\Vert U^2\Vert\Psi'\rangle\rvert^2$	0.0008	$^5D_0 \rightarrow {}^7F_2$	$\lvert\langle\Psi\Vert U^2\Vert\Psi'\rangle\rvert^2$	0.0032
$^5L_6 \leftarrow {}^7F_1$	$\lvert\langle\Psi\Vert U^6\Vert\Psi'\rangle\rvert^2$	0.0090	$^5D_0 \rightarrow {}^7F_4$	$\lvert\langle\Psi\Vert U^4\Vert\Psi'\rangle\rvert^2$	0.0023
$^5L_6 \leftarrow {}^7F_0$	$\lvert\langle\Psi\Vert U^6\Vert\Psi'\rangle\rvert^2$	0.0155	$^5D_0 \rightarrow {}^7F_6$	$\lvert\langle\Psi\Vert U^6\Vert\Psi'\rangle\rvert^2$	0.0002
$^5D_4 \leftarrow {}^7F_0$	$\lvert\langle\Psi\Vert U^4\Vert\Psi'\rangle\rvert^2$	0.0011			

6.4 Emission Spectra

In the following we give typical examples of luminescence spectra of the Ln^{III} ions, with emission from the main luminescent levels. Depending on the chemical environment of the ion, the shape of the spectra may differ substantially (for instance, the relative intensity and CF splitting of the bands), but the energy of the transitions remains relatively insensitive; in addition vibronic transitions as well

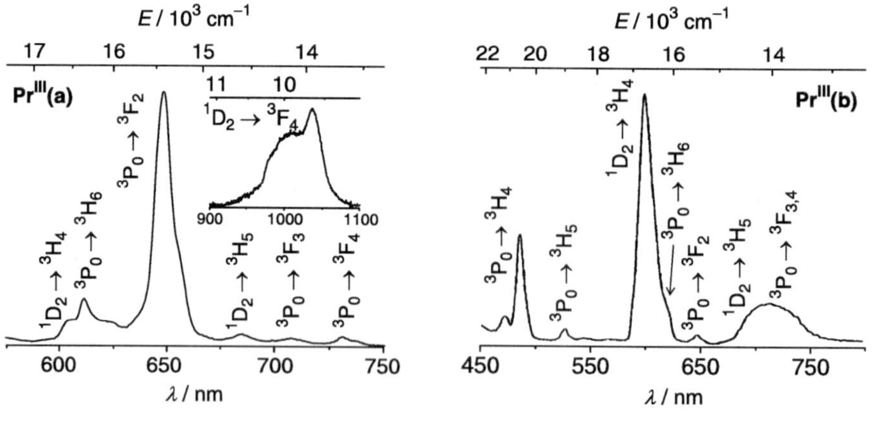

Fig. 17 (continued)

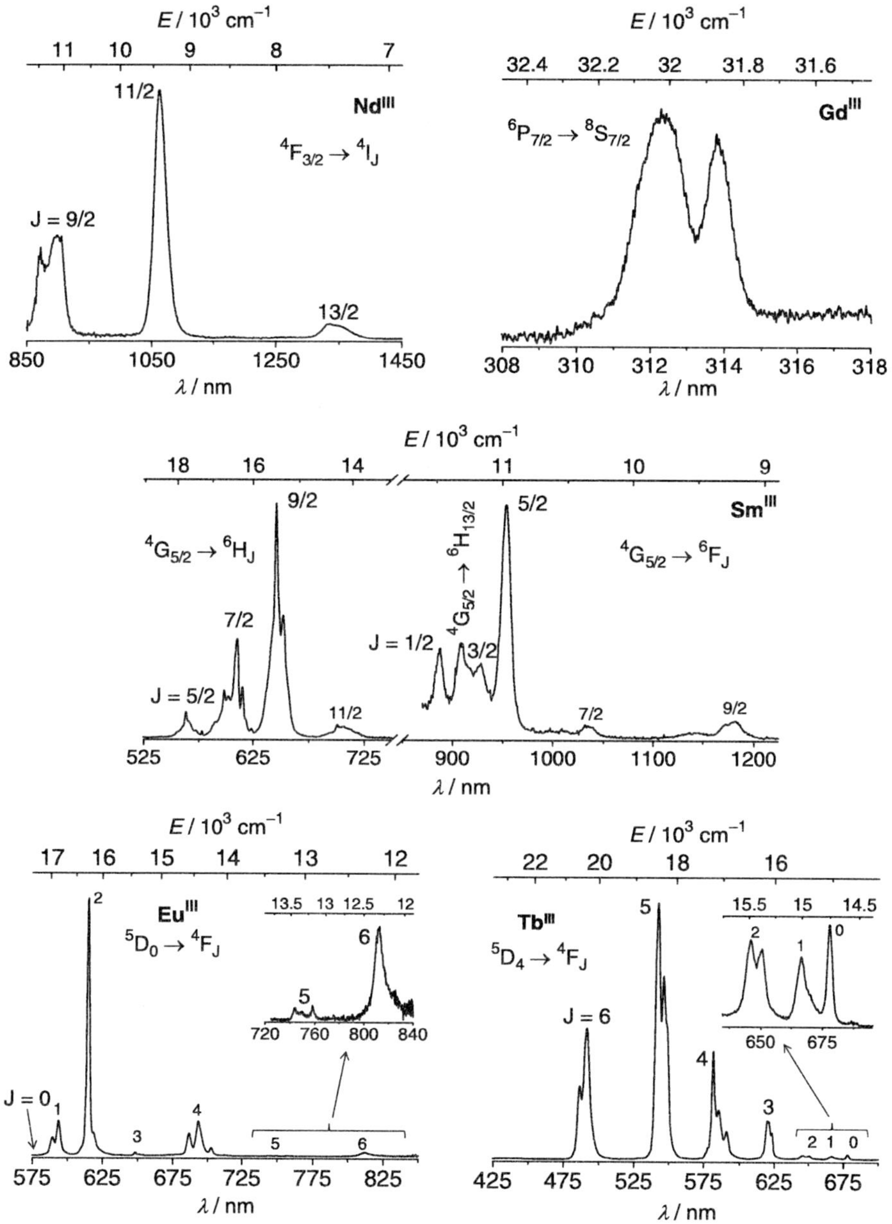

Fig. 17 (continued)

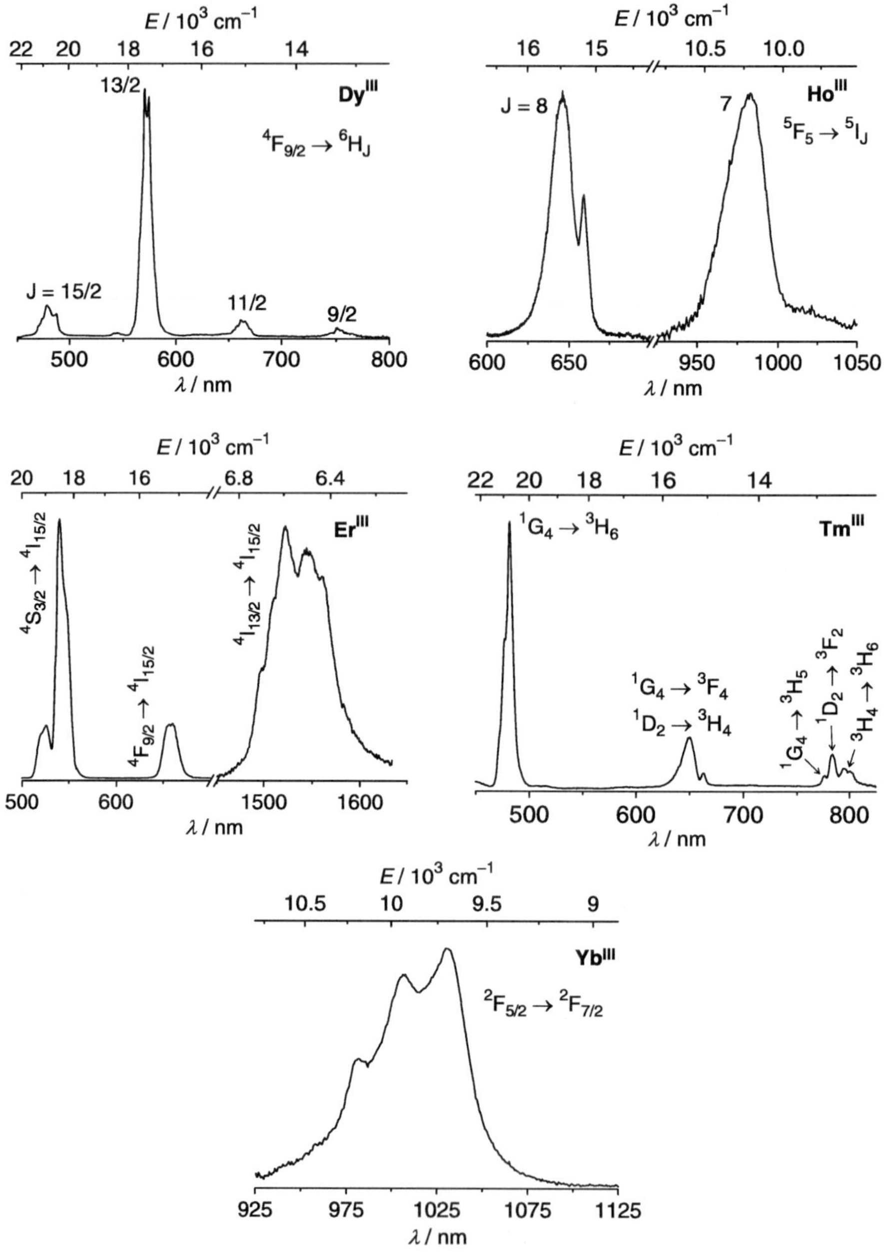

Fig. 17 (continued) Typical examples of Ln^{III} emission spectra under ligand excitation (320–340 nm): microcrystalline samples of dimeric $[Ln(hfa)_3(L)]_2$ (L = 4-cyanopyridine *N*-oxide) for Ln = Pr(a), Nd, Sm, Dy, Ho, Tm, Yb [49]; solutions of $[Pr(b)(L^1)_2(NO_3)]$ in CH_2Cl_2 (L^1 = dihydrobis-[3-(2-pyridyl)pyrazolyl]borate) [50] and $Cs_3[Ln(dpa)_3]$ $\approx$ 2–3.7 $\times$ 10^{-2} M in Tris–HCl 0.1 M (pH 7.4), Ln = Eu, Tb [7]. Emission of Gd is measured on a microcrystalline sample of Gd_2O_3 and of Er (up-conversion, λ_{exc} = 980 nm) on a doped sample of $NaYF_4$:Yb/Er (18/4 %) (this work). All spectra are recorded at room temperature (except for Tm, Ho, and Pr(a) in the visible range, 77 K), corrected for the instrumental function, and normalized

as emission from other levels may also show up. In the case of Pr^{III}, the two spectra shown are quite different because different emissive states are sensitized.

References

1. Liu G, Jacquier B (2005) Spectroscopic properties of rare earths in optical materials. Tsinghua University Press & Springer, Beijing & Heidelberg
2. Hüfner S (1978) Optical spectra of transparent rare earth compounds. Academic, New York
3. Carnall WT, Goodman GL, Rajnak K, Rana RS (1989) A systematic analysis of the spectra of lanthanides doped into single crystal LaF_3. J Chem Phys 90:3443–3457
4. Peijzel PS, Meijerink A, Wegh RT, Reid MF, Burdick GW (2005) A complete $4f^n$ energy level diagram for all trivalent lanthanide ions. J Solid State Chem 178(2):448–453
5. Görller-Walrand C, Binnemans K (1998) Spectral intensities of f-f transitions. In: Gschneidner KA Jr, Eyring L (eds) Handbook on the physics and chemistry of rare earths, vol 25. Elsevier BV, Amsterdam, Ch. 167
6. Walsh BM (2006) Judd–Ofelt theory: principles and practices. In: Di Bartolo B, Forte O (eds) Advances in spectroscopy for lasers and sensing. Springer Verlag, Berlin, pp 403–433
7. Aebischer A, Gumy F, Bünzli JCG (2009) Intrinsic quantum yields and radiative lifetimes of lanthanide tris(dipicolinates). Chem Phys Phys Chem 11:1346–1353
8. Carnall WT (1979) The absorption and fluorescence spectra of rare earth ions in solution. In: Gschneidner KA Jr, Eyring L (eds) Handbook on the physics and chemistry of rare earths, vol 3. Elsevier BV, Amsterdam, Ch. 24
9. Dorenbos P (2000) The f-d transitions of the trivalent lanthanides in halogenides and chalcogenides. J Lumin 91:91–106
10. Shionoya S, Yen WM (1999) Principal phosphor materials and their optical properties. In: Shionoya S, Yen WM (eds) Phosphor handbook. CRC, Boca Raton, Ch. 235
11. Comby S, Bünzli JCG (2007) Lanthanide near-infrared luminescence in molecular probes and devices. In: Gschneidner KA Jr, Bünzli JCG, Pecharsky V (eds) Handbook on the physics and chemistry of rare earths, vol 37. Elsevier BV, Amsterdam, Ch. 24
12. Bredol M, Kynast U, Ronda C (1991) Designing luminescent materials. Adv Mater 3: 361–367
13. Brunet E, Juanes O, Sedano R, Rodriguez-Ubis JC (2002) Lanthanide complexes of polycarboxylate-bearing dipyrazolylpyridine ligands with near-unity luminescence quantum yields: the effect of pyridine substitution. Photochem Photobiol Sci 1:613–618
14. Kleinerman M, Choi SI (1968) Exciton-migration processes in crystalline lanthanide chelates I. Triplet exciton migration in lanthanide chelates of 1, 10-phenantholine. J Chem Phys 49:3901–3908
15. Bünzli JCG, Petoud S, Moret E (1999) Luminescent properties of the Hekis(nitrito)europate (III) ion. Spectrosc Lett 32:155–163
16. Charbonnière LJ, Balsiger C, Schenk KJ, Bünzli JCG (1998) Complexes of *p-tert*-Butylcalix [5]arene with lanthanides: synthesis, structure and photophysical properties. J Chem Soc Dalton Trans 505–510
17. Werts MHV, Jukes RTF, Verhoeven JW (2002) The emission spectrum and the radiative lifetime of Eu^{3+} in luminescent lanthanide complexes. Phys Chem Chem Phys 4:1542–1548
18. Piguet C, Bünzli JCG (2010) Self-assembled lanthanide helicates: from basic thermodynamics to applications. In: Gschneidner KA Jr, Bünzli JCG, Pecharsky V (eds) Handbook on the physics and chemistry of rare earths, vol 40. Elsevier BV, Amsterdam, Ch. 247
19. Kleinerman M (1969) Energy migration in lanthanide chelates. J Chem Phys 51:2370–2381
20. de Sà GF, Malta OL, Donega CD, Simas AM, Longo RL, Santa-Cruz PA, da Silva EF (2000) Spectroscopic properties and design of highly luminescent lanthanide coordination complexes. Coord Chem Rev 196:165–195

21. Bünzli JCG, Chauvin AS, Vandevyver CDB, Song B, Comby S (2008) Lanthanide bimetallic helicates for *in vitro* imaging and sensing. Ann NY Acad Sci 1130:97–105
22. Pandya S, Yu JH, Parker D (2006) Engineering emissive lanthanide complexes for molecular imaging and sensing. Dalton Trans 2757–2766
23. Hemmilä I, Mukkala VM (2001) Time-resolution in fluorometry technologies, labels, and applications in bioanalytical assays. Crit Rev Clin Lab Sci 38:441–519
24. Latva M, Takalo H, Mukkala VM, Matachescu C, Rodriguez-Ibis JC, Kankare J (1997) Correlation between the lowest triplet state energy level of the ligand and lanthanide(III) luminescence quantum yield. J Lumin 75:149–169
25. Archer RD, Chen HY, Thompson LC (1998) Synthesis, characterization, and luminescence of europium(III) Schiff base complexes. Inorg Chem 37:2089–2095
26. Sato S, Wada M (1970) Relations between intramolecular energy transfer efficiencies and triplet state energies in rare earth β-diketonates. Bull Chem Soc Jpn 43:1955–1962
27. Deiters E, Song B, Chauvin AS, Vandevyver CDB, Bünzli JCG (2009) Luminescent bimetallic lanthanide bioprobes for cellular imaging with excitation into the visible. Chem Eur J 15:885–900
28. Eaton DF (1988) Reference materials for fluorescence measurements. J Pure Appl Chem 60:1107–1114
29. Melhuish WH (1961) Quantum efficiencies of fluorescence of organic substances: effect of solvent and concentration of the fluorescent solute. J Phys Chem 65:229–235
30. De Mello JC, Wittmann HF, Friend RH (1997) An improved experimental determination of external photoluminescence quantum efficiency. Adv Mater 9:230–232
31. Bünzli JCG, Choppin GR (eds) (1989) Lanthanide probes in life, chemical, and earth sciences: theory and practice. Elsevier, Amsterdam
32. Frey ST, Horrocks WdeW Jr (1995) On correlating the frequency of the $^5D_0 \leftarrow {}^7F_0$ transition in Eu^{3+} complexes with the sum of nephelauxetic parameters for all of the coordinating atoms. Inorg Chim Acta 229:383–390
33. Bünzli JCG, Pradervand GO (1986) The Eu(III) ion as luminescent probe: laser-spectroscopic investigation of the metal ion sites in an 18-crown-6 complex. J Phys Chem Phys 85: 2489–2497
34. Görller-Walrand C, Binnemans K (1996) Rationalization of crystal field parameters. In: Gschneidner KA Jr, Eyring L (eds) Handbook on the physics and chemistry of rare earths, vol 23. Elsevier BV, Amsterdam, Ch. 155
35. Puntus LN, Chauvin AS, Varbanov S, Bünzli JCG (2007) Lanthanide complexes with a calix [8]arene bearing phosphinoyl pendant arms. Eur J Inorg Chem 2315–2326
36. Supkowski RM, Horrocks WdeW Jr (2002) On the determination of the number of water molecules, *q*, coordinated to europium(III) ions in solution from luminescence decay lifetimes. Inorg Chim Acta 340:44–48
37. Beeby A, Clarkson IM, Dickins RS, Faulkner S, Parker D, Royle L, de Sousa AS, Williams JAG, Woods M (1999) Nonradiative deactivation of the excited states of europium, terbium and ytterbium complexes by proximate energy-matched OH, NH and CH oscillators: an improved luminescence method for establishing solution hydration states. J Chem Soc Perkin Trans 2:493–503
38. Kimura T, Kato Y (1998) Luminescence study on determination of the inner-sphere hydration number of Am(III) and Nd(III). J Alloys Compd 271:867–871
39. Kimura T, Kato Y (1998) Luminescence study on hydration states of lanthanide(III) polyaminocarboxylate complexes in aqueous solution. J Alloys Compd 275–277:806–810
40. Horrocks WdeW Jr, Sudnick DR (1981) Lanthanide ion luminescence probes of the structure of biological macromolecules. Acc Chem Res 14:384–392
41. Eliseeva SV, Bünzli JCG (2010) Lanthanide luminescence for functional materials and biosciences. Chem Soc Rev 39:189–227
42. Bünzli JCG, Piguet C (2005) Taking advantage of luminescent lanthanide ions. Chem Soc Rev 34:1048–1077

43. Bünzli JCG (2009) Lanthanide luminescent bioprobes (LLBs). Chem Lett 38:104–109
44. Xiao M, Selvin PR (2001) Quantum yields of luminescent lanthanide chelates and far-red dyes measured by resonance energy transfer. J Am Chem Soc 123:7067–7073
45. Horrocks WdeW Jr, Arkle VK, Liotta FJ, Sudnick DR (1983) Kinetic parameters for a system at equilibirium from the time course of luminescence emission: a new probe of equilibrium dynamics. Excited-state europium(III) as a species label. J Am Chem Soc 105:3455–3459
46. Ermolaev VL, Gruzdev VP (1984) Novel spectral-kinetic method for investigation of ligand exchange in labile metal complexes in solutions. Inorg Chim Acta 95:179–185
47. Song B, Vandevyver CDB, Deiters E, Chauvin AS, Hemmilä I, Bünzli JCG (2008) A versatile method for quantification of DNA and PCR products based on time-resolved Eu^{III} luminescence. Analyst 133:1749–1756
48. Bünzli JCG, Vuckovic MM (1984) Solvation of neodymium(iii) perchlorate and nitrate in organic solvents as determined by spectroscopic measurements. Inorg Chim Acta 95:105–112
49. Eliseeva SV, Ryazanov M, Gumy F, Troyanov SI, Lepnev LS, Bünzli JCG, Kuzmina NP (2006) Dimeric complexes of lanthanide(III) hexafluoroacetylacetonates with 4-cyanopyridine *N*-oxide: synthesis, crystal structure, magnetic, and photoluminescent properties. Eur J Inorg Chem 4809–4820
50. Davies GM, Pope SJA, Adams H, Faulkner S, Ward MD (2005) Photophysical properties of Pr(III) and Er(III) complexes of poly-(pyrazolyl)borates. Photochem Photobiol Sci 4:829–834

Stable Luminescent Chelates and Macrocyclic Compounds

G. Mathis and H. Bazin

Abstract This review is focused on the lanthanide probes usable for time-resolved luminescence resonance energy transfer experiment and in the development of bioassays. The basic principle of heterogeneous time-resolved fluorescence (TRF) assays and homogeneous TRF assays are summarized. The criteria that should fulfill a lanthanide luminescent probe to be useful in the design of bioassays in diagnostic or drug discovery (high throughput screening) are defined as brightness, absorption wavelength, luminescence decay, instrumentation crosstalk, stability, lipophilicity/hydrophilicity, photobleaching, quenching phenomenon, conjugation chemistry, and synthesis practicability. The photophysical properties and the fulfillment of the above criteria are commented for the most representative structures insisting on the available stability data. Two main groups of molecules are described: (1) the luminescent stable chelates and (2) the macrocyclic-based ligands. The stable chelates are based on EDTA, DTPA, podant-like scaffold, and peptide scaffolds. The macrocyclic compounds described are macrocycles, macrocycles with pendant groups, and macropolycyclic cage ligands (cryptands). Applications of lanthanides complexes to cell based assays as well as time-resolved microscopy and imaging are discussed.

Keywords Fluoroimmunoassays · High throughput screening · Homogeneous time-resolved fluorescence · Lanthanide chelates · Lanthanide cryptates

Contents

G. Mathis and H. Bazin (✉)
CISbio Bioassays, BP 84175, 30204 Bagnols-sur-Céze Cedex, France
e-mail: hbazin@cisbio.com

P. Hänninen and H. Härmä (eds.), *Lanthanide Luminescence: Photophysical, Analytical and Biological Aspects*, Springer Ser Fluoresc (2011) 7: 47–88, DOI 10.1007/4243_2010_5,

Abbreviations

Φ	Quantum yield
BP	Bipyridine
BSA	Bovine serum albumin
DMEM	Dulbecco's modified Eagle's medium
DO3A	1,4,7,10-Tetraazacyclododecane-*N,N′,N″*-triacetic acid
DOTA	Tetraazacyclododecane tetraacetic acid
EDTA	Ethylene diamine tetraacetic acid
EGTA	Ethylene glycol tetraacetic acid
ELISA	Enzyme-linked immunosorbent assay
FCS	Fetal calf serum
FITC	Fluorescein isothiocyanate
FRET	Förster resonance energy transfer
GFP	Green fluorescent protein
HSA	Human serum albumin
HTRF	Homogeneous time-resolved fluorescence
HTS	High throughput screening
K_{OW}	Octanol/water partition coefficient
LANCE	Lanthanide chelate excitation
LBT	Lanthanide-binding tag
LRET	Luminescence resonance energy transfer
MRI	Magnetic resonance imaging
nH_2O	Number of water molecules

NHS	*N*-hydroxysuccinimide
NIR	Near-infrared
PMT	Photomultiplier tube
RPMI	Roswell Park Memorial Institute medium
TATP	1,4,8,9-Tetra-aza-triphenylene
TEAAc	Triethylammonium acetate
TRF	Time-resolved fluorescence
TR-FRET	Time-resolved Förster resonance energy transfer
Tris	Trishydroxymethylaminométhane

1 Introduction and Scope

Today, fluorescence is the method of choice in many cases for biomolecular labeling in life science research and development, where it is continuously replacing radio-labeling. This trend was particularly observed in the domain of high throughput screening (HTS). However, not all fluorescence techniques are amenable to HTS, where miniaturization is playing a central role, especially when reagent and test compound savings are crucial. When using bulk fluorescence techniques, robust assay systems can be miniaturized down to 10 μl volumes and are still compatible with the requirements for ultra-HTS (uHTS). Prominent bulk fluorescence techniques use total fluorescence [1], fluorescence polarization (FP) [2, 3], luminescence decay [4], and time-resolved fluorescence (TRF) also known as time-gated fluorescence [5, 6]. TRF make use of fluorophores with long luminescence decay (μs to ms), such as rare earth elements (lanthanides), especially europium (Eu)/terbium (Tb) chelates or cryptates. Samarium and dysprosium can also be used, with some limitations due to their characteristic shorter luminescence decay. In TRF measurements, a delay of several hundred of microseconds separates the pulsed excitation of the fluorophore from the detection of fluorescence emission. This cuts out short-lived fluorescence and reduces significantly any auto-fluorescence or background signal (light scattering), allowing sensitive and specific signal detection. In addition, the very large Stokes shift (>200 nm) of the lanthanide and their sharp emission peak (full width at half maximum ~10 nm) contribute to the low background signal [7]. The TRF mode was initially exemplified by heterogeneous assays such as DELFIA® (Dissociation Enhanced Lanthanide Fluoroimmunoassay) developed during the 1970s [8] and marketed by LKB/Wallac. In DELFIA®-based binding assays, a ligand is labeled with a non-photoactive lanthanide chelate. The amount of specifically bound labeled ligand is analyzed following addition of an "enhancement solution" after wash steps that remove unbound label. The addition of enhancement solution is a critical step required to obtain detectable signal from the lanthanide ion. The low pH of the enhancement solution is required for efficient release of the lanthanide from the non-fluorescent chelate. The presence of a non-ionic detergent serves to form a micelle around the organic complex thereby preventing quenching of the signal by excluding water molecule from the lanthanide complexation sphere. On addition of the enhancement solution, a fluorescent chelate is formed, enhancing the lanthanide fluorescence by up to 10^7-fold. Using a LEADseeker reader (Amersham) in TRF mode, a lower limit of detection < 0.5 pM can be obtained (europium in 100 μl total volume in DELFIA® fluorescence enhancement solution, 96-well black plates, 20-s imaging). Interestingly, in a comparative study, DELFIA® was shown to give less variability than ELISA [9].

The needs for homogeneous TRF assay more amenable to automation required in the HTS domain, and to a lesser extent in the diagnostic field, brought the development of simple fully homogeneous assay, known as HTRF®, using TRF mode and lanthanide cryptates [5, 10]. A similar method using lanthanide chelates known as LANCE™ was developed [11]. Such homogeneous techniques make use of luminescence resonance energy transfer (LRET). Energy transfer can be measured with essentially no contaminating background and donor–acceptor distances of up to 10 nm can be measured. By temporal and spectral discrimination, donor emission and acceptor emission (both intensity and luminescence decay) can be independently measured. This leads to dramatically improved signal to background compared to FRET. Specifically, in LRET the acceptor emission due only to energy transfer can be measured with no background, in contrast to FRET. Contaminating background in FRET when measuring energy transfer via an increase in acceptor fluorescence, arises from two sources: direct excitation of the acceptor by the excitation light and donor emission at wavelengths where one looks for acceptor emission. In LRET, both sources are eliminated. Homogeneous TRF assays are particularly advantageous in the HTS domain [12], since many compounds belonging to the screened libraries are strongly fluorescent and are used at a high concentration (usually 10–100 μM in the microplate wells) inducing high fluorescence background which can only be eliminated because of the gated mode. This allow HTS assays to be run using sub-nanomolar lanthanide complex labeled biomolecules giving a robust signal undisturbed by the fluorescence background. Although alternative FP assays have been developed for HTS, due to the lower sensitivity and to limit interference from fluorescent compound of the screened library, a higher concentration of labeled biomolecule has to be used. Typically, 300 fmol per well of labeled peptide should be used in an optimized FP HTS assay [13], in contrast with 2 fmol per well usually required in a lanthanide based homogeneous TRF assay. This lower amount of labeled entity can yield important savings in a HTS campaign. In addition, the consumption of enzyme, as in a typically kinase assay, is five time higher in an FP assay compared to TR-FRET. This is important to consider since the kinase itself could be either difficult to obtain in active form, hence being a limiting factor, or a costly element of the assay [14]. For HTS kinase assay, addition of EDTA (~10–20 mM final concentration in wells) to stop the kinase activity is needed. The behavior of various commercially available lanthanide complexes was tested in the presence of EDTA [14]; for some europium complexes the initial fluorescence dropped by a factor of ten after 8-h incubation. This is important to consider since the pharmaceutical industry needs robust assays exhibiting a signal stable over time, allowing for eventual storage of microtiter plates after the final revelation step and before reading.

Although some ready-to-use Eu-labeled antibodies are commercially available as part of “toolboxes” it is also desirable to label a biomolecule of interest using a simple one step immunochemistry with a reactive Eu complex ensuring the conservation of the biological activity (as immunoreactivity) in the labeled molecule. The coupling protocol often preferred is the *N*-hydroxysuccinimide (NHS) activated ester method which can be used in aqueous buffer for selective N-terminal peptide labeling (pH 7) or more extensive labeling (pH 8–9) of larger proteins as antibodies. Therefore, ideally, a lanthanide complex should be amenable for functionalization as NHS ester. This can be a handicap for many lanthanide complexes containing multiple carboxylic functions as chelator since in this case, selective functionalization as NHS ester and stability in acidic conditions (needed for purification step) can be challenging.

It is important to note that, to increase signal intensity, multiple lanthanide labeling of even small molecules is possible because lanthanide luminescence, in contrast to organic dye fluorescence, [15] does not undergo concentration quenching. This can be attributed to the lack of overlap between the absorption and excitation spectra due to the large Stokes shift.

The use of multiple, simultaneously acquired parameters (multiplexing) should allow more complex assemblies of biomolecular interactions to be studied and time and reagent savings in the HTS context. In the case of europium, the choice for FRET acceptor is limited to near-infrared (NIR) fluorophores as allophycocyanine derived acceptors or indocyanine acceptor dyes. Allophycocyanines are characterized by a high quantum yield and a high molar absorbance ensuring a strong spectral overlap with europium emission and thus an optimal Förster radius [7]. Furthermore, in some cases the high molecular weight can be an advantage by limiting the extent of dynamic FRET [16] if high concentration of acceptor is required in a particular assay. NIR acceptors belonging to the indocyanine series are convenient provided that the multiple labeling necessary to insure optimal labeling of large proteins (as antibodies) has no detrimental effect on the photophysics as self-quenching phenomenon [15].

The terbium complexes allow more flexibility in the choice of suitable acceptors and more particularly of multiple acceptors in the scope of multiplexing, since the terbium emission lines are more evenly disposed in the emission spectrum (compared to europium emission) displaying lines at 490, 545, 585, and 620 nm. Thus, either fluorescein, rhodamin, or indocyanines derived acceptors can be used, their respective emission falling in workable “windows” between successive emission lines or in the NIR window. Furthermore, green fluorescent protein (GFP) or GFP-like acceptors have been recently used to design assay involving a fusion protein substrate [17].

For bioanalytical applications (in vivo or in vitro) high thermodynamic stability, kinetic inertness, and a saturated coordination sphere are required. Given the lability of lanthanide ions (formation of insoluble hydroxides) and their need for high coordination numbers, this poses a real challenge to design an efficient and robust probe; therefore, several approach and scaffolds have been used. Most of the bioanalytical applications are using "Bifunctional Chelators" (ligands bearing two functions, one used to covalently attach the ligand to biological material and one which strongly coordinates to the metallic ion); therefore, we will concentrate mostly on these aspects – "sensors" being discussed in another chapter.

This chapter will focus on the most recent work, since some recent reviews have been covering the subject up to 2007 [18, 19].

2 Criteria for Optimum and Robust Luminescent Probe for Bioanalytical Applications

Over the years, in our laboratory, we have currently tested most of the lanthanide complexes which were available either as marketed compounds or as samples coming from academic research groups. Since the first published luminescent lanthanide complexes, many research groups both in academia or industry have been working for the design of a "perfect" luminescent probe which could fulfill all the bioanalytical applications. For an industrial application, the compound must be well characterized; its synthesis and the final analytical characteristics must be reproducible and should not vary upon storage and further handling. This section summarizes the requirements and parameters to evaluate the practical usefulness of a given complex. The following sections will review the current lanthanide complexes literature focusing on the complexes fulfilling most of this criteria and seen through the prism of their true potential or industrial value. Since some commercial TRF readers are equipped with a nitrogen laser excitation allowing short pulse width (to improve the S/N ratio), the absorbance at 337 nm is an important criterion or more precisely the brightness, which is the product of the molar absorbance (or absorption cross-section, ε in $M^{-1}\ cm^{-1}$) at the excitation wavelength and the fluorescence quantum yield (Φ). The brightness, symbolized as B (expressed in $M^{-1}\ cm^{-1}$), represents the overall efficiency of the luminophore and is an important parameter useful in comparing complexes of different structures. According to our experience, one can set a low limit ~300 $M^{-1}\ cm^{-1}$ for the brightness characterizing an "efficient" complex. Care should be taken concerning the quantum yields quoted in the literature; some authors report the overall quantum yield (which is the value used to compute the brightness) and sometime the intrinsic lanthanide quantum yield of luminescence (Φ_{Ln}).

2.1 *The Excitation Wavelength Maximum*

The excitation wavelength maximum must be above 300 nm to have some chances to fulfill the brightness criterion. If the excitation wavelength is below 300 nm the inner filter effect, either from proteins or from screened compounds would reduce dramatically the luminescence output signal. In that case, the use of a ratiometric

[10] measurement for filter effect correction would be of limited efficiency. The λ_{max} is not the only factor to consider; some complexes present λ_{max} around 300 nm but display a shallow line-shape on the long wavelength side of the absorbance spectrum so they can present enough brightness upon 337-nm excitation. The shorter the light wavelength necessary to excite the chromophore, the stronger will be the filter effect from the medium. This is particularly prominent in the presence of "chemical library compounds" under screening, which are added at 10 μM final concentration in wells for primary screening. This is also the case for experiments on living cell in a "minimum", but already complex, medium such as DMEM. The absorbance in DMEM, for a 1-cm light path, are the following (in parentheses the % transmission): $A_{337} = 0.2$ (63%), $A_{290} = 0.58$ (26%), $A_{280} = 1.23$ (5.9%); furthermore, working with 0.1% BSA, often required to avoid the unspecific binding of peptides or lipophilic ligands, results in an increased absorbance of $A_{280} = 1.89$ (1.3%). It is clear from this above example that attempts to use a complex characterized by a λ_{max} ~280 would have serious limitations since most of the exciting light will be absorbed due to the medium inner filter effect. This is also why a ratiometric measurement is of major importance for proper measurement of fluorescence intensity [10]. Furthermore, wavelengths above ~320 nm are often preferred for practical and economical reason since they allow to use non-quartz optics (A_{320} ~0.2 for 1-mm thickness glass microscopy slide).

2.2 *The Luminescence Decay*

The luminescence decay, preferably mono-exponential, must be long enough to have an efficient rejection of the noise, as light scattering, coming from the excitation pulse. Using a flash lamp excitation the residual scattered light can last upto 50–100 μs until reaching negligible value. Using a nitrogen laser, even if the pulse is characterized by a steeper descending front, due to the much higher initial intensity the residual scattered light last up to ~10 μs. Therefore, this consideration sets the minimum workable luminescence decay to ~50 μs. Furthermore, the luminescence decay value (τ_D) measured for the luminescent donor not engaged in a FRET process is shortened if the luminescent lanthanide is involved in a FRET process. In case of high FRET efficiency (i.e., more than 50%), this shortening can be a problem since the donor luminescence decay would drop down to more than half of the initial value and therefore if the initial luminescence decay is ~100 μs the actual luminescence decay would be less than 50 μs. In this case, most of the luminescence would be emitted before the gating time and would result in low output signal, and decreasing the gating time would increase the signal and also the background. Therefore, for the isolated lanthanide, a τ_D ~200 μs luminescence decay can be set as an absolute minimum and ideally it should be in the 1-ms range so that the luminescence decay τ_{DA} (in the presence of an acceptor) would be long enough in case of high FRET efficiency. For more details see the first chapter

(J.C. Bünzli), as well as a very good review article describing all the technical aspects of the TR-FRET application to HTS [7].

2.3 The Donor Contribution in the Acceptor Channel

The donor contribution in the acceptor channel (crosstalk) should be as low as possible; the impact of this contribution on a bioassay is not obvious to anticipate starting from a lanthanide complex emission spectrum, since many instrumental factors, such as the filter settings (bandpass width), have to be considered. The intensity distribution between the emission lines is critical, particularly for europium complexes, with a strong impact of the ligand structure and symmetry (for terbium complexes, this impact is reduced). Care must be exercised in comparing published emission spectra, since many of the published spectra are not corrected for the photomultiplicator sensitivity (which falls off rapidly between 650 and 800 nm even using a "red PMT"). The consequence is that the 690-nm ($^5D_0 \rightarrow {}^7F_4$) band seems much smaller than its true value. Some articles do indeed show spectra corrected for the sensitivity of the detection system (which contains contributions from the PMT, but also from the monochromators and optics). Whenever such corrections have been applied, this is usually indicated in the experimental section of the article.

2.4 Complexes Stability in the Presence of Competing Cations

Many important enzymes assayed in HTS (as kinases) need Mn^{2+} or Mg^{2+} as cofactors, this later being particularly detrimental on many lanthanide chelates even on those considered as "stable chelates." Ideally, a complex should resist several hours in the presence of millimolar amounts of Mn^{2+} or Mg^{2+}. Additionally and in particular for the design cellular bioassays, the complex must be stable in the presence of ~120 mM Ca^{2+}, as well as many vitamin and cofactors present in the culture media.

2.5 Complexes Stability in the Presence of Competing Complexing Agent (Ligand Exchange)

EDTA is routinely used to stop enzymatic reactions, the final EDTA concentration in wells being ~10 mM. Therefore, a complex should withstand at least 10 mM EDTA for several hours. Additionally, the complex must be stable in the presence of citrate and phosphate (present in many current buffers as PBS and culture media). This later leads to a precipitation of lanthanide ions arising from decomplexation. For special application, an additional criterion can be thermal stability (use of thermostable enzymes) [21].

A stability/robustness test can be easily performed using small amounts of complexes. Preferentially, the test should be performed on nanomolar concentration range for the complex (rather than micromolar concentration) since for practical analytical application the complexes would be used at nanomolar concentrations. The samples are prepared in 100 μl in 96-well black microtiter plates using phosphate and HEPES buffer alone and in admixture with the competitors as described in Table 1. The fluorescence intensity decreases (expressed in percentage) are measured in comparison with a reference compound of established stability. To perform such an experiment one can use any dedicated time-resolved reader such as BMG Rubystar (www.bmglabtech.com) or Perkin–Elmer Wallac 1420 VICTOR2™ (http://las.perkinelmer.com) or multi-mode microplate reader such as BMG Pherastar FS, Berthold Mithras LB940 (www.Berthold.com), Biotek Synergy-2 (www.biotek.com), TECAN safire II, Infinite® F500 (www.tecan.com) or their equivalent. Some readers present specific advantages, for instance, lifetime measurement can be performed using the "advanced mode" on Rubystar or the "decay curve monitoring" mode on Pherastar FS (either with flash lamp or laser excitation) recording fluorescence intensity decay every 10 μs over a 2-ms period and reprocessing the data on a flowchart. This allows, if required, a more in-depth study by measuring the specific decay time of a luminescent complex in various conditions. Other readers such as TECAN safire II allow to record excitation and emission spectra in 100 μl 96-well microtiter plates in the nanomolar range. A complete stability/robustness test performed on a 96-well 100 μl format requires less than a nanomol of complex. Besides saving valuable samples, the clear advantage is that the tests can be run at the nanomolar concentration rather than the micromolar for "cuvette" spectrofluorimeter, therefore, at a more relevant complex concentration. Concerning the effect of cell culture medium we found that the most detrimental is the RPMI containing Phenol Red, and to a less extent the DMEM. The DMEM (without Phenol Red) causing by itself a 50% intensity quenching (filter effect) due to a strong absorbance in the range of 300–340 nm. In addition, the advantage of the TRF mode is obvious in this medium due to the presence of micromolar amount of fluorescent species such as folic acid and proteins, the background fluorescence perturbing prompt fluorescence measurement in the steady state mode.

Table 1 Minimal medium or competitor concentration for screening lanthanides complexes

Signal decrease (%)	PO_4 0.05 M pH7 0.1% BSA	HEPES 0.05 M pH7 0.1% BSA	+120 mM $CaCl_2$	+1 mM $MgCl_2$	+20 mM EDTA	+RPMI 1/1	+FCS 1/1
Within minutes	<5	<5	<5	<5	<10	<20	<30
Overnight at 20°C	<5	<5	<5	<5	<20	<30	<40

2.6 Balanced Lipophilicity–Hydrophilicity

The ligand must be obviously lipophilic enough in order to shield the lanthanide from water molecules (ensuring a low non-radiative deactivation) but should not stick to the glassware and plastic container (as microtiter plates). A nanomolar or sub-nanomolar solution of the lanthanide complex should present stable photophysical characteristics (as emission intensity) over 24 h, in case of non-specific binding, use of additive (as BSA) is usually helpful.

Additionally for cellular bioassays for which a permeant probe is desired, the complex should fulfill the rule established for optimum cell penetration (charge, K_{OW}). This topic will be discussed later and is exemplified in a recent article [22].

2.7 Photobleaching

This is usually not a problem (compared to conventional fluorophores) since many europium probes have proven to be robust and resistant to Photobleaching [23].

2.8 Quenching from Metabolites

This question is not addressed in most of the articles describing the synthesis and photophysical studies of lanthanide complexes, although this is a key issue in traditional diagnostic application involving analysis in body fluids. This is even more crucial considering the growing field of time-resolved microscopy application, cellular and intracellular imaging on living cell. It is known that species such as ascorbate ($E_{1/2} = +0.30$ V*), iodide (+0.54 V*), urate (+0.59 V*), and bromide (+1.07 V*) are able to deactivate the lanthanide excited state, leading to a reduction in both the emission luminescence decay and intensity [10, 24, 25] (* potentials measured versus Ag/AgCl reference electrode with saturated KCl). This dynamic quenching process is characterized by a Stern–Volmer quenching constant, representing the concentration of reductant needed to reduce the luminescence decay or emission intensity to 50% of its original value. The Stern–Volmer quenching constants values for the quenching by iodide, ascorbate, and urate have been measured for a series of structurally related macrocyclic ligands, based on various chromophore [24, 25]. From these studies, it was observed that (1) the Tb complexes are more sensitive to quenching than the Eu analogs as expected from the higher excited state energy of the Tb $^5D_4$ state and (2) urate quenching is much more effective than anticipated on the basis of its one-electron oxidation potential. The quenching requires collisional encounter, and the luminescence decay of the encounter complex is likely to be determined by steric factors and by the local electrostatic potential gradient and so is very dependent of the lanthanide environment. Therefore, the evaluation of the quenching by metabolites model is of the utmost importance as a criterion for the validation of a complex to be of use for the design of robust bioassays. Noteworthily this quenching effect, basically undesired, can be turned into an advantage to design homogeneous TRF assays for hybridization [26] and assay for uric acid [25].

Adjuvants such as fluoride ions, were shown to suppress the quenching effect from uric acid present in serum sample [10]. Additionally, a lanthanide complex designed for cell-based assay should be insensitive to luminescence quenching by azide [27], which is a common internalization inhibitor used in ligand–receptor binding studies [28].

2.9 Conjugation Chemistry

Once a lanthanide complex possessing the optimum photophysical characteristics (or at least presenting the best compromise for a given application) and except in the case of ion or metabolites probes, the complex will be ultimately conjugated to a biomolecule of interest. This step can be a hurdle since in some cases the coupling chemistry or the biomolecule itself can modify the initial photophysical characteristics. The NHS activation of an aliphatic carboxylic group is the best choice in terms of chemical stability (purification and storage), reactivity at pH 7–8, and biological integrity of the labeled molecules. It is noteworthy that this is the preferred chemistry for marketed organic dyes. Iodoacetyl or isothiocyanate needing higher pH which can be detrimental for the biological activity, fluorescein-NHS is nowadays preferred to FITC. Ideally, the complex should also be available as a primary amino-alkyl derivative which gives the largest possibilities for further immunochemistry, since it can be converted at will to maleimide or thiol using commercially available heterobifunctional reagents.

2.10 Practicability of the Synthesis and Purification

This seems to be obvious and is particularly important in view of the industrial applications; nevertheless, this is often hampered in many publications. Among the complexes described in literature we have tested, some were very promising in respect of all the photophysical data, but totally useless since the purification of the lanthanide complex itself, or its conjugates was not compatible with industrial chromatographic process.

3 Luminescent Stable Chelates

To fulfill the above criteria a number of lanthanide complexes comprising a chromophore (antenna) and a chelating subunit were designed using various strategies (see [19] for more details). A first class is represented by the so-called stable chelates which are described in the following section and enter into two broad categories: the linear polydentate type ligands and the podants type ligands.

A second class is represented by the macrocyclic compounds comprising the macropolycyclic ligands as cryptates will be presented in a further section.

3.1 Linear Polydentate Ligands

3.1.1 EDTA-Like Scaffold

With logK in the range 15–20, $[Ln(edta)]_2$ (ethylenetriaminepentaacetic acid) complexes are quite stable and consequently, the aminocarboxylate complexing unit has been grafted on numerous scaffolds and chromophores to generate stable fluorescent chelates previously described in the literature [29–31] and in patents [32–35]. Ligands based on EDTA scaffold are shown in Fig. 1.

An EDTA based chelate bearing a suitably substituted pyridine such as the heptadentate [2,2′,2″,2‴-({4-[(4-isothiocyanatophenyl) ethnyl] pyridine-2,6 diyl} bis-(methylenenitrilo)) tetrakis(acetato) europium (III)] (**1**) was marketed (initially by Wallac and then by Perkin–Elmer) under the name Eu-W1024-ITC chelate activated as an isothiocyanate derivative. This chelate presents a maximum excitation around 340 nm (ε_{310} ~27,000 M^{-1} cm^{-1}) and a maximum emission at 613 nm. This chelate is used in the LANCE™ homogeneous assays developed by Hemmilä and the Wallac's group in 1996 and further marketed by Perkin–Elmer [35]. According to Perkin–Elmer application note (www.perkinelmer.com/lifesciences, "Stability of the Wallac LANCE™ Eu-chelates"), the Eu-W1024 chelate is relatively stable although it loses about half of the fluorescence within 2 h either upon incubation in the presence of 50 mM EDTA or in a 50 mM pH 5 succinate buffer. The manganese effect is even more drastic since 80–90% of the fluorescence is lost within 5 min incubation in the presence of 500 μM Mn^{2+}. Interestingly, this effect is partially canceled in the presence of a calculated amount of EDTA. For experiments requiring high EDTA concentrations (>20 mM EDTA), a high temperature, or a low pH the W8044 Europium chelate (vide infra) is preferred because of its higher stability.

The more stable chelates are composed of a nonadentate chelating ligand, such as terpyridine or a terpyridine analog with a five-membered ring as triazole (vide infra). The former is exemplified by 4′-(3-amino-4-methoxyphenyl)-6,6″-bis[*N*,*N*-bis(carboxymethyl) aminomethyl]-2,2′:6′,2″-terpyridine **2** (trivial name, terpyridine-bis(methylenamine)tetraacetic acid: TMT) [36]. The Eu-TMT complex has high molar absorbance (ε_{319} ~15,200 M^{-1} cm^{-1}; pH 9.0). TMT forms a nonadentate complex with Eu^{3+}, occupying all Eu^{3+} coordination sites resulting in a high binding constant. The aromatic terpyridyl system also provides the chelated Eu^{3+} ion with a relatively hydrophobic environment necessary for maximal fluorescence and luminescence decay. The reported luminescence decay is $\tau = 1.4$ ms (in 50 mM Tris–HCl pH 7.4). The nonadentate complexation brings an effective stability since the Eu-TMT fluorescence does not decrease in pH 3 citrate

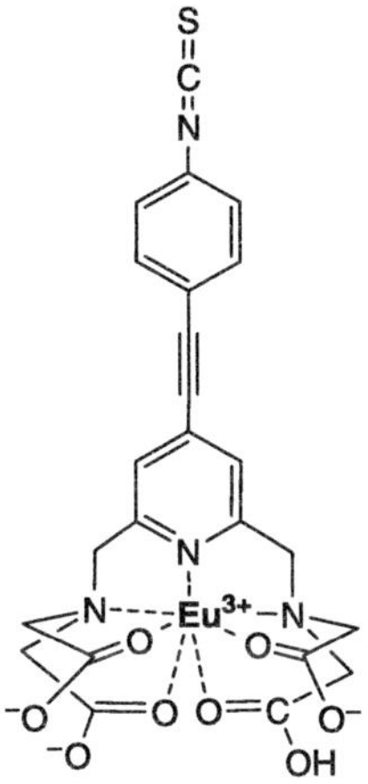

Eu-W1024-ITC (**1**)

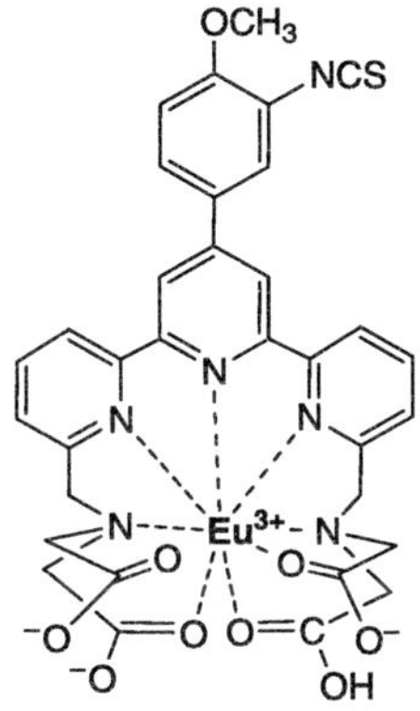

Amersham Eu(TMT) ITC (**2**)

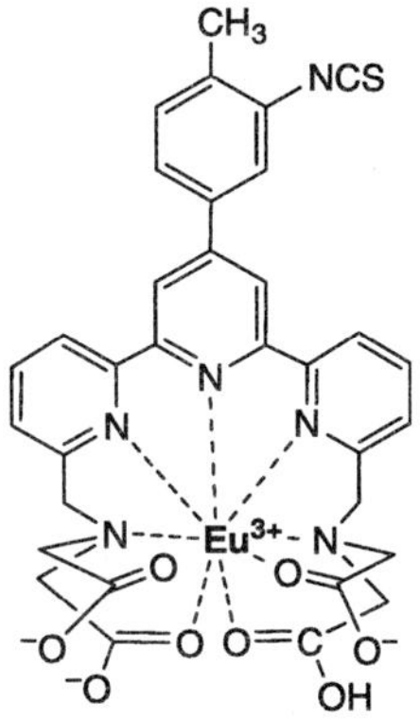

Eu-TMT (**3**)

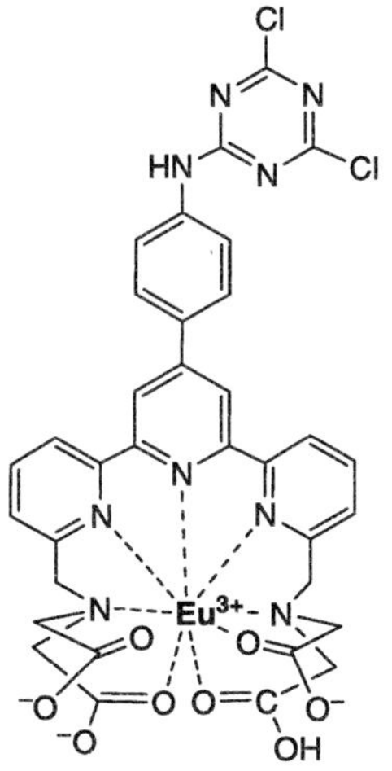

Eu-W8044 DTA (**4**)

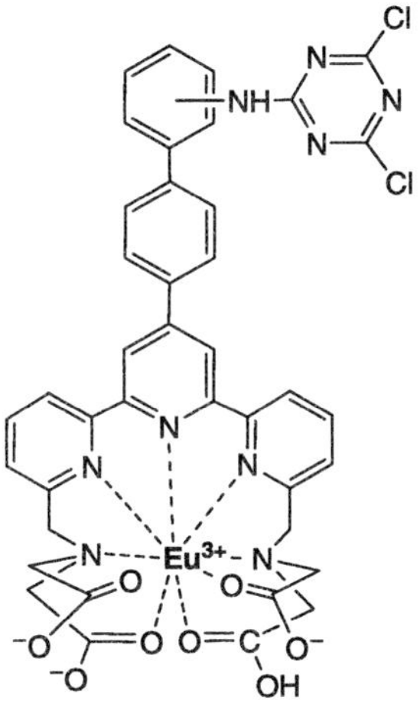

Eu-(DTBTA)(**5**)

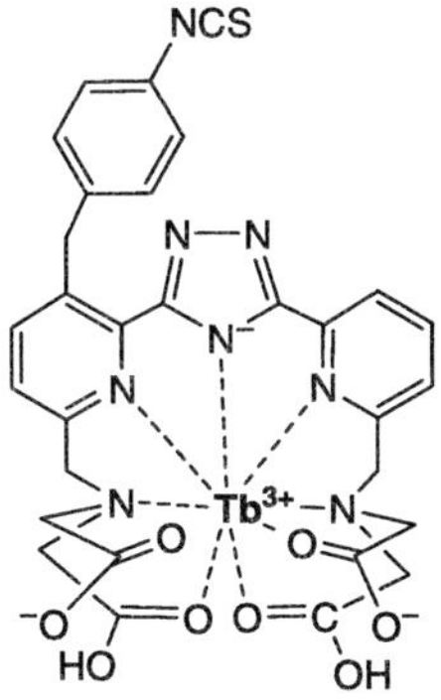

Tb-W14016 (**6**)

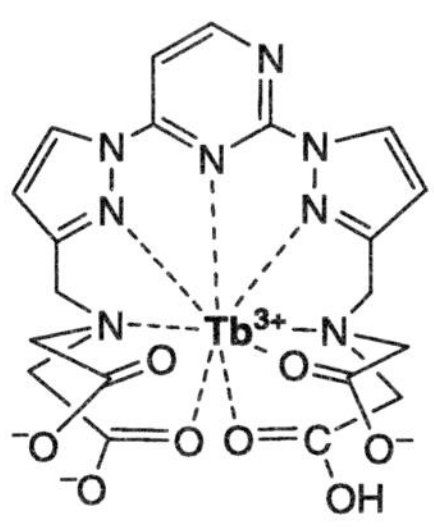

Tb-PzPMPz (**7**)

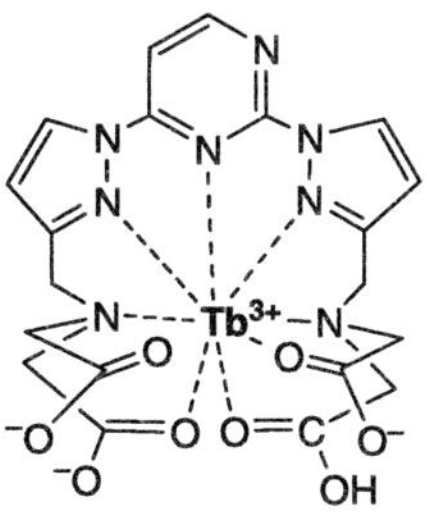

Tb-PzPyPz (**8**)

Fig. 1 Ligands based on EDTA scaffold

buffer. This Eu-TMT complex has been used to design time-resolved resonance energy, transfer assays using cyanine (CY5) conjugates as acceptor, and was evaluated in comparison with the diethylenetriaminepentaacetic acid (DTPA)-cs124/CY3 FRET couple [37].

It is now marketed under the name of "Amersham Eu (TMT) isothiocyanate" by GE Healthcare (www.gelifesciences.com). The luminescence intensity of the Eu (TMT) was reported (either water or Tris buffer) in the presence of various additives: the addition of Ca^{2+}, Mg^{2+}, and EGTA had no effect on intensity. However, when either iron or manganese ions were added, the intensity dropped to zero almost immediately.

A closely related Eu-TMT structure **3** (one methyl group replacing a methoxy group) was developed [38]: $\lambda_{max} = 295$ and 333 nm (ratio 1: 0.56; no molar absorbance quoted), Em = 615 nm, τ (H_2O) = 0.9 ms, nH_2O ~0.28, no reported value for quantum yield [38].

A similar terpyridine based chelate is marketed (Perkin–Elmer) as Eu-W8044 DTA chelate **4** and has a dichlorotriazinyl group as a reactive arm [39]. Again the nonadentate complexation increases the stability, the fluorescence is stable at pH 5 and decreases by ~20% within 2 h incubation in 50 mM succinate pH 4, but 90% of the fluorescence is lost at pH 3 (the difference of behavior with the Eu-TMT *vide supra* could be due to the difference in buffer composition). The Eu-W8044 is also more stable in the presence of EDTA and can withstand 2 h in 250 mM EDTA pH 7.8, but manganese effect is still detrimental since 80% of the fluorescence is lost within 30 min in the presence of 100 μM Mn^{2+}; as described above for the heptadentate Eu-W1024 chelate, a calculated amount of EDTA cancels the manganese effect (www.perkinelmer.com/lifesciences, "Stability of the Wallac LANCE™ Eu-chelates").

Similar nonadentate complexes such as {2,2′,2″,2‴-{[4′-(aminobiphenyl-4-yl)-2,2′:6′,2″-terpyridine-6,6′-diyl] bis(methylenenitrilo)}-tetrakis(acetato)-europium (III)} (Eu-ATBTA) have been activated as 4,6-dichloro-1,3,5-triazinyl **5** (DTBTA) or succinimidyl (NHS-ATBTA) derivatives and coupled to streptavidin and oligodeoxynucleotides [40]. The photophysical data are: $\lambda_{max} = 297$ nm and 335 nm (ε_{335} ~31,000 M^{-1} cm^{-1}; pH 9.0) $\Phi_{tot} = 0.09$ and $\tau = 1.02$ ms, the fluorescence does not decrease below 1 mM EDTA (10 mM Tris–HCl pH 8); interestingly, this chelate is stable in conditions simulating PCR (30 min at 90°C). This Eu(III) chelate was recently used for time-resolved microscopy [41].

A nonadentate bipyridino-triazole fluorescent Tb chelate (activated as isothiocyanate) was marketed by Wallac (product W14016). The basicity of the five-membered unsaturated heterocycles (e.g., pKa for pyridinium is 5.25 and for imidazolium is 6.95) makes them better donors than pyridine with respect to lanthanide complexation. Moreover, in some of the chelates the aromatic structure is negatively charged (e.g., compounds containing 1,2,4-triazol-3,5-ylene). Not only does this stabilize the chelate, it additionally changes their adsorption properties diminishing unspecific binding to column materials and plastics. Tb(III) chelate 3-{6′-[*N*,*N*-bis(carboxymethyl) aminomethyl]-4′-(*p*-isothiocyanatobenzyl)-2′-pyridyl}-5-{6″-[*N*,*N*-bis(carboxy methyl)aminomethyl]-2″-pyridyl}-1,2,4-triazole **6** has been used for bioassays [42] and the photophysical characteristics are available in the literature [43] $\lambda_{max} = 300$ nm, ($\varepsilon_{300} = 9{,}000$ M^{-1} cm^{-1}), $\Phi_{tot} = 0.05$, τ (H_2O) = 1.25 ms. No data are available concerning the effect of EDTA and potentially interfering cations. The brightness at 300 nm being already close to the minimum criterion, the data for 337-nm excitation is not available, but one can anticipate that this chelate is better suited for measurement on a flash lamp reader but is not well adapted for a reader equipped with a nitrogen laser. Recently, a similar chelate bearing a bis-pyrazolyl-pyrimidine **7** was described [44]. The bis-pyrazolyl pyrimidine chromophore was chosen since the replacement of one carbon by nitrogen was expected to overcome a steric problem arising in previously studied bis-pyrazolyl-pyridine **8** [45].

The photophysical characteristics are documented $\lambda_{max} = 300$ nm [ε = 13,700 M^{-1} cm^{-1}, τ (H_2O) = 2.74 ms, residual water molecules (nH_2O < 0.5)], Φ_{tot} (H_2O) ~0.39 ($\Phi_{Eu} = 0.60$) and triplet state energy of gadolinium complex (26,666 cm^{-1}). The terbium complex of 2,6-bis{3-[*N*,*N*-bis(carboxymethyl)aminomethyl] pyrazol-1-yl}pyridine **8** displayed remarkable luminescence properties: $\lambda_{max} = 313$ nm ($\varepsilon = 7{,}800$ M^{-1} cm^{-1}), Φ_{tot} (H_2O) = 0.60, τ (H_2O) = 2.75 ms. The photophysical characteristics are well documented (luminescence decay in D_2O and at 77 K) residual water molecules (nH_2O ~0.2–0.4) and triplet state energy of gadolinium complex (25,150 cm^{-1}) however the actual expected brightness upon laser excitation cannot be compared with other complexes since the absorbance at 337 nm is not reported. The conjugation of the complex with biomolecules was performed by its derivatization (as isothiocyanate) in relatively low yield using a tedious synthetic protocol. Unfortunately, for either the bis-pyrazolylpyridine or the bis-pyrazolylpyrimidine lanthanide complexes, no data are available concerning the stability in the presence of EDTA or competing ions (Mn^{2+}, Ca^{2+}) or in potentially troublesome but "basic" buffer as phosphate buffer; therefore, it is hard to evaluate the real usefulness of these complexes in a real bioassay, as in the HTS context.

3.1.2 DTPA-Like Scaffold

The ligands based on DTPA scaffold are displayed in Fig. 2. The DTPA was coupled to a chromophore, the 7-amino-4-methylquinolin-2(1*H*)-one (carbostyril-124, cs124) [46]. The DTPA structure (as di-anhydride) is easy to functionalize as a bifunctional chelate allowing the coupling to a biomolecule of interest. The DTPA-cs124 **9** itself is a good sensitizer for terbium with favorable molar absorbance $\varepsilon_{337} = 9{,}600$ M^{-1} cm^{-1} and luminescence decay τ (H_2O) = 1.55 ms. In spite of nH_2O ~1.1 inducing a partial deactivation, the quantum yield $\Phi_{tot} = 0.32$ is acceptable. One drawback is that the coupling to a biomolecule usually induces on the tethered (Tb^{3+})DTPA-cs124 **10** a drop in the quantum yield (0.2–0.25) and in the luminescence decay characterized by typical bi-exponential decay [46] illustrating the presence of several emitting species. The amide bond

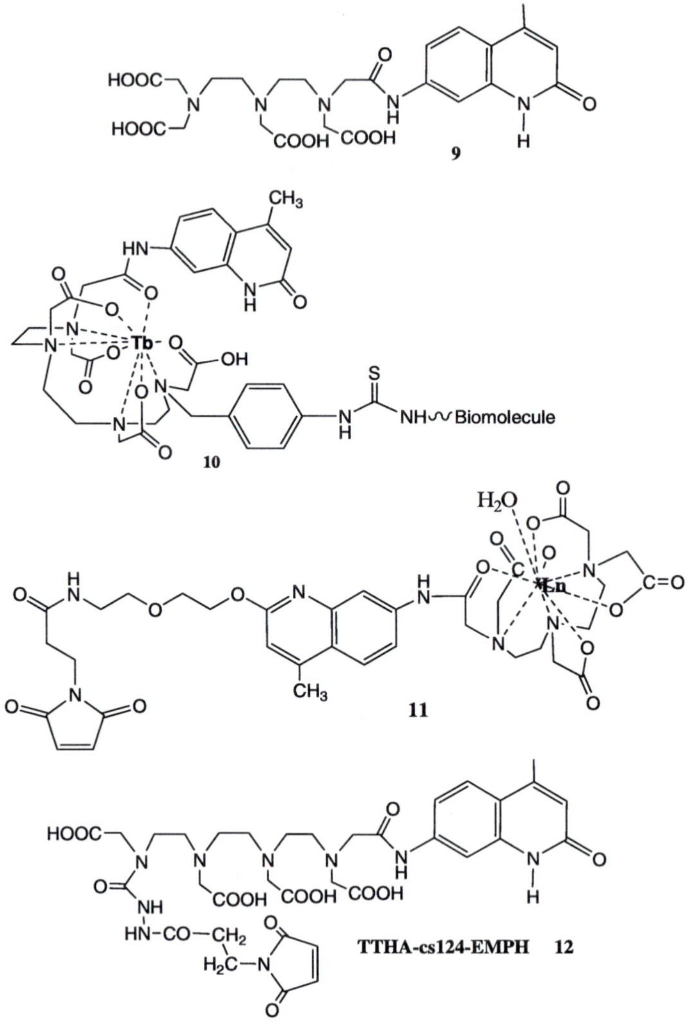

Fig. 2 Ligands based on DTPA and TTHA scaffold

formed from one of the carboxyl groups is able to maintain coordination to the lanthanide [46] and the binding constant is sufficiently high for most practical purposes. Nevertheless, care must be exercised in the labeling, conjugates purification steps, and further handling since the lanthanide DTPA-cs124 complexes are sensitive to competing ions and complexing agents. The tethered (Tb^{3+})DTPA-cs124 complexes **10** is stable up to 10 mM Mg^{2+} but the luminescence intensity decreases twofold in the presence of 500 μM Mn^{2+} and between 50 and 100 mM EDTA. The (Tb^{3+})DTPA-cs124 labeling reagents are reported to be sensitive to phosphate buffer, which induces the precipitation of the terbium, and is not compatible to acidic medium such as formic or trifluoracetic acid medium traditionally used for peptides RP-HPLC purification. Although TEAAc based eluant can be used for RP-HPLC purification, this buffer is well adapted to (Tb^{3+})DTPA-cs124 labeled ODN its efficiency is very sequence-dependent in the case of labeled peptides (peak broadening). A (Tb^{3+}) DTPA-cs124 is marketed as Lanthascreen™ (www.invitrogen.com) under the form of amine-reactive, thiol-reactive, as well as bioconjugates. Many example for the use of such (Tb^{3+})DTPA-cs124 **9** in time-resolved resonance energy transfer assays are reported [34].

Recently, DTPA was coupled to a 2-alkoxy-4-methylquinolin-7-amine [47], which is an O-substituted (enol form) derivative of carbostyril-124 known to exist under two tautomeric forms (enol and keto form) [48]. For this lanthanide chelate **11**, partial photophysical data were reported: $\lambda_{max} = 330$ nm (ε not quoted), Eu complex τ (H_2O) $= 0.61$ ms (mono-exponential). The Tb complex τ (H_2O) $= 1.24$ ms (mono-exponential). Both complexes exhibited Φ_{tot} (H_2O) ~60% of the respective Eu or Tb DTPA-CS124 complex taken as reference. Both Eu and Tb complexes were converted into maleimide derivatives. The advantage compared to the previously reported DTPA-CS124 is that the linkage to maleimide (and thus to the biomolecule) does not involve a second carboxylate from the DTPA scaffold, and so the lanthanide is expected to be octacoordinated. No quantitative data were reported about stability, but one could anticipate a sufficient stability since the labeling of a His-tagged protein was performed directly onto Ni-NTA beads (4°C in the presence of TCEP) followed by imidazol elution. It is reported that the lanthanide was not liberated in this process. Luminescence decay for the conjugated complexes: Eu τ (H_2O) $= 0.75$ ms, Tb τ (H_2O) $= 1.17$ ms (both 95% mono-exponential). Nevertheless, the labeling efficiency does not seem to be optimum since a 30-fold excess of maleimide derivative resulted in 70% labeling of the single cysteine residue of the protein. The coupling procedure involving TCEP during the reaction could be the reason for such low efficiency since TCEP, usually considered as inert towards maleimide, was shown to interfere [49].

3.1.3 TTHA-Like Scaffold

The ligands based on triethylenetetraaminehexanoic acid (TTHA) scaffold are shown in Fig. 2. A TTHA scaffold connected to carbostyril-124 was described **12** [50, 51]. This 10-dentate ligand was selected for minimizing the lanthanide hydration and indeed it has shown no coordinated water molecules. (Eu)TTHA-cs124: τ (H_2O) $= 1.19$ ms, τ (D_2O) $= 1.79$, $nH_2O = 0.3$, relative brightness (relative to (Eu)DTPA-cs124) $= 2.7$. (Tb)TTHA-cs124: τ (H_2O) $= 2.10$ ms, τ (D_2O) $= 2.31$, $nH_2O = 0.19$, brightness (relative to (Tb)DTPA-cs124) $= 0.31$. The luminescence intensity and luminescence decay measurements were carried out in the presence of excess $TbCl_3$ or $EuCl_3$ (1:1.2 molar ratio) added to the chelate prior to experiment. This precaution, together with the observation that emission intensity is concentration-dependent and luminescence decay is constant, indicated that the binding stability constant is unexpectedly small. Even if the new chelates provide better protection to the lanthanide ions from solvent limiting the accessibility of water molecules to the metal center, no data were given concerning the sensitivity to competing ions or ligands.

3.1.4 Podand-Type Ligand

Tripodal Ligands

Original tripod ligand (Fig. 3) built from two 6′-carboxy-6-methylene-2,2′-bipyridine chromophoric arms linked to amino group of a glutamate skeleton **13** were developed by Ziessel's group [52]. The reported photophysical properties are globally favorable, Eu(III) complex: $\lambda_{max} = 308$ nm ($\varepsilon_{308} = 19{,}700\ M^{-1}\ cm^{-1}$), Φ_{tot} (H_2O) = 0.08, τ (H_2O) = 0.62 ms; Tb(III) complex: ($\varepsilon_{308} = 20{,}800\ M^{-1}\ cm^{-1}$), Φ_{tot} (H_2O) = 0.31, τ (H_2O) = 1.48 ms ; ligand triplet excited state $E = 22{,}100\ cm^{-1}$ ($1{,}700\ cm^{-1}$ above the Tb^{3+} emitting state). The published emission spectrum of the terbium complex display a prominent $^5D_0 \rightarrow {}^7F_2$ transition ideal for a sensitive and selective detection in TR-FRET mode (reduced crosstalk). It is clear from the published absorption spectra that because of the absorption falling to baseline at wavelengths longer than 330 nm, these complexes can only be used on a flash lamp reader but are not compatible with nitrogen laser excitation (337 nm). Stability studies showed that the complexes are stable for several days in common buffers used as Tris–HCl (10 mM, pH 7.0) and PBS (20 mM PO_4 150 mM NaCl, pH 7.4). No stability data in the presence of EDTA and competing cations are available. Some more complexes based on the same scaffold were published by the same group [53].

Recently, a pyrazole-pyridine Tb(III) tripod **14** was synthesized [54]. The reported photophysical data ($\Phi_{tot} < 0.001$ and $\tau \sim 0.9$ ms in methanol) together with the comparison between experimental and calculated absorption electronic spectra [52], suggest that the terbium tripod complex presents a Tb[podand]·$4H_2O$ structure, where the podand ligand was coordinated to Tb^{3+} ion by six oxygen atoms from the carboxylate group. Thus, the terbium being coordinated by the carboxylate rather than the chromophore nitrogen, the terbium is not shielded from water and the intramolecular energy transfer process is unfavorable. Furthermore, the UV absorption spectrum [55] showed an abrupt edge below 320 nm and almost no absorbance at 337 nm, and therefore the chromophore is not well fitted for laser excitation.

Tetrapodal Ligand

A series of octadentate complexes containing 2-hydroxyisophthalamide (salicylamide) chelating units linked to 1,1,2,2-tetraamino-propyl-1,2-diaminoethane backbone (symbolized as H(2,2)-) were synthesized in Raymond's group [56] (Fig. 3).

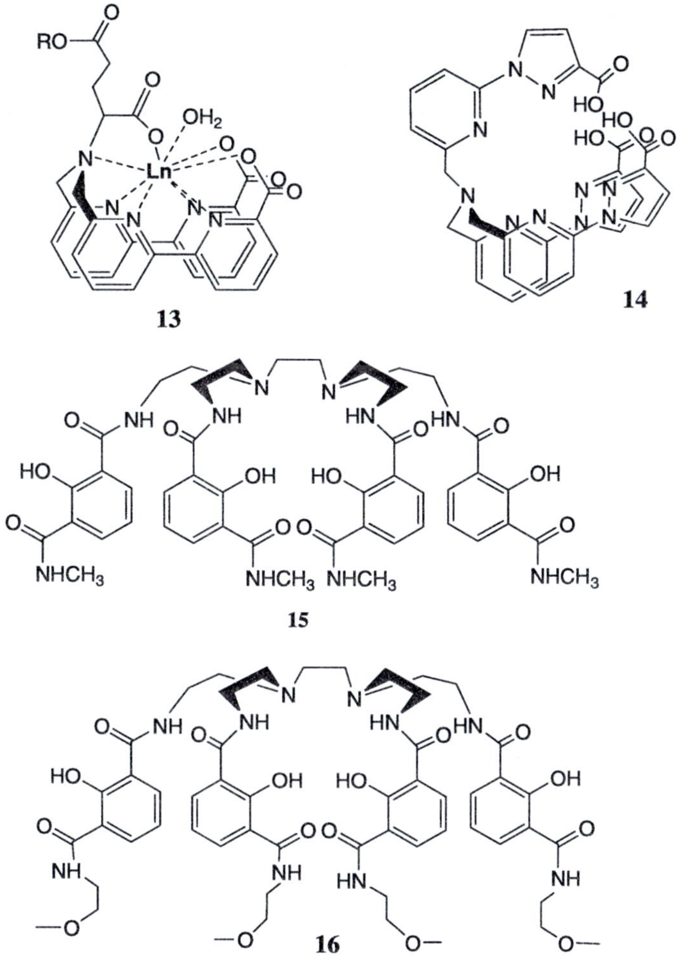

Fig. 3 Ligand based on tripod and tetrapod scaffold

The photophysical properties of **15** are: $\lambda_{max} = 350$ nm ($\varepsilon_{350} = 26{,}800$ M^{-1} cm^{-1}) (10 mM Tris buffer pH 8.5), Φ_{tot} (H_2O) $= 0.59$. So the brightness B $= 15{,}800$ cm^{-1} which is about 20 times the value of commercially available complexes. Unfortunately, the luminescence decay time was not quoted in the article. The 2-hydroxyisophthalamide group is a very good ligand for Ln^{3+} cations and also provides excellent sensitization of the Tb^{3+} luminescence through a particularly efficient ligand-to-lanthanide energy transfer process [56]. The complexes are highly soluble and stable in water at physiological pH and in Tris or phosphate buffers. These properties, combined with the ease of synthesis, make these ligands very attractive, provided that analogous ligands containing a suitable linker for covalent labeling of biomolecule would be available. Indeed, analogous ligands functionalized with a linker were patented [57] and structurally similar 2-hydroxyisophthalamide ligand derivatives that utilize the H(2,2)-backbone have been recently commercialized and utilized as a luminescent probe for high sensitivity homogeneous time-resolved fluorescence (HTRF) technology [58]. Although no solubility issues were mentioned in the original paper [53], more recently a new series of ligands was designed [59] incorporating methoxyethyl units to increase the water solubility (Fig. 3). The more interesting ligand is (Tb)**16** ($n = 0$ named H(2,2)-IAM-MOE) Φ_{tot} (H_2O) = 0.56, τ (H_2O) = 2.63 ms, τ (D_2O) = 3.27 ms, nH_2O = 0.1. From low temperature phosphorescence on Gd(**16**) complex, the triplet states was measured $E = 23{,}170\ cm^{-1}$, which is about 2,750 cm^{-1} higher in energy than the Tb(III) $^5D_4$ emitting state, and therefore in the range proposed for optimal ligand-to-Tb(III) energy transfer [43]. The only data available concerning the stability of Tb(**15**) are qualitative: micromolar solutions of complex in Tris and phosphate buffer (pH 7.4) can be kept for a long period of time without losing luminescence. The observation that Tb(**15**) complex in phosphate buffer (0.01 M) gives detectable luminescence down to 10^{-15} M as quoted [56] was surprising and it was shown later [59] that this was due to contamination and the detection limit was rather between 10^{-10} and 10^{-11} M. Stability constants were measured for Tb(**16**) ensuring stability at nanomolar concentrations. This is stable enough for a use in immunoassays, although additional data concerning stability in the presence of EDTA, Mn^{2+}, Mg^{2+}, and urate are desirable to evaluate the true potential of this complex.

3.1.5 Lanthanide-Binding Tags

Lanthanide-binding tags (LBTs) (Fig. 4) are short peptides (17–20 amino acids) that selectively and avidly bind lanthanide ions [60]. Originally based on Ca^{2+}-binding loops, LBTs have been optimized for Tb^{3+} binding using synthetic peptide libraries. In these motifs, the indole ring of a strategically positioned tryptophan (Trp) residue (**17**) sensitizes Tb. Recently, two LBTs containing unnatural amino acids as potential lanthanide sensitizers were studied [61]. The first **18** is based on carbostyril-124 which has been used as a sensitizer of both Tb and Eu luminescence in the context of polyaminocarboxylate chelates as described above. The second **19**

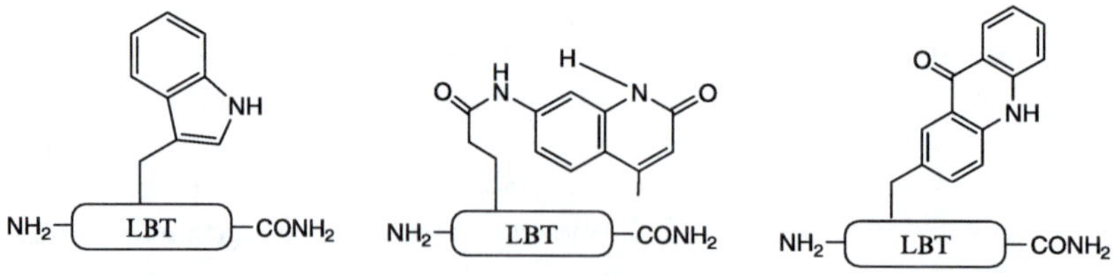

LBT(Trp) **17** LBT(cs124) **18** LBT(Acd) **19**

H_2N-FITDNNDGXIEGDELLLEEG-$CONH_2$

LBT peptide equence, the aminoacid bearing the sensitizer side-chain is denoted X

Fig. 4 Lanthanide-binding tags (LBTs)

is based on the acridone fluorophore (Acd), which has been previously used to sensitize Eu luminescence. The peptides were synthesized on solid phase using an acid labile linkage to the resin; thus, a mild cleavage released the peptide still protected on the side chains but bearing a free carboxylic function on C-ter which could further be converted into thioester. The peptide was then coupled using Natural Chemical Ligation to the *N*-ter cysteine of a protein bearing a SH2 domain. The LBT(cs124) tethered to the protein presented a K_D ~46 nM for Tb^{3+} and ~69 nM for Eu^{3+}. The luminescence decay data were not given but from their value in H_2O and D_2O a nH_2O ~0 was determined. One drawback was mentioned concerning the photobleaching of the CS124 LBT under continuous excitation.

4 Macrocyclic Compounds

A first large class of macrocyclic-based lanthanide ligands was designed taking advantage of all the knowledge coming from the gadolinium complexes used for magnetic resonance imaging (MRI) as DOTA derivatives. The second class is formed by the macropolycyclic compounds as cryptates. The synthesis of macrocyclic ligands with pendant groups such as DO3A derivative, which represent the majority of the compounds which received some practical applications, is usually straightforward although more demanding than linear polydentate ligands. Usually, the synthetic scheme involves a protection, alkylation, deprotection, and alkylation sequence. Amines function of a preformed azamacrocycle (cyclen) can be temporarily protected by suitable protective groups before the alkylation with an halogeno derivative of the chromophore. For a recent review see [62, 63].

4.1 Macrocyclic Ligands

Lanthanide chelates based on a macrocyclic Schiff base (Fig. 5) can be prepared from 2,6-diacetylpyridine and diaminoalkane by using a lanthanide ion as a template. Although Schiff bases are normally labile, the lanthanide ion stabilizes the macrocyclic moiety if the structure formed is an 18-membered macrocycle **20** [64]. Due to limited chelate stability, assays have been performed in the presence of excess lanthanide ion; nevertheless, the stability of Eu^{3+} chelation in these macrocyclic chelates does not permit effective and reliable dissociation enhancement [8, 65]. Detection sensitivity has been enhanced by using the "lanthanide enhanced luminescence" LEL (also known as RETEL effect or cofluorescence) in association with gadolinium chelates [66]. Nevertheless, care must be taken in LEL experiments since some background may arise from the traces of europium present in the gadolinium complex used as enhancer. Such macrocyclic Schiff base is commercialized (Research Organics Inc.; http://www.resorg.com) as Quantum Dye® [67] functionalized as isothiocyanate. Reported photophysical characteristics for the

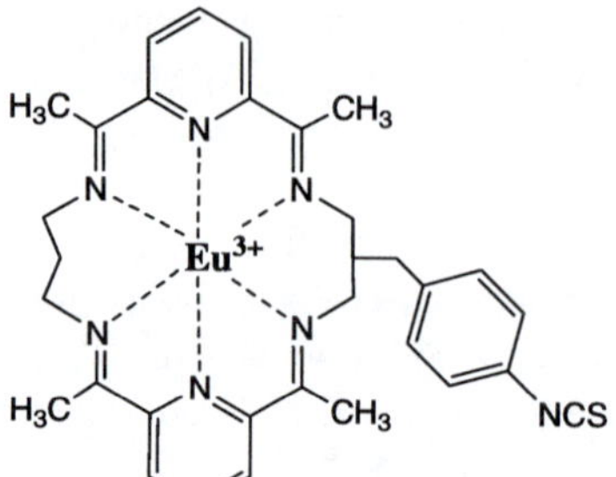

20

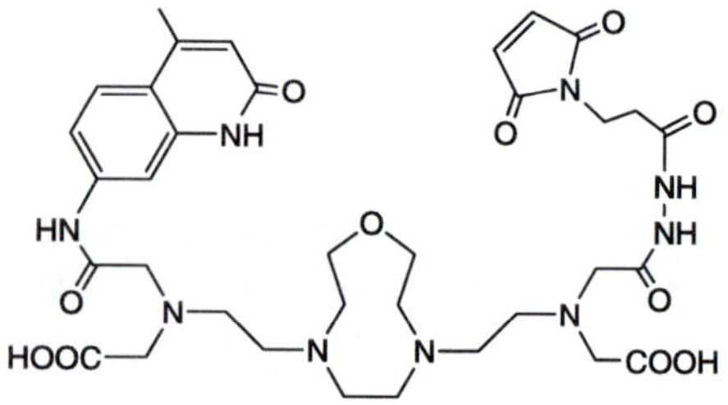

21 (N_2O-cs124)

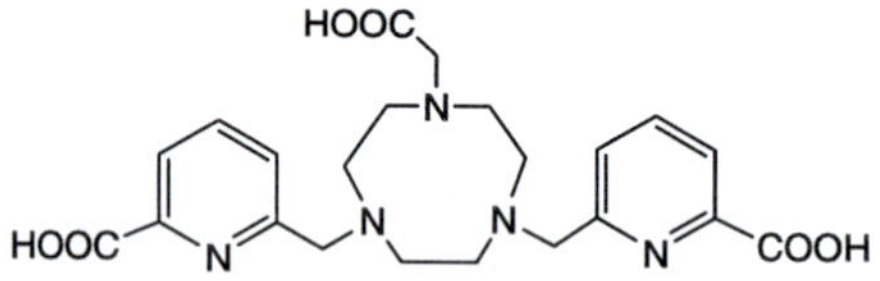

22 H_3bpatcn

Fig. 5 Ligands based on macrocyclic Schiff base and macrocycles with pendant groups

non-functionalized Eu (III) Hexaaza Macrocyclic [68] are as follows: $\lambda_{max} = 298$ nm (ε_{298} ~11,200 M^{-1} cm^{-1}), 375 nm (ε_{337} ~400 M^{-1} cm^{-1}, f–f absorption band of Eu^{3+}), τ (H_2O) ~0.70 ms, τ (D_2O) ~2.0 ms Φ_{tot} (D_2O) = 0.006. For the functionalized complex, τ (H_2O) ~0.427 ms Φ_{tot} (H_2O) = 0.006, and τ (H_2O) ~0.43 ms. The absorption spectrum reaching baseline at 331-nm excitation of this complex is not compatible with 337 nm laser but only to readers equipped with flash lamp. To circumvent the intrinsically low quantum yield, peptide carriers labeled with multiple lanthanide complexes and containing a reactive functionality that could be linked to a biomolecule have been used [69]. So far only a few applications in the field bioassays [70, 71] and one application concerning time-gated flow cytometry [72] have been reported.

4.2 *Macrocyclic Ligands with Pendant Groups*

Lanthanide macrocyclic ligands are mostly polyaza macrocycles bearing pendant groups used either for complexation purposes (mostly carboxylic acetate and methylphosphonates groups) or for sensitization purposes (chromophoric groups). In a few cases the chromophoric group is part of the macrocycle itself. In the following section we will classify the various complexes found according to the increasing size of the macrocyclic rings and using the common acronyms:

1. NOTA (1,4,7-triazacyclononane-*N*,*N′*,*N″*-triacetate)
2. DO3A (1,4,7,10-tetraazacyclododecane-1,4,7,-triacetate)

3. DOTA (1,4,7,10-tetraazacyclododecane-1,4,7,10-tetraacetate)
4. TETA (1,4,8,11-tetraazacyclotetradecane-1,4,8,11-tetraacetate)
5. PCTA(3,6,9,15-tetraazabicyclo[9.3.1]pentadeca-1(15),11,13-triene-3,6,9-triacetate).

A comprehensive article [73] discusses a comparison of the complexation behavior and the metal ion selectivity of macrocycles and open-chain ligands. The reported stability constants for lanthanide complexes (Table 2) show that DOTA scaffold is expected to give the more stable complexes.

One should also mention an interesting article [74] describing the photophysical properties of the macrocyclic "DOTA-like" europium complexes upon direct excitation of the emissive europium ion at $\lambda_{exc} = 394$ nm.

4.2.1 ODCN Scaffold

A series of 9 and 10-dentate chelates were recently reported [81]. The (Tb) 1-oxa-4,7-diazacyclononane based chelate (N2O-cs124) **21** bearing three carboxylic groups and carbostyril 124 as sensitizer (Fig. 5) provided the best protection to the lanthanide ions from solvent molecule, and forms a stable lanthanide coordination compounds.

(Tb)N_2O-cs124: τ (H_2O) = 1.89 ms τ (H_2O) = 1.91 ms (slight excess of ligand, no experimental details for buffer), nH_2O ~0.02, relative brightness relative to (Tb)DTPA-cs124 = 1.3. (Eu)N_2O-cs124: Φ_{tot} (H_2O) = τ (H_2O) = 1.0 ms τ (H_2O) = 1.39 ms (slight excess of ligand, no experimental details for buffer), nH_2O ~0.29, relative brightness relative to (Eu)DTPA-cs124 = 1.22. The molar absorbance is not given but should be of the same order as for DTPA-cs124 (ε_{337} = 9,600 M^{-1} cm^{-1}). The (Tb)N_2O-conjugated to proteins displays a bi-exponential decay $\tau_1 = 1.34$ (75%), $\tau_2 = 0.22$ (25%). No stability data in the presence of competing ions, EDTA, or metabolites were given, but at least the complex is stable in the presence of 5 mM Mg^{2+} which was present in the coupling reaction (4°C overnight in 20 mM HEPES, pH 7.4).

4.2.2 NOTA Scaffold

A ligand (H_3bpatcn) **22** in which two picolinate and one acetate arm are connected to a 1,4,7-triazacyclonane macrocycle (TCN) was synthesized in Mazzanti group

Table 2 Stability Constants ([ML]/[M][L]) of the Ln(NOTA), Ln(DO3A), Ln(DOTA), Ln (TETA), Ln(PCTA), and DTPA

M^{3+}	NOTA [75]	DO3A [76]	DOTA [75]	TETA [77]	PCTA	DTPA [80]
Tb^{3+}	–	–	24.4	14.8	–	22.7
Eu^{3+}	13.9	–	23.5	14.0	20.3 [78]	22.4
Gd^{3+}	14.3	21.1	24.7	13.8	20.4 [79]	22.5

[82] (Fig. 5). A complete photophysical study was reported: Eu(bpatcn) $\lambda_{max} = 273$ nm ($\varepsilon = 9{,}050\ M^{-1}\ cm^{-1}$) Φ_{tot} (H_2O) = 0.05, τ (H_2O) = 0.54 ms (Tris buffer pH 7.4), nH_2O ~1.2 ; Tb(bpatcn) $\lambda_{max} = 273$ nm ($\varepsilon = 9{,}100\ M^{-1}\ cm^{-1}$) Φ_{tot} (H_2O) = 0.43, τ (H_2O) = 1.49 ms. The stability constant $LogK_{GdL} = 15.8$ comparing favorably with NOTA itself, but no data were given concerning its behavior in the presence of competing ions and ligand as EDTA, as well as possible quenching by metabolites. Although the luminescence decay values and brightness (B ~450 for Eu complex and ~3,900 for Tb complex) are compatible with a use as probe, the absorption maximum at 273 nm and the absorbance falling to near zero above 300 nm (due to the pyridine chromophore) presents a serious handicap, with respect to inner filter effect, for practical use in a diagnostic or screening context. Furthermore, this complex would have to be functionalized for coupling to biomolecules.

4.2.3 DOTA/DO3A Scaffold

The ligand derived from DO3A structure is shown in Fig. 6 These scaffolds offer the possibility to built tetrapodal ligands holding one chromophore; when the ligand is linked to the macrocycle replacing one of the carboxylic functions the ligand is truly a DO3A analog. Considering the known stability of derivatives such as Gd (DOTA) and Gd(DO3A), which are used in vivo as MRI contrast agent, exhibit a high kinetic inertness under physiological conditions and should be a good platform to design stable luminescent complexes. Indeed, many complexes are reported incorporating various chromophores. A closely related scaffold is the 1,4,7,10-tetrakis (carbamoylmethyl)-1,4,7,10-tetraazacyclododecane (TCMC) an amide derivative of DOTA. In some complexes, carboxylic groups are replaced (partially or fully) by methylphosphonate groups. Parker's group in Durham did a tremendous amount of work on this class of lanthanide complexes including several innovative chromophores; over sixty different europium and terbium complexes have been synthesized. The complete photophysical studies are being carried out including evaluation of stability interference with metabolites (citrate, ascorbate, urate). The most prominent aspects in terms of use for bioassays including the TRF microscopy applications, for which there is a renewed interest [83–87], are listed in Table 3. An azathioxanthone europium chelate belonging to this series received a successful application in the design on a europium luminescence assay of lactate and citrate in biological fluids [88]. This assay allows the titration of citrate in microliter of sample; a decrease in citrate is used here as a biomarker for prostate cancer. A very recent review [22] summarizes the main trends of selected cell-penetrating lanthanide complexes used as optical probes for microscopy on living cells. For the more promising in vivo applications (criterion of reduced interference with metabolites), additional toxicity studies on cell culture have been carried out. The more promising complexes are (Tb)**32** presenting slow cell uptake and relatively low toxicity (IC_{50} ~58 μM) and (Eu)**33** presenting fast cell uptake and low toxicity ($IC_{50} > 175$ μM). Interestingly, (Eu)**33** complex presented a very large decrease in Eu emission intensity upon binding to HSA (suggesting that the chromophore

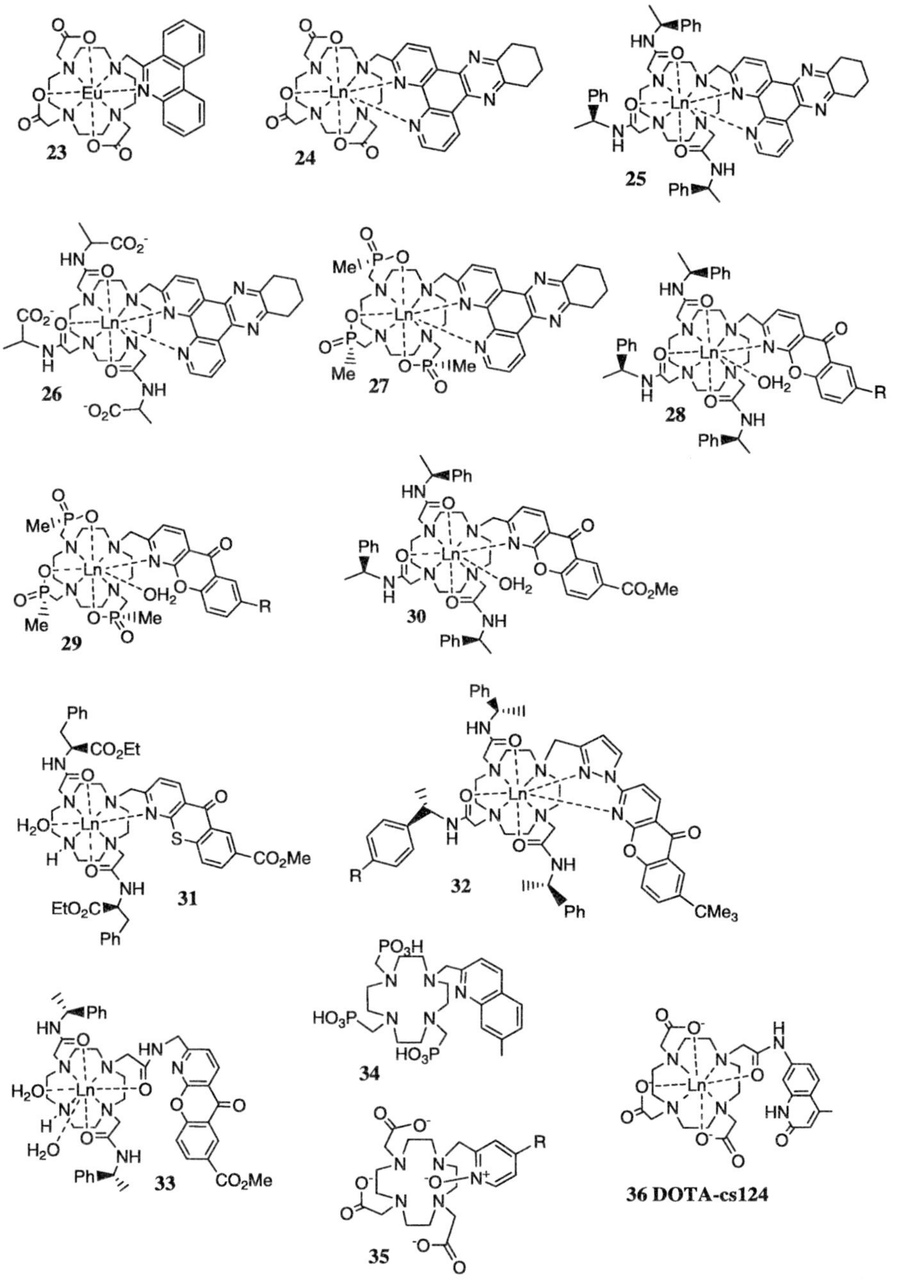

Fig. 6 Ligands based on DO3A scaffold

Table 3 Chromophore, structure numbering, and photophysical properties of DO3A scaffold based lanthanide complexes

Chromophore structure	Exc (nm)	Luminescence decay (ms)	Φ_{tot} (H_2O)	Ref.
Phenanthridine **23**	355	1.00 (Eu)	0.06 (Eu)	[87]
TATP **24**	348	1.06 (Eu)	0.21 (Eu)	[89]
		1.46 (Tb)	0.36 (Tb)	
TATP **25**	348	1.04 (Eu)	0.16 (Eu)	[25, 90, 91]
		1.56 (Tb)	0.40 (Tb)	
TATP **26**	348	0.96 (Eu)	0.18 (Eu)	[24, 90, 91]
		1.49 (Tb)	0.36 (Tb)	
TATP **27**	348	1.14 (Eu)	0.21 (Eu)	[90, 91]
		2.90 (Tb)	0.49 (Tb)	
Azaxanthone R = H **28**	338	1.82 (Tb)	0.37 (Tb)	[22]
		0.56 (Eu)	0.08 (Eu)	
Azaxanthone **29**	337	1.16 (Eu)	0.18 (Eu)	[92]
		3.57 (Tb)	0.43 (Tb)	
Azaxanthone-CO_2Me **30**	335	1.65 (Tb)	0.37 (Tb)	[92]
		0.54 (Eu)	0.08 (Eu)	
Azathioxanthone **31**	384	0.3 (Eu)	0.09 (Eu)	[93]
Pyrazolyl-azaxanthone **32**	348	2.22 (Tb)	0.61 (Tb)	[94]
Azaxanthone-CO_2Me **33**	335	1.17 (Tb) 0.26 (Eu)	n.d	[95]
Methylquinoline **34**		0.62 (Eu)	n.d	[96, 97]

excited state was being quenched by a change transfer process), although the complex gave a bright labeling within the cell. Most of the complexes were shown to concentrate in lysosomes, nucleoli, or mitochondria. Cell imaging using lanthanide chelates is still in its infancy and it will need time to be able to correlate structure and fate of the molecule within the subcellular compartments.

A DO3A phosphonate analog bearing a methylquinoline chromophore **34** was designed to be used as bimodal imaging agent either as Gd complex for MRI or Eu complex for luminescence [96, 97]. The (Eu)**34** was also tethered to a benzodiazepine receptor ligand (PK11195) moiety. Experiments on glioblastoma cells and displacement experiment with unlabeled PK11195, fluorescence imaging showed intracellular uptake. This suggested that the complex was resistant to quenching. Although no quantitative toxicity (IC_{50}) data were published, the compounds seem relatively non-toxic since concentrations of 25 μM were used for fluorescence microscopy with (Eu)**34**-PK11195 and up to 1 mM total concentration for a (Eu)**34**-(Gd)**34**-PK11195 cocktail. Preliminary studies indicate that (Ln)-PK11195 is localized in the same region of the cell as the mitochondrial dye JC-1. No quantitative data were given concerning quantum yield and luminescence decay, the λ_{max} = 320 nm (ε not given) [96], and the emission spectrum showed a low symmetry (strong 680-nm band) not favorable for LRET experiments. Later the same group reported a conjugate of PK11195 and lissamine dye [98] since they consider the excitation of the sensitizing methylquinoline chromophore at 320 nm as a limiting factor considering the attenuation through glass optics of standard microscopes.

A DOTA scaffold bearing a pyridine-*N*-oxide **35** was recently reported [99]. The pyridine could be either unsubstituted (R = H) or bearing a 4-carboxylic group. The luminescence decays were mono-exponential (Eu)**35**, τ (H_2O) = 0.624 ms,

τ (D_2O) = 1.90 ms. The quantum yield was not given nor was the absorbance. The complexes seemed stable since they could be analyzed on RP-HPLC using an acetonitrile gradient in 0.1% TFA.

A DO3A scaffold connected to carbostyril-124 **36** was described [50, 51] together with the linear TTHA analog **12**. ($\varepsilon_{337} = 10^4$ M^{-1} cm^{-1}), (Tb)**36** τ (H_2O) = 1.54 ms, τ (D_2O) = 2.61 ms, $\Phi_{tot} \sim 0.32$, Φ_{tot} (H_2O) = 0.32. (Eu)**36** τ (H_2O) = 0.62 ms, τ (D_2O) = 2.25 ms, brightness (relative to (Eu)DTPA-cs124) = 0.57, Φ_{tot} (H_2O) = 0.057, $n H_2O$ ~1.1. The quantum yield was not substantially increased in this DOTA derivative compared to previously reported acyclic complexes (DTPA-cs124). One important feature of this article is the discussion about the lanthanide quantum yield (Φ_{Ln}). This parameter has been often poorly determined; however, it is a key parameter in LRET experiment since the distance R, inferred from energy transfer experiments via donor luminescence decay or intensity measurements, depends on knowing the donor (Φ_{Ln}) quantum yield: $R \propto \Phi_{Ln}^{1/6}$. Very often this has been explicitly or implicitly assumed to be unity in D_2O. The authors showed this assumption is not rigorously correct. Using diffusion-enhanced resonance energy transfer (DERET) between Tb DTPA-cs124 and fluorescein (i.e., measuring the dynamic transfer through the decrease of donor luminescence decay as a function of the acceptor concentration) the Q_{Tb} in DTPA-cs124 was determined to be 0.486 (versus QLn in D_2O = 0.818). Similarly, by transfer between Eu-DTPA-cs124 and SulfoRhodamine 101, the lanthanide quantum yield Φ_{Eu} in Eu-DOTA-cs124 was determined to be 0.137 (versus Q_{Ln} in D_2O = 0.65). Nevertheless, the errors introduced in calculating Förster radii are relatively small (<7% for Eu-DTPA-cs124 since it is $Q_{Ln}^{1/6}$).

4.3 *Macrocyclic Ligands with Endocyclic Chromophore*

As pointed out in a recent article [100] and as reviewed herein, to optimize the stability and the luminescence properties of Eu and Tb complexes in aqueous solutions many ligands have been synthesized and numerous chromophores have been investigated for the sensitization of luminescence of these ions. However, as a consequence of these requirements of optimal kinetic inertness and luminescence, only a few viable lanthanide labels (essentially based on Eu complexes) have been developed and tested up to now. Whereas numerous examples of photoactive lanthanide complexes based on macrocyclic ligands where a chromophoric unit is located on the dangling arms are described (*vide supra*), only a few reports are based on lanthanide binding by macrocyclic compounds incorporating both an endocyclic chromophoric unit and pendant carboxylate groups. Such macrocyclic ligands built from a complexing moiety associated with an endocyclic chromophore group and an exocyclic carboxylate group represent an alternative as displayed in Fig. 7. Polyaminocarboxylates have excellent water solubility and a high binding constant for lanthanides, whereas the macrocyclic structure can increase the thermodynamic and (or) kinetic stability of the complexes ("macrocyclic effect").

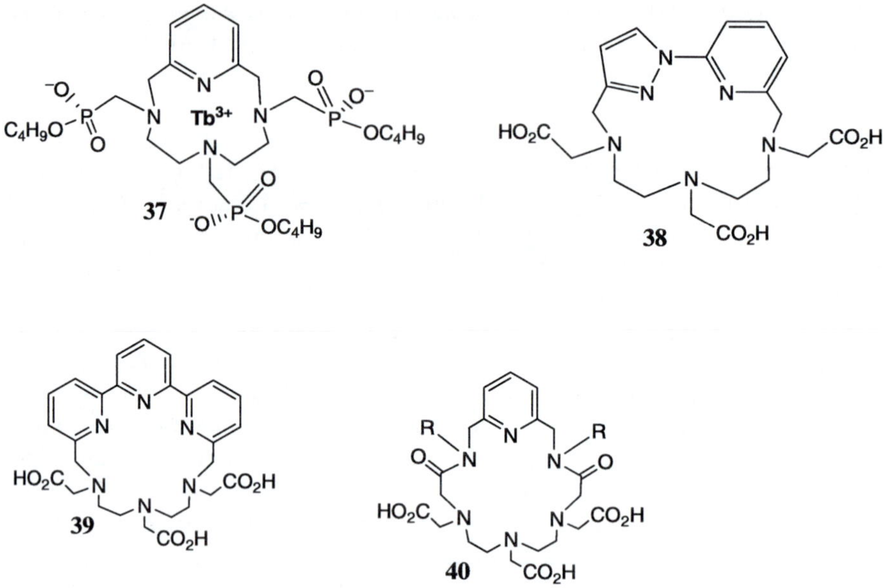

Fig. 7 Macrocyclic ligands with endocyclic chromophore

Moreover, in such systems, the antenna-to-lanthanide sensitization step occurs between partners in a rigid conformation that can improve the intramolecular energy transfer rates.

4.3.1 PCTA Scaffold

PCTA is a tetraazamacrocyclic ligand bearing a pyridine chromophore and three carboxylic functions. It offers seven potential donor atoms able to coordinate a lanthanide. The stability constant of 20.3 for (Eu^{3+})PCTA is acceptable in order to work in the presence of competing ions or ligands [101]. A well known complex **37** is based on a "phosphonate equivalent" of DOTA: Tb(III)-3,6,9-tris(methylene phosphonic acid *n*-butyl ester)-3,6,9,15-tetraazabicyclo[9.3.1]-pentadeca-1 (15),11,13-triene (Tb-PCTMB) [102, 103] $\lambda_{max} = 270$ nm, $\varepsilon_{269} = 3{,}000\ M^{-1}\ cm^{-1}$, τ (H_2O) = 4.98 ms [104] Φ_{tot} (H_2O) = 0.51. As one could expect from a single pyridine chromophore the absorption is rather low and the absorption falls below the workable window (300–340 nm): $\lambda_{max} = 269$ nm, $\varepsilon_{269} = 4{,}600\ M^{-1}\ cm^{-1}$.

4.3.2 Pentaazacyclopentadecane Scaffold

Pentaazamacrocyclic ligand incorporating either an intracyclic bipyridine (BP) [105] or a pyrazolylpyridine **38** [100] chromophore was recently synthesized in Picard's group in Toulouse. The pyrazolylpyridine unit of **38** exhibits promising properties for photon collection and transfer to terbium cations since its triplet level (26,100 cm^{-1}) is well situated over the terbium emissive state.

(Eu)**38**: λ_{max} = 294 nm (ε ~11,500 M^{-1} cm^{-1}) Φ_{tot} (H_2O) = 0.05, τ (H_2O) = 0.68 ms (Borate buffer + 0.4 M KF), τ (H_2O) = 0.68 ms (Borate buffer pH 8.6), $n H_2O$ ~0.82; (Tb)**38**: λ_{max} = 294 nm (ε ~11,500 M^{-1} cm^{-1}) Φ_{tot} (H_2O) = 0.49, τ (H_2O) = 2.67 ms (Borate buffer + 0.4 M KF), τ (H_2O) = 3.16 and 0.68 ms (Borate buffer + 0.4 M KF), $n H_2O$ ~0.82. The stability of the complex was determined by luminescent experiments, monitoring the disappearance of the 545-nm peak as a function of time. The emission properties of the Tb complex in aerated solution at room temperature in various buffers (borate buffer, pH 8.6; Tris buffer, pH 7.3, phosphate buffer, pH 7.4), highlights the kinetic inertness of the complex in aqueous media; the fluorescence properties remained unchanged for several days in the examined buffers. The Tb complex resists very well dynamic quenching by urate; the measured luminescence decay for (Tb)**38** (10 μM) in the absence and in the presence of urate (50 and 500 μM) is nearly constant (1.85, 1.82, and 1.76 ms, respectively). This complex is also resistant to dissociation in the presence of ethylenediaminetetraacetic acid (EDTA); competitive titration of Tb complex with EDTA (up to tenfold excess of EDTA, pH 8.6) showed no evidence of ligand exchange after 2.5 days. This is an improvement over previously reported Tb complexes derived from the corresponding acyclic ligand, for which 53% dissociation was observed after 2.5 days under the same conditions. Less than a 5% time-dependent change was observed over a 24-h period when the Tb complex (3 μM) was incubated in a mixture of Tris buffer/human serum (2:1) at pH 7.4 and the luminescence decay (1.74 ms mono-exponential decay) remained unchanged. The only drawback is the sharp fall below 325 nm observed on the excitation spectrum, which would make a 337-nm laser excitation inefficient; nevertheless, this complex should be compatible with readers equipped with a flash lamp such as Pherastar (BMG Labtech) or Victor V (Perkin–Elmer Life Sciences). This compound is indeed promising and as pointed out by the authors the synthesis of the corresponding functionalized compound remains to be done. This complete photophysical study should be recognized as a good example to be followed since it includes the study of complex stability in the presence of competing ions, EDTA, and serum.

4.3.3 Hexaazacyclohexadecane (Hexacyclen) Scaffold

Tetraazamacrocyclic ligands incorporating an intracyclic terpyridine chromophore **39** was described [106, 107], as expected the absorption maxima (335 nm) was more advantageous compared to **38**. The physical characteristics were described: τ (H_2O) = 1.06 ms, Φ_{tot} (H_2O) = 0.18 (Eu complex), τ (H_2O) = 1.11 ms, Φ_{tot} (H_2O) = 0.21 (Tb complex). The relative lack of variation (H_2O versus D_2O) in the luminescence decays indicated that no water molecule is coordinated to the metal ion suggesting that the nine binding sites provided by the ligand are coordinating the lanthanide. No decomposition of these complexes was observed after 1 day in the presence of a fivefold excess of EDTA and DTPA. These results indicate high stability constants between ligand and lanthanide ions (log >20). The molar absorbance is not reported but should be ~12,000 M^{-1} cm^{-1} at 335 nm, and therefore the brightness should be competitive with commercial chelates.

4.3.4 Hexaazacyclooctadecane Scaffold

A hexaazacyclooctadecane incorporating an endocyclic pyridine chromophore **40** [108] was used either as gadolinium complex for MRI experiments or as terbium complex for fluorescence imaging. The basic structure consisting of a DTBA unit cyclized with a 2,6-bis[(*N*,*N*′-alkyl)-aminomethyl]pyridine derivative allows to obtain easily a series of homologous ligands and thus lanthanide complexes possessing various lipophilicity. The ligands are named according to the alkyl chain R = butyl (**40-C4**), R = decyl (**40-C10**), and R = dodecyl (**40-C12**). The cellular uptake was evaluated by measuring the luminescence retained on the cells after incubation with 10 μM Tb complex and washing, using time-resolved luminescence (λ_{em} = 545 nm, λ_{exc} = 262 nm, delay time = 0.1 ms, and gate time = 1 ms). The **40-C4** complex showed almost no retention; in contrast, about 10–12% of the **40-C10** and **40-C12** complexes were bound to the cells within 40 min. The authors described interesting experiments of DERET [16] using as acceptor a fluorescein derivative (calcein). The acceptor is either a permeant calcein-Acetoxymethyl ester (calcein-AM, hydrolyzed by cellular esterases into calcein) or the impermeant calcein itself in order to determine the localization of the bound lanthanide chelate. Cells loaded with (Tb)**40-C10**, washed several times and then incubated with calcein showed, upon 262-nm excitation, a calcein emission (~520 nm) in time-resolved mode. Obviously, this signal is due to energy transfer from (Tb)**40-C10** since adding more calcein to the cell suspension increased the calcein emission intensity and decreased the intensity of Tb emission at 490 nm correspondingly. In contrast, when calcein was delivered inside cells using calcein-AM, calcein emission was only detected in the steady state fluorescence mode, but not in time-resolved mode. The intracellular localization of calcein was also confirmed by the fluorescence microscopy. Since calcein emission was undetectable in the time-resolved mode, it suggests that no energy transfer has occurred between intracellular calcein and (Tb) **40-C10**. On the basis of the results from DEFRET, the authors concluded that cells take up (Tb)**40-C10** mainly through hydrophobic interactions between cell membranes and two alkyl chains of the lanthanide complex. Once taken up by cells, (Tb) **40-C10** adopts a configuration in which decyl alkanes insert into the out leaflet of membrane lipid layer, and the hydrophilic Tb chelate faces the extracellular medium.

4.4 Cryptands

Cryptands are ligands, mainly represented by macrobicyclic or macrotricyclic structure, capable of ion encapsulation due to their cage-like structures (see Fig. 8). The inclusion compound formed by a metal ion encapsulated within a

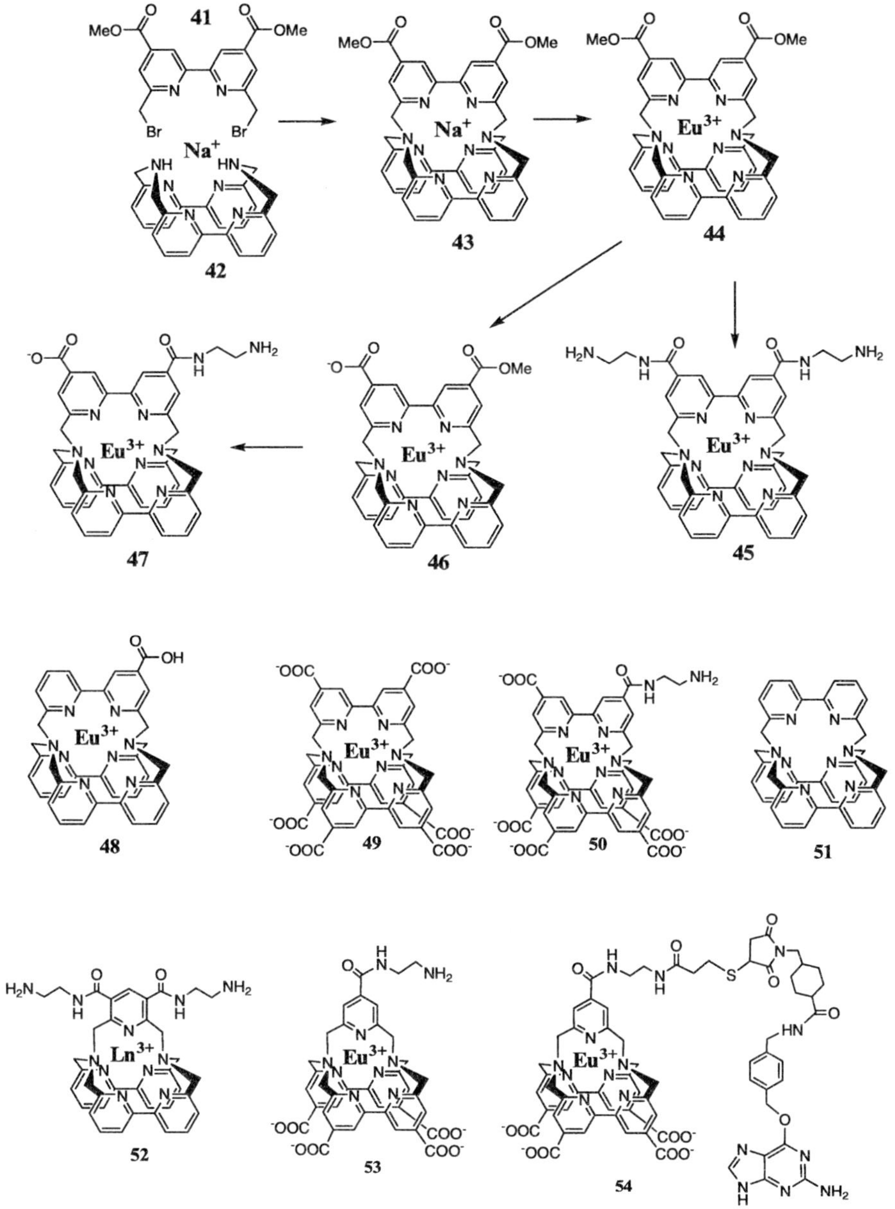

Fig. 8 Cryptands structures based on bipyridine units

three-dimensional cage is named cryptate (also named "clathrochelates"). Concerning the cryptate nomenclature, "short names" are generated by an extension of the macrocycle nomenclature. For example, a bicyclic cage formed by three branches linking two bridgehead nitrogens, each branch formed by a triethylene glycol unit, will be named [2.2.2] (IUPAC name: 1,10-diaza-4,7,13,16,21,24-

hexaoxabicyclo[8.8.8] hexacosane) and the corresponding lanthanide cryptate will be symbolized as $Ln^{3+} \subset [2.2.2]$. In the case of a cryptate based on heterocyclic units, as BPs, linking two bridgehead nitrogens, the lanthanide cryptate will be symbolized as $Ln^{3+} \subset$ [BP.BP.BP]. For unsymmetrical ligands as for example the one formed by two BP units and one pyridine unit, the lanthanide cryptate will be symbolized as $Ln^{3+} \subset$ [Py.BP.BP]. The following section describes the latest development in the field of lanthanide macrocyclic cage-like complexes. There are several review articles dealing with lanthanide cryptates [109–113]. A large number of cryptands have been reported. The polyether cryptants exhibit affinities for "hard" metal ions such as the alkali and alkaline earths; the latter ions bind via electrostatic interaction with the dipolar ether functions, replacing solvent molecules from the coordination sphere of the metal. Relative to their single-ring analogs, bicyclic cryptands usually exhibit higher formation constants for metal binding that can involve both favorable entropy and/or enthalpy terms a phenomenon known as the "Cryptate Effect" [114] characterized by high stability and selectivity, slow exchange rates and efficient shielding of the bound substrate from the environment [115]. To achieve optimal binding properties, it is important to obtain just the right balance between flexibility (which lowers the energy barrier to complexation and decomplexation) and rigidity (which confers selectivity, but can result in inhibition of the ingress of a metal ion into the cavity) [11, 116]. Generally, the metal ion whose ionic crystal radius best matches the radius of the cryptand cavity will form the most stable complex. The correspondence between cavity size and complex stability is more pronounced with the cryptands than with single macrocycle ligands [110]. The selectivity and stability of cryptates are also influenced by the cryptand structural flexibility, the number and type of cryptand donor atoms, and the solvation energy of the metal ion. The high stability of the lanthanide complexation is best documented for the $Eu^{3+} \subset [2.2.2]$ cryptate [117] and the Eu and Tb [BP.BP.BP] cryptates [118]. The higher stability of lanthanide cryptate, in contrast to chelates, is also exemplified by the observation that no precipitation of $Ln(OH)_3$ occurs upon prolonged incubation in strong alkaline solution. Furthermore, the addition of high concentrations of complexing agent such as EDTA does not lead to ligand exchange reaction as observed for stable lanthanide chelates (vide supra). Although the cage structure has a shielding effect, the influence of anions such as iodide and phosphate ions on a series of lanthanide cryptates has been investigated [119]. It has been found that addition of iodide anions to aqueous solution of $Eu^{3+} \subset [2.2.1]$ results in the quenching of the luminescence of the complex. However, the quenching effect of urate or azide on the lanthanide fluorescence is very dependent of the cage structure (vide infra) and on the environment, since the presence of fluoride can suppress the quenching effect [10]. The synthesis of a cryptand is usually more demanding than macrocyclic or open-chain chelates. In some cases, as for non-functionalized symmetrical cryptand, the synthesis can be achieved in one step through template aided macrobicyclization [120]. Bifunctional lanthanide cryptate as **43**, **44** and **45,** are synthesized either by bicyclization of a bis-bromomethyl derivative **41** on a preformed azamacrocycle **42** [121] or alternatively by bicyclization of a bis-aminomethyl derivative (suitably functionalized) with two equivalents of a bis-bromomethyl derivative.

4.4.1 Tribranched Macrocyclic Cage

Trisbipyridine Cryptates Scaffold

The functionalized derivatives of $Eu^{3+} \subset$ [BP.BP.BP] cryptate represent the structure which has received the largest biomedical applications as luminescent probe. The BP unit itself has been incorporated in number of ligands [122] and particularly cryptates [121]. The luminescent properties of cryptates, as well as other lanthanides complexes, have been reviewed [122] with respect to biomedical applications. The amine functionalized derivative **45** was initially synthesized in Lehn's group in Strasbourg [123]; the photophysical properties have been thoroughly studied [124, 125]: τ (H_2O) = 0.34 ms, τ (D_2O) = 1.53 ms. Low temperature luminescence showed that in this complex the thermal deactivation is very limited and the luminescence in D_2O showed the presence of two water molecules. In the presence of fluoride ions, the luminescence decay is increased to 1.00 ms [10] and the quantum yield to $\Phi_{tot} = 0.10$. The $\lambda_{max} = 305$ nm ($\varepsilon \sim 18{,}000\ M^{-1}\ cm^{-1}$) could be considered as sub-optimal considering laser excitation at 337 nm. Nevertheless, the shape of the absorbance/excitation spectrum presents a shallow edge on the long wavelength side and the brightness ($\sim 360\ M^{-1}\ cm^{-1}$) upon 337-nm excitation is high enough for an efficient excitation. The absorption spectrum of the di-functionalized $Eu^{3+} \subset$ [BP.BP.BP] cryptate is a convolution of the BPs absorption and of the bipyridine-4,4′-dicarboxylate absorption which presents a bathochromic shift compared to the BP units.

Recently, a mono-functionalized version **48** of $Eu^{3+} \subset$ [BP.BP.BP] was synthesized incorporating a bipyridine-4-carboxylate [126]. The photophysical properties have not yet been reported, but must be very close to **45** although the loss of one carboxylate should slightly decrease the 337-nm absorbance and thus the brightness.

Alternatively, the monofunctionalization can be achieved through mono-saponification of **44** yielding the mono methyl ester **46** which can be aminolized into **47**; the single amine function can be converted to a suitably activated derivative using hetero-bifunctional cross-linking reagents for further coupling to various biomolecules.

Numerous bioassays format have been designed using derivatives of the mono-functionalized europium cryptate such as **47** (commercially available from CISbio, see http://www.htrf.com). Since the last reviews [127] some new applications $Eu^{3+} \subset$ [BP.BP.BP] cryptate were developed through the HTRF technology, as caspase assay [128], telomerase assay [129], leukotriene b4 assay [130], inositol-1-phosphate assay [131], minisequencing [132], and MutS-DNA interaction [133]. Recently, the HTRF technology has been used in the study of protein–protein interactions and the authors demonstrated that measured K_D values compare favorably with those calculated from independent experiments [134].

In the design of TR-FRET bioassays, the direct labeling of small molecules is not always recommended and in some cases the indirect labeling through an "affinity pair" (as biotin/streptavidine) gives better results. As pointed out earlier a too high FRET efficiency, due to very close proximity between donor and acceptor, leads to a reduction in the luminescence decay of the sensitized emission, and thus a deterioration of the signal/noise ratio. In that case an indirect labeling increasing the distance reduces the FRET efficiency and finally improves the bioassay analytical performances [135].

In the scope of shifting the absorption spectrum towards longer wavelength, new europium trisbipyridine cryptate bearing multiple carboxylate substituents as **49** [136] and its mono-functionalized derivative **50** were synthesized. Indeed, the absorption λ_{max} was shifted to 320 nm and the absorbance at 337 nm and thus the brightness was increased roughly by a factor of two. The luminescence decay in H_2O, D_2O, and at low temperature were very close to those of the unsubstituted europium trisbipyridine cryptate showing that this substitution does not modify the other photophysical properties [136]. The behavior in the presence of serum is similar to cryptates **45** and **47** and the luminescence decay in the presence of fluoride is also in the millisecond range.

Pyridine Bis-Bipyridine Cryptates Scaffold

The removal of one of the pyridine unit from the [BP.BP.BP] cryptant **51** leads to a new series of ligands namely the pyridine bis-bipyridine cryptates scaffold symbolized as [PY.BP.BP].

The Tb $\subset$ **51** displayed a very intense luminescence at low temperature, but at room temperature the luminescence was much weaker and the luminescence decay was relatively short [τ (H_2O) = 0.45 ms]. It was concluded that a strong (temperature-dependent) energy back transfer from terbium to bipyridines was responsible for this situation [134]. This back transfer is due to a short gap between the cryptand bipyridine triplet level (22,100 cm^{-1}) [137] and the terbium excited level (Tb^{3+} first excited level $^5D_4$ lies at 20,500 cm^{-1}) [43]. Surprisingly, in the Tb^{3+} $\subset$ **52** cryptate this phenomenon is not observed [138], although the chromophores are also BP units. A photophysical study of the Gd^{3+} $\subset$ **52** phosphorescence showed a triplet level energy of 23,640 cm^{-1} [138]. Thus, in this complex the triplet level lies higher compared to **51** and the energy gap (3,100 cm^{-1}) is enough to minimize the back transfer. The quantum yield Φ_{tot} (H_2O) = 0.25 and the luminescence decay τ (H_2O) = 0.85 ms are suitable for TR-FRET applications. The corresponding Eu^{3+} $\subset$ **52** cryptate showed a quantum yield Φ_{tot} (H_2O) = 0.14 and a luminescence decay τ (H_2O) = 0.77 ms. The sensitivity of Tb^{3+} $\subset$ **52** cryptate to quenching in serum is a limitation, although this complex has been used as a luminescent probe to label microcystin, a hepatotoxin produced by freshwater cyanobacteria, and to build a time-resolved fluoroimmunometric assay (TR-FIA) to quantify the presence of the toxin in water [139]. One drawback of the cryptate **52** is that the absorption spectrum ($\lambda_{max} = 291$ nm, $\varepsilon \sim 18{,}000\ M^{-1}\ cm^{-1}$) is not adapted to 337-nm laser excitation, although the shoulder around 310 nm allows an efficient excitation on readers equipped with a flash lamp [139]. In order to shift the absorbance towards 337 nm, a europium pyridine bis-pypyridine cryptate **53** bearing four carboxylate groups was synthesized [140]. This resulted indeed in a bathochromic shift ($\lambda_{max} = 322$ nm, $\varepsilon \sim 18{,}000\ M^{-1}\ cm^{-1}$) and an increased absorption ($\varepsilon_{337} \sim 13{,}300\ M^{-1}\ cm^{-1}$). Furthermore, the cryptate **53** was almost insensitive to serum quenching and thus can be used without addition of fluoride. This is particularly interesting for in vivo applications since cells hardly tolerate fluorides.

The cryptate **53** has been used for TRF microscopy imaging of $GABA_B$ receptors on living cells [86]. This cryptate was conjugated to anti-tag antibodies and used to monitor, through resonance energy transfer, the heterodimerization of $GABA_B$ receptors expressed at the surface of living cells [141, 142]. The cryptate **53** was also conjugated to a linear pseudopeptide antagonist of the vasopressin V_{1a} receptor which was used to develop a HTRF binding assay on the surface of living cells [143]. The cryptate **53** was conjugated to a functionalized O^6-benzylguanine moiety and the resulting conjugate **54** was used as substrate for a suicide enzyme evolved from guanine *O*-alkyl transferase and known as "SNAP-Tag" [144]. The cryptate labeled substrate **54** was used for specific labeling of an engineered $GABA_B$ receptor expressed at the surface of living cells [145] and to probe protein–protein interaction at the cell surface using HTRF technology.

Cryptands Containing a Tetralactam Unit

Cryptands based on a macrobicycles tetralactam scaffold bearing either a BP unit **55** or a bis-pyrazolo-pyridine unit **56** as chromophore have been described [146, 147] (Fig. 9). They showed encouraging physicochemical and photophysical properties, but to the best of our knowledge they were not further studied.

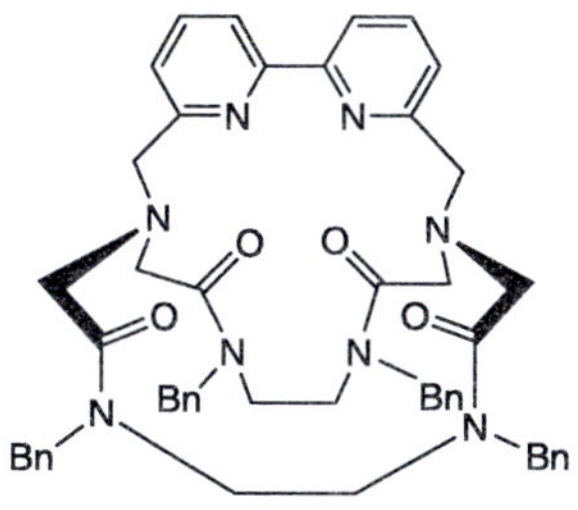

55

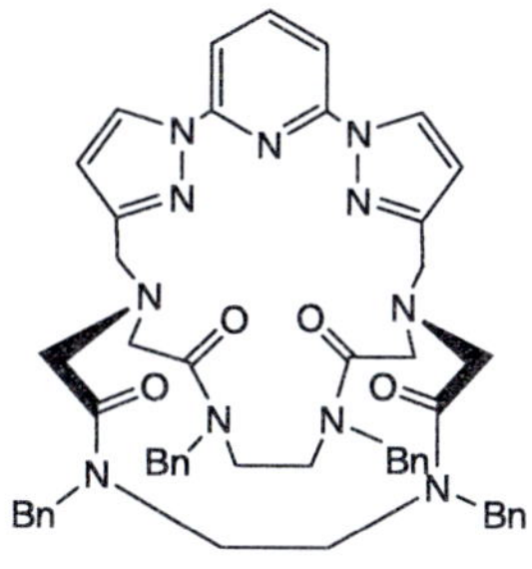

56

Fig. 9 Cryptands based on tetralactam unit

Cryptates Containing a *N*-oxides

This series of cryptand have been recently reviewed [148]. Although Eu cryptates based on a trisbipyridine di-*N*-oxide structure [149] had interesting photophysical properties in terms of quantum yield [150], their application was not investigated further.

4.4.2 Tetrabranched Macrocylic Cage

Macrotricyclic cage structure obtained by capping a tetrapodal scaffold as **15** (Fig. 3) has been described recently [151] by Raymond's group. The presence of four chromophores is expected to allow an efficient light harvesting and effective coordination (octadentate).

5 Conclusion

Even though the first generation of lanthanide cryptates and chelates are widely used for research, diagnostic, and HTS applications, the need for more sophisticated bioassays to answer new questions or challenging biological processes foster the research for new lanthanide luminescent probes [152]. The constant need for higher sensitivity drives the research for new antenna chromophores and molecular scaffolds. The demand from the biologists to mimic more closely the "real world" requires bioassays involving the use of living cells and thus is leading to new constraints in term of stability and quenching issues. New trends consist in the use of new chemical tools for biomolecular imaging [153] and specific in vivo labeling of proteins in their biological context [144] as an alternative to labeling with GFP.

Imaging using time-resolved microscopy is under development in several research groups to visualize biological events within living cells by using various luminescent lanthanide complexes [22]. Alternatively, two-photon microscopy imaging using lanthanide complexes is also studied in several groups [154, 155]. The specificity brought by LRET experiment where two or more partners of a biological interaction are labeled can even replace the need for fluorescence microscopy in some applications [145]. An interesting new application resides in the selective quenching to provide a multiparametric detection [156]. The TBP cryptate **47** luminescence is quenched by a monoclonal antibody, whereas a carboxylate substituted cryptate (as **50** or **53**) is not quenched.

An alternative approach to multiparametrism makes use of lanthanide complexes in conjunction with quantum dots as acceptors [157]. Another growing field for applications of lanthanide complexes is the design of luminescent sensors responding to various analytes such as pH, anions (phosphate, carbonates), metal ions, hydrogen peroxide, and ATP. Since many enzymatic reactions are associated with the consumption or formation of small molecules like hydrogen peroxide,

ATP, pyrophosphate, or phosphate such sensors are useful for the determination of enzyme activities and the screening of enzyme regulators both in clinical chemistry and in drug discovery [158]. The development of time-gated CCD camera using pulsed flash lamp as *Imagex-TGi* camera (Photonic Research Systems Ltd; www.prsbio.com) allows the detection of cryptate luminescence on DNA arrays. Nowadays, commercially available time-resolved fluorimeter, nitrogen laser, and flash lamp are the main sources used for excitation of lanthanide complexes. Considering the potential of simpler more robust light emitting diodes (LED) and the availability of ultraviolet LED some equipments such as time-gated luminescence flow cytometer have been developed [159]. Lanthanide complexes [160] containing 2-hydroxyisophthalamide chromophores present a molar absorbance ε_{365} ~12,000 M^{-1} cm^{-1} compatible with an efficient excitation using a 365-nm LED (but not with a 385-nm LED), opening new horizons in the field of time-resolved instrumentation. Recently, published lanthanide luminescent helicates $[Eu_2(LC_2(CO_2H))_3]$ present absorption properties ($\varepsilon_{344} = 57{,}900\ M^{-1}\ cm^{-1}$) compatible with 365-nm LED excitation and interesting photophysical data (long luminescence decay time 2.45 ms). Although detail stability studies were not given, the complex seems promising since it was successfully used for cell staining and imaging using a TRF microscope [161].

Finally, a possible application is the design of luminescent probes for bioimaging in animals, which is nowadays being carried out using lanthanide nanoparticles. For these applications the in-depth excitation of the luminescent probes is challenging and leads to the development of two-photon excitation [162]. This amount of new potential applications will still drive the synthesis of new photoactive lanthanide complexes and continue to stimulate the inventiveness of researchers in this field.

References

1. Pope AJ et al (1999) Drug Discov Today 4:350–362
2. Kolb JM, Yamanaka G, Manly SP (1996) J Biomol Screen 1:203–210
3. Burke TJ, Loniello KR, Beebe JA, Ervin KM (2003) Comb Chem High Throughput Screen 6:183–194
4. Turconi S et al (2001) Drug Discov Today 1:27–39
5. Mathis G (1999) J Biomol Screen 4:309–314
6. Karvinen J et al (2002) J Biomol Screen 7:223–231
7. Bünzli JC, Piguet C (2005) Chem Soc Rev 34:1048–1077
8. Marshall NJ, Dakubu S, Jackson T, Ekins RP (1981) Pulsed light, time resolved fluoroimmunoassay. In: Albertini A, Ekins RP (eds) Monoclonal antibodies and developments in immunoassay. Elsevier, Amsterdam, pp 101–108
9. Kadkhodayan S, Elliott LO, Mausisa G, Ackerly Wallweber H, Deshayes K, Feng B, Fairbrother WJ (2007) Assay Drug Dev Technol 5:501–514
10. Mathis G (1993) Clin Chem 39:1953–1959
11. Hemmilä I, Webb S (1997) Drug Discov Today 2:373–381
12. Trinquet E, Mathis G (2006) Mol Biosyst 2:380–387

13. Simeonov A, Yasgar A, Jadhav A, Lokesh GL, Klumpp Ca, Michael S, Austin CP, Natarajan A, Inglese J (2008) Anal Biochem 375:60–70
14. Klumpp M, Boettcher A, Becker D, Meder G, Blank J, Leder L et al (2006) J Biomol Screen 11:617–633
15. Berlier JE, Rothe A, Buller G, Bradford J, Gray DR, Filanoski BJ, Telford WG, Yue S, Liu J, Cheung CY, Chang W, Hirsch JD, Beechem JM, Haugland RP, Haugland RP (2003) J Histochem Cytochem 51:1699–1712
16. Thomas DD, Carlsen WF, Stryer L (1978) Proc Natl Acad Sci U S A 75:5746–5750
17. Riddle SM, Vedvik KL, Hanson GT, Vogel KW (2006) Anal Biochem 356:108–116
18. Hemmilä I, Laitala V (2005) J Fluoresc 15:529–542
19. Leonard JP, Nolan CB, Stomeo F, Gunnlaugsson T (2007) Top Curr Chem 281:1–43
20. Pope AJ, Haupts UM, Moore KJ (1999) Drug Discov Today 4:350–362
21. Alpha-Bazin B, Bazin H, Guillemer S, Sauvaigo S, Mathis G (2000) Nucleosides Nucleotides Nucleic Acids 19:1463–1474
22. Montgomery C, Murray B, New EJ, Pal R, Parker D (2009) Acc Chem Res 42:925–937
23. Gudgin-Dickson EF, Pollak A, Diamandis EP (1995) J Photochem Photobiol B Biol 27:3–19
24. Poole RA, Kielar F, Richardson SL, Stenson PA, Parker D (2006) Chem Commun 39: 4084–4086
25. Kielar F, Montgomery CP, New EJ, Parker D, Poole RA, Richardson SL, Stenson PA (2007) Org Biomol Chem 5:2975–2982
26. Bazin H, Guillemer S, Mathis G (2002) J Fluoresc 12:245–248
27. Lis S (2002) J Alloys Comp 341(1–2):45–50
28. Horuk R, Huang JJ, Covington M, Newton RC (1987) J Biol Chem 262(34):16275–16278
29. Mukkala VM, Sund C, Kwiatkowski M, Pasanen P, Högberg M, Kankare J, Takalo H (1992) Helv Chim Acta 75:1621–1632
30. Mukkala VM, Helenius M, Hemmilä I, Kankare J, Takalo H (1993) Helv Chim Acta 76:1361–1378
31. Takalo H, Hemmilä I, Sutela T, Latva M (1996) Helv Chim Acta 79:789–802
32. Takalo H, Mukkala V-M. Luminescent lanthanide chelates. WO/1993/011433
33. Kankare J, Takalo H, Hänninen E, Helenius M, Mukkala V-M. Terpyridine derivatives. WO/1990/000550
34. Toner JL, Hilborn DA, Murray BJ, Hossain TZ, Snow RA, Saha AK, Philion R, Shearman CW, Shah C (1998) US Patent 5,707,603
35. Tuomola M, Cooper KM, Lahdenperä S, Baxter GA, Elliott CT, Kennedy DG, Lövgren T (2002) Analyst 127:83–86
36. Saha AK, Kross K, Kloszewski ED, Upson DA, Toner JL, Snow RA, Black CDV, Desai VC (1993) J Am Chem Soc 115:11032–11033
37. Jones SG, Lee DY, Wright JF, Jones CN, Teear ML, Gregory SJ, Burns DD (2001) J Fluoresc 11:13–21
38. Cooper ME, Sammes PG (2000) J Chem Soc Perkin Trans 28:1695–1700
39. Takalo H, Mukkala VM (1993) WO 93/11433
40. Nishioka T, Yuan J, Yamamoto Y, Sumitomo K, Wang Z, Hashino K, Hosoya C, Ikawa K, Wang G, Matsumoto K (2006) Inorg Chem 45:4088–4096
41. Kimura H, Mukaida M, Watanabe M, Hashino K, Nishioka T, Tomino Y, Yoshida K, Matsumoto K (2008) Anal Biochem 372:119–121
42. Ylikoski A, Hellman J, Matikainen T, Käkönen SM, Karp M, Väänänen HK, Lövgren T, Pettersson K (1998) J Bone Miner Res 13:1183–1190
43. Latva M, Takalo H, Mukkala VM, Matachescu C, Rodriguez-Ubis JC, Kankare J (1997) J Luminesc 75:149–169
44. Brunet E, Juanez O, Sedano R, Rodríguez-Ubis JC (2007) Tetrahedron Lett 7:1091–1094
45. Brunet E, Juanes O, Sedano R, Rodríguez-Ubis JC (2002) Photochem Photobiol Sci 7:613–618
46. Selvin PR, Jancarik J, Li M, Hung LW (1996) Inorg Chem 35:700–705
47. Kim SH, Ge P, Katzenellenbogen JA (2009) Chem Commun 2:183–5

48. Saroja G, Sankaran NB, Samanta A (1996) Chem Phys Lett 249:392–398
49. Shafer DE, Inman JK, Lees A (2000) Anal Biochem 282(1):161–164
50. Li M, Selvin PR (1995) J Am Chem Soc 117:8132–8138
51. Xiao M, Selvin PR (2001) J Am Chem Soc 123:7067–7073
52. Weibel N, Charbonnière L, Guardigli M, Roda A, Ziessel R (2004) J Am Chem Soc 126:4888–4896
53. Charbonnière LJ, Weibel N, Retailleau P, Ziessel R (2006) Chem Eur J 13:346–58
54. Brunet E, Juanes O, Rodríguez-Blasco MA, Vila-Nova SP, Rodríguez-Ubis JC (2007) Tetrahedron Lett 48(8):1353–1355
55. Freire RO, Vila-Nova SP, Brunet E, Juanes O, Rodríguez-Ubis JC, Alves S Jr (2007) Chem Phys Lett 443(4–6):378–382
56. Petoud S, Cohen SM, Bünzli JCG, Raymond KN (2003) J Am Chem Soc 125:13324–13325
57. Raymond KN (2009) US20090036537A1
58. Trinquet E, Gregor N, Degorce F, Tardieu JL, Seguin P (2008) Poster 11023, SBS. http://www.htrf.com/files/resources/poster_ip-one-tb_sbs_08_ss-couple.pdf
59. Samuel APS, Moore EG, Melchior M, Xu J, Raymond KN (2008) Inorg Chem 47:7535–7544
60. Macmanus JP, Hogue CW, Marsden BJ, Sikorska Sikorska M, Szabo AG (1990) J Biol Chem 265:10358–10366
61. Reynolds AM, Sculimbrene BR, Imperiali B (2008) Bioconj Chem 19:588–91
62. Suchý M, Hudson RHE (2008) Chem Eur J 29:4847–4865
63. Hermann P, Kotek J (2008) Ten-membered rings or larger with one or more nitrogen atoms. In: Katritzky AR, Ramsden CA, Scriven EFV, Taylor RJK (eds) Comprehensive heterocyclic chemistry III. Elsevier, Oxford, pp 613–666
64. De Cola L, Smailes DL, Vallarino LM (1986) Inorg Chem 25:1729–1732
65. Saito K, Lee RT, Lee YC (1998) Anal Biochem 258(2):311–314
66. Leif RC, Vallarino LM, Becker MC, Yang S (2006) Cytometry A 69:940–946
67. Vallarino LM, Leif RC (1997) Macrocycle complexes of yttrium, the lanthanides and the actinides having peripheral coupling functionalities continuation-in-part. US Patent 5,696,240
68. Sabbatini N, De Cola L, Vallarino LM, Blasse G (1987) J Phys Chem 91:4681–4685
69. Leif RC, Vallarino LM, Becker MC, Yang S (2006) Cytometry A 69A:767–778
70. Lee YC, Kawasaki N, Lee RT, Suzuki N (1998) Glycobiology 8:849–856
71. Leif RC, Becker MC, Bromm AJ Jr, Vallarino LM, Williams JW, Williams SA, Yang S (2002) Optimizing the luminescence of lanthanide(III) macrocyclic complexes for the detection of Anti5BrdU. In: Farkas DL, Leif RC (eds) Optical diagnostics of living cells V. SPIE Proceedings, vol 4622, pp 250–261
72. Jin D, Piper JA, Leif RC, Yang S, Ferrari BC, Yuan J, Wang G, Vallarino LM, Williams JW (2009) J Biomed Opt 14 (in press). doi: 10.1117/1.3103770
73. Hancock R (1992) J Chem Educ 69:615–621
74. Táborský P, Svobodovál I, Hnatejko Z, Lubal P, Lis S, Försterová M, Hermann P, Lukeš I, Havell J (2005) J Fluoresc 15:507–512
75. Cacheris WP, Nickel SK, Sherry AD (1987) Inorg Chem 26:958–960
76. Kumar K, Chang CA, Tweedle MF (1993) Inorg Chem 32:587–593
77. Wu SL, Horrocks WD Jr (1997) J Chem Soc Dalton Trans 9:1497–1502
78. Aime S, Botta M, Crich SG, Giovenzana GB, Jommi G, Pagliarin R, Sisti M (1997) Inorg Chem 36:2992–3000
79. Siaugue JM, Segat-Dioury F, Favre-Reguillon A, Wintgens V, Madic C, Guy A (2003) J Photochem Photobiol A 156:23–29
80. Foos J, Moeller T, Thompson LC (1962) J Inorg Nucl Chem 24:499
81. Ge P, Selvin PR (2008) Bioconj Chem 19:1105–1111
82. Nonat A, Gateau C, Fries PH, Mazzanti M (2006) Chem Eur J 12:7133–7150
83. Vereb G, Jares-Erijman E, Selvin PR, Jovin TM (1998) Biophys J 74:2210–2222

84. Charbonniere ZR, Guardigli M, Roda A, Sabbatini N, Cesario M (2001) J Am Chem Soc 123:2436–2437
85. Hanaoka K, Kikuchi K, Kobayashi S, Nagano T (2007) J Am Chem Soc 129:13502–13509
86. Ghose S, Trinquet E, Laget M, Bazin H, Mathis G (2008) J Alloys Comp 451:35–37
87. Beeby S, Botchway SW, Clarkson IM, Faulkner S, Parker AW, Parker D, Williams JA (2000) J Photochem Photobiol B 57:83–89
88. Pal R, Parker D, Costello LC (2009) Org Biomol Chem 7:1525–1528
89. Poole RA, Bobba G, Cann MJ, Frias JC, Parker D, Peacock RD (2005) Org Biomol Chem 3:1013–1024
90. Poole RA (2006) PhD Thesis, Durham University (UK)
91. Poole RA, Montgomery CP, New EJ, Congreve A, Parker D, Botta M (2007) Org Biomol Chem 5:2055–2062
92. Atkinson P, Findlay KS, Kielar F, Pal R, Parker D, Poole RA, Puschmann H, Richardson SL, Stenson PA, Thompson AL, Yu J (2006) Org Biomol Chem 4:1707–1722
93. Yu J, Parker D, Pal R, Poole RA, Cann MJ (2006) J Am Chem Soc 128:2294–2299
94. Montgomery CP, Parker D, Lamarque L (2007) Chem Commun 37:3841–3843
95. Murray BS, New EJ, Pal R, Parker D (2008) Org Biomol Chem 6(12):2085–94
96. Manning HC, Goebel T, Thompson RC, Price RR, Lee H, Bornhop DJ (2004) Bioconj Chem 15:1488–95
97. Manning HC, Goebel T, Marx JN, Bornhop DJ (2002) Org Lett 4:1075–1078
98. Manning HC, Goebel T, Marx JN, Bornhop DJ (2006) Bioconj Chem 17:735–740
99. Polášek M, Kotek J, Hermann P, Císařová I, Binnemans K, Lukeša I (2009) Inorg Chem 48:466–475
100. Nasso I, Bedel S, Galaup C, Picard C (2008) Eur J Inorg Chem:2064–2074
101. Tircsó G, Kovács Z, Sherry AD (2006) Inorg Chem 45:9269–9280
102. Bornhop DJ, Hubbard DS, Houlne MP, Adair C, Kiefer GE, Pence BC, Morgan DL (1999) Anal Chem 71:2607–2615
103. Kiefer GE, Jackson L, Bornhop DJ (1999) US Patent 5,928,627
104. Hubbard DS (1998) Master of Science, University of Texas
105. Nasso I, Galaup C, Havas F, Tisnès P, Picard C, Laurent S, Vander Elst L, Muller RN (2005) Inorg Chem 44:8293–8305
106. Galaup C, Couchet JM, Picard C, Tisnès P (2001) Tetrahedron Lett 42:6275–6278
107. Galaup C, Couchet JM, Bedel S, Tisnès P, Picard C (2005) J Org Chem 70:2274–2284
108. Zheng Q, Dai H, Merritt ME, Malloy C, Pan CY, Li WH (2005) J Am Chem Soc 127:16178–16188
109. Sabbatini N, Guardigli M, Lehn JM (1993) Coord Chem Rev 123:201–228
110. Izatt RM, Pawlak K, Bradshaw JS, Bruening RL (1995) Chem Rev 95:2529–2586
111. Alexander V (1995) Chem Rev 95:273–342
112. Yam VWW, Lo KKW (1999) Coord Chem Rev 184:157–240
113. Fallis IA (2002) Annu Rep Prog Chem Sect A: Inorg Chem 98:321–368
114. Lehn JM (1977) Pure Appl Chem 47:857–870
115. Lehn JM (1978) Acc Chem Res 2:49–57
116. Adam KR, Atkinson IM, Kim J, Lindoy LF, Matthews OA, Meehan GV, Raciti F, Skelton BW, Svenstrup N, White AH (2001) J Chem Soc Dalton Trans 16:2388–2397
117. Shestakova AK, Chertkov VA, Schneider HJ (2000) Tetrahedron Lett 41:6753–6756
118. Cross JP, Dadabhoy A, Sammes PG (2004) J Luminesc 110(3):113–124
119. Sabbatini N, Guardigli M, Lehn JM, Mathis G (1992) J Alloys Comp 180:363–367
120. Rodriguez-Ubis JC, Alpha B, Plancherel D, Lehn JM (1984) Helv Chim Acta 67:2264–2269
121. Kaes C, Katz A, Hosseini MW (2000) Chem Rev 100:3553–3590
122. Faulkner S, Matthews JL (2003) Fluorescent complexes for biomedical applications. In: McCleverty JA, Meyer TJ (eds) Comprehensive coordination chemistry II. Pergamon, Oxford, pp 913–944
123. Alpha B, Lehn JM, Mathis G (1987) Angew Chem Int Ed Engl 26:266–267

124. Alpha B, Balzani V, Lehn JM, Perathoner S, Sabbatini N (1987) Angew Chem Int Ed Engl 26:1266–1267
125. Alpha B, Ballardini R, Balzani V, Lehn JM, Perathoner S, Sabbatini N (1990) Photochem Photobiol 52:299–306
126. Havas F, Danel M, Galaup C, Tisnès P, Picard C (2007) Tetrahedron Lett 48:999–1002
127. Bazin H, Trinquet E, Mathis G (2002) J Biotechnol 82:233–50
128. Préaudat M, Ouled-Diaf J, Alpha-Bazin B, Mathis G, Mitsugi T, Aono Y, Takahashi K, Takemoto H (2002) J Biomol Screen 7:267–74
129. Gabourdes M, Bourgine V, Mathis G, Bazin H, Alpha-Bazin B (2004) Anal Biochem 333:105–13
130. Liang AM, Claret E, Ouled-Diaf J, Jean A, Vogel D, Light DR, Jones SW, Guilford WJ, Parkinson JF, Snider RM (2007) J Biomol Screen 12:536–545
131. Trinquet E, Fink M, Bazin H, Grillet F, Maurin F, Bourrier E, Ansanay H, Leroy C, Michaud A, Durroux T, Maurel D, Malhaire F, Goudet C, Pin JP, Naval M, Hernout O, Chretien F, Chapleur Y, Mathis G (2006) Anal Biochem 358:126–35
132. Lopez-Crapez E, Bazin H, Chevalier J, Trinquet E, Grenier J, Mathis G (2005) Hum Mutat 25(5):468–75
133. Lopez-Crapez E, Malinge JM, Gatchitch F, Casano L, Langlois T, Pugnière M, Roquet F, Mathis G, Bazin H (2008) Anal Biochem 383:301–6
134. Newton P, Harrison P, Clulow S (2008) J Biomol Screen 13:674–82
135. Vogel KW, Vedvik KL (2006) J Biomol Screen 11:439–443
136. Guillaumont D, Bazin H, Benech JM, Boyer M, Mathis G (2007) Chem Phys Chem 8:480–488
137. Blasse G, Dirksen GJ, Sabbatini N, Perathoner S, Lehn JM, Alpha B (1988) J Phys Chem 92:2419–2422
138. Vila-Nova SP, Pereira GAL, Albuquerque RQ, Mathis G, Bazin H, Autiero H, de Sá GF, Alves S Jr (2004) J Luminesc 109:173–179
139. Oliveira EJA, Vila Nova SP, Alves-Jr S, Santa-Cruz P, Molica RJR, Teixeira A, Malageño E, Lima Filho JL (2006) J Braz Chem Soc 17:243–250
140. Autiero H, Bazin H, Mathis G (2006) US Patent 7,087,384
141. Maurel D, Kniazeff J, Mathis G, Trinquet E, Pin JP, Ansanay H (2004) Anal Biochem 329:253–62
142. Albizu L, Teppaz G, Seyer R, Bazin H, Ansanay H, Manning M, Mouillac B, Durroux T (2007) J Med Chem 50:4976–85
143. Albizu L, Breton BMN, Pin JP C, Manning M, Mouillac B, Barberis C, Durroux T (2006) Mol Pharmacol 70:1783–91
144. Johnsson N, Johnsson K (2007) Chem Biol 2:31–38
145. Maurel D, Comps-Agrar L, Brock C, Rives ML, Bourrier E, Ayoub MA, Bazin H, Tinel N, Durroux T, Prézeau L, Trinquet E, Pin JP (2008) Nat Methods 5:561–7
146. Azéma J, Galaup C, Picard C, Tisnès P, Ramos P, Juanes O, Rodríguez-Ubis JC, Brunet E (2000) Tetrahedron 56:2673–2681
147. Galaup C, Azéma J, Tisnès P, Picard C, Ramos P, Juanes O, Brunet E, Rodríguez-Ubis JC (2002) Helv Chim Acta 85:1613–1625
148. Pietraszkiewicz M, Karpiuk J, Pietraszkiewicz O (2000) J Alloys Comp 300:141–146
149. Paul-Roth CO, Lehn JM, Guilhem J, Pascard C (1995) Helv Chim Acta 78:1895–1903
150. Sabbatini N, Guardigli M, Manet I, Ungaro R, Casnati A, Ziessel R, Ulrich G, Asfari Z, Lehn JM (1995) Pure Appl Chem 67:135–140
151. Raymond KN, Corneillie TM, XU J (2008). Luminescent macrocyclic lanthanide complexes. WO 2008 063721A3
152. Hovinen J, Guy PM (2009) Bioconj Chem 20:404–421
153. Miller LW, Cornish VW (2005) Curr Opin Chem Biol 9:56–61
154. Connally RE, Piper JA (2008) Ann N Y Acad Sci 1130:106–116
155. Picot A, Baldeck D'Aléo A, PL GA, Duperray A, Andraud C, Maury O (2008) J Am Chem Soc 130:1532–1533

156. Mathis G, Trinquet E, Alberti P, Laget M (2009) Method for suppressing a FRET signal, FRET signal suppressor agents and use in a method for multiplexing biological events. US20090162861A1
157. Hildebrandt N, Löhmannsröben H-G (2007) Curr Chem Biol 1:167–186
158. Spangler CM, Spangler C, Schäerling M (2008) Ann N Y Acad Sci 1130:138–48
159. Jin D, Connally R, Piper J (2007) Cytometry A 71A:797–808
160. Petoud S, Muller G, Moore EG, Xu J, Sokolnicki J, Riehl JP, Le UN, Cohen SM, Raymond KN (2007) J Am Chem Soc 129:77–83
161. Fernández-Moreira V, Song B, Sivagnanam V, Chauvin AS, Vandevyver CDB, Gijs M, Hemmilä I, Lehr HA, Bünzli JCG (2010) Analyst 135:42–52
162. Hänninen P, Soukka J, Soini JT (2008) Ann N Y Acad Sci 1130:320–326

Lanthanide Nanoparticules as Photoluminescent Reporters

Tero Soukka and Harri Härmä

Abstract Luminescent lanthanide nanoparticles provide a combination of high luminescence intensity, long luminescence lifetime, sharp emission peaks with narrow bandwidth, and a large Stokes' shift, leading to a high-performance reporter technology for bioanalytical assays. The high specific activity of luminescent lanthanide nanoparticles enables improved assay sensitivities compared to the conventional fluorescent reporters, and the different luminescent lanthanide ions with well-separated emission wavelengths facilitate multiparametric assays. The versatility of lanthanide chelate-doped polymeric nanoparticles as reporters is further increased by the use of the most effective organic light-harvesting ligands optimized for enhanced luminescence and the availability of inexpensive high-power solid-state light sources. Moreover, entrapping within polymeric nanoparticles allows the use of the lanthanide complexes with a weak water solubility or a low thermodynamic stability. In this review, we discuss the effects of nanoparticle composition, particle size, synthesis, surface modification, and biomolecule conjugation on the assay performance in addition to applications of these reporters in both heterogeneous and homogeneous assays. In the future, additional value may be provided by hybrid lanthanide nanoparticles, which combine, e.g., luminescent and magnetic properties for detection and controlling the assay kinetics and washing efficiency.

Keywords Bioanalytical assays · Immunoassays · Lanthanide nanoparticles · Luminescence resonance energy transfer · Time-resolved fluorometry

---

T. Soukka (✉)
Department of Biotechnology, University of Turku, 20520 Turku, Finland
e-mail: tejoso@utu.fi

H. Härmä
Laboratory of Biophysics, University of Turku, 20520 Turku, Finland
e-mail: harri.harma@utu.fi

P. Hänninen and H. Härmä (eds.), *Lanthanide Luminescence: Photophysical, Analytical and Biological Aspects*, Springer Ser Fluoresc (2011) 7: 89–114, DOI 10.1007/4243_2010_11,

Contents

1 Introduction

Europium(III) chelate-dyed nanoparticles as particulate photoluminescent reporters were first introduced for bionalytical analysis already in late 1970s [1]. At that time, the radioactive isotopes were the prevalent reporter technology in immunoassays [2] or other ligand-binding assays [3]. Methodological improvements such as labeled antibody-based assays and noncompetitive two-site assay principle [4–6] had been introduced, and the advantages of monoclonal antibodies [7] were about to be recognized [8–10]. The advantages of nonisotopic high specific activity labels in noncompetitive methods were also explained theoretically [11–13], supporting the development of numerous nonisotopic labels and detection technologies such as time-resolved fluorometry of the lanthanide chelates [14–16]. However, fluorescent reporters had been recognized as preferred strategy to replace radioisotopes earlier [17]; they were an inexpensive, stable, and a nontoxic alternative enabling rapid detection compared to radioisotopes. Long signal acquisition time of radiolabels, a limited shelf-life, and special precaution to radiation safety led to development of nonradioactive labeling technologies. The problems related to scattered excitation light and background fluorescence (also referred as autofluorescence) from both solid-phase and biological material, however, severely limited the detection of conventional fluorescent reporters [18, 19]. Elimination of short-lifetime background by utilization of long-lifetime fluorescent reporters and temporal resolution in detection was proposed as a preferred strategy towards improved sensitivity in luminescence-based immunoassays [17]. Chelate complexes of rare earth metals, primarily europium(III) and terbium(III), were known to produce long-lifetime luminescence up to milliseconds [20], enabling feasible time-resolved detection instrumentation. Yet the development of a fluorescent metal chelate suitable for conjugation to antibodies turned out to be challenging.

To circumvent the challenge of developing conjugatable, intrinsically luminescent rare earth chelates for antibody labeling, two separate approaches were introduced. An elegant solution was provided by coupling the metal ion to antibodies in a nonluminescent complex utilizing a derivative of ethylenediaminetetraacetate (EDTA)-like chelating structure. After completing the binding assay, dissociation of the lanthanide ion from the chelate was realized by lowering pH. This enabled the formation of another highly luminescent complex with newly added light-harvesting ligands [15, 21]. The method known as dissociation enhanced lanthanide fluoroimmunoassay (DELFIA™) was developed to a successful commercial product using the formation of secondary chelating structure with multiple lanthanide ions (europium(III), samarium(III), and terbium(III)) [22]. The second approach utilized the incorporation of luminescent rare earth metal chelates in polymeric particles coated with multiple antibodies [1]. Ultraviolet-excited inorganic lanthanide-doped phosphors [23] and conjugatable, intrinsically luminescent lanthanide chelates [24–26] were introduced later as an alternative approach for labeling with long-lifetime luminescent reporters. The lanthanide chelated-dyed particulate reporters, however, remained with little use in immunoassay applications until early 2000s, when their advantages were demonstrated again [27]. Thereafter, they have been widely applied in biomedical research including bioaffinity assays and imaging applications.

Luminescence techniques are often used in detection of binding events in biochemical assays. There is an increasing need to accomplish signal measurement on small quantities of sample containing a limited number of optical signaling probes. The challenge is to analyze small volumes or small spatial areas in samples with low concentration of analyte molecules. Nanostructures interfacing with biological systems have attracted widespread interest in biology and medicine. Fluorescent nanoparticles have distinct advantages over conventional dye molecules in detection of biomolecules because of their high signal amplification, reproducible physical features, and photostability. Dye nanoparticles, dye-doped nanoparticles, and semiconductor and metal nanoparticles have been identified as promising particulate luminescent reporters in bioanalysis. Here, we review the advances in the development of lanthanide-based nanoparticles for bioanalysis and their requirements for detection instrumentation.

2 Particulate Lanthanide-Based Reporters

Lanthanide-doped nanoparticles represent a promising new class of luminescent probes. This is due to their excellent chemical and optical properties: availability of multiple lanthanide ions providing different, well-separated emission wavelengths, size- and shape-independent luminescent properties, large effective Stokes' shifts, sharp emission peaks with band-widths of 10–20 nm, low photobleaching, stability to photochemical degradation, low toxicity, and the absence of blinking. These unique properties make lanthanide-doped nanoparticles highly suitable luminescent probes for multicolor labeling applications.

During the last 10 years, particulate photoluminescent lanthanide reporters [28–31] have gained an increasing research interest. This has been linked to the development of other particulate reporters, especially fluorescent quantum dots [32, 33], which have been shown to provide some significant advantages over conventional fluorescent labels; e.g., size-dependent emission wavelength and minimal photobleaching have rendered quantum dots a preferable label technology for applications such as multiplexing or fluorescence-based imaging [34]. Particulate reporters can also provide an extreme specific activity, which is needed especially in miniaturized assay systems, where a smaller number of the label molecules have to be detected. It is obvious that the particulate reporters do not fulfill the requirement of small molecular size for a universal or an ideal label, but they can provide improved stability, speed and convenience of detection, or lower overall cost of the label, detection reagents, and detection system, which are at least equally important in the development of new bioanalytical assay technologies [35–37].

2.1 *Dye-Doped Nanoparticles*

Dye-doped nanoparticles consist of organic or inorganic dye molecules dispersed inside a solid matrix. Incorporation of dyes inside the matrix increases the luminescence signal of a single label unit and protects the dye from interacting with the surrounding environment. Cross-linking of the matrix leads to low porosity, preventing partially oxygen diffusion and therefore photobleaching of the dye. Various matrices have been prepared for incorporation of lanthanide labels, and two main strategies for the preparation of lanthanide nanoparticles can be identified: physical adsorption or incorporation and covalent binding of the dye to the particle.

When physical incorporation of the lanthanide chelates is applied, the chelating structures interact through hydrophobic forces with hydrophobic polymers, preventing the dye molecules leaching from the particle. Frank and Sundberg prepared polystyrene nanoparticles embedded with thenoyltrifluoroacetonate lanthanide chelates complexed with tri-*n*-octylphosphine oxide in the late 1970s [1]. Latex nanoparticles containing encapsulated terbium(III) ions were described in the early 1990s. However, these particles were not intrinsically luminescent, requiring the addition of dipicolinic acid to extract the lanthanide ions and to enable the formation of luminescent complexes in solution [38]. More than 20 years after the original work by Frank and Sundberg, lanthanide-embedded polystyrene nanoparticles were finally shown to give superior sensitivity in immunoassays [27]. Thereafter, multiple lanthanide-based polymer nanoparticles have been prepared and applied for assay purposes. Europium (III)- and terbium(III)-comprising nanoparticles have been fabricated through free-radical dispersion copolymerization of styrene, poly(ethylene glycol) (PEG) macromonomer, and the lanthanide complex coordinated with 1,10-phenanthroline possessing acryl group [39]. Nanospheres copolymerized with luminescent polymerizable β-diketone or 2,2′,2″,2‴-{(6-methacroylamidohex-1-yl-imino)bis(methylene)-bis[4-(thiophen-2-yl)pyridine-6,2-diyl]bis(methylenenitrilo)}tetrakis-(acetic acid) europium(III) chelates have been reported [40]. Multiplexed polystyrene nanosized particles embedded with Eu(III), Tb(III), Sm(III), and Dy(III) chelates were prepared through copolymerization with acrylic acid [41]. These chelates were physically incorporated into the prepared nanoparticles.

Silica is an alternative host for lanthanide chelates. Synthesis of nanosized silica particles by well-established hydrolysis of tetraethylorthosilicate (TEOS) with ammonium hydroxide is a common route for particle preparation. The transparent inert silica material is an excellent matrix for dye-embedded nanolabels as less spectral interferences are expected from the host material. Multiple lanthanide particles have been prepared for bioanalysis. Nanoparticles have been synthesized by hydrolysis of TEOS with ammonium hydroxide in a microemulsion-containing aqueous solution of *N*,*N*,N^1,N^1-[2,6-bis(3′-aminomethyl-1′-pyrazolyl)-phenylpyridine]tetrakis(acetate)-Tb(III) (BPTA-Tb^{3+}) chelate. In the microemulsion, the aqueous phase containing the lanthanide chelate formed nanodroplets acting as nanoreactors for the synthesis of nanoparticles. Since the size of the nanoparticles is dependent on the size of the droplets, it can be controlled, e.g., by the type of surfactants and by adjusting the concentration of the reactants [42–46].

Dual-lanthanide-chelated silica nanoparticles have been prepared using a 2,9-bis[*N*,*N*-bis(carboxymethyl)aminomethyl]-1,10-phenanthroline ligand binding both Eu(III) and Tb(III). Covalent linkage of 3-aminopropyl(triethoxy) silane and the chelate on silica nanoparticles in the presence of precisely controlled ratio of Eu(III) and Tb(III) formed luminescent particles, which could be excited by a single wavelength and having two distinctive emission wavelengths [47]. In a similar fashion, the Eu(III) complex with 4,4-bis(1,1,2,2,3,3-heptafluoro-4,6-hexanedion-6-yl)chlorosulfo-*o*-terphenyl (BHHCT) or Tb(III) chelate with *N*,*N*,N^1,N^1-[2,6-bis(3-aminomethyl-1-pyrazolyl)-phenylpyridine] (BPTA) were covalently coupled on aminated silica nanoparticles [48].

2.2 *Dye Nanoparticles*

Particles prepared within a host are not necessarily ideal as the volume is largely embedded in the matrix and less volume is available for dye incorporation. Therefore, dye nanoparticles constructed of dye molecules and a protected shell is an interesting approach for preparing nanosized lanthanide particles. The large Stokes' shift of the lanthanide chelates allows the packing of chelating structures close to one another without severe loss of luminescent properties through self- or concentration quenching (i.e., inner filter effect) [16]. Dye nanoparticles have been prepared from tris(dibenzoylmethane)-mono(phenanthroline) and tris (dibenzoylmethane)-mono(5-aminophenanthroline) europium(III) chelates as well as from europium(III) naphtoyltrifluoroacetone:trioctylphosphineoxide complexes [49, 50]. Obviously, clustered lanthanide chelates are not stable and require stabilizing structures to protect them both from dissolution and adverse environmental effects in aqueous phase. Therefore, stabilization through silica has been applied [49, 50] (see preceding section). The particle size can be varied with careful adjustment of synthesis conditions, and small-sized nanoparticles can be formed. Such small-sized particles approaching the size of proteins have been prepared from Eu (III) complex of tris(thenoyltrifluoroacetonato)-2-(*N*,*N*-diethylanilin-4-yl)-4,6-bis (3,5-dimethylpyrazol-1-yl)-1,3,5-triazine [51].

2.3 *Inorganic Nanoparticles*

Inorganic phosphors are potential labels for time-resolved luminescence staining and assays in aqueous environment [23, 52–55]. The lanthanide phosphors have essentially infinite shelf life, no toxicity, no photobleaching, and are unaffected by environmental conditions such as pH, temperature, enzymatic reactions, or solvent effects. Their major drawback is that the luminescence per lanthanide ion is significantly less than from the dye-doped or dye nanoparticles due to the weak absorption of individual ions partly compensated by their higher number. Inorganic nanoparticles, however, can be prepared readily in large quantities with relatively simple methods. The size of the nanoparticles can be controlled from low nanometer scale to several hundred nanometers with a narrow size distribution.

By far, the most reported lanthanide nanoparticles are those with inorganic host structure. They have been prepared through milling of crude materials and synthesis by wet-chemistry or combustion methods. Inorganic phosphors have been ball-milled from large microparticles to a size between 100 and 300 nm, stabilized with polycarboxylic acids and coated with biomaterials for bioaffinity assay purposes [23, 52, 56]. Ball-milling is not an optimal approach as it limits the particle size obtained to approximately 100 nm. In bioaffinity assay applications, nanoparticle label sizes extending below the limits of the ball-milling technique are often preferred. Therefore, a number of wet-chemical syntheses or combustion methods have been studied. A number of lanthanide-doped phosphate or vanadate nanoparticles and nanowires have been prepared [57–64]. Lanthanide oxide nanoparticles have also been under extensive research [65–69]. Lanthanide-based upconversion nanoparticles have been widely studied due to their unique luminescent properties [70–76], and the properties of these nanostructures are discussed more closely under chapter [150].

2.4 *Hybrid Nanoparticles*

Hybrid materials have obtained a large interest in nanomaterials research and, recently, lanthanide-based materials have been combined with magnetic and metal nanomaterials in order to construct nanoparticles with advanced features for bioanalysis. Nanoscale magnetic/luminescent core–shell particles have been prepared for DNA quantification. Fe_3O_4/Eu:Gd_2O_3 core–shell nanoparticles of 200–400 nm in diameter were synthesized by spray pyrolysis [77].

Hybrids of metal nanoparticles and lanthanides have brought up new lanthanide-based sensors and improved luminescence signal of lanthanide chelates. Adsorption of a thiolated heptadentate macrocyclic Eu(III) cyclen conjugate to gold nanoparticles has formed the basis of water soluble sensor nanoparticle vehicles. A ternary complex of Eu(III) gold nanoparticle and a β-diketone antenna gave rise to the formation of a highly luminescent nanoparticle complex. This nanoparticle was applied to the detection of biologically relevant phosphates such as flavin monophosphate through the displacement of the antenna [78].

Lakowicz et al. have reviewed luminescence enhancement of lanthanide chelates on the surface of metal nanostructures [79]. When the Eu(III) chelate with TTA (thenoyltrifluoroacetylacetonate) was deposited on silver islands, the signal intensity increased 5-fold while the lifetime sustained a 100-fold decrease. The lifetime reduction was explained by electromagnetic coupling between the metal ion and the silver islands. In a similar approach using silver nanoparticles and Eu(III) dipicolinic acid, Wang et al. attributed the signal enhancement to a strong coupling between the radiation transition and surface plasmon resonance [80]. Eu(III) pyridine-3,5-dicarboxylate and Eu(III) β-diketonate aggregates on silver nanoparticles have also been reported to enhance luminescent properties of the Eu(III)-complex [81–83].

2.5 *Biological Nanoparticles*

Biological nanoparticles can serve as hosts to lanthanide-based materials. Liposomes and proteins can be readily produced in large quantities having precise predetermined particle nanosize. Liposomes have been loaded with Eu(III) chelated with diethylenetriamine pentaacetate (DTPA) and shown to function as a nanolabel in a model assay system. The europium(III) chelate was not directly luminescent and required a signal enhancement step in a separation assay [84, 85]. Improved Eu(III) chelate containing liposome nanolabels have been studied by incorporating luminescent chelating structures directly into liposomes. Homogeneous separation-free model assay has been introduced based on fluorescent liposomal structures [86, 87]. Recently, protein-based biological apoferritin nanoparticles loaded with europium(III) ions have been prepared [88]. These biological nanoparticle labels require an additional enhancement step to extract the ions from the particles and to form luminescent complexes in solution.

2.6 *Luminescence Characteristics*

Both the lanthanide chelate-incorporated nanoparticles [27] and the inorganic lanthanide-doped nanocrystals [28] share some common characteristics of lanthanide luminescence: multiple, lanthanide-specific narrow emission bands at visible wavelengths, long luminescence lifetime, and well-separated excitation and emission wavelengths. The mechanism of excitation, however, is entirely different between the two particulate lanthanide reporters: the lanthanide ions complexed to light harvesting ligands are typically excited through absorption by ligand and energy transfer from an excited state of the ligand (typically the triplet state) to the ion, whereas the inorganic lanthanide-doped nanocrystals utilize directly the absorption bands of the emissive lanthanide ion (activator) or energy-transfer from another absorptive co-dopant ion (sensitizer). This results in entirely different excitation spectra and renders the inorganic lanthanide-doped nanocrystals dependent on short-wavelength ultraviolet excitation providing typically lower luminescence efficiency. The molar absorptivity of the lanthanide ions at the direct excitation bands can be over 10,000-fold lower compared to the molar absorptivity of the ions complexed to the light harvesting ligands. The higher photoluminescence output per a single lanthanide ion renders the lanthanide chelate-incorporated polymeric nanoparticle labels preferred over inorganic particles in many applications. In case of Er(III)- and Yb(III)-doped inorganic upconverting nanocrystals, the weak absorptivity is partly compensated by availability of low-cost, high-power laser diode excitation in the near-infrared spectral range.

Each of the luminescent lanthanides produces emission lines at wavelengths characteristic to the ion (Table 1), i.e., different lanthanides, samarium(III), europium(III), terbium(III), and dysprosium(III), produce emission peaks at distinct wavelengths when chelated with a light-harvesting ligand. Concentration-dependent quenching of dyes in nanostructures is expected to occur at orders of magnitude higher concentration for lanthanide-based materials compared to conventional organic fluorophores due to a large Stokes' shift of the lanthanide chelating structures. Thus, lanthanide chelates can be densely packed into polymeric nanoparticles [41] to provide highly photoluminescent particulate reporters.

Molecular lanthanide chelates need to be designed to be kinetically and thermodynamically stable, water-soluble, and biocompatible for coupling to biomolecules, whereas lanthanide chelates incorporated into nanoparticles can be selected for their luminescence properties only. This is an advantage since the stability and water solubility requirements of the molecular lanthanide chelates are sometimes achieved at the cost of optimal photoluminescence properties. Moreover, lanthanide

Table 1 Emission characteristics of various lanthanides and their chelates

Lanthanide	Principal emission band (nm)	Transition	Typical lifetime	Detectability[a] (mol/L)
Samarium(III)	643/598[b]	$^5G_{5/2} \rightarrow {}^6H_{7/2}, {}^6H_{9/2}$	50–100 μs	250×10^{-15}
Europium(III)	613	$^5D_0 \rightarrow {}^7F_2$	0.5–1 ms	30×10^{-15}
Terbium(III)	545	$^5D_4 \rightarrow {}^7F_5$	0.1–2 ms	25×10^{-15}
Dysprosium(III)	573	$^4F_{9/2} \rightarrow {}^6H_{13/2}$	Below 10 μs	750×10^{-15}

[a]Specifications for DELFIA technology with PerkinElmer Victor X

[b]From [89]; other spectral characteristics from [90]

chelate incorporated into nanoparticle reporters can enable a sensitive, quadruple label technology for applications requiring multiplexing and are more versatile labels for bioanalytical assays than luminescent molecular lanthanide chelates [91]. Another advantage of particulate lanthanide reporters is that they are potentially insensitive to the environment [92] unlike some of the intrinsically luminescent molecular chelates, which respond to other molecules in their proximate environment [26, 93].

Inorganic lanthanide-doped nanoparticles provide typically lower luminescence emission intensity compared to lanthanide-chelate dyed nanoparticles. However, they have additional advantages by being insensitive to photobleaching. Polymer-based lanthanide chelate nanoparticles display improved stability towards pH and quenching of metal ions [92], but they are still moderately susceptible to photobleaching. Thus, inorganic lanthanide-doped nanoparticles should be preferred for luminescence imaging applications [28, 94], for which fluorescent nanocrystals such as quantum dots have been already shown to present significant advantages [34] over conventional fluorescent dyes. Inorganic lanthanide-doped nanophosphors possess exceptional stability at high temperature and under harsh conditions; the luminescence intensity of europium(III)-activated yttrium oxysulfide is only slightly affected by temperature change from −20°C to 100°C and pH change from 5 to 8 [23].

2.7 *Surface Chemistry*

Lanthanide chelate-dyed polystyrene particles containing carboxylic acid groups have been covalently coated with antibodies, streptavidin [95], and nucleic acids [96] for different biomolecular-binding assays. Particulate labels can be coated with antibodies and used as such, but the obvious steric and kinetic problems associated with the large molecular size can be particularly avoided by indirect detection of the bound antibodies using, e.g., biotin–streptavidin interaction [97, 98]. Several research groups have synthesized silica nanoparticles doped with europium(III) or terbium(III) chelates and surface-coated with organosilanes featuring primary amino groups for conjugation of streptavidin or antibodies [44, 46, 99]. Modification of the silica surface with dextran layer before conjugation of antibodies resulted in an exceptionally low nonspecific binding enabling highly sensitive assays [48]. In another approach, the nanoparticle surface has been coated with a layer of bovine serum albumin (BSA) before further conjugating to the antibody [100].

Surface chemistry and surface modification play an important role in the conjugation of proteins to inorganic lanthanide-doped nanoparticles [28]. Inorganic nanophosphors, such as europium-activated yttrium oxide and oxysulfide, typically do not possess any groups available for covalent coupling, and passive adsorption of proteins may not be efficient and stable enough – although it has been utilized in early applications together with polycarboxylic acids [23, 52]. Polycarboxylic acids strongly adsorb on the lanthanide phosphors giving a sufficient colloidal stability in water [94]. Adsorbed polyacrylic acids can be further activated by carbodiimides for covalent conjugation to biomolecules through primary amino groups. Recently, europium(III)-activated gadolinium oxide nanoparticles have been coated with streptavidin and used to detect biotinylated nucleic acids [101].

3 Heterogeneous Bioanalytical Assays

The nanoparticle reporter and the optimal nanoparticle–antibody bioconjugate should be individually designed for heterogeneous bioanalytical assays, taking into account the particulate size and shape, luminescence properties, colloidal stability (monodispersivity) and, in addition, specific properties of the assay such as nonspecific binding and specific binding affinity of the bioconjugate. Here, we briefly discuss the current understanding of the preferred nanoparticle characteristics in heterogeneous two-site assay format.

3.1 High Assay Sensitivity

There has been an increasing need for assay miniaturization and the detection of reduced numbers of analyte molecules. Thus, concentration beyond the current limit of detection is required for the detection of, e.g., viral antigens, antiviral antibodies, or circulating tumor cells, for discovering new clinically valuable analytes, and for replacing the current nucleic acid amplification or cell culture-based assays for detecting viruses, bacteria, and other pathogens (for a review, see [102]). Although the detection limits required for viruses and bacteria, e.g., 10–1,000 molecules/mL ($1 \times 10^{-20} - 1 \times 10^{-18}$ mol/L), may seem to be beyond the measurable range of any immunoassay, it should be noted that these analytes are large macromolecules, containing tens or hundreds of identical epitopes on the surface, which can significantly improve their detectability in noncompetitive two-site immunoassays. The high luminescence intensity of particulate lanthanide reporters favors their application in immunoassays, where extreme sensitivity is required. The same objective has also been addressed using lanthanide labeled streptavidin macroconjugates of thyroglobulin [103], polyvinylamine [104, 105], polyaminoacids [106], or precisely assembled conjugates of fluorescent proteins [107]. However, the high specific activity does not solely provide an improved sensitivity in a two-site immunoassay. For extreme sensitivity, an affinity larger than that of the labeled molecular binders is required. The high specific activity of the label and the high affinity allow reducing the number of labels in the reaction, which in turn decreases the background signal. The association rate of antibody-coated nanoparticles is dependent on the surface area covered by antibodies (the number of active binding sites), whereas the dissociation rate is independent of the covered area [108]. Since the antibody association rate increases with increasing number of coated antibodies, the nanoparticle–antibody binding affinity also increases, resulting eventually in an improved monovalent affinity compared to individual antibody. The improved monovalent binding affinity of nanoparticle–antibody bioconjugates further enables recognition of analytes using smaller concentration of bioconjugates and improves the obtainable signal-to-background ratio. Thus, it is the combination of the specific activity, the improved affinity, and the reduced number of assay components that leads to high assay sensitivity in a two-site immunoassay nanoparticle-label concepts.

Both lanthanide chelate-doped nanoparticles and inorganic lanthanide-doped nanoparticles can provide the common advantages of lanthanide luminescence to bioassays. These advantages, however, have been erroneously also attributed to quantum dots [109]. The europium(III) chelate-dyed nanoparticles [27] enable supersensitive immunoassay for prostate specific antigen (0.040 ng/L, 1.2 fmol/L, i.e., 730,000 molecules/mL), where the detectability of the label is not anymore a limiting factor [110]. In this assay and the other microtitration-well-based assays utilizing particulate lanthanide reporters [95, 98, 111], the signal has been measured directly from solid-phase bound reporters (Fig. 1) using time-resolved mode of the commercially available plate fluorometer. Direct surface-readout removes the need for separate dissociation enhancement or substrate incubation step associated with conventional DELFIA [22] and ELISA methods in microtitration wells. Commercial europium(III) chelate-dyed polystyrene particles [112–114] have been recently utilized in lateral flow assays, in an optical waveguide-based immunosensor and in protein array on surface of nanoporous silicon [115]. Best reported assay

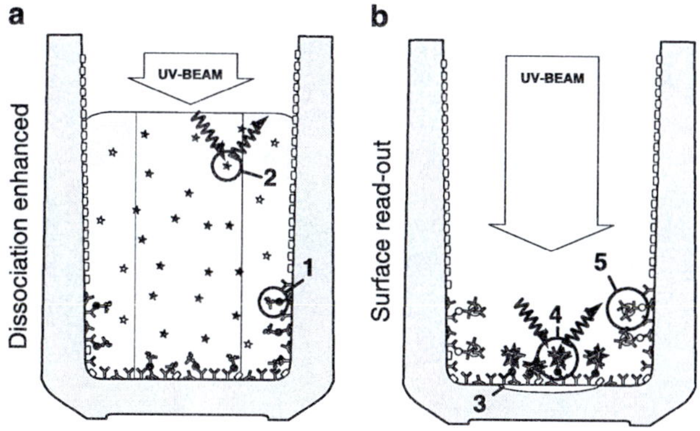

Fig. 1 Two different detection principles in a microtitration well: **(a)** dissociation enhanced measurement of nonfluorescent europium(III) chelates and **(c)** surface read-out of bound nanoparticle–antibody bioconjugates from the bottom of the microtitration well. *Explanation*: **(a)** Europium(III) ions are released from the nonluminescent chelates attached to the labeled antibody **(1)** to enhancement solution establishing highly luminescent complexes **(2)**. **(b)** The nanoparticle–antibody bioconjugates bound on an illuminated area **(3)** on the bottom of the well generate a signal **(4)**, while the bioconjugates on the sides **(5)** are omitted in the surface read-out

performance, however, has been described with in-house prepared lanthanide-chelate-labeled silica nanoparticles [48] due to incredibly low nonspecific binding.

In current assay concepts, lanthanide nanoparticles typically possess sufficient specific activity and luminescence signal for not limiting the assay sensitivity. Therefore, further improvements in assay sensitivity can be mainly directed to decreasing nonspecific binding of nanoparticle–bioconjugates to solid surfaces. In recent studies, the nonspecific binding issue in lanthanide-based nanoparticle assays has been addressed to elucidate its origin [116, 117]. In a two-site immunoassay study, either the particle size or the size of the antibody or its fragments at the capture, or the nanoparticle label surface was shown not to have a predominant impact on nonspecific binding. In another study, nonspecific binding was found not to originate from any microtiter plate solid-phase related defects or denaturation of antibodies on solid phases. Therefore, further investigation is required to clarify the reasons for nonspecific binding in order to fully benefit from the high specific activity and high affinity of the lanthanide nanoparticle labels.

3.2 Reporter Size

The small nanosized particulates reporters are potential labels for bioanalysis but may conflict with the required specific activity of the reporters as individual label moieties can only be packed to a certain density inside the particulate. The luminescence intensity (Fig. 2) of lanthanide chelate-dyed polystyrene nanoparticles increases proportionally to the particle volume. Apparently, the volume-dependent increase in specific activity is also applicable to inorganic nanoparticule labels. With respect to size alone, excluding any effect on assay performance, a large particulate size is preferred as higher specific activity per single reporter is reached. However, a large size of the particulates is associated with problems in heterogeneous immunoassay: Indeed, the large diameter of particle reporters causes steric restrictions and can prevent the formation of a sandwich pair between the solid-phase antibody:analyte complex and the bioconjugated nanoparticle label [117]. A large size of the nanoparticle reporter limits ultimately also the dynamic range of the heterogeneous solid-phase immunoassay at high analyte concentrations due to the packing and steric issues (Fig. 3). The diffusion of large particulates is slower than that of antibodies labeled with molecular reporters, and the particle concentration cannot be excessively increased to compensate this without increasing the assay background noise and thus loosing the available potential sensitivity. Inevitably, the reporter size cannot be excessively increased and thus the specific activity of the particulate label must be compromised for an optimal assay performance. A large nanoparticle label size is also believed to cause highly nonspecific binding due to an increase in surface contact area. However, a recent report shows that this is not the case and, in fact, quite similar nonspecific binding was demonstrated using europium chelated nanoparticles of 47–202 nm in diameter [117]. On the contrary, size-dependent problems related to nanoparticles much smaller than

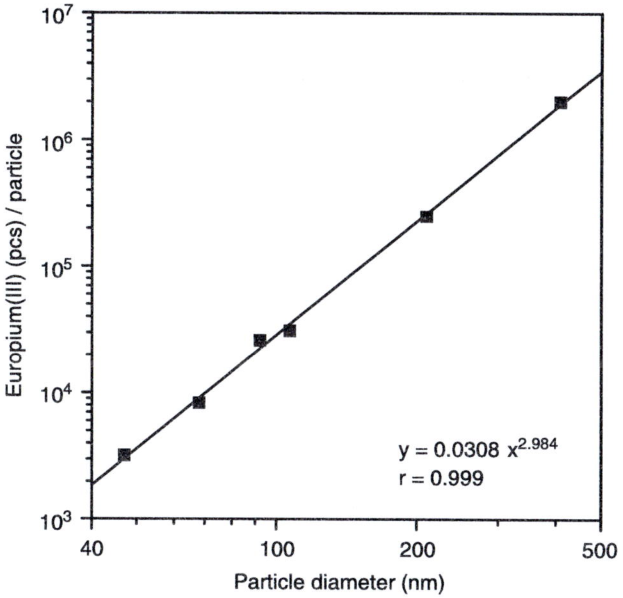

Fig. 2 Size-dependent luminescence intensity of commercially available europium(III) chelate-dyed nanoparticles (obtained from Seradyn, IN) in 1 g/L Triton X-100 solution. The luminescence intensity of particulate lanthanide reporters is defined as the equivalent number of europium(III) ions in DELFIA enhancement solution resulting in a signal corresponding to that originating from a single particle. The commercial europium(III) chelate-dyed nanoparticles obtained from Seradyn, Inc. contain europium(III) tris(2-naphthoyl trifluoroacetonate)

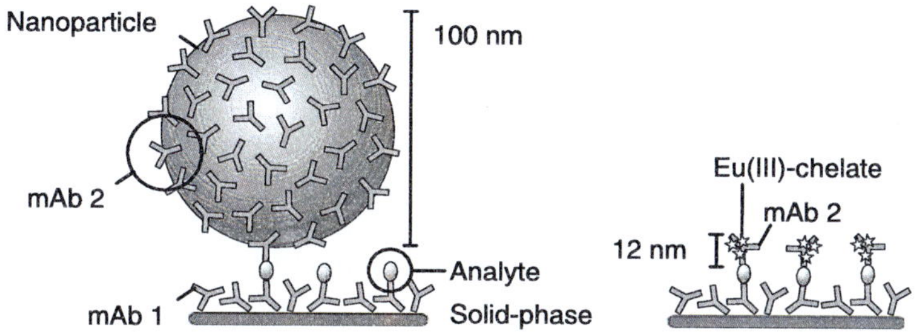

Fig. 3 Principle of a noncompetitive two-site immunoassay using (*left*) a europium(III) chelate-dyed nanoparticle label and (*right*) a conventional europium(III) chelate. The passively coated mAb1 (monoclonal antibody) recognizes one epitope on the analyte and the mAb2 recognizes another, nonoverlapping epitope. In the nanoparticle-based assay, the nanoparticle–antibody bioconjugates of mAb1 measure analyte correctly only at low concentrations, when the analyte molecules are sparsely distributed on the surface. In the reference immunoassay, the analyte is measured using the conventional molecular europium(III)-chelate-labeled mAb2

100 nm in diameter have been related to colloidal instability [117]. The potential problems associated with the large dimensions of the particulate reporters, however, can be partially circumvented and predicted [108, 118]. In practice, particle

diameters of 100 nm or below, preferably less than 60 nm, are selected in most applications to result in an optimal performance if colloidal stability of the nanoparticle suspension can be retained.

3.3 Advanced Features

Hybrid materials may provide an additional dimension to heterogeneous assay development as the functionality of the nanoparticle labels can be increased. An interesting possibility is to use magnetic Co:Nd:Fe_2O_3 (core) luminescent Eu: Gd_2O_3 (shell) core–shell nanoparticles with diameter 200–400 nm in immunoassays, enabling manipulation of nanoparticles by an external magnetic field during the separation and washing steps [119, 120]. The magnetic field can be utilized to improve the washing efficiency for removing weakly bound nanoparticles (nonspecific binding) from the solid phase. Similar nanoparticles with magnetic Fe_2O_3 core have been utilized in nucleic acid hybridization assays [77]. The active manipulation of particulate reporters by magnetic field can be an advantage to improve assay kinetics and washing efficiency. Inorganic lanthanide-doped nanoparticles such as $LaPO_4$:Ce^{3+}, Tb^{3+} can also be synthesized with a diameter as small as 10 nm and coupled to proteins with different conjugation strategies [121]. New interesting opportunities are available by utilizing multiplexing based on nanophosphors doped with different lanthanides. Inorganic lanthanide-doped nanophosphors are also well-suited for applications such as microarray imaging [122] – the absence of photobleaching renders the arrays with inorganic particulate lanthanide reporters readable practically infinitely.

4 Homogeneous Bioanalytical Assays

Separation-free, homogeneous bioanalytical assays offer significant advantages over heterogeneous assay protocols, enabling simplified automation and faster assays. The common drawbacks of homogeneous assays are a narrower dynamic range and a lower sensitivity. We discuss here the potential and preferred characteristics of lanthanide nanoparticles in luminescence resonance energy transfer (LRET)-based homogeneous assays.

4.1 Luminescence Resonance Energy Transfer

Particulate lanthanide reporters can be utilized as donors in resonant energy transfer to short-lifetime fluorescent acceptors, enabling homogeneous assay technology. Temporal resolution eliminates autofluorescence and the direct fluorescence of an acceptor, providing the same advantages as molecular lanthanide chelates [123, 124] or cryptates [125]. The sharp and narrow emission peaks related to lanthanide luminescence minimize direct donor emission at the acceptor-specific emission wavelength. This is essential for sensitive measurement of the energy-transfer excited acceptor emission. An obvious problem in resonance energy transfer assays based on particulate reporters is the distance restriction between donor and acceptor set to below 10 nm by Förster radius. Efficient use of particulate lanthanide reporters as donors requires relatively small nanoparticles as only the lanthanide donors located within the Förster radius (that is in the outer surface layer of the particulate) from the surface-bound acceptors are able to contribute to the energy transfer. The high specific activity of a particulate is thus only partially exploited in energy transfer as the donor emission originating from the core of the nanoparticle can be absorbed by nonbound acceptors generating background fluorescence with an apparent long lifetime at the acceptor specific emission wavelength. According to current knowledge [126] Förster's or energy-transfer radius for europium(III) and Cy5 [127] is less than 70 Å, which would result in an optimal size for particulate luminescent donors less than 20 nm in diameter.

Referring to the time-resolved emission spectrum of europium(III) chelate nanoparticles, it appears that the optimum wavelengths for the measurement of energy transfer are 730 nm and 770–790 nm in order to eliminate the long-lifetime emission of the europium complex (Fig. 4). In the case of terbium(III) chelate nanoparticles, the situation is slightly better since no emission peak exists at wavelengths larger than 700 nm. The long lifetime emission of europium(III) chelate nanoparticles would generate cross-talk at the wavelength of commonly employed acceptor dyes such as Cy5 and APC (allophycocyanin) with 665 nm emission and therefore limit the sensitivity of the homogeneous assay. In principle, LRET from europium(III) with an emission peak at 615 nm and measurement of the sensitized emission at 730 nm would require acceptor dyes with an exceptionally large Stokes' shift, over 100 nm, although this is not necessarily required with luminescent lanthanides as efficient energy transfer has been demonstrated with conventional far-red-emitting fluorophores. It has been proposed and shown that with long-lifetime fluorescent donors, FRET is efficient even when the donor emission and the acceptor absorption spectrum do not overlap significantly (non-overlapping FRET) [128–130]. In cases where maximum overlap of the absorption spectra of the donors with the major emission band of europium(III) at 615 nm is desirable, several acceptor dyes, e.g., AlexaFluor 680 and 700 (Invitrogen) and Cy5 (GE Healthcare), are commercially available.

Inorganic europium(III) ion doped oxide nanoparticles with an average diameter of approximately 35 nm (wide size distribution) have been used as donors in FRET experiments with Cy5 acceptor fluorophore [131]. These nanoparticles were further size-fractionated by centrifugation and the supernatant was retained to produce nanoparticles with an average diameter of 17 nm. The smaller nanoparticles contained more Eu(III) ions close to the surface resulting in better luminescence quenching of the donor. Improved performance has been obtained in resonance energy transfer to fluorescein acceptor by utilizing inorganic $LaPO_4$:Ce(III) (core),

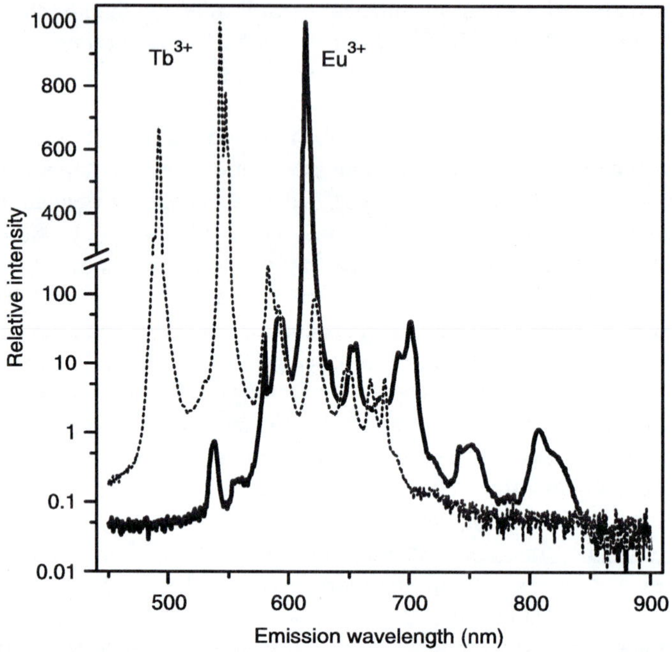

Fig. 4 Time-resolved emission spectrum of luminescent europium(III) chelate-dyed nanoparticles (*solid line*) and terbium(III) chelate (*broken line*). Time-resolved luminescence emission spectra of europium(III) chelate-dyed nanoparticles (107 nm in diameter, obtained from Seradyn, IN) in 1 g/L Triton X-100 solution (excitation at 340 nm) and terbium(III) ions in DELFIA enhancement solution and DELFIA enhancer (PerkinElmer LAS, MA) (excitation at 320 nm) were measured with a Cary Eclipse fluorescence spectrophotometer (Varian Inc., Mulgrave, Australia)

Tb(III) (shell) core–shell nanoparticles as donor, in which the emissive terbium(III) ions are located in the outer layer of the nanophosphor [132]. Similar arrangement is also possible with modified liposome membranes dyed with europium(III) 2-naphthoyltrifluoroacetone chelate as donor in energy transfer to APC [86, 87].

Recently, europium(III) chelate dyed nanoparticles with a diameter of 107 nm have been utilized as donors in a FRET-based competitive immunoassay designed for measuring 17β-estradiol concentration [133]. The nanoparticles were coated with recombinant antiestradiol antibody Fab fragments to shorten the average distance between the europium(III) chelates inside the particle and the acceptor dye conjugated to estradiol. AlexaFluor 680 was chosen as an acceptor to minimize the direct crosstalk of europium(III) emission at 730 nm. The study showed that high acceptor density on the surface results in less energy-transfer excited signal from a single acceptor. This may be associated with a smaller number of luminescent europium(III) ions able to transfer energy to a single acceptor and self-quenching via acceptor–acceptor energy transfer. The developed assay was later employed to screen enzyme inhibitors by measuring estradiol concentration [134, 135] and for comparing the efficiency of individual europium(III) chelates versus chelate-dyed nanoparticles [136]. In the last article, similar nanoparticles with a diameter of 68 nm have been also utilized to confirm the theoretical hypothesis that better assay performance can be obtained with smaller nanoparticles as a relatively smaller fraction of the particle volume is outside Förster's radius from the surface-bound acceptors. In theory, the optimal size of the particulate reporter in homogeneous FRET-based assays is less than 20 nm in diameter, which enables donors in the entire particle volume to participate in resonance energy transfer to surface-bound acceptors. When the particulate reporter is significantly larger, the excess of donors only contribute to background fluorescence due to reabsorption of the donor emission by nonbound acceptors. Core–shell type particulate reporters, in which the donors are located only in the shell layer, can thus provide an improved assay performance due to lower background luminescence.

Specific bioaffinity assays are always accompanied with nonspecific binding interactions and this has been seen as an obstacle in assay development. However, nonspecific binding in combination with core–shell europium(III) and europium (III)-dyed polymeric nanoparticles has been recently exploited in FRET-based assays for total protein concentration measurement [137, 138]. The method was based on the competitive adsorption of sample proteins and acceptor-labeled proteins to polymeric Eu(III) nanoparticles. High sensitivity and low, subnanomolar limit of detection was partially reached due to a short Förster's radius as the acceptor label was brought in direct contact with the surface.

5 Detection Instrumentation

Lanthanide chelate-doped polymeric nanoparticles and inorganic lanthanide-doped nanocrystals require a dedicated time-resolved fluorometer with pulsed ultraviolet excitation [16]. The excitation light source is typically either a broad-band Xe flash lamp with appropriate filtering [21, 139] or a 337 nm nitrogen (N_2) laser [125, 140, 141] able to provide higher energy excitation pulses. Nitrogen laser is optimal for most lanthanide labels, but the laser is itself bulky and it renders the detection instrument inevitably more expensive. Detection instruments based on 360 nm light-emitting diode (LED) excitation have been described [142], but the light output power of UV-LEDs at typical excitation wavelengths (below 340 nm) is yet suboptimal for efficient excitation [113, 114]. The emission intensity of lanthanide chelates is lower than that of equivalent concentration of conventional fluorescent dyes because of the longer turnaround time [143]. Each chelated lanthanide ion emits a single photon in a period of several hundred microseconds, whereas with conventional fluorescent, the equivalent period is nanoseconds. This results in requiring more sensitive light detection and, in practice, photomultiplier tube-based detection (PMT) is required as photodiodes are currently not sensitive enough. The pulsed excitation and time-resolved detection can be achieved either with electronic triggering and gating alone or combined with a mechanical chopper. In principle, the mechanical chopper provides optimum performance by simplifying the temporal gating of the detector. It has been applied especially in time-resolved fluorescence microscopy with Peltier-cooled charge-coupled device (CCD) cameras for detection [23, 27, 144, 145]. The higher emission intensity of particulate lanthanide reporters facilitates their use in CCD-based time-resolved luminescence imaging by compensating the weaker detection sensitivity of CCD compared to PMT. Lanthanide nanoparticles yet require temporal gating for sensitive detection, rendering them dependent on time-resolved fluorescence imaging devices.

Currently, lanthanide-based time-resolved fluorometry is compatible with small hand-held analyzers comprising an integrated detector [146]. If the sensitivity issue can be solved, UV-LED excitation and/or photodiode-based detection will render the actual time-resolved luminescence detector module significantly smaller and also produce a significant cost advantage. Since lanthanide chelated-dyed nanoparticles contain a large number of chelates, they can potentially solve the sensitivity problem – at least partially. Some europium(III) chelate-dyed nanoparticles can even be excited above 400 nm wavelength [51, 147, 148], which, together with the development of UV-LED technology, may change the situation in the near future. Further advantage of UV-LED (and also N_2 laser) excitation is the rapid decay of the excitation pulse which, in the case of Xe flash lamps, can limit the detection of the very rapidly decaying emission of dysprosium(III) and samarium(III) [89] or of the energy-transfer excited emission with short apparent decay time in TR-FRET (time-resolved fluorescence resonance energy-transfer) applications [149].

6 Outlook

In bioanalytical applications, both lanthanide chelate-incorporated and inorganic lanthanide-doped nanoparticles provide significant advantages in bioanalysis over their conventional counterparts including quantum dots. The future challenges stand in the development of small-sized chelate-incorporated nanoparticles and inorganic nanophosphors with improved luminescence characteristics as well as their commercial availability for research use. Time-resolved plate fluorometers are now available from several manufacturers, facilitating the implementation of lanthanide-based nanoparticle bioanalytical technologies.

References

1. Frank DS, Sundberg MW (1978) Fluorescent labels comprising rare earth chelates. US Patent 4,283,382
2. Yalow RS, Berson SA (1960) Immunoassay of endogenous plasma insulin in man. J Clin Invest 39:1157–1175

3. Ekins RP (1960) The estimation of thyroxine in human plasma by an electrophoretic technique. Clin Chim Acta 5:453–459
4. Addison GM, Hales CN (1971) Two site assay of human growth hormone. Horm Metab Res 3:59–60
5. Miles LE, Hales CN (1968) Labelled antibodies and immunological assay systems. Nature 219:186–189
6. Wide L, Bennich H, Johansson SG (1967) Diagnosis of allergy by an in-vitro test for allergen antibodies. Lancet 2:1105–1107
7. Köhler G, Milstein C (1975) Continuous cultures of fused cells secreting antibody of predefined specificity. Nature 256:495–497
8. David, G S and Greene, H E (1984) Immunometric assays using monoclonal antibodies. US Patent 4,486,530
9. Ekins RP (1989) A shadow over immunoassay. Nature 340:256–258
10. Sevier ED, David GS, Martinis J et al (1981) Monoclonal antibodies in clinical immunology. Clin Chem 27:1797–1806
11. Ekins RP, Chu F, Micallef J (1989) High specific activity chemiluminescent and fluorescent markers: their potential application to high sensitivity and 'multi-analyte' immunoassays. J Biolumin Chemilumin 4:59–78
12. Jackson TM, Ekins RP (1986) Theoretical limitations on immunoassay sensitivity. Current practise and potential advantages of fluorescent Eu^{3+} chelates as non-radioisotopic tracers. J Immunol Methods 87:13–20
13. Jackson TM, Marshall NJ, Ekins RP (1983) Optimisation of immunoradiometric (labelled antibody) assays. In: Hunter WM, Corrie JET (eds) Immunoassays for clinical chemistry. Churchill Livingstone, Edinburgh
14. Ekins RP, Dakubu S (1985) The development of high sensitivity pulsed light, time-resolved fluoroimmunoassays. Pure Appl Chem 57:437–482
15. Siitari H, Hemmilä I, Soini E et al (1983) Detection of hepatitis B surface antigen using time-resolved fluoroimmunoassay. Nature 301:258–260
16. Soini E, Lövgren T (1987) Time-resolved fluorescence of lanthanide probes and applications in biotechnology. CRC Crit Rev Anal Chem 18:105–154
17. Soini E, Hemmilä I (1979) Fluoroimmunoassay: present status and key problems. Clin Chem 25:353–361
18. Ekins RP (1985) Current concepts and future developments. In: Collins WP (ed) Alternative immunoassays. Wiley, New York
19. Hemmilä I (1985) Fluoroimmunoassays and immunofluorometric assays. Clin Chem 31:359–370
20. Haddad PR (1977) The application of ternary complexes to spectrofluorometric analysis. Talanta 24:1–13
21. Soini E, Kojola H (1983) Time-resolved fluorometer for lanthanide chelates–a new generation of nonisotopic immunoassays. Clin Chem 29:65–68
22. Hemmilä I, Dakubu S, Mukkala VM et al (1984) Europium as a label in time-resolved immunofluorometric assays. Anal Biochem 137:335–343
23. Beverloo HB, van Schadewijk A, van Gelderen-Boele S et al (1990) Inorganic phosphors as new luminescent labels for immunocytochemistry and time-resolved microscopy. Cytometry 11:784–792
24. Alpha B, Balzani V, Lehn J-M et al (1987) Luminescence probes: The Eu^{3+}- and Tb^{3+} -cryptates of polypyridine macrobicyclic ligands. Angew Chem Int Ed Engl 26:1266–1267
25. Takalo H, Mukkala VM, Mikola H et al (1994) Synthesis of europium(III) chelates suitable for labeling of bioactive molecules. Bioconjug Chem 5:278–282
26. Takalo H, Mukkala V-M, Meriö L (1997) Development of luminescent terbium(III) chelates for protein labelling: effect of triplet-state energy level. Helv Chim Acta 80:372–387
27. Härmä H, Soukka T, Lövgren T (2001) Europium nanoparticles and time-resolved fluorescence for ultrasensitive detection of prostate-specific antigen. Clin Chem 47:561–568

28. Dosev D, Nichkova M, Kennedy IM (2008) Inorganic lanthanide nanophosphors in biotechnology. J Nanosci Nanotechnol 8:1052–1067
29. Escribano P, Julian-Lopez B, Planelles-Arago J et al (2008) Photonic and nanobiophotonic properties of luminescent lanthanide-doped hybrid organic- inorganic materials. J Mater Chem 18:23–40
30. Seydack M (2005) Nanoparticle labels in immunosensing using optical detection methods. Biosens Bioelectron 20:2454–2469
31. Yuan JL, Wang GL (2006) Lanthanide-based luminescence probes and time-resolved luminescence bioassays. Trends Anal Chem 25:490–500
32. Bruchez M Jr, Moronne M, Gin P et al (1998) Semiconductor nanocrystals as fluorescent biological labels. Science 281:2013–2016
33. Chan WC, Nie S (1998) Quantum dot bioconjugates for ultrasensitive nonisotopic detection. Science 281:2016–2018
34. Michalet X, Pinaud FF, Bentolila LA et al (2005) Quantum dots for live cells, in vivo imaging, and diagnostics. Science 307:538–544
35. Ekins RP (1997) Immunoassay design and optimization. In: Price CP, Newman DJ (eds) Principles and practice of immunoassay. Macmillan, London
36. Kricka LJ (1994) Selected strategies for improving sensitivity and reliability of immunoassays. Clin Chem 40:347–357
37. Kricka LJ (1999) Nucleic acid detection technologies – labels, strategies, and formats. Clin Chem 45:453–458
38. Guo C-Y, Shankar RR, Abe S et al (1992) Functionalized, probe-vontaining, latex nanoparticles. Anal Biochem 207:241–248
39. Tamaki K, Shimomura M (2002) Fabrications of luminescent polymeric nanoparticles containing lanthanide (III) ion complexes. Int J Nanosci 1:533–537
40. Hakala H, Mukkala VM, Sutela T et al (2006) Synthesis and properties of nanospheres copolymerised with luminescent europium(III) chelates. Org Biomol Chem 4:1383–1386
41. Huhtinen P, Kivela M, Kuronen O et al (2005) Synthesis, characterization, and application of Eu(III), Tb(III), Sm(III), and Dy(III) lanthanide chelate nanoparticle labels. Anal Chem 77:2643–2648
42. Chen Y, Chi YM, Wen HM et al (2007) Sensitized luminescent terbium nanoparticles: preparation and time-resolved fluorescence assay for DNA. Anal Chem 79:960–965
43. Chen Y, Lu ZH (2007) Dye sensitized luminescent europium nanoparticles and its time-resolved fluorometric assay for DNA. Anal Chim Acta 587:180–186
44. Hai XD, Tan MQ, Wang G et al (2004) Preparation and a time-resolved fluoroimmunoassay application of new europium fluorescent nanoparticles. Anal Sci 20:245–246
45. Sharma P, Brown S, Varshney M et al (2008) Surfactant-mediated fabrication of optical nanoprobes. Adv Polymer Sci 218:189–233
46. Ye ZQ, Tan MQ, Wang GL et al (2004) Preparation, characterization, and time-resolved fluorometric application of silica-coated terbium(Ill) fluorescent nanoparticles. Anal Chem 76:513–518
47. Zhang H, Xu Y, Yang W et al (2007) Dual-lanthanide-chelated silica nanoparticles as labels for highly sensitive time-resolved fluorometry. Chem Mater 19:5875–5881
48. Xu Y, Li QG (2007) Multiple fluorescent labeling of silica nanoparticles with lanthanide chelates for highly sensitive time-resolved Immunofluorometric assays. Clin Chem 53: 1503–1510
49. Härmä H, Graf C, Hänninen P (2008) Synthesis and characterization of core-shell europium (III)-silica nanoparticles. J Nanopart Res 10:1221–1224
50. Härmä H, Keränen AM, Lövgren T (2007) Synthesis and characterization of europium(III) nanoparticles for time-resolved fluoroimmunoassay of prostate-specific antigen. Nanotechnology 18:075604
51. Wen XF, Li MY, Wang Y et al (2008) Colloidal nanoparticles of a europium complex with enhanced luminescent properties. Langmuir 24:6932–6936

52. Beverloo HB, van Schadewijk A, Zijlmans HJMAA et al (1992) Immunochemical detection of proteins and nucleic acids on filters using small luminescent inorganic crystals as markers. Anal Biochem 203:326–334
53. Faris GW, Wright WH, Pati S et al (1996) Upconverting reporters for biomedical diagnostics: Applications in antibody and DNA detection. In: Sevick-Muraca E, Benaron D (eds) Trends in optics and photonics series. Optical Society of America, Washington DC
54. Hampl J, Hall M, Mufti NA et al (2001) Upconverting phosphor reporters in immunochromatographic assays. Anal Biochem 288:176–187
55. Zijlmans HJMAA, Bonnet J, Burton J et al (1999) Detection of cell and tissue surface antigens using up-converting phosphors: a new reporter technology. Anal Biochem 267:30–36
56. Soukka T, Kuningas K, Rantanen T et al (2005) Photochemical characterization of up-converting inorganic lanthanide phosphors as potential labels. J Fluoresc 15:513–528
57. Buissette V, Moreau M, Gacoin T et al (2004) Colloidal synthesis of luminescent rhabdophane LaPO4: Ln(3+)center dot xH(2)O (Ln = Ce, Tb, Eu; x approximate to 0.7) nanocrystals. Chem Mater 16:3767–3773
58. Fan WL, Bu YX, Song XY et al (2007) Selective synthesis and luminescent properties of monazite- and zircon-type LaVO4: Ln (Ln = Eu, Sm, and Dy) nanocrystals. Cryst Growth Des 7:2361–2366
59. Hebbink GA, Stouwdam JW, Reinhoudt DN et al (2002) Lanthanide(III)-doped nanoparticles that emit in the near-infrared. Adv Mater 14:1147–1150
60. Kömpe K, Borchert H, Storz J et al (2003) Green-emitting CePO4:Tb/LaPO4 core-shell nanoparticles with 70% photoluminescence quantum yield. Angew Chem Int Ed Engl 42:5513–5516
61. Riwotzki K, Meyssamy H, Kornowski A et al (2000) Liquid-phase synthesis of doped nanoparticles: Colloids of luminescing LaPO4: Eu and CePO4: Tb particles with a narrow particle size distribution. J Phys Chem B 104:2824–2828
62. Song HW, Yu LX, Lu SZ et al (2004) Remarkable differences in photoluminescent properties between LaPO4: Eu one-dimensional nanowires and zero-dimensional nanoparticles. Appl Phys Lett 85:470–472
63. Stouwdam JW, van Veggel FCJM (2004) Improvement in the luminescence properties and processability of LaF3/Ln and LaPO4/Ln nanoparticles by surface modification. Langmuir 20:11763–11771
64. Wang F, Xue XJ, Liu XG (2008) Multicolor tuning of (Ln, P)-Doped YVO4 nanoparticles by single-wavelength excitation. Angew Chem Int Ed Engl 47:906–909
65. Dosev D, Nichkova M, Liu M et al (2005) Application of fluorescent Eu: Gd2O3 nanoparticles to the visualization of protein micropatterns. Proc SPIE 5699:473–481
66. Gordon WO, Carter JA, Tissue BM (2004) Long-lifetime luminescence of lanthanide-doped gadolinium oxide nanoparticles for immunoassays. J Lumin 108:339–342
67. Karmaoui M, Ferreira RAS, Mane AT et al (2006) Lanthanide-based lamellar nanohybrids: Synthesis, structural characterization, and optical properties. Chem Mater 18:4493–4499
68. Nichkova M, Dosev D, Gee SJ et al (2005) Microarray immunoassay for phenoxybenzoic acid using polymer encapsulated Eu: Gd2O3 nanoparticles as fluorescent labels. Anal Chem 77:6864–6873
69. Nichkova M, Dosev D, Perron R et al (2006) Eu3+-doped Gd2O3 nanoparticles as reporters for optical detection and visualization of antibodies patterned by microcontact printing. Anal Bioanal Chem 384:631–637
70. Buissette V, Giaume D, Gacoin T et al (2006) Aqueous routes to lanthanide-doped oxide nanophosphors. J Mater Chem 16:529–539
71. Heer S, Kömpe K, Güdel H-U et al (2004) Highly efficient multicolour upconversion emission in transparent colloids of nanoparticle-doped NaYF4 nanocrystals. Adv Mater 16:2102–2105
72. Heer S, Lehmann O, Haase M et al (2003) Blue, green, and red upconversion emission from lanthanide-doped LuPO4 and YbPO4 nanocrystals in a transparent colloidal solution. Angew Chem Int Ed Engl 42:3179–3182

73. Hirai T, Orikoshi T (2004) Preparation of yttrium oxysulfide phosphor nanoparticles with infrared-to-green and -blue upconversion emission using an emulsion liquid membrane system. J Colloid Interf Sci 273:470–477
74. Holm BA, Bergey EJ, De T et al (2002) Nanotechnology in biomedical applications. Mol Cryst Liq Cryst 374:589–598
75. Li L, Jiang WG, Pan HH et al (2007) Improved luminescence of lanthanide(III)-doped nanophosphors by linear aggregation. J Phys Chem C 111:4111–4115
76. Yu MX, Li FY, Chen ZG et al (2009) Laser scanning up-conversion luminescence microscopy for imaging cells labeled with rare-earth nanophosphors. Anal Chem 81:930–935
77. Son A, Dosev D, Nichkova M et al (2007) Quantitative DNA hybridization in solution using magnetic/luminescent core-shell nanoparticles. Anal Biochem 370:186–194
78. Massue J, Quinn SJ, Gunnlaugsson T (2008) Lanthanide luminescent displacement assays: The sensing of phosphate anions using Eu(III)-cyclen-conjugated gold nanoparticles in aqueous solution. J Am Chem Soc 130:6900
79. Lakowicz JR (2001) Radiative decay engineering: biophysical and biomedical applications. Anal Biochem 298:1–24
80. Wang YH, Zhou XR, Wang T et al (2008) Enhanced luminescence from lanthanide complex by silver nanoparticles. Mater Lett 62:3582–3584
81. Nabika H, Deki S (2003) Surface-enhanced luminescence from Eu3+ complex nearby Ag colloids. Eur Phys J D 24:369–372
82. Sun YY, Jiu HF, Zhang DG et al (2005) Preparation and optical properties of Eu(III) complexes J-aggregate formed on the surface of silver nanoparticles. Chem Phys Lett 410:204–208
83. Sun YY, Zheng Z, Yan Q et al (2006) Effects of Ag colloidal nanoparticles on luminescent properties of Eu(III) beta-diketone. Mater Lett 60:2756–2758
84. Laukkanen ML, Orellana A, Keinanen K (1995) Use of genetically engineered lipid-tagged antibody to generate functional europium chelate-loaded liposomes. Application in fluoroimmunoassay. J Immunol Methods 185:95–102
85. Orellana A, Laukkanen ML, Keinanen K (1996) Europium chelate-loaded liposomes: a tool for the study of binding and integrity of liposomes. Biochim Biophys Acta 1284:29–34
86. Okabayashi Y, Ikeuchi I (1998) Liposome immunoassay by long-lived fluorescence detection. Analyst 123:1329–1332
87. Pihlasalo S, Hara M, Hanninen P et al (2009) Liposome-based homogeneous luminescence resonance energy transfer. Anal Biochem 384:231–237
88. Jääskeläinen A, Harinen RR, Lamminmaki U et al (2007) Production of apoferritin-based bioinorganic hybrid nanoparticles by bacterial fermentation followed by self-assembly. Small 3:1362–1367
89. Hakala H, Liitti P, Puukka K et al (2002) Novel luminescent samarium(III) chelates. Inorg Chem Commun 5:1059–1062
90. Hemmilä I, Mukkala V-M (2001) Time-resolution in fluorometry technologies, labels, and applications in bioanalytical assays. Crit Rev Clin Lab Sci 38:441–519
91. Hemmila I (1995) Luminescent lanthanide chelates – a way to more sensitive diagnostic methods. J Alloys Comp 225:480–485
92. Kokko L, Lovgren T, Soukka T (2007) Europium(III)-chelates embedded in nanoparticles are protected from interfering compounds present in assay media. Anal Chim Acta 585:17–23
93. Nurmi J, Ylikoski A, Soukka T et al (2000) A new label technology for the detection of specific polymerase chain reaction products in a closed tube. Nucleic Acids Res 28:e28
94. Beverloo HB, van Schadewijk A, Bonnet J et al (1992) Preparation and microscopic visualization of multicolor luminescent immunophosphors. Cytometry 13:561–570
95. Härmä H, Soukka T, Lönnberg S et al (2000) Zeptomole detection sensitivity of prostate-specific antigen in a rapid microtitre plate assay using time-resolved fluorescence. Luminescence 15:351–355

96. Huhtinen P, Vaarno J, Soukka T et al (2004) Europium(III) nanoparticle-label-based assay for the detection of nucleic acids. Nanotechnology 15:1708–1715
97. Diamandis EP, Christopoulos TK (1991) The biotin-(strept)avidin system: principles and applications in biotechnology. Clin Chem 37:625–636
98. Huhtinen P, Soukka T, Lovgren T et al (2004) Immunoassay of total prostate-specific antigen using europium(III) nanoparticle labels and streptavidin-biotin technology. J Immunol Methods 294:111–122
99. Tan MQ, Wang GL, Hai XD et al (2004) Development of functionalized fluorescent europium nanoparticles for biolabeling and time-resolved fluorometric applications. J Mater Chem 14:2896–2901
100. Ye ZQ, Tan MQ, Wang GL et al (2005) Development of functionalized terbium fluorescent nanoparticles for antibody labeling and time-resolved fluoroimmunoassay application. Talanta 65:206–210
101. Son A, Nichkova M, Dosev D et al (2008) Luminescent lanthanide nanoparticles as labels in DNA microarrays for quantification of methyl tertiary butyl ether degrading bacteria. J Nanosci Nanotechnol 8:2463–2467
102. Iqbal SS, Mayo MW, Bruno JG et al (2000) A review of molecular recognition technologies for detection of biological threat agents. Biosens Bioelectron 15:549–578
103. Morton RC, Diamandis EP (1990) Streptavidin-based macromolecular complex labeled with a Europium chelator suitable for time-resolved fluorescence immunoassay applications. Anal Chem 62:1841–1845
104. Scorilas A, Bjartell A, Lilja H et al (2000) Streptavidin-polyvinylamine conjugates labeled with a europium chelate: applications in immunoassay, immunohistochemistry, and microarrays. Clin Chem 46:1450–1455
105. Scorilas A, Diamandis EP (2000) Polyvinylamine-streptavidin complexes labeled with a europium chelator: a universal detection reagent for solid-phase time resolved fluorometric applications. Clin Biochem 33:345–350
106. Qin QP, Lövgren T, Pettersson K (2001) Development of highly fluorescent detection reagents for the construction of ultrasensitive immunoassays. Anal Chem 73:1521–1529
107. Russell JC, Colpitts TL, Holets-McCormack SR et al (2002) Solid phase assembly of defined protein conjugates. Bioconjug Chem 13:958–965
108. Soukka T, Härmä H, Paukkunen J et al (2001) Utilization of kinetically enhanced monovalent binding affinity by immunoassays based on multivalent nanoparticle-antibody bioconjugates. Anal Chem 73:2254–2260
109. Azzazy HME, Mansour MMH, Kazmierczak SC (2006) Nanodiagnostics: a new frontier for clinical laboratory medicine. Clin Chem 52:1238–1246
110. Soukka T, Paukkunen J, Härmä H et al (2001) Supersensitive time-resolved immunofluorometric assay of free prostate-specific antigen with nanoparticle label technology. Clin Chem 47:1269–1278
111. Soukka T, Antonen K, Härmä H et al (2003) Highly sensitive immunoassay of free prostate-specific antigen in serum using europium(III) nanoparticle label technology. Clin Chim Acta 328:45–58
112. Cummins CM, Koivunen ME, Stephanian A et al (2006) Application of europium(III) chelate-dyed nanoparticle labels in a competitive atrazine fluoroimmunoassay on an ITO waveguide. Biosens Bioelectron 21:1077–1085
113. Rundstrom G, Jonsson A, Martensson O et al (2007) Lateral flow immunoassay using europium (III) chelate microparticles and time-resolved fluorescence for eosinophils and neutrophils in whole blood. Clin Chem 53:342–348
114. Song XD, Knotts M (2008) Time-resolved luminescent lateral flow assay technology. Anal Chim Acta 626:186–192
115. Järås K, Tajudin AA, Ressine A et al (2008) ENSAM: Europium nanoparticles for signal enhancement of antibody microarrays on nanoporous silicon. J Proteome Res 7:1308–1314

116. Näreoja T, Määttänen A, Peltonen J et al (2009) Impact of surface defects and denaturation of capture surface proteins on nonspecific binding in immunoassays using antibody-coated polystyrene nanoparticle labels. J Immunol Methods 347:24–30
117. Näreoja T, Vehniäinen M, Lamminmaki U, et al (2009) Study on nonspecificity of an immuoassay using Eu-doped polystyrene nanoparticle labels. J Immunol Methods 345:80–89
118. Swift JL, Cramb DT (2008) Nanoparticles as fluorescence labels: is size all that matters? Biophys J 95:865–876
119. Dosev D, Nichkova M, Dumas RK et al (2007) Magnetic/luminescent core/shell particles synthesized by spray pyrolysis and their application in immunoassays with internal standard. Nanotechnology 18:55102
120. Nichkova M, Dosev D, Gee SJ et al (2007) Multiplexed immunoassays for proteins using magnetic luminescent nanoparticles for internal calibration. Anal Biochem 369:34–40
121. Weiler V, Haase M, Walter C et al (2007) Luminescent nanoparticles for molecular medicine. Biophotonics 2007: Optics in life. Science 6633:J6332
122. Shen J, Sun LD, Yan CH (2008) Luminescent rare earth nanomaterials for bioprobe applications. Dalton Trans 5687–5697
123. Blomberg K, Hurskainen P, Hemmilä I (1999) Terbium and rhodamine as labels in a homogeneous time-resolved fluorometric energy transfer assay of the beta subunit of human chorionic gonadotropin in serum. Clin Chem 45:855–861
124. Stenroos K, Hurskainen P, Eriksson S et al (1998) Homogeneous time-resolved IL-2-IL-2R alpha assay using fluorescence resonance energy transfer. Cytokine 10:495–499
125. Mathis G (1993) Rare earth cryptates and homogeneous fluoroimmunoassays with human sera. Clin Chem 39:1953–1959
126. Wu P, Brand L (1994) Resonance energy transfer: methods and applications. Anal Biochem 218:1–13
127. Selvin PR, Rana TM, Hearst JE (1994) Luminescence resonance energy transfer. J Am Chem Soc 116:6029–6030
128. Hemmilä I, Hurskainen P, Blomberg K, et al (1998) Homogeneous luminescence energy transfer assays. PCT International Application WO 98/15830
129. Kokko T, Kokko L, Soukka T (2008) Terbium(III) chelate as an efficient donor for multiple-wavelength fluorescent acceptors. J Fluoresc 19:159–164
130. Laitala V, Hemmila I (2005) Homogeneous assay based on anti-Stokes' shift time-resolved fluorescence resonance energy-transfer measurement. Anal Chem 77:1483–1487
131. Casanova D, Giaume D, Gacoin T et al (2006) Single lanthanide-doped oxide nanoparticles as donors in fluorescence resonance energy transfer experiments. J Phys Chem B 110:19264–19270
132. Meyer C, Haase M, Hoheisel W, et al (2004) Core/shell nanoparticles suitable for (F)RET-assays. PCT International Application WO 2004/096944 A1
133. Kokko L, Sandberg K, Lövgren T et al (2004) Europium(III) chelate-dyed nanoparticles as donors in a homogeneous proximity-based immunoassay for estradiol. Anal Chim Acta 503:155–162
134. Kokko L, Jaakohuhta S, Lindroos P et al (2006) Improved homogeneous proximity-based screening assay of potential inhibitors of 17 beta-hydroxysteroid dehydrogenases. Assay Drug Dev Tech 4:671–678
135. Kokko L, Johansson N, Lovgren T et al (2005) Enzyme inhibitor screening using a homogeneous proximity-based immunoassay for estradiol. J Biomol Screen 10:348–354
136. Kokko L, Kokko I, Lovgren T et al (2008) Particulate and soluble Eu(III)-chelates as donor labels in homogeneous fluorescence resonance energy transfer based immunoassay. Anal Chim Acta 606:72–79
137. Härmä H, Dahne L, Pihlasalo S et al (2008) Sensitive quantitative protein concentration method using luminescent resonance energy transfer on a layer-by-layer europium(III) chelate particle sensor. Anal Chem 80:9781–9786

138. Valanne A, Suojanen J, Peltonen J et al (2009) Multiple sized europium(III) chelate-dyed polystyrene particles as donors in FRET – an application for sensitive protein quantification utilizing competitive adsorption. Analyst 134:980–986
139. Prat O, Lopez E, Mathis G (1991) Europium(III) cryptate – a fluorescent label for the detection of DNA hybrids on solid support. Anal Biochem 195:283–289
140. Dechaud H, Bador R, Claustrat F et al (1986) Laser-excited Immunofluorometric assay of prolactin, with use of antibodies coupled to lanthanide-labeled diethylenetriaminepentaacetic acid. Clin Chem 32:1323–1327
141. Diamandis EP (1988) Immunoassays with time-resolved fluorescence spectroscopy – principles and applications. Clin Biochem 21:139–150
142. Connally R, Dekker P, Piper J (2006) A novel luminescence analyser for europium chelates using solid-state excitation and a gated photomultiplier. Proc SPIE 6371:U133–U142
143. Lövgren T, Pettersson K (1990) Time-resolved fluoroimmunoassay, advantages and limitations. In: Van Dyke K, Van Dyke R (eds) Luminescence immunoassay and molecular applications. CRC, Boca Raton, FL
144. Connally RE, Piper JA (2008) Time-gated luminescence microscopy. Ann N Y Acad Sci 1130:106–116
145. Seveus L, Vaisala M, Syrjanen S et al (1992) Time-resolved fluorescence imaging of europium chelate label in immunohistochemistry and in situ hybridization. Cytometry 13: 329–338
146. Smith MC, Steimle G, Ivanov S et al (2007) An integrated portable hand-held analyser for real-time isothermal nucleic acid amplification. Anal Chim Acta 598:286–294
147. Werts MHV, Duin MA, Hofstraat JW et al (1999) Bathochromicity of Michler's ketone upon coordination with lanthanide(III) beta-diketonates enable efficient sensitisation of Eu^{3+} for luminescence under visible light excitation. Chem Commun 1999:799–800
148. Wu J, Wang GL, Jin DY, et al (2008) Luminescent europium nanoparticles with a wide excitation range from UV to visible light for biolabeling and time-gated luminescence bioimaging. Chem Commun 365–367
149. Laitala V, Hemmila I (2005) Homogeneous assay based on low quantum yield Sm(III)-donor and anti-Stokes' shift time-resolved fluorescence resonance energy-transfer measurement. Anal Chim Acta 551:73–78
150. Wang H-Q, Thomas N (2010) Upconverting nanoparticles. Springer Ser Fluoresc doi: 10.1007/4243_2010_8

Upconverting Nanoparticles

Hai-Qiao Wang and Thomas Nann

Abstract Upconverting nanoparticles have attracted a large amount of interest, primarily due to their attractive potential applications in bioimaging and therapy. This chapter reviews different methods of synthesis for the preparation of upconverting nanoparticles, covering the theory of nucleation and growth, as well as a wide range of methods, for example, arrested precipitation, hot injection, microemulsion and microwave-assisted techniques. Finally, some examples for bio/medical applications of upconverting nanoparticles are discussed.

Keywords Bioimaging · Colloids · Photodynamic therapy · Synthesis methods · Upconverting nanoparticles

Contents

H.-Q. Wang
Institute Materials for Electronics and Energy Technology (i-MEET), Friedrich-Alexander-Universität Erlangen-Nürnberg, Martensstrasse 7, D-91058 Erlangen, Germany

T. Nann (✉)
Ian Wark Research Institute, University of South Australia, Mawson Lakes Blvd, Mawson Lakes, SA 5095, Australia
e-mail: thomas.nann@unisa.edu.au

P. Hänninen and H. Härmä (eds.), *Lanthanide Luminescence: Photophysical, Analytical and Biological Aspects*, Springer Ser Fluoresc (2011) 7: 115–132, DOI 10.1007/4243_2010_8,

Photoluminescence upconversion can be observed when a luminophor is excited with light of lower energy than the light emitted (anti-Stokes emission). The difference between multi-photon excitation and upconversion luminescence is discussed below. Multi-photon excitation requires excitation pulses with femtosecond frequency, whereas the strength of the absorption depends on the square of the light intensity. In contrast to that, typical upconversion mechanisms involve the excitation of an ion in the ground state by one photon and subsequent energy transfer to another ion in the ground state under absorption of one or more photons and de-excitation of the first ion. Upconversion luminescence can be excited by continuous wave irradiation. To date, upconversion luminescence has not been observed in organic fluorophores. Apart from very few dubious reports, all papers on upconversion photoluminescence deal with transition metal or rare-earth-doped systems.

The effect of upconversion luminescence was first discovered in the late 1950s and studied intensively during the 1960s. The history and physical fundamentals of upconversion phosphors have been reviewed by Auzel in 2004 [1]. Until 2003, upconversion luminescence was observed exclusively in bulk, transition metal or rare-earth-doped phosphors. Applications for these phosphors were suggested in the fields of displays [2], detectors [3, 4] and laser materials [5]. Güdel et al. [6] were the first to publish the synthesis and characterisation of upconverting nanoparticles in 2003. Nanoparticles with diameters in the lower nanometre size regime can be synthesised under mild conditions, form transparent colloidal dispersions and behave to some degree like molecules. Thus, Güdel's work has paved the way for many interesting applications, which could not be realised – due to the lack of upconverting molecules – so far. Bioimaging and medical applications are of particular interest in this field and will be discussed later.

This chapter focuses on the synthesis, characterisation, derivatisation and application of rare-earth-doped, upconverting nanoparticles. Principles and applications of bulk upconverting phosphors have been reviewed recently [7]. The same applies to physical and theoretical details [1]. Different synthesis methods for rare-earth-doped, upconverting nanoparticles will be introduced (Sect. 2), and their effect on the properties of the resulting nanoparticles or nanocrystalline material will be discussed. As nanoparticles need to be derivatised to be used in almost any application, a special emphasis will be given to the derivatisation and phase-transfer of upconverting nanoparticles (Sect. 3). Finally, some potential applications of these nanocrystals will be discussed (Sect. 4).

1 Types of Upconverting Nanoparticles

Upconverting nanoparticles have been synthesised from basically all upconverting materials known to date. This includes oxides, phosphates and vanadates, as well as yttriumfluoride-based and various mixed systems. The exact state and properties of the resulting nanoparticles or nanocrystalline material depend to a large degree on the method of their synthesis (cf. Sect. 2). However, as upconverting (nano)crystals differ mainly in their host lattice, they can be grouped on this basis. Table 1 shows an overview of upconverting nanoparticles and nanocrystalline material prepared so far. It has to be noted that only colloidal nanoparticles can be used for applications such as bio-labelling.

2 Synthesis Methods

Several methods have been used to synthesise upconverting nanoparticles. The choice of method depends to a large degree on the material of the nanoparticles to be synthesised. Furthermore, the properties of the resulting nanoparticles or

Table 1 Upconverting nanoparticles, synthesis methods and ligands

Material	Synthesis method	Ligands/state	References
$CaS:Sm^{3+}$	Sol-gel	–/Nanocrystalline	[8]
Y_2O_2S:Yb,Er (or Ho, Tm)	Solid–gas reaction	2-Methyl-2-ethylheptanoic acid /nanocrystalline	[9]
$Lu_2O_3:Er^{3+}$	Combustion	–/Nanocrystalline	[10]
$Y_2O_3:Nd^{3+}$	Combustion	–/Nanocrystalline	[11]
$Y_2O_3:Er^{3+}$	Vapour phase		[10, 12]
$Y_2O_3:Yb^{3+},Er^{3+}$	Arrested precipitation		[13]
(or Ho^{3+}, Tm^{3+})			[14]
$Y_2O_3:Li^{+},Er^{3+}$			[15]
$Y_2O_3:Yb^{3+},Er^{3+},Li^{+}$			[16]
$ZrO_2:Er^{3+}$	Sol-gel	–/Nanocrystalline	[17]
$ZrO_2:Yb^{3+},Er^{3+}$	Combustion and sol-gel	–/Nanocrystalline	[18]
$Gd_2O_3:Tm^{3+}$ (or Er^{3+} or Yb^{3+})	Sol-gel	–/Nanocrystalline	[19]
$Gd_3Ga_5O_{12}:Er^{3+}$	Combustion	–/Nanocrystalline	[20]
$Gd_3Ga_5O_{12}:Pr^{3+}$			[21]
$LuPO_4:Tm^{3+}$	Arrested precipitation	Dodecylamine/colloidal	[6]
$LuPO_4:Yb^{3+},Tm^{3+}$			
$YbPO_4:Er^{3+}$			
$LaPO_4$:Yb,Er	Arrested precipitation	CTAB/nanocrystalline	[22]
$LaPO_4$:Yb@Er (core–shell)	Solvothermal		
$NaYF_4:Yb^{3+},Er^{3+}$ or Yb^{3+},Tm^{3+}	Arrested precipitation	EDTA/colloidal	[23, 24]
$NaREF_4$ (RE: Pr to Lu, Y)	Thermal decomposition	Oleic acid/colloidal	[25]
$NaYF_4:Yb^{3+}$, Er^{3+}	Thermal decomposition	Oleic acid/colloidal	[26–28]
or Yb^{3+}/Tm^{3+}	Microwave synthesis		[29]
$NaGdF_4:Yb^{3+}/Er^{3+}$	Arrested precipitation	–/Nanocrystalline	[30]

nanocrystalline material depend on the synthesis method as well. Some methods result in nanocrystals that do not bear any ligands and are thus not stabilised in dispersion (cf. Table 1). These nanocrystals are generally difficult to be used in applications where aggregation is a problem (e.g., bio-labelling). Some methods can be used to synthesise stabilised nanoparticles, but have limitations regarding the material.

Nanocrystal nucleation and growth follow the basic principles of crystallisation: a nucleation event precedes the growth of nanocrystals and eventually bulk crystals. This mechanism was studied by Victor LaMer and is therefore referred to as *LaMer mechanism* frequently [31, 32]. Scheme 1 depicts a LaMer plot which shows the concentration of dissolved reactants as nucleation and growth proceeds. The reaction can be divided into three phases: first, the concentration of reactant gradually increases and eventually exceeds solubility (Scheme 1, phase I). Second, the concentration of reactants reaches the critical limit of supersaturation and rapid nucleation occurs. This nucleation burst results in a sudden decrease in reactant concentration (Scheme 1, phase II). Finally, nuclei grow slowly as the reaction solution depletes in reactants (Scheme 1, phase III). As growth is usually thermodynamically favoured over nucleation, nanoparticles can be grown monodispersely when phase II can be limited to a short period of time by suitable choice of reactant concentration and temperature. The LaMer mechanism follows general principles and applies to the vast majority of nanocrystal and -particle synthesis methods.

Although the LaMer mechanism describes the different stages of nanocrystal nucleation and growth, it does not account for the kinetics of these processes. The kinetics of nanocrystal nucleation can be easily derived on the assumption that nuclei are spherical and grow isotropically. Then, the magnitude of surface free energy required to create a nanoparticle of radius r is:

$$\Delta G_S = 4\pi r^2 \gamma \tag{1}$$

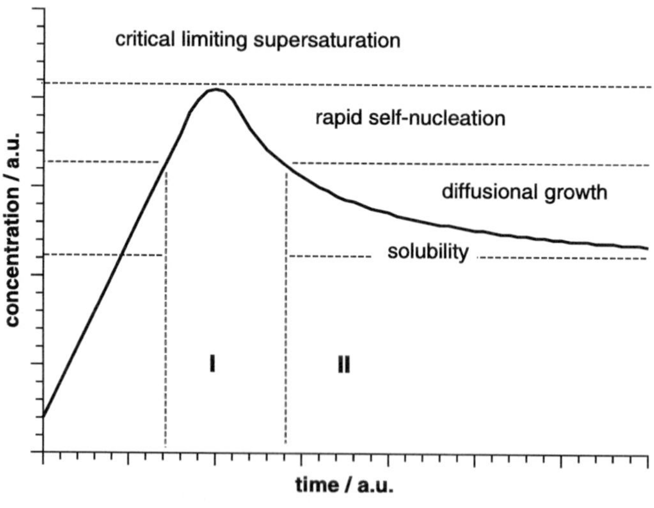

Scheme 1 LaMer plot

with $4\pi r^2$ as the nanoparticles' surface and γ as surface tension. ΔG_S is a positive quantity and therefore favours the back-reaction (viz. dissolution of the nanoparticle). The magnitude of the volume free energy for the same process can be written as:

$$\Delta G_V = \frac{4}{3}\pi r^3 \Delta G_v \tag{2}$$

where $4/3\pi r^3$ is the volume of the particle and ΔG_V the free energy change of the crystallisation per unit volume (typically the lattice energy). ΔG_V is a negative value and thus advances spontaneous nucleation. The overall free energy change of formation of a nanocrystal is the sum of ΔG_S and ΔG_V. Scheme 2 shows Gibbs free energies of a nanoparticle at different radii.

The magnitude of the overall free energy change of nucleation is positive for small nuclei, reaches a maximum when nuclei have a critical radius r_c and becomes negative for larger particles. Therefore, nuclei with radii smaller than r_c are likely to dissolve, whereas nuclei with larger radii r_c start to grow. The magnitude of the critical radius can be calculated by differentiation of ΔG with respect to r and subsequent setting to zero:

$$\frac{\mathrm{d}\Delta G}{\mathrm{d}r} = 8\pi r_c \gamma + 4\pi r_c^2 \Delta G_v = 0 \tag{3}$$

Rearrangement results in

$$r_c = -\frac{2\gamma}{\Delta G_v} \tag{4}$$

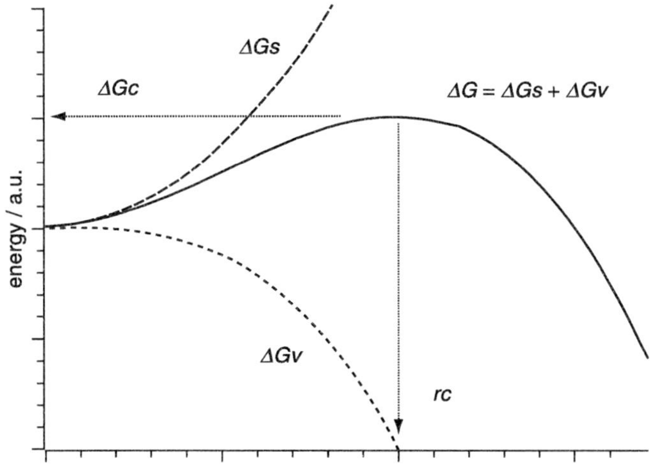

Scheme 2 Gibbs free energy change of the formation of differently sized nanocrystals (nucleation)

and insertion in ΔG gives the critical free energy change of nucleation:

$$\Delta G_c = \frac{16\pi\gamma^3}{3\Delta G_v^2} = \frac{4}{3}\pi\gamma r_c^2 \tag{5}$$

The critical radius r_c in (5) can be replaced using the Kelvin equation:

$$\ln S = \frac{2\gamma V_m}{r RT} \tag{6}$$

with S as supersaturation of reactants and V_m as molecular volume. Insertion of (6) into (5) results in:

$$\Delta G_c = \frac{16\pi\gamma^2 V_m^2}{3(RT \ln S)^2} \tag{7}$$

The critical free energy change can be seen as activation energy for the formation of stable nuclei. Applying this to the Arrhenius equation enables us to estimate the temperature dependent rate of nucleation:

$$\ln k_n = \ln A - \frac{\Delta G_c}{RT} = \ln A - \frac{16\pi\gamma^3 V_m^2}{3(RT)^3 (\ln S)^2} \tag{8}$$

In (8), all quantities are constant for a given synthesis, except for the temperature T and the supersaturation S. It can be seen that the rate of nucleation increases on increasing temperature and supersaturation. Reactants deplete (usually) when nuclei are formed, and thus the reaction rate decreases. Monodisperse nanoparticles can be synthesised by choosing the reaction temperature so that it is just above the nucleation temperature at moderate supersaturation. Then, a so-called nucleation burst occurs (rapid nucleation followed by a sudden drop in nucleation rate), followed by slow growth of the nuclei. This generic mechanism is applicable to the majority of synthesis methods.

2.1 *Arrested Precipitation*

Arrested precipitation denotes a technique where a poorly soluble product is precipitated within a template by mixing solutions of the respective ions. The template might be a microemulsion, surface ligand solution, mesoporous material (e.g., anodised alumina), polymer or dendrimer, or any other system that provides a confined space for the precipitation. Historically, arrested precipitation was the first method used to synthesise semiconductor nanoparticles that were used to study quantum size effects systematically [33] (and thus paved the way for the whole field of nanosciences).

Arrested precipitation can be performed in many different ways: most simply, two solutions containing ions of a poorly soluble salt are mixed within a template. Alternatively, ions (or reactants) can be released slowly as described in stage I of the LaMer mechanism. Subsequently, nucleation and growth take place within the template. Furthermore, arrested precipitation can be combined with other methods of synthesis such as the sol-gel method. In that case, a gel is "precipitated" within the template and condensed in a second step. All of these variations have been used to synthesise upconverting nanoparticles.

Güdel et al. have mixed lanthanide chlorides and phosphoric acid in a high-boiling mixture of surface ligands [6]. Upconverting nanoparticles were obtained after heating the mixture. The nanocrystals were highly crystalline and colloidally stable. The same group did use a variation in their former protocol to synthesise $NaYF_4$ and $NaGdF_4$-based upconverting nanocrystals as well [23, 30]. Again, the nanoparticles were highly crystalline and luminescent. Figure 1 shows a typical example of a transmission electron micrograph (TEM) of upconverting nanocrystals obtained by arrested precipitation. The spectroscopic properties of these nanoparticles did not differ significantly from those of the corresponding bulk crystals. However, it seems to be imperative to precipitate the nanocrystals at high temperature to observer luminescence. Guo et al. have synthesised $NaYF_4$-based nanocrystals by arrested precipitation at low temperatures and could only observe luminescence after annealing of the particles at 400–600°C [24].

The groups of Song and Zhang et al. have used a combination of a sol-gel method and arrested precipitation to prepare nanocrystalline Y_2O_3-based upconverters [15, 16]. They first prepared a gel from yttrium and lanthanide nitrates and ammonia hydroxide that was "arrested" by citrate ligands. Then, they calcinated the dried gel at 800°C and obtained upconverting nanocrystalline material. It has to be stated that methods that involve annealing/calcination always result in nanocrystalline material rather than colloidally stable nanoparticles.

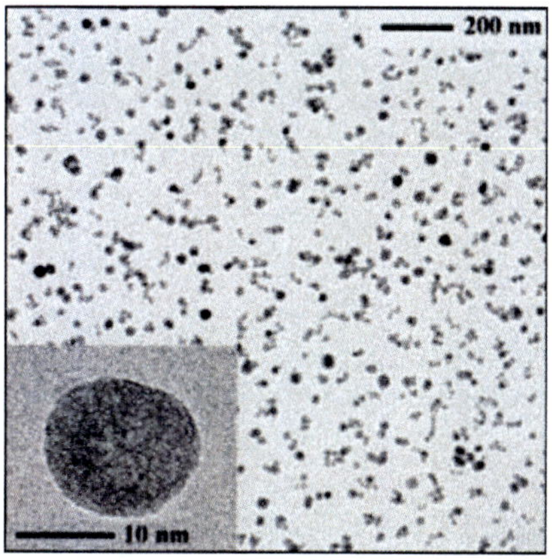

Fig. 1 Transmission electron micrograph of $NaYF_4$:20% Yb,2%Ernanocrystals. Copyright Wiley-VCH Verlag GmbH & Co. KGaA. Reproduced with permission from [23]

2.2 Combustion Synthesis

The idea of the combustion synthesis method is based on the assumption that nanoparticles can be formed by a very rapid heat "pulse" (the combustion or explosion). A typical reaction mixture consists of metal nitrates and a propellant. On combustion, the mixture yields the desired products and volatile by-products. The reaction scheme for such a reaction might be:

$$\begin{aligned} &6M(NO_3)_3 + 10NH_2CH_2COOH + 18O_2 \rightarrow \\ &\quad 3M_2O_3 + 5N_2 + 18NO_2 + 20CO_2 + 25H_2O \end{aligned} \tag{9}$$

whereas glycine serves as propellant for the reaction. Typical products of combustion syntheses are nanocrystalline metal oxide powders. To date, there is no combustion synthesis method that leads to colloidal nanoparticles, as ligands would be burnt during combustion.

Y_2O_3 has been the most popular host material for upconverting nanocrystals prepared by combustion synthesis [11–13, 34]. Capobianco et al. have been the first to publish nanocrystalline, upconverting Y_2O_3:Er^{3+} in 2000 [12]. They studied the crystallographic and spectroscopic properties of their material and found that the nanocrystalline state allows for the adsorption of CO_2, which introduced additional relaxation pathways to the system. Later, the same group intensified its studies on the Y_2O_3:Yb^{3+}, Er^{3+} system with emphasis on the spectroscopic properties [13, 34]. A very similar study was performed on nanocrystalline Lu_2O_3:Er^{3+} [10]. Some minor differences in the spectroscopic properties were found as compared with Y_2O_3:Er^{3+}, which were attributed to the larger nanocrystal size in case of Lu_2O_3:Er^{3+}.

Finally, the same group of researchers has studied several nanocrystalline upconverters on the basis of a $Gd_3Ga_5O_{12}$ host lattice [20, 21, 35]. The focus of their studies was on the spectroscopic properties of the nanomaterial again, and they found some minor differences as compared with bulk crystals.

2.3 Sol-Gel Methods

Sol-gel methods involve typically three steps: first, dissolution of metal salts in a mostly aqueous solvent (formation of the sol), second, formation of a gel, and third, condensation of the gel. The products of sol-gel methods resemble those of combustion syntheses as the final step involves typically sintering at high temperatures. Therefore, the same limitations regarding the application of nanocrystalline upconverting nanoparticles apply for both methods.

Nanocrystalline upconverting material on the basis of CaS [8], ZrO_2 [17, 18] and Gd_2O_3 [19] has been prepared by sol-gel synthesis. The focus of the researchers was on the spectroscopic properties of this material again, and it was found that only slight differences as compared with bulk crystals could be found.

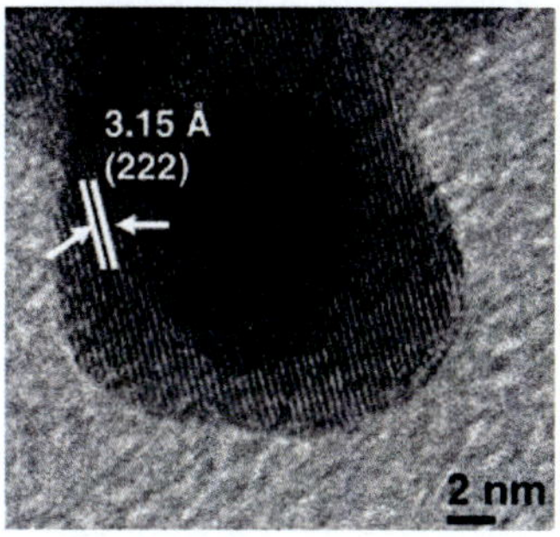

Fig. 2 High-resolution transmission electron micrograph (HRTEM) of $LaPO_4$:Yb,Er core/shell nanocrystals. Copyright ACS. Reproduced with permission from [22]

2.4 Solvothermal Methods

A solvothermal synthesis method involves reaction between reactants in a subcritical or supercritical solvents. Solvothermal syntheses are usually performed within an autoclave at high temperature and pressure.

We are aware of only one publication where a solvothermal synthesis method has been used for the preparation of upconverting $LaPO_4$:Yb,Er nanorods [22]. This publication is outstanding because beyond being unique in terms of the synthesis method, it is the only one showing a preparation method for upconverting nanorods and core/shell nanoparticles.

Figure 2 shows a high-resolution transmission electron micrograph (HRTEM) of core/shell $LaPO_4$:Yb,Er nanocrystals. The cores of these particles have been doped with 1% of Er^{3+} and the shells with approximately 1% of Yb^{3+}. The particles did crystallise in a monoclinic monazite structure. Core and shell can be clearly distinguished in the HRTEM. It was observed that the upconversion luminescence intensity of core/shell nanoparticles is larger as compared with homogeneous ones.

Another interesting aspect of the solvothermal synthesis of $LaPO_4$:Yb,Er nanocrystals includes the fact that this method allows for the synthesis of anisotropic particles. Ghosh et al. trace the anisotropic growth back to a diffusion model, whereas the surfactant cetyltriammonium bromide (CTAB) blocks certain crystallographic faces of the nuclei selectively [22]. It has to be mentioned that Ghosh et al. have calcined their particles at 900°C, which removed all ligands from the particles' surfaces. However, colloidal nanoparticles can be prepared by solvothermal synthesis methods as well.

2.5 Thermal Decomposition/Hot Injection

The thermal decomposition of organometallic precursors is probably the most important method for the synthesis of colloidal, inorganic nanocrystals in general (this method is sometimes called "hot injection method"). Typically, a solution of surface ligands (surfactants) is heated in a high-boiling solvent. One or more metal salts may be already present. When the temperature of the solution has reached the decomposition temperature of the organometallic precursor, the latter is swiftly injected. Reactants are suddenly released on decomposition of the precursor. The injection causes a nucleation burst and a sudden drop in temperature, because the precursor solution is usually colder than the ligand solution. The reaction profile follows roughly the LaMer scheme as displayed in Scheme 1. The major difference is that the scheme starts at the maximum of phase II when using the hot injection method.

To date, primarily alkali lanthanide tetrafluorides such as $NaYF_4$ and related crystals have been prepared by thermal decomposition. The majority of publications in this field use a method first published by Mai et al [25]. These researchers used rare-earth trifluoroacetates as reactants. The trifluoroacetates decomposit in the heat under release of fluoride and volatile by-products. Mai et al. have focussed on the synthesis of differently sized and shaped nanocrystals but did not show any upconversion data. The first upconverting $NaYF_4$-based nanocrystals synthesised using the Mai method have been published by Boyer et al. [27, 28]. The method has been modified by Ehlert et al. to realise four different, spectroscopically distinguishable upconverting nanocrystal species [26]. Figure 3 shows the upconversion luminescence of these nanocrystals on excitation with a 980 nm laser diode.

Nanocrystals synthesised by thermal decomposition are usually colloidally stable and can be manipulated on a single particle basis. Ehlert et al. have synthesised an additional silica shell on top of their nanocrystals and thus rendered them water-dispersible and bio-compatible. Figure 4 shows a TEM micrograph of these core/shell nanoparticles. The “frog-spawn” structure gives evidence that the upconverting nanoparticles were found in the exact centre of the silica shell. The optical properties of the core/shell nanoparticles did not change as compared with un-shellednanocrystals.

2.6 *Microwave-Assisted Synthesis*

Microwave irradiation (MWI) is an electromagnetic irradiation in the frequency range of 0.3–300 GHz. Magnetrons were developed by Randall and Booth at the

Fig. 3 Photographic picture of upconverting nanocrystals with four different *colours* (no filters or overexposure were used). Excitation: 980 nm laser diode. Reprinted with permission from [26]. Copyright 2008 American Chemical Society

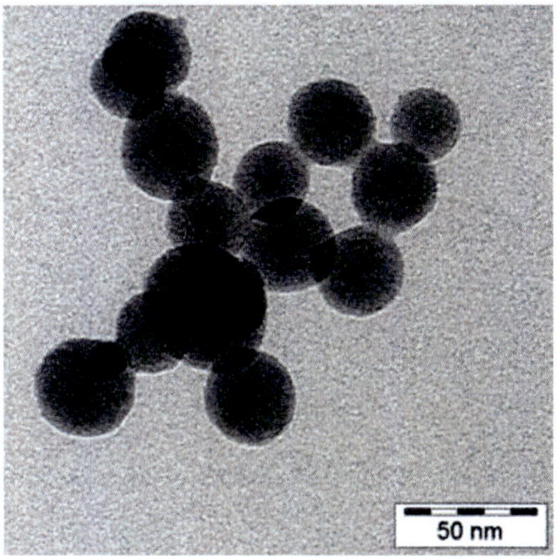

Fig. 4 TEM micrograph of $NaYF_4$:Yb,Er/silica core/shell nanoparticles. Reprinted with permission from [26]. Copyright 2008 American Chemical Society

University of Birmingham during Word War II [36]. Microwave heating is a well-established method not only for the domestic preparation of meals but also for the industrial heating of rubber, wood, paper and agricultural products. Even though MWI was first used for organic synthesis by Gedye et al. in 1986 [37, 38], the method became popular for chemical synthesis only in the past few years [39].

The effective heating of a substance by MWI depends on the ability of a material (solvent or reagent) to absorb microwave energy and convert it into heat. This is mainly based on two principal mechanisms: first, rotation of dipoles and, second, ionic conduction, i.e., by means of reversals of solvent dipoles and the resulting displacement of charged ions of a solute in the solvent [40]. The exertion of MWI on a substance results in alignment of the dipoles or ions with the applied electric field. As the field oscillates, the dipoles or ions attempt to realign themselves with the alternating electric field, and energy is converted into heat through molecular friction and dielectric loss. The amount of heat generated by this process is directly related to the ability of the matrix to align itself with the frequency of the applied field [39].

Microwave-assisted synthesis has some advantages as compared with conventional (conductive) heating: reaction rates are accelerated, milder reaction conditions can be used, higher chemical yield, lower energy consumption, different reaction selectivities, high reproducibility and many others. The first application of microwave-assisted synthesis for the preparation of nanoparticles has been published by Komarneni et al. in 2002 [41]. Other articles by various groups on the microwave-assisted synthesis of different nanocrystals followed rapidly (for example [42–44]). It is difficult to use a hot injection method in combination with microwave-assisted synthesis for practical reasons. Therefore, nucleation and growth of nanocrystals follow strictly the LaMer mechanism as discussed above: polar precursors decomposit under release of reactants (phase I). Then, a nucleation burst occurs (phase II) and eventually nanocrystals start to grow (phase III).

So far, only one publication from our own group deals with the microwave-assisted synthesis of upconverting nanocrystals [29]. It was found that alkali yttrium tetrafluoride nanocrystals can be synthesised extremely monodispersely within a short period of time. Figure 5 shows a TEM micrograph of $NaYF_4$:Yb^{3+}/Er^{3+} nanocrystals that were prepared by microwave-assisted synthesis.

The microwave-assisted synthesis method allows for the preparation of differently shaped nanocrystals as well. Figure 6 shows a TEM micrograph of cube-shaped $NaYF_4$:Yb^{3+}/Tm^{3+} nanocrystals.

Finally, it has been shown that the luminescence intensity of upconverting nanocrystals can be increased significantly using a mixed $NaYF_4$, $LiYF_4$ host lattice.

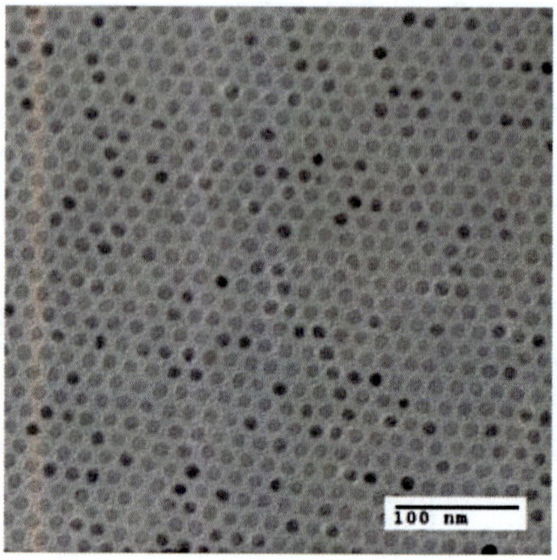

Fig. 5 TEM micrograph of $NaY_{0.78}F_4$:$Yb_{0.2}{}^{3+}$, $Er_{0.02}{}^{3+}$ nanocrystals. Average diameter of the nanoparticles was about 10 nm

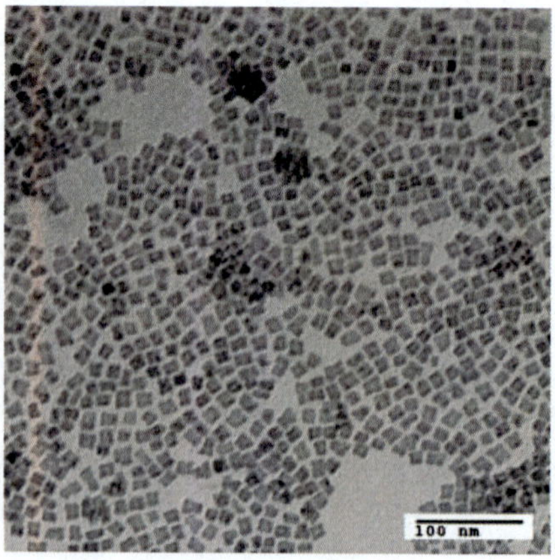

Fig. 6 TEM micrograph of cube-shaped $NaY_{0.78}F_4$: $Yb_{0.2}{}^{3+}$, $Tm_{0.02}{}^{3+}$ nanocrystals. Cube size was approximately 10 × 14 nm

3 Applications

To date, the vast majority of applications using upconverting nanoparticles or nanocrystalline material are to be found in the field of bioanalytics or medical diagnostics. This is not surprising because synthesis methods of upconverting nanocrystals are still in their infancy, and the advantages of using these nanocrystals are most obvious in the life sciences.

3.1 Bioanalytical and Medical Applications

Using upconverting nanoparticles for bio-labelling has some distinct advantages: infrared light causes minimal damage to tissue, it penetrates deep, it does not excite any biological fluorophores (practically no background luminescence), high signal to noise ratio and synthesis of these particles can be made very inexpensive. Bio-labelling was probably one of the major motivators for synthesising upconverting nanocrystals in the past decade.

In 1999, Zijlmans and co-workers first reported on the detection of cell and tissue surface antigens using upconverting crystals (diameter approximately 0.2–0.4 μm) as luminescent reporters [45]. Even though this report did show the potential of using upconverters for bio-labelling, the particles were too big for most applications in the field. In recent years, as the synthesis of high quality (small and highly luminescent) upconverting nanocrystals and their surface modification improved, more achievements have been accomplished in the field of life sciences, including in vivo imaging, in vitro detection, and molecular sensing based on fluorescence resonance energy transfer (FRET).

In 2005, Li et al. developed a FRET system with biotinylated $NaYF_4$:Yb/Er upconverting nanoparticles (diameter ca. 50 nm) as energy donor and biotinylated gold nanoparticles as energy acceptors [46]. They applied it to detect trace amounts of avidin, and a detection limit of 0.5 nM was achieved.

Zhang et al. designed a highly sensitive and specific nucleotide sensor and quantitatively detected the perfectly matched target DNA at a detection limit of 1.3 nM in 2006 [47] (results shown in Fig. 7). They did not provide any detailed information about the upconverting nanoparticles they used; other than that they were based on Er^{3+}-doped $NaYF_4$ upconverting particles as donor and carboxytetramethyl rhodamine (TAMRA) as acceptor.

In the same year, Lim et al. reported on the in vivo and scanning electron microscopy imaging of upconversion nanophosphors (50–200 nm) in *Caenorhabditis elegans* [48]. The viability of the nanoparticles for biological imaging was confirmed by imaging the digestive system of the nematode worm (cf. Fig. 8). The upconversion nanoparticles have been identified by a scanning electron microscopy at high spatial resolution and energy-dispersive X-ray analysis.

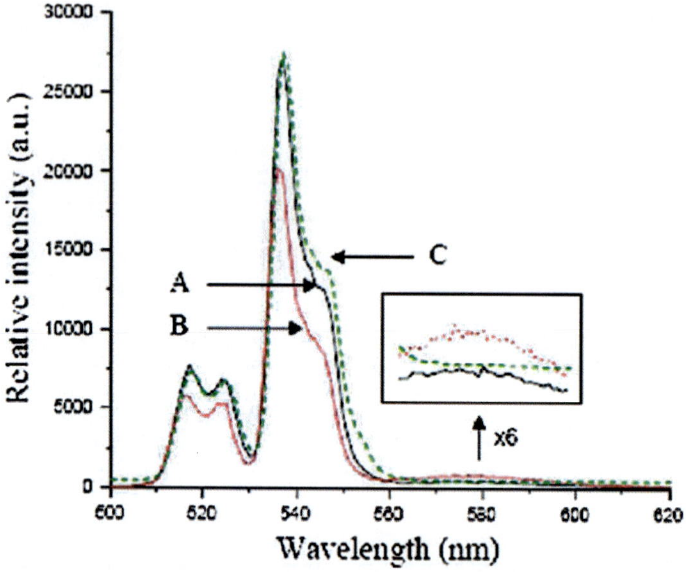

Fig. 7 Photoluminescence spectra of a mixture of DNA1-photon upconverting particles and DNA2-TAMRA at 975 nm excitation: (**A**) before addition of DNA; (**B**) after addition of target DNA; (**C**) after addition of mismatched DNA. Reprinted with permission from [47]. Copyright 2006 American Chemical Society

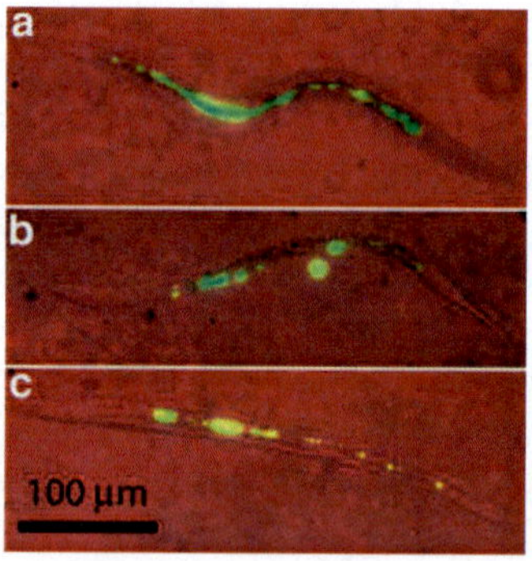

Fig. 8 False *colour* two-photon images of *Caenorhabditis elegans* at 980 nm excitation with red representing the bright field and green for the phosphor emission. Reprinted with permission from [48]. Copyright 2006 American Chemical Society

Again in 2006, Abrams et al. developed a microfluidic device for detection of pathogens in oral samples by combining a microfluidic system and upconverting phosphor particles (diameter ca. 400 nm, excited with 980 nm IR light) [49]. The utilisation of upconverting phosphor technology for a reporter provided an enhancement in sensitivity in the range of three orders of magnitude as compared with conventional ethidium bromide-stained agarose gels.

All of the publications discussed above did use upconverting (nano)particles of sizes far beyond the mesoscopic scale. Li et al. published the first application of upconverting nanocrystals for bioanalytics where the nanocrystals were fairly uniform in size and had diameters below 50 nm [50]. The group combined magnetite nanoparticles with upconverting nanocrystals ($NaYF_4$:Yb^{3+}/Er^{3+}) for the sensitive detection of DNA by a magnetic field-assisted bioseparation and concentration technology and obtained some preliminary positive results.

In 2008, Rantanen et al. have published a sensitive enzyme activity assay where they used upconverting donors (average size 340 nm) and a Black-Berry 650 quencher [51]. Again, the nanoparticles used were comparatively large.

Xu et al. reported on the autofluorescence insensitive imaging in a tissue phantom using $NaYF_4$:Yb^{3+}/Tm^{3+} upconverting nanocrystals illuminated with near-infrared radiation in 2008 [52]. An exceptionally high contrast was obtained compared to the result of downconverting fluorophores.

Eventually, Zhang et al. used upconverting PEI/$NaYF_4$:Yb^{3+}, Er^{3+} nanoparticles of about 50 nm diameter for bioimaging [53] and photodynamic therapy (PDT) [54].

Figure 9 depicts an example of bioimaging using upconverting nanoparticles of HT29 cells and a live rat. The comparison with quantum dots shows clearly the advantages of using upconverting nanocrystals for this purpose. However, these

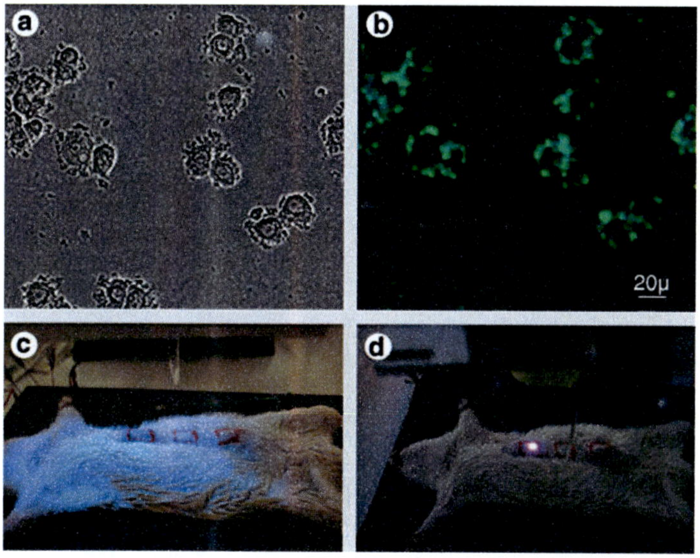

Fig. 9 Bright field (**a**) and confocal-fluorescence (**b**) images of human colonic adenocarcinoma cells (HT29) incubated with folic acid functionalised PEI/$NaYF_4$:Yb,Er nanoparticles. Intradermal injection of PEI/$NaYF_4$:Yb,Er nanoparticles (to an approximate depth of 5 mm) on shaved rat abdominal skin (**d**) shows fluorescence on excitation with near-infrared laser. Similar injection of quantum dots (**c**) does not show fluorescence on excitation with UV. Reprinted with permission from [54]. Copyright 2008

results have to be treated with care, as the bright "spot" in Fig. 9d might be also caused by scattered excitation light (typical CCD cameras are sensitive at 980 nm). Furthermore, the group showed that these particles can be used as nanotransducers for PDT in deep tissue. The results showed that the upconverting nanoparticles could be excited even after deep intramuscular injection into rats and suggested the potential clinical use of upconverting nanoparticles in imaging and PDT of cancer in deep tissues.

It can be concluded that upconverting nanoparticles show significant advantages when used in the field of life sciences. However, only very preliminary work has been done so far. It can be expected that this technology will advance rapidly as high quality upconverting nanoparticles as described, for example, in Sect. 2.6 will be exploited for bioanalytics and medical diagnostics and therapy.

3.2 Non-Life Sciences Applications

Recently, Li et al. have reported on a light-switchable optical system [55]. They have coated colloidal upconverting LaF_3:Yb,Ho nanocrystals with photochromic diarylethene derivatives. The luminescence of the upconverting nanocrystals is quenched when the diarylethene has been photochemically cyclicised. On photochemical ring-opening, the quanching is waivered and the nanocrystal luminescence restored.

In 2008, Laine et al. reported on a method of processing selected rare-earth ions-doped Y_2O_3 upconverting nanophosphors to produce transparent polycrystalline upconverting nanoceramics, and discussed its potential application in 3D displays [56]. Figure 10 shows an example of an Er-doped sample and the corresponding red upconversion luminescence.

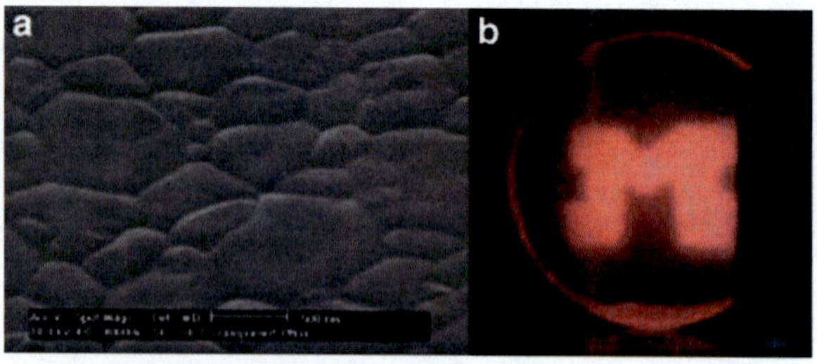

Fig. 10 (a) Microstructure of polished and thermally etched surface of HIPped Er-doped material. (b) Red emission (662 nm) on exposure to 980 nm light. Copyright Wiley-VCH Verlag GmbH & Co. KGaA. Reproduced with permission from [56]

4 Conclusions

The synthesis and exploitation of upconverting nanocrystals are still in their infancy. However, significant progress has been made during the past several years in both synthesis of highly monodisperse and colloical nanoparticles, and in – primarily bioanalytical – applications of upconverters. It can be expected that considerable progress will be made shortly and that many new fields of exploitation of upconverting nanoparticles will be seen in this rapidly emerging field.

References

1. Auzel F (2004) Chem Rev 104:139–173
2. Downing E, Hesselink L, Ralston J, Macfarlane R (1996) Science 273:1185–1189
3. Bloembergen N (1959) Phys Rev Lett 2:84–85
4. Breteau JM, Ayral JL, Micheron F, Auzel F (1990) J Appl Phys 67:1102–1107
5. Johnson LF, Guggenheim HJ (1971) Appl Phys Lett 19:44–47
6. Heer S, Lehmann O, Haase M, Güdel H (2003) Angew Chem Int Ed 42:3179–3182
7. Gamelin DR, Güdel HU (2001) Top Curr Chem 214:1–56
8. Gong X, Chen WJ, Wu PF, Chan WK (1998) Appl Phys Lett 73:2875–2877
9. Hirai T, Orikoshi T (2004) J Colloid Interface Sci 273:470–477
10. Capobianco JA, Vetrone F, Boyer JC, Speghini A, Bettinelli M (2002) Opt Mater 19:259–268
11. Tessari G, Bettinelli M, Speghini A, Ajo D, Pozza G, Depero LE, Allieri B, Sangaletti L (1999) Appl Surf Sci 144–145:686–689
12. Capobianco JA, Vetrone F, D'Alesio T, Tessari G, Speghini A, Bettinelli M (2000) Phys Chem Chem Phys 2:3203–3207
13. Vetrone F, Boyer JC, Capobianco JA, Speghini A, Bettinelli M (2003) J Phys Chem B 107:1107–1112
14. Glaspell G, Anderson J, Wilkins JR, El-Shall MS (2008) J Phys Chem C 112:11527–11531
15. Bai Y, Yang K, Wang Y, Zhang X, Song Y (2008) Opt Commun 281:2930–2932
16. Chen G, Liu H, Liang H, Somesfalean G, Zhang Z (2008) J Phys Chem C 112:12030–12036
17. Jia R, Yang W, Bai Y, Li T (2006) Opt Mater 28:246–249
18. Hyppänen I, Hölsä J, Kankare J, Lastusaari M, Pihlgren L (2007) J Nanomater 2007: Article ID 16391, 8 p
19. Guo H, Dong N, Yin M, Zhang W, Lou L, Xia S (2004) J Phys Chem B 108:19205–19209
20. Vetrone F, Boyer J, Capobianco JA, Speghini A, Bettinelli M (2003) J Phys Chem B 107:10747–10752
21. Naccache R, Betrone F, Boyer JC, Capobianco JA, Speghini A, Bettinelli M (2004) J Nanosci Nanotech 4:1025–1031
22. Ghosh P, Oliva J, Rosa EDL, Haldar KK, Solis D, Patra A (2008) J Phys Chem C 112: 9650–9658
23. Heer S, Kömpe K, Güdel H, Haase M (2004) Adv Mater 16:2102–2105
24. Yi G, Lu H, Zhao S, Ge Y, Yang W, Chen D, Guo L (2004) Nano Lett 4:2191–2196
25. Mai HX, Zhang YW, Si R, Yan ZG, Sun LD, You LP, Yan CH (2006) J Am Chem Soc 128:6426–6436
26. Ehlert O, Thomann R, Darbandi M, Nann T (2008) ACS Nano 2:120–124
27. Boyer JC, Vetrone F, Cuccia LA, Capobianco JA (2006) J Am Chem Soc 128:7444–7445
28. Boyer JC, Cuccia LA, Capobianco JA (2007) Nano Lett 7:847–852
29. Wang H, Nann T (2009) ACS Nano 3:3804–3808

30. Aebischer A, Heer S, Biner D, Krämer K, Haase M, Güdel HU (2005) Chem Phys Lett 407:124–128
31. LaMer VK, Dinegar RH (1950) J Am Chem Soc 72:4847–4854
32. Pound GM, LaMer VK (1952) J Am Chem Soc 74:2323–2332
33. Steigerwald ML, Brus LE (1989) Annu Rev Mater Sci 19:471–495
34. Vetrone F, Boyer JC, Capobianco JA, Speghini A, Bettinelli M (2004) J Appl Phys 96:661–667
35. Boyer JC, Vetrone F, Capobianco JA, Speghini A, Bettinelli M (2004) Chem Phys Lett 390:403–407
36. Michael D, Mingos P, Baghurst DR (1991) Chem Soc Rev 20:1–47
37. Gedye R, Smith F, Westaway K, Ali H, Baldisera L, Laberge L, Rousell J (1986) Tetrahedron Lett 27:279–282
38. Giguere RJ, Bray TL, Duncan SM, Majetich G (1986) Tetrahedron Lett 27:4945–4948
39. Kappe CO (2004) Angew Chem Int Ed 43:6250–6284
40. Kubrakova I, Toropchenova E (2008) Inorg Mater 44:1509–1519
41. Komarneni S, Li D, Newalkar B, Katsuki H, Bhalla AS (2002) Langmuir 18:5959–5962
42. Pang Q, Shi J, Liu Y, Xing D, Gong M, Xu N (2003) Mater Sci Eng B 103:57–61
43. Gerbec JA, Magana D, Washington A, Strouse GF (2005) J Am Chem Soc 127:15791–15800
44. Panda AB, Glaspell G, El-Shall MS (2007) J Phys Chem C 111:1861–1864
45. Zijlmans HJMAA, Bonnet J, Burton J, Kardos K, Vail T, Niedbala RS, Tanke HJ (1999) Anal Biochem 267:30–36
46. Wang L, Yan R, Huo Z, Wang L, Zeng J, Bao J, Wang X, Peng Q, Li Y (2005) Angew Chem Int Ed 44:6054–6057
47. Zhang P, Rogelj S, Nguyen K, Wheeler D (2006) J Am Chem Soc 128:12410–12411
48. Lim SF, Riehn R, Ryu WS, Khanarian N, Tung C, Tank D, Austin RH (2006) Nano Lett 6:169–174
49. Abrams WR, Barber CA, McCann K, Tong G, Chen Z, Mauk MG, Wang J, Volkov A, Bourdelle P, Cortjens PLAM, Zuiderwijk M, Kardos K, Li S, Tanke HJ, Niedbala RS, Malamud D, Bau H (2007) Ann NY Acad Sci 1098:375–388
50. Wang L, Li Y (2006) Chem Commun:2557
51. Rantanen T, Järvenpää M, Vuojola J, Kuningas K, Soukka T (2008) Angew Chem Int Ed 47:3811–3813
52. Xu CT, Svensson N, Axelsson J, Svenmarker P, Somesfalean G, Chen G, Liang H, Liu H, Zhang Z, Andersson-Engels S (2008) Appl Phys Lett 93:171103
53. Chatterjee DK, Rufaihah AJ, Zhang Y (2008) Biomaterials 29:937–943
54. Chatterjee DK, Yong Z (2008) Nanomedicine 3:73–82
55. Zhiguo Z (2008) Chem Commun 2008:4786–4788
56. Hinklin TR, Rand SC, Laine RM (2008) Adv Mater 20:1270–1273

Near-Infrared Luminescent Labels and Probes Based on Lanthanide Ions and Their Potential for Applications in Bioanalytical Detection and Imaging

Martinus H.V. Werts

Abstract The use of near-infrared (NIR) light in bioanalytical detection and biological imaging presents certain advantages over UV–visible light in terms of penetration depth, reduction of background signals and decreased phototoxicity. Under suitable conditions, complexes of ytterbium(III)-, neodymium(III)- and erbium (III)-containing organic chromophores may display relatively bright NIR luminescence upon excitation with visible light. This contrasts with the more commonly studied visibly luminescent europium(III) and terbium(III) complexes, which generally need UV or blue light for excitation. We discuss the current knowledge on NIR luminescence of lanthanide complexes (including holmium(III), praseodymium(III) and thulium(III)) in solution. It is suggested that among the NIR luminescent lanthanide complexes, ytterbium(III) complexes are likely to be the most promising candidates for biophotonic applications. The study of NIR luminescence and its application in bioanalytical detection are enabled by progress in detector and light source technology, which are also addressed. Recent developments in lanthanide-based NIR luminescent materials, including complexes and doped nanoparticles, are discussed in the light of their potential for bioanalytical application.

Keywords Biophotonics · Lanthanide complexes · Near-infrared luminescence

Contents

M.H.V. Werts
Centre National de la Recherche Scientifique, Ecole Normale Supérieure de Cachan/Bretagne, SATIE (CNRS UMR8029), Campus de Ker Lann, 35170 Bruz, France
e-mail: martinus.werts@bretagne.ens-cachan.fr

P. Hänninen and H. Härmä (eds.), *Lanthanide Luminescence: Photophysical, Analytical and Biological Aspects*, Springer Ser Fluoresc (2011) 7: 133–160, DOI 10.1007/4243_2010_9,

Abbreviations

CCD	Charge-coupled device
DNA	Deoxyribonucleic acid
DTPA	Diethylenetriamine pentaacetic acid (or any of its conjugate acetate ions)
ECL	Electrochemiluminescence
EDTA	Ethylenediamine tetraacetic acid (or any of its conjugate acetate ions)
FRET	Förster resonance energy transfer, a form of excitation energy transfer (sometimes referred to with the incorrect term “fluorescence resonance energy transfer”)
NIR	Near-infrared
OD	Optical density
PMT	Photomultiplier tube
UV	Ultraviolet
Tris	Tris(hydroxymethyl)aminomethane

Symbols

ε	Molar absorption coefficient
$\varepsilon_{\mathrm{Ln}}$	Molar absorption coefficient of the lanthanide ion
$\varepsilon_{\mathrm{sens}}$	Molar absorption coefficient of the sensitiser
$\varepsilon\Phi$	Photoluminescence brightness (product of molar absorption coefficient and photoluminescence quantum yield)

η_{sens} Excitation energy transfer efficiency from the sensitiser to the lanthanide ion in a luminescent lanthanide complex
τ_{obs} Observed photoluminescence decay time
τ_{rad} Radiative (or "natural") lifetime of the luminescent transition
Φ_{tot} Overall photoluminescence quantum yield of a sensitiser-modified luminescent lanthanide complex (excitation via the sensitiser)
Φ_{Ln} Intrinsic photoluminescence quantum yield of the lanthanide ion in a luminescent lanthanide complex

The use of near-infrared (NIR) light in luminescent bioanalytical detection and biological imaging presents certain advantages over the use of ultraviolet (UV) and visible light. Biological media are often not completely limpid, and scattered excitation light may have a negative effect on the signal-to-background contrast in photoluminescence detection. The scattering of light emitted inside the sample leads to loss of signal and blurring of spatial information. Light scattering decreases rapidly with increasing wavelength (think of blue skies and red sunsets) [2]. Therefore, the use of longer wavelengths may increase the sensitivity of luminescence detection, provided that other parameters such as the brightness of the labels and the detector sensitivities do not change.

A second and probably a more important reason for the usefulness of NIR light in bioanalytics and bio-imaging is the prevalence in biological samples of endogenous chromophores and fluorophores with radiative transitions in the visible and ultraviolet light. The autofluorescence associated with the endogenous fluorophores can result in a strong background, which can be reduced by working at long excitation and detection wavelengths [3]. On the long wavelength side, the range of accessible excitation and emission wavelengths is limited by the onset of absorptions due to vibrational overtones of water and organic molecules. Depending on the sample thickness and composition, the NIR optical window may extend from 700 to over 1,200 nm (Fig. 1).

The signal for visibly luminescent Eu^{3+} and Tb^{3+} complexes can be separated from biological background fluorescence by time-delayed detection (see ['Imaging lanthanide luminescence by time-resolved microscopy', Tanke]), but such complexes generally need ultraviolet excitation, which calls for the use of relatively expensive and bulky excitation sources, and may require special quartz optics.

Thus, it is interesting to generate options for using longer excitation and detection wavelengths. The investigation of NIR luminescent lanthanide complexes is a relatively recent activity. This is a result from the lack of availability of suitable detectors and the lack of knowledge of the proper conditions (ligands, solvents and photosensitisers) for observing the luminescence of NIR luminescent lanthanide ions.

Here, we will give an overview on the work on NIR labels and probes based on lanthanide ions. The emphasis will be on the NIR luminescent ions (Yb^{3+}, Nd^{3+}, Er^{3+}, Pr^{3+}, Ho^{3+}, Tm^{3+}), which may be incorporated in complexes that can be excited by ultraviolet, visible or possibly NIR light. Alternatively, multiphoton excitation and lanthanide-based upconversion make possible the use of NIR

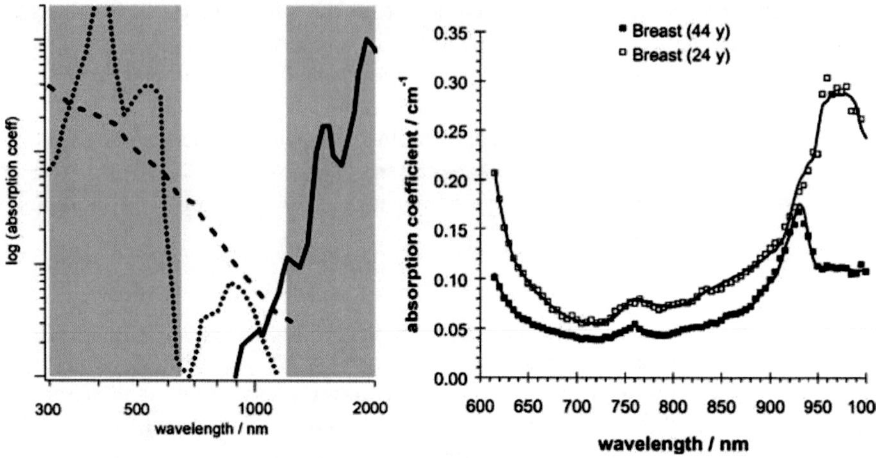

Fig. 1 *Left*: Absorption profiles (*logarithmic scale*) of water (*solid line*) and two typical pigments found in biological tissue (*dotted line*: oxyhaemoglobin, *dashed line*: melanin), demonstrating the "optical window" existing in the near-infrared (NIR). *Right*: Example of NIR transparency in biological tissue. Absorption spectra of the female breast on 44 year (*filled squares*) and 24 year (*open squares*) old volunteers. Reproduced from [1] with permission of The Royal Society of Chemistry for the European Society for Photobiology, the European Photochemistry Association and the RSC

excitation to generate visible lanthanide luminescence. This possibility will be addressed briefly. Attention will also be given to the recent development in lanthanide-doped inorganic nanoparticles.

1 Near-Infrared Luminescence of Lanthanide Ions in Aqueous and Organic Media

Most of the work on luminescent lanthanide complexes and the luminescence of lanthanide ions in organic, aqueous and liquid environments have been concerned with the visibly luminescent Eu^{3+} and Tb^{3+} ions. Two other visibly luminescent ions, Dy^{3+} and Sm^{3+} have also received attention, albeit to a lesser extent. This is likely related to their inferior luminescence quantum yields in these media as a result of the more efficient non-radiative relaxation of their excited states. In these two ions, smaller gaps exist between the lowest luminescent state and the lower lying energy levels, which lead to faster non-radiative relaxation (see ['Basics of Lanthanide Photophysics', Bünzli, Eliseeva]). Gadolinium(III) is known to possess very high luminescence quantum yields in solution, but it emits far into the UV (310 nm), which makes it less interesting for applications, either in biological detection or elsewhere.

The luminescence of the other lanthanide ions, ytterbium(III), neodymium(III), erbium(III), praseodymium(III), holmium(III), thulium(III), cerium(III) and promethium(III), in organic and aqueous media has long been considered to be too

Table 1 Approximate number of publications dealing with luminescent lanthanide complexes obtained searching for instance "ytterbium AND luminesc* AND complexes"

Ion	Number of publications
Yb^{3+}	168
Nd^{3+}	198
Er^{3+}	143
Ho^{3+}	14
Pr^{3+}	37
Tm^{3+}	11
Ce^{3+}	(44)[a]
Pm^{3+}	0

[a]None of these publications involve infrared cerium(III) *f–f* luminescence

weak to be studied conveniently. Nevertheless, this situation has changed recently, particularly as a result of the emergence of sensitive NIR detection systems and the interest for detection in biological media. Complexes of some of these lanthanides have been known for quite some time to exhibit NIR luminescence, e.g. Crosby and Kasha [4] described in 1958 the photosensitised emission of Yb^{3+} chelates, whereas solutions of neodymium(III) complexes were studied for laser action in the 1960s [5]. However, it is true that the weak emission of these complexes in a spectral region where sensitive detectors were lacking has impeded studies of NIR luminescent lanthanide ions.

A swift survey of the literature (Table 1) indicates that the lanthanide ions other than Eu, Tb, Dy and Sm may be classified in three groups, according to the approximate number of publications dealing with the NIR luminescence of their complexes. A correlation between this number and the luminescence quantum yields of these complexes may be anticipated. Especially the Yb–Nd–Er group has received considerable attention, in relation with their luminescence being conveniently detectable by current detection technologies. The Ho–Pr–Tm group, which – as we will see below – has very weak emissions, and studies of their luminescent complexes have begun recently.

As has been described in ['Basics of Lanthanide Photophysics', Bünzli, Eliseeva] and in ['Stable Luminescent Chelates and MacrocyclicCompounds', Mathis, Bazin], luminescent lanthanide complexes contain an "antenna" chromophore that serves to efficiently absorb excitation light and transfer this energy to the incorporated lanthanide ion that can then exhibit its photoluminescence. In this case, the overall luminescence quantum yield, Φ_{tot}, is given by the product of the photosensitisation efficiency η_{sens}, which gives the overall antenna-to-lanthanide energy transfer efficiency, and the intrinsic photoluminescence quantum yield of the lanthanide ion, Φ_{Ln} [6].

$$\Phi_{tot} = \eta_{sens}\Phi_{Ln} \tag{1}$$

The sensitisation efficiency η_{sens} can readily be optimised to approach unity [7, 8], particularly by minimising the distance between the antenna chromophore and the lanthanide ion. In particular for NIR luminescent lanthanide ions, the overall luminescence quantum yield is mainly limited by the intrinsic luminescence quantum yield of the lanthanide ion Φ_{Ln}. The intrinsic quantum yield is difficult to measure directly but can be deduced from the observed luminescence decay times, τ_{obs}, once a reliable estimate of τ_{rad} is known (see also ['Basics of Lanthanide Photophysics', Bünzli, Eliseeva]).

$$\Phi_{Ln} = \frac{\tau_{obs}}{\tau_{rad}} \tag{2}$$

In order to obtain an idea of the luminescence decay times and the intrinsic luminescent quantum yields of NIR luminescent lanthanide complexes, Table 2 compiles some typical decay times obtained in recent work. The decay times are much shorter than the luminescence decay times for Eu(III) and Tb(III), which are often on the order of hundreds of microseconds.

The lifetime data contains a range of different ligand systems, in different solvents, and is only intended as an indication of usually observed values. As will be explained below, the surrounding ligand and solvent have profound effects on the non-radiative deactivation of NIR luminescent lanthanide ions.

The observed luminescence lifetimes in Table 2 can be translated into order-of-magnitude estimates for the intrinsic luminescence quantum yield of the lanthanide ions using Eqn. 2 and values for τ_{rad} from Table ['Ground state (G), main emissive...', table 6] in ['Basics of Lanthanide Photophysics', Bünzli, Eliseeva]. These estimates are collected in Table 3, together with some measured luminescence quantum yields for these ions in luminescent complexes. The quantum yields are very low compared to those of Eu^{3+} and Tb^{3+} complexes.

Table 2 Examples of typically observed luminescence decay times for the NIR transitions in luminescent lanthanide complexes in solution

Ion	Transition	Wavelength (nm)	τ_{obs} (typical)	References
Yb^{3+}	$^2F_{5/2} \rightarrow {}^2F_{7/2}$	980	1.91 μs; 10.4 μs	[7]
			0.7 μs ... 10.9 μs	[9]
Nd^{3+}	$^4F_{3/2} \rightarrow {}^4I_{11/2}$	1,060	250 ns; 280 ns	[7]
			700 ns; 6.3 μs	[8]
Er^{3+}	$^4I_{13/2} \rightarrow {}^4I_{15/2}$	1,530	1.2 μs; 1.6 μs	[10]
	$^4I_{13/2} \rightarrow {}^4I_{15/2}$	1,530	1.46 μs	[7]
Pr^{3+}	$^1D_2 \rightarrow {}^3F_4$	1,020	110 ns; 73 ns; 56 ns; 54 ns	[10]
	$^1D_2 \rightarrow {}^3F_4$	1,030	8.0 ns	[11]
	$^1D_2 \rightarrow {}^3H_4$	605	8.8 ns	[11]
Ho^{3+}	$^5F_5 \rightarrow {}^5I_6$	1,479	–	[12]
	$^5F_5 \rightarrow {}^5I_7$	975	–	[12]
	$^5F_5 \rightarrow {}^5I_7$	980	6.5 ns	[11]
	$^5F_5 \rightarrow {}^5I_8$	650	6.4 ns	[11]
Tm^{3+}	$^3F_4 \rightarrow {}^3H_4$	1,465	–	[12]

Lifetimes and quantum yields are sensitive to ligand structure and solvent. The values presented here serve to illustrate the typical luminescence lifetimes attainable for complexes in solution

Table 3 Typical quantum yields for typical, reasonably optimised lanthanide complexes in solution, estimated Φ_{Ln} from luminescence lifetime data, and observed Φ_{obs} reported in the literature

Ion	Φ_{Ln} (estimate)	Φ_{obs}	References
Yb^{3+}	10^{-2}	2×10^{-2}	[12]
		5×10^{-3}	[7]
Er^{3+}	10^{-4}	3×10^{-4}	[12]
		2×10^{-4}	[7]
Nd^{3+}	10^{-3}	4×10^{-3}	[12]
		3×10^{-2}	[13]
		4×10^{-4}	[7]
Pr^{3+}	10^{-4}		
Ho^{3+}	10^{-5}	2×10^{-5}	[12]
Tm^{3+}	–	6×10^{-5}	[12]

In spite of the generally low luminescence quantum yields, NIR luminescent lanthanide complexes, and particularly those of Yb^{3+}, are of interest for bioanalytical applications, since (1) their luminescence will be less hindered by endogenic fluorescence from biological material, (2) antenna chromophores with high extinction coefficients may be incorporated into the complexes that compensate for the low quantum yield and (3) powerful visible and NIR excitation sources can be used. In the following, we will outline the work done on the development of NIR luminescent lanthanide complexes.

1.1 Neodymium(III), Erbium(III) and Ytterbium(III) Complexes

In 1958, Crosby and Kasha [4] described the photosensitised emission of Yb^{3+} chelates, whereas solutions of neodymium(III) complexes were studied for laser action in the 1960s [5]. Until the early 1990s, there have been some reports of Yb^{3+}–porphyrin complexes [14, 15] and the ability of coloured metal indicators to act as photosensitisers for Nd^{3+} and Yb^{3+} [16, 17]. Systematic studies of NIR luminescence in organic and aqueous media began in the middle of the 1990s with reports on Nd(III) [18–20], Yb(III) [9, 18] and Er(III) [18, 21]. For erbium(III), of importance for optical telecommunications, these were the first reports of its luminescence in an organic environment opening possibilities for the development of Er^{3+} doped polymers for plastic photonics. It was demonstrated that organic dyes such as fluorescein, eosin [18] and others [22] could be incorporated as successful antenna chromophores for Nd^{3+}, Yb^{3+} and Er^{3+} into the ligand architecture (see e.g. Fig. 2). Other types of antenna chromophores were identified such as organometallic complexes [25].

The efficiency of the sensitisation processes depends critically on the distance between the ion and the sensitiser: the antenna chromophore needs to be in close contact with the lanthanide to ensure efficient transfer, as has been demonstrated for the fluorescein-type of antenna chromophore [7, 18]. The intensity of the ligand

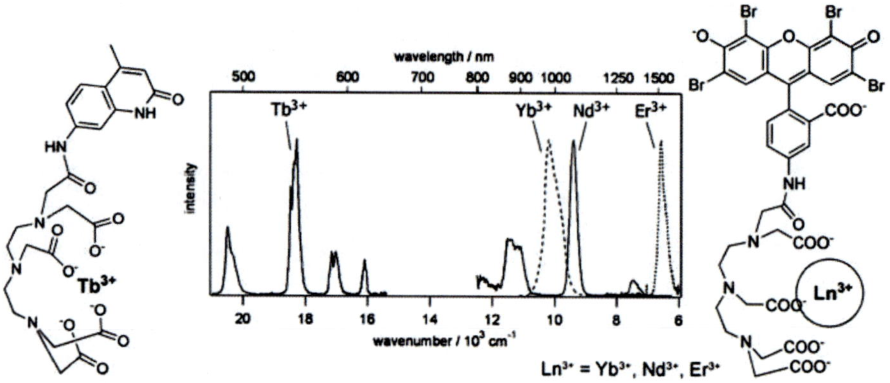

Fig. 2 Emission spectra of sensitiser-modified DTPA complexes. The visibly luminescent terbium (III) complex [23, 24] contains a carbostyril sensitiser, excited at 320 nm. The NIR luminescent ytterbium(III), neodymium(III) and erbium(III) complexes [18] are based on an eosin antenna chromophore, excited at 520 nm. *DTPA* diethylenetriaminepentaacetic acid

fluorescence and the sensitivity of the sensitised lanthanide luminescence towards the presence of oxygen in solution are effective indicators for the efficiency of the sensitisation process. When in close contact with the energy-accepting lanthanide ion (Nd^{3+}, Er^{3+} or Yb^{3+}), the fluorescence of the fluorescein chromophore is completely quenched, and no sensitivity towards the presence of oxygen is found, indicating optimal sensitisation [7]. Complexes in which the chromophore is situated at approximately 0.5 nm of the ion still display relatively intense ligand-centred fluorescence and sensitivity to dissolved oxygen [18].

The non-radiative decay channels in Yb^{3+}-based complexes in solution have been characterised quantitatively [26], showing that not only directly bound water molecules quench their luminescence but that quenching is also caused by closely diffusing water molecules. This is an important information for the development of Yb^{3+} complexes for application in biological systems where water is omnipresent as it stresses the need of encapsulating the luminescent ion in a large protecting ligand. The reduction in non-radiative deactivation of the luminescent state of NIR lanthanide ions is crucial for their application, be it as luminophores or as elements for light amplification. Molecular vibrations in the ligand that encapsulates the ion, particularly C–H, N–H and O–H, are efficient quenchers of NIR lanthanide luminescence. An illustrative example of efforts to reduce this problem is the complex developed by Hasegawa et al. [13] where all of the C–H oscillators have been replaced by the lower-energy C–F group (Fig. 3, left). The experimentally determined luminescence quantum yield of this complex reaches a record value of 3%, which is obviously low compared to lanthanide ions in glasses and crystals or organic dyes in solution whose quantum yields can often exceed 50%. A recent, supplementary strategy for maintaining NIR luminescence is the encapsulation of the ions in aromatic shells, which rigourously prevents quenching solvent molecules from approaching the lanthanide ion [28, 29].

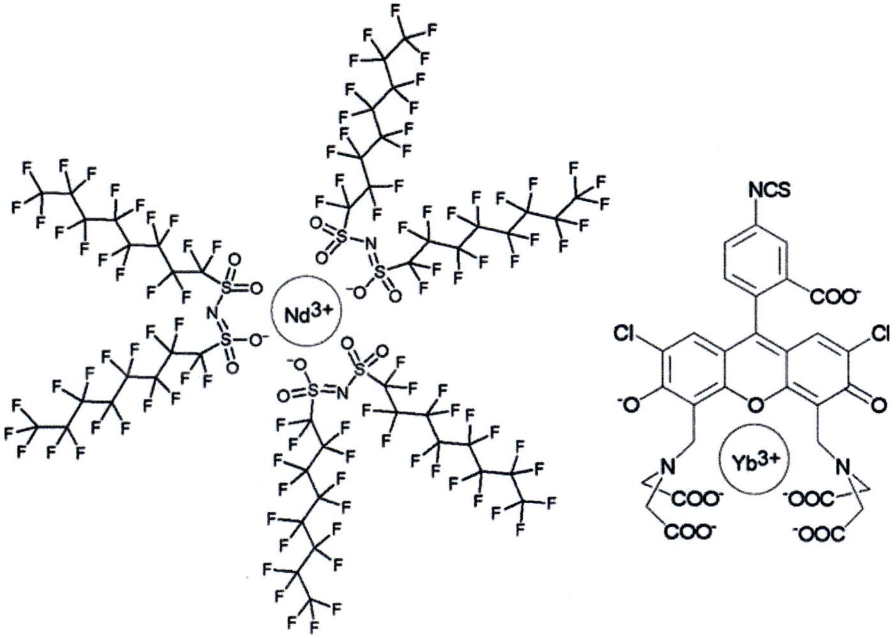

Fig. 3 Two examples of NIR luminescent lanthanide complexes. *Left*: Nd^{3+} complexes with reduced non-radiative deactivation due to replacement of C–H by C–F oscillators [13]. *Right*: Yb(FxITC), an NIR luminescent label for the labelling of biological molecules [27]

A sub-optimal photoluminescence quantum yield is not necessarily a major problem. Other factors come into play when developing applications. In medical diagnostic and imaging applications, the luminescence signal generated is proportional to the product of the extinction coefficient of the complex and the quantum yield, the brightness $\varepsilon\Phi$, which means that also the extinction coefficient is an important factor. Additional factors come into play as well: the optical transparency of the embedding matrix, which is much better in the NIR for biological material. Especially Yb^{3+} should be a good ion for luminescent complexes in biological applications as it has all of its emission concentrated in one narrow spectral band around 980 nm, where an average biological medium is at its most transparent. Use of an Yb^{3+} complex, Yb(FxITC) (Fig. 3, right), in a prototypical fluoroimmunoassay has been demonstrated [27].

If ytterbium(III) complexes are to be used in in vivo cellular imaging instead of in in vitro diagnostics, the potential toxicity [30] of lanthanide ions becomes an issue, and the complexes should display low dissociation constants in order to limit the amount of free intracellular lanthanide ions. For instance, the dissociation constant of the Yb^{3+}–fluorexon complex was found to be comparable to that of the corresponding EDTA complex [7], being higher than DTPA and DOTA complexes, but still limiting the presence of free Yb^{3+}. However, the cytotoxicity of this luminescent lanthanide complex and its derivatives is not known.

1.2 Praseodymium(III), Holmium(III) and Thulium(III) Complexes

Concerning the luminescence of the second group of NIR luminescent lanthanide ions, the ions Pr^{3+}, Tm^{3+} and Ho^{3+} display even weaker luminescence in solution than Yb^{3+}, Nd^{3+} and Er^{3+}, and the luminescence of their complexes has only recently been measured [12, 31].

Praseodymium(III) luminescence has been observed from two different excited states, $^3P_0$ and $^1D_2$ [31]. The latter level gives rise to NIR luminescence at 890 and 1,070 nm, together with an emission at 605 nm. The quantum yield of β-diketonate complexes of Pr^{3+} in toluene was estimated to be significantly less than 10^{-3}. Interesting, a publication by the same group also reports chemiluminescence of the β-diketonate complex $Pr(fod)_3$, induced by the decomposition of a dioxetane with subsequent energy transfer to the Pr^{3+} ion [32].

Luminescence from Pr^{3+} has been observed for pyrazolylborate complexes in organic solution [10, 33]. NIR luminescence of Pr^{3+} in aqueous solution has also been reported (Fig. 4). The extremely short luminescence decay times observed (8 ns) indicate a very low luminescence quantum yield in this environment [11] and therefore, Pr^{3+} is expected to have less favourable characteristics for application in NIR luminescent labels.

Thulium(III) complexes are mainly known for their combined weak visible/red/NIR emission from the $^1G_4$ state, which lies at 21,200 cm^{-1} and thus requires near-UV

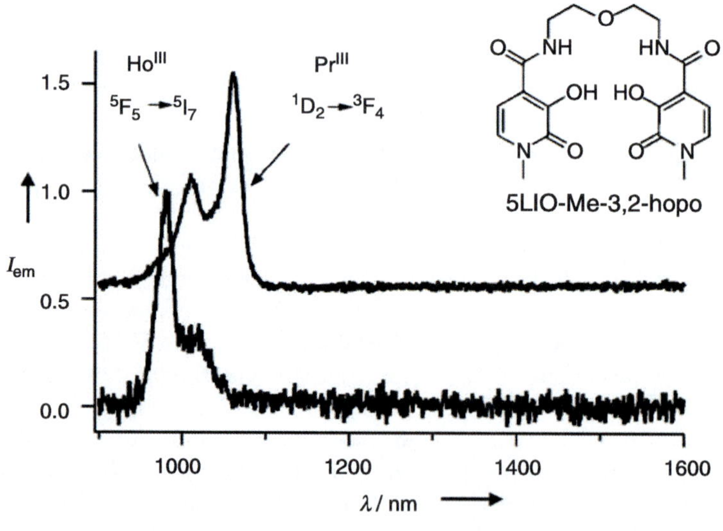

Fig. 4 The NIR emission spectra [11] (excitation at 345 nm) of holmium(III) and praseodymium (III) complexes with hydroxypyridinone-based ligand "5LIO-Me-3,2-hopo" in buffered aqueous solution (TRIS, 0.1 M, pH 7.4). The emission spectra of the Pr(III) complex are vertically offset for clarity. (Reproduced from [11] with permission. Copyright Wiley-VCH Verlag GmbH & Co. KGaA)

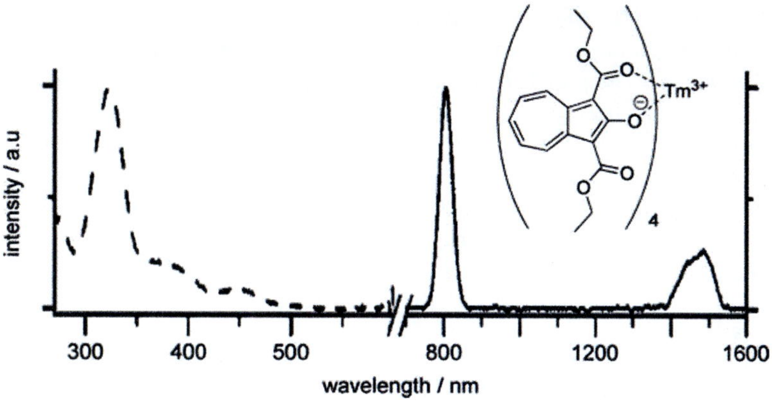

Fig. 5 NIR photoluminescence excitation (*dashed line*, emission monitored at 800 nm) and emission spectra (*solid line*, excitation at 380 nm) of a thulium(III) complex (10^{-5} M in acetonitrile). Data adapted from [36]

absorbing antenna chromophores [34, 35]. The emission bands from this state are situated at 480, 650 and 770 nm. Recently, emission has been detected from the $^3F_4$ state with emission bands at 790 and 1,465 nm [12]. These $^3F_4$ emissions can be sensitised through visibly absorbing tropolonate ligands that have triplet energies that are too low for sensitisation of the $^1G_4$ emissions. This NIR thulium(III) can also be sensitised by azulene-type chromophores (Fig. 5), which have a moderate absorption in the red part of the visible light [36].

NIR holmium(III) luminescence in solution has been detected (Fig. 4) but found to be exceedingly weak (quantum yield ~10^{-5}) [11, 12]. The low luminescence quantum yields, combined with the limited perspectives for improvements, make holmium(III) a less desirable candidate for luminescent bioanalytical applications.

1.3 Cerium(III) and Promethium(III) Complexes

Luminescence emanating from *f–f* transitions of lanthanide complexes in solution has been observed, and studied, for all trivalent lanthanide ions, except for Ce^{3+} and Pm^{3+}. For each of these two ions, the reason for the absence of data is different. Cerium(III) has only one *f–f* transition, $^2F_{7/2} \rightarrow {}^2F_{5/2}$, which lies too far in the infrared (~5,000 nm) [37] to have a probability to display any detectable emission in molecular media. Instead of this single *f–f* transition, complexes of cerium(III) can display *f–d* transitions, which have considerable extinction coefficients and are sometimes fluorescent. However, the associated emissions are situated in the UV/blue part of the spectrum and do not display the long luminescence decay lifetimes associated with the *f–f* transitions lanthanide ions [38, 39]. This makes cerium(III) complexes less interesting for biological labelling, although energy transfer from cerium(III) to terbium(III) complexes may be used to probe supramolecular interactions in soft matter [40].

The element promethium only exists as unstable isotopes, and does not occur naturally, although lines presumably corresponding to Pm have been detected in the spectra of certain stars [41, 42]. Seventeen isotopes are known, of which the longest half life (17.7 years) belongs to ^{145}Pm. The isotope that is "usually" employed is ^{147}Pm (3.7 years). This isotope is one of the fission products of ^{238}U. This means that its first nucleosynthesis by man occurred probably as a by-product of the experiments in the 1930s and early 1940s that finally led to the realisation of the Manhattan Project. The identification and discovery of the element was however only reported in 1947 [43].

Promethium-147 can also be obtained from the decay of ^{147}Nd (which in turn can be produced by bombarding a uranium carbide target with energetic protons from a particle accelerator). Mixed with ZnS (a well-known light emitting material), Pm(III) chloride was used in the glow-in-the-dark paint used in analogue watches: the particles emitted by the decay of the Pm excite the ZnS, which displays its luminescence.

Owing to the relatively long half life and the relative ease of obtaining ^{147}Pm, the chemical and spectral properties of Pm^{3+} have been studied, although – understandably – not to a great extent. Interestingly, Krupke et al. have been able to demonstrate laser action of YAG (yttrium–aluminium garnet) doped with Pm^{3+} ions [44]. The light emission and laser characteristics of Pm^{3+}-doped materials are reminiscent of their neodymium(III) homologues, although the dominant emission transition for Pm^{3+} is at 920 nm instead of 1,060 nm.

The luminescence of Pm^{3+} and its chelates in solution has not been reported. Since the spectroscopy is similar to Nd^{3+}, it may very well be that sensitisers that have already been shown to be effective with Nd^{3+} will also work with promethium (III). Judging from the energy gap between the emissive state and the ground state of this ion, its sensitivity towards quenching by proximate vibrational oscillators should be lower than that of neodymium(III). Promethium(III) may prove to be an interesting NIR-emitting ion. Its radioactivity, however, is a likely impediment for most applications (remember that luminescent lanthanide complexes were proposed as replacements for radioactive iodine in radio-immunoassays in the first place) [45]. It might be that – if chosen properly – the sensitiser will also act as a scintillator, in which case the decay of one Pm may excite the luminescence of an other one. Taking this argument further: the β-decay of Pm, which converts it into Sm, might lead to the simultaneous excitation of an included scintillator-sensitiser, leading to autogenerated samarium(III) emission. This would present a very exotic example of radiation photochemistry.

2 Particularities of the Architecture and Molecular Engineering of Near-Infrared Luminescent Lanthanide Complexes

The guidelines applied in the molecular engineering of visibly luminescent Eu^{3+}, Tb^{3+}, Dy^{3+} and Sm^{3+} complexes also apply to the NIR lanthanide complexes, concerning shielding of the ions from the exterior and antenna chromophore, particularly concerning complex stability, and compatibility with the biological media where they find their intended application (see Sect. 3.1). Many of the ligand systems already developed may readily be used with NIR lanthanide ions, which are facilitated by the virtually identical coordination chemistry of all trivalent lanthanide ions (of course, subtle differences between the ions exist!). The reader is referred to ['Basics of Lanthanide Photophysics', Bünzli, Eliseeva] and ['Luminescent chemical and physical sensors based on lanthanide complexes', Schäferling], and [Stable Luminescent Chelates and MacrocyclicCompounds, Mathis, Bazin] and ['Lanthanide assemblies and polymetallic complexes', Faulkner] that deal with the design of coordinating ligand systems for luminescent lanthanide complexes, particularly for visibly luminescent lanthanide complexes.

When optimising ligands for use with NIR luminescent lanthanide ions, it is advantageous to take advantage of the greater freedom in the choice of antenna chromophores. Chromophores can be used that absorb at longer wavelengths and consequently have larger maximum extinction coefficients. We have already mentioned different classes of sensitisers: well-known organic dyes [18, 46, 47], metal indicators [7, 16], organometallics [25] and porphyrins [14, 48–50]. Many visibly absorbing organic and organometallic chromophores can be sensitisers for lanthanide luminescence, provided that the energy of their lowest excited (triplet) state is sufficiently high to transfer energy efficiently and irreversibly to the accepting lanthanide ion.

One recent example of the sensitisers proposed specifically for NIR luminescent lanthanide ions is azulene, which is a chromophore that has long fascinated photochemists [36]. Its use may have been inspired by scientific curiosity as azulene has only a modest absorption in the red, and more powerful sensitisers do exist. Another recent example is the curcuminoids [51], which are molecules of natural origin (constituents of *curcuma* or turmeric). The use of naturally occurring chromophores with the NIR luminescent ions is reminiscent of the application of naturally occurring dipicolinic acid and salicylic acid with Eu^{3+} and Tb^{3+}, which were among the very first chromophores to be used for the photosensitisation of lanthanide ions.

Porphyrins and related compounds are also photophysically interesting sensitisers of NIR lanthanide luminescence [14, 48–50]. Their so-called Soret absorption band, typically located in the blue region of the visible spectrum (400–430 nm), has extremely high molar absorption coefficients (well over 10^5 M^{-1} cm^{-1}). However, the Soret band is relatively narrow, which limits the choice of excitation wavelengths. Excitation of the porphyrin complexes using wavelengths farther into the red may be preferred, but there, the molar absorption coefficients are not as high. Another potential problem in using porphyrin complexes may be the limited solubility in aqueous environments, although the balance between hydrophobic and hydrophilic may be tuned to suit particular applications [52, 53].

Organic chromophores are very effective as sensitisers, provided that the energy transfer from excited chromophores to the lanthanide ions is efficient. Their sensitising power comes from their large cross-sections. The brightness

($\varepsilon\Phi$) of a lanthanide complex in which the lanthanide ion is excited directly is given by:

$$(\varepsilon\Phi) = \varepsilon_{\mathrm{Ln}}\Phi_{\mathrm{Ln}} \quad (3)$$

When using a sensitising chromophore to excite the lanthanide luminescence, the brightness becomes:

$$(\varepsilon\Phi) = \varepsilon_{\mathrm{sens}}\Phi_{\mathrm{tot}} = \varepsilon_{\mathrm{sens}}\eta_{\mathrm{sens}}\Phi_{\mathrm{Ln}} \quad (4)$$

With η_{sens} approaching unity, the overall brightness of the complex can become over 10^5 times larger, since $\varepsilon_{\mathrm{sens}}$ (values in the $10^4 \ldots 10^5\ \mathrm{M}^{-1}\ \mathrm{cm}^{-1}$ range) can easily be 10^5 as large as Φ_{Ln} (typically $0.1 \ldots 1\ \mathrm{M}^{-1}\ \mathrm{cm}^{-1}$). Moreover, absorption bands of organic chromophores are much larger than the narrow transitions of lanthanide ions, displaying high extinction coefficients over a wide range of wavelengths.

For high sensitisation efficiencies, it is crucial to minimise the distance between the antenna chromophore and the lanthanide ion, since the efficiency of antenna-to-lanthanide energy transfer and thus, η_{sens} depends critically on distance. Energy transfer from a chromophore to the lanthanide ion proceeds via the Dexter mechanism [54], which is only operable at very short distance, contrary to the Förster resonance energy transfer mechanism (FRET) that can extend over many Angstroms, depending on the photophysical properties of the participating energy transfer partners. The reason for the predominance of the Dexter mechanism in luminescent lanthanide complexes is the fact that the energy accepting lanthanide ion only has very weak transitions between its electronic energy levels (forbidden/weakly allowed transitions, see Chap. 2).

In the case of ytterbium(III), an original, alternative photosensitisation mechanism has been described for the combination of this ion with an indole chromophore [55]. The excited indole chromophore is quenched by Yb^{3+} through electron transfer, yielding Yb^{2+}, which is relatively stable, and the indole radical cation. The photoinduced charge separation is followed by a charge recombination (restoring both the indole and Yb^{3+}) that frees enough energy to leave the Yb^{3+} in an excited, luminescent state. This mechanism would explain the photosensitisation of Yb^{3+} luminescence by indole despite the negligible spectral overlap between indole and Yb^{3+}. This "internal redox" process may be operational in certain Yb^{3+} complexes, especially those with chromophores with high-lying excited states (UV). However, in most NIR luminescent lanthanide complexes, the Dexter mechanism is operational.

The strong distance dependence of the lanthanide luminescence sensitisation efficiency is clearly illustrated in Fig. 6, in which two fluorescein-based ytterbium(III) complexes are compared. Both complexes were measured in water. Even though the ligand structures are slightly different, the intrinsic lanthanide luminescence quantum yields (Φ_{Ln}) are comparable, judging from the very comparable luminescence decay times [7, 56] of both complexes (around 1 μs in aqueous buffer, 8...10 μs in D_2O). Since in both cases the same chromophore is used, the enormous difference is luminescence intensity can therefore be attributed to a change in η_{sens}.

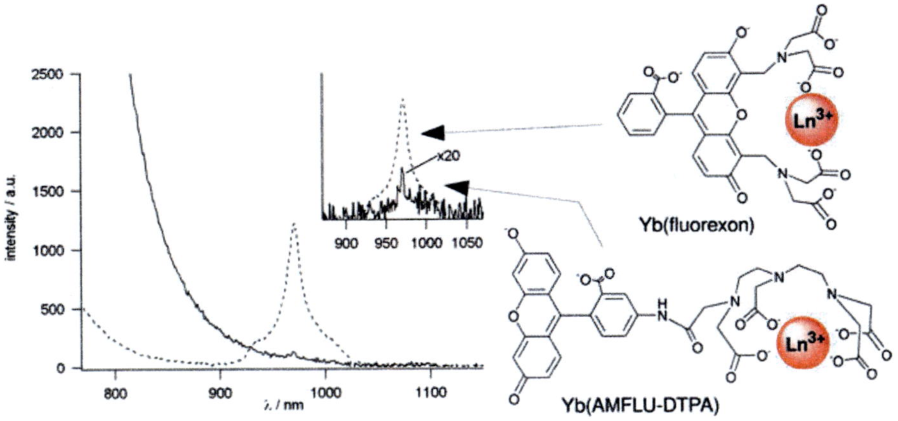

Fig. 6 The dramatic effect of antenna-lanthanide separation on the overall sensitised NIR luminescence intensity in fluorescein-based ytterbium(III) complexes. In the Yb–fluorexon complex (*dashed line*), the lanthanide ion is much closer to the chromophore than in Yb-AMFLU-DTPA (*solid line*). The emission spectra (excitation at 490 nm) are a direct comparison of the luminescence intensity of equally concentrated aqueous buffer solutions (TRIS, pH 8.3). The inset shows a close-up of the $Yb3^{+}$ luminescence, after subtraction of the contribution of the antenna fluorescence

In both complexes, the sensitising chromophore is based on fluorescein, and the actual part of the molecule that is responsible for the photophysical properties of fluorescein is the annelated three-ring system [57, 58]. The attached phenyl group has little influence on the photophysical properties, since it is not conjugated with the three-ring system. In the case of Yb(AMFLU-DTPA), the substituted phenyl group mainly acts as a spacer that further separates the sensitising chromophore and the lanthanide ion, slowing down energy transfer from the chromophore to the lanthanide ion. Moreover, the longer distance diminishes the lanthanide heavy atom effect that enhances intersystem crossing in the chromophore that converts the initial excited singlet state into a long-lived triplet state that is able to transfer energy to the lanthanide ion. In Yb(AMFLU-DTPA), the ligand fluorescence is much stronger than in Yb(fluorexon), indicative of less efficient triplet formation or, in all cases, a less efficient lanthanide sensitisation. Ligand fluorescence is a process competing with the generation of NIR lanthanide luminescence.

An efficiently luminescent lanthanide complex is therefore a complex that displays negligible fluorescence from the sensitising organic ligand. At first, it may therefore be surprising that antenna chromophores based on notoriously fluorescent ligands such as fluorescein can still be efficient sensitisers for lanthanide luminescence [7]. Fluorescein has been found to be more efficient as a sensitiser than its derivatives eosin and erythrosin in spite of the higher triplet yield of the latter two. This is attributed to the higher triplet energy of fluorescein, compared to eosin and erythrosin, giving less energy back-transfer from the lanthanide to the sensitiser [59].

The initial preference for the selection of sensitising chromophores would be on chromophores with a high triplet yield and weak fluorescence [60]. However, as it turns out, lanthanide ions can effectively suppress ligand fluorescence in luminescent complexes, particularly by a combination of the heavy-atom effect and their paramagnetism, which both enhance singlet–triplet intersystem crossing [7]. The extra yield of triplets leads to a more efficient sensitisation of lanthanide luminescence. Additionally, under favourable conditions, lanthanide ions with many accepting energy levels may be able to receive their excitation energy directly from the antenna chromophore's singlet state [46].

It is interesting to consider the limits on the excitation wavelength for NIR luminescent complexes, i.e. what is the longest wavelength a sensitising chromophore can absorb and still be a potent photosensitiser for NIR lanthanide luminescence? Such estimates have been made in the past for visible luminescent lanthanide ions [60]. In order to exclude back-energy transfer from the excited lanthanide ion to the triplet state of the sensitiser, this triplet state should be well above the lanthanide luminescent state (~3,500 cm^{-1}). In the case of organic chromophores, which only display significant light absorption for excitation in their singlet excited state, the singlet–triplet energy separation of the chromophore should be added to this gap. Although the initial estimate [60] assumed that an energy gap of 8,500 cm^{-1} need to exist between the absorption of the sensitising chromophore and the luminescent level of the lanthanide ion, it appears from experimental observations [61] that a gap (between the singlet state of the sensitiser and the luminescent state of the lanthanide ion) of 6,000–6,500 cm^{-1} may be sufficient, depending on the chromophore. For efficiently luminescent sensitiser-modified Yb^{3+} complexes (luminescent state near 10,200 cm^{-1}), this means that the limit for ligand absorption will be in the 600–670 nm range.

Although on the sensitiser side, NIR luminescent lanthanide complexes provide a certain amount of architectural and photophysical freedom compared to visibly luminescent lanthanide complexes, NIR luminescent lanthanide ions are much more prone to non-radiative deactivation by nearby molecular oscillators, particularly O–H, N–H and even C–H. Therefore, the ligand should offer suitable protection against this non-radiative deactivation by keeping quenching vibrations (in particular, water, urea) at a distance. Naturally, it should contain a minimal amount of quenching capability. This is an interesting challenge since NIR luminescent lanthanide ions are even quenched by C–H oscillators [26, 62], which are not easily avoided in organic molecules.

Current strategies [8] for the optimisation of the lanthanide luminescence quantum yield (Φ_{Ln}) for NIR luminescent ions consists in replacing C–H vibrations by C–F or C–D vibrations in the ligand (see e.g. Fig. 3, right). Moreover, the ligand shell around the luminescent ion is made tight, hydrophobic and bulky (Fig. 7), in order to keep any diffusion quenchers (such as water molecules) at a distance [28, 29]. An open challenge is to combine this ion-protecting architecture with the integration of an efficient and visible-light absorbing sensitiser at close proximity of the ion.

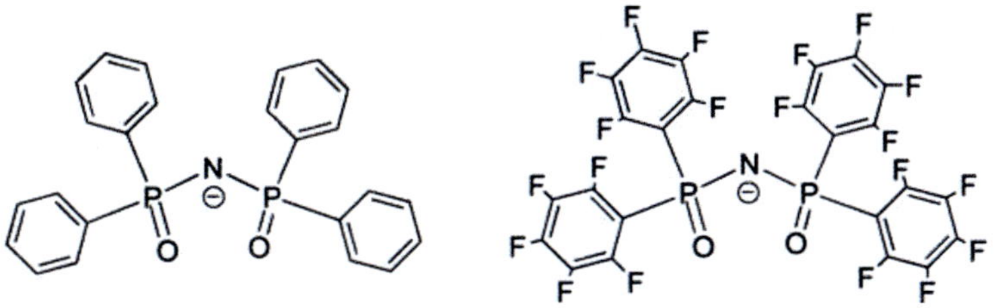

Fig. 7 Imidodiphosphinate "shell"-type ligands [28, 29] providing a very tight shielding of the lanthanide ion against quenching molecules. When used with NIR luminescent lanthanide ions, the perfluorinated imidophosphinate ligand yields very long luminescence lifetimes and – presumably – high photoluminescence quantum yields

In the optimisation of every class of luminescent materials, it is crucial to have quantitative data on the luminescence brightness of each member of the class. This quantitative brightness, in molecular terms, is the product of the extinction coefficient and the overall luminescence quantum yield (see Chap. 2). More detailed information on the photophysical performance of luminescent lanthanide complexes may be obtained by dissecting this brightness into the individual contributions of extinction coefficient, energy transfer efficiency and intrinsic lanthanide luminescence yield.

Without at least relative values of brightnesses and quantum yields, claims of "bright", "brilliant", "intensely luminescent" NIR luminescent lanthanide complexes are of little value. Absence of noise in an emission spectrum may indeed mean that the measured complex has a high brightness but may also be the result of a very efficient light detection system and/or a powerful excitation source.

Since the detailed study of NIR luminescent lanthanide complexes has begun relatively recently, little reference data on the luminescence of these complexes exist, in contrast to references for visible luminescence and fluorescence [63]. However, luminescence quantum yield data from different research groups generally converge to values of similar order of magnitude. Several of the complexes described and characterised in the literature may be at least used as semi-quantitative standards. The quantum yield data reported in the literature, however, has as yet not been completely checked for its coherence.

Another important parameter is the photostability of the luminophores as this determines the total number of photons that will be emitted by the complex before it is destroyed by a photoinduced process. Data on the photostability of lanthanide complexes is very scarce. The photostability of lanthanide complexes is expected to be superior to that of organic fluorophores since the excitation energy only remains on the organic ligand a very short time before it is transferred to the lanthanide ion, which usually does not undergo any photochemical transformations.

3 Near-Infrared Luminescent Lanthanide Ions in Inorganic Nanoparticles

Inorganic hosts such as glasses and crystals have less effective non-radiative deactivation channels for excited lanthanide ions compared to molecular hosts. NIR luminescence, emission from higher excited states and luminescence upconversion are therefore more readily observed for lanthanides in such inorganic hosts. For biological detection and for the doping of lanthanide ions into polymeric hosts, it is interesting to have objects with the luminescent properties of inorganic phosphors but with sizes comparable to molecules (only several nanometers) and with similar solubility and physicochemical behaviour. Such Ln^{3+} doped "nanophosphors" are currently under active development, especially using "wet chemical" synthetic methods, which are often relatively low cost and easily scaled up. We will focus here on particles that are dispersible (soluble) in organic and aqueous solvents since those are most interesting for application in either bio-imaging or as photonically active elements in polymer-based waveguides. Moreover, we will mainly focus on NIR luminescence and compare nanoparticles to molecular complexes. ['Lanthanide nanoparticules as photoluminescent reporters', Soukka, Härma] and ['Upconverting nanoparticles', Wang, Nann] give a broader and more detailed overview of luminescent lanthanide-doped nanoparticles.

In a pioneering work in this field [64], the micrometre-sized grains of an industrial europium(III)-doped phosphor powder were ground into smaller phosphor particles. These smaller particles were coated with negatively charged polyelectrolytes and subsequently with proteins (such as avidin) to render them biocompatible. Imaging was achieved using pulsed excitation and delayed detection in a conventional fluorescence microscope, which suppresses the autofluorescence of the biological sample, making use of the fact that the decay times of biofluorescence (on the order of nanoseconds) are orders of magnitude shorter than the decay of lanthanide luminescence (approx. 1 ms). Specifically designed upconverting phosphors of composition Y_2O_2S:(Yb, Er) have been used for upconversion luminescence detection in nucleic acid microarrays (see also ['Upconverting nanoparticles', Wang, Nann]) [65]. Very sensitive detection of the upconverted green luminescence can be achieved under 980 nm NIR excitation. The particles used are on the order of several hundred nanometers in size, which can still be considered very large compared to molecular dimensions.

Smaller particles have been investigated recently: Ln^{3+}-doped $LaPO_4$ [66] and LaF_3 [67] nanoparticles with sizes in the 5–15 nm range. Indeed, NIR luminescence from the Nd^{3+} and Er^{3+}-doped particles appear to suffer less non-radiative deactivation than in complexes with organic ligands, although efficient quenching still occurs near the surface of the particles . In 12 nm sized $NaYF_4$:(Yb,Er) and $NaYF_4$:(Yb,Tm) particles, upconversion luminescence is observed under NIR excitation [68, 69].

This type of doped inorganic nanoparticles lack the excellent light absorption properties of organic chromophores, necessitating the weak direct excitation of lanthanide ions, although Ce^{3+} has been used to enhance the ultraviolet absorption of Tb^{3+} particles [66]. A different strategy to obtain Ln^{3+}-doped nanoparticles with efficient light absorption is the incorporation of lanthanide ions into semiconductor nanoparticles, which has been achieved recently [70]. The overall photoluminescence quantum yield of these doped quantum dots was found to be very low ($<10^{-3}$), but ample room exists for the improvement of these materials through advances in the control of size-distribution and the introduction of protecting capping layers. Further functionalisation of the nanoparticle surface is necessary to make them biocompatible [71, 72]. Especially in the controlled surface modification of inorganic nanoparticles by organic molecules, much work remains to be done to improve colloidal stability and biological compatibility.

Recently also, silica materials doped with sensitiser-modified NIR luminescent lanthanide complexes have been described [73] as well as NIR luminescent mesoporous materials [74] doped with such complexes. These reports do not describe the behaviour of these materials as nanoparticles in suspension, but the synthesis procedure might be adapted to obtain such materials suitable for use in biological media. The encapsulated lanthanide complexes appear to have luminescent characteristics very similar to the complexes in solution (e.g. luminescence decay times), although the silica-based ytterbium(III) and neodymium(III) materials show a roughly twofold increase in NIR luminescence intensity when they are dried for 3 weeks at 50°C. No indications exist that the encapsulation in these materials drastically improves the photoluminescence quantum yield of NIR luminescent lanthanide complexes.

Each type of host for lanthanide ions has its strengths and weaknesses for the construction of luminescent materials. Organic ligands generally induce efficient non-radiative deactivation of lanthanide ions, particularly for highly excited states, but offer tunability of their physicochemical and biological behaviour through molecular chemistry. Inorganic hosts give rise to richer lanthanide photophysics through reduced non-radiative deactivation channels but are not as easily processed as organic materials and are often bio-incompatible. As we have seen in this review, the traditional weaknesses of the different types of hosts are currently being addressed: surface-modified nanoparticles may eventually be used in organic and biological media for polymer optical amplifiers or biological imaging, and luminescent lanthanide complexes may generate light in electroluminescent devices or work efficiently as NIR emitters.

4 Working Towards Applications of Near-Infrared Luminescent Lanthanides in Biosensing, In Vitro Diagnostics and Imaging

A few demonstrations of the use of NIR luminescent lanthanide-based materials in bioanalytics and bio-imaging have appeared in recent years. Advances in optoelectronics have brought light sources and detectors that are well-suited for working with NIR luminescent materials, and improvements in sensitivity for NIR-based detection methods are still to be expected. The relative transparency of biological material in the NIR gives an impetus to the research on NIR luminescent lanthanide-based materials.

In this part, we will sketch the opportunities that exist for the application of NIR luminescent lanthanide materials and review several examples of the use of these materials in actual/proof-of-principle bio-applications. Our discussion will include some of the instrumental aspects of the detection of NIR luminescence as it is in this area that enabling technologies for the application of NIR luminescent lanthanide complexes have appeared.

4.1 Instrumental Aspects of the Generation and Detection of Near-Infrared Luminescence

The application of NIR luminescent materials depends on the availability of robust, cost-effective excitation sources and light detectors for such materials. For NIR luminescent lanthanide complexes, it is especially the relatively limited choice of accessible detectors that has slowed their investigation and further development. Most standard spectrofluorimetric and microscopic equipment are equipped with detectors that are mainly sensible in the visible. An extension of the spectroscopic sensitivity of this equipment farther into the red usually does not go to wavelengths longer than 850 nm (the limit of typical multi-alkali photocathodes). We will briefly discuss some recent developments in excitation sources and detectors that are applicable in particular to NIR luminescent lanthanide complexes.

4.2 Excitation of Near-Infrared Luminescent Lanthanide Ions and Complexes

Current optical imaging and photoluminescence detection methods mainly use "classical" light sources such as mercury and xenon lamps and "conventional" lasers such as argon-ion and helium–neon. Diode-pumped solid-state lasers, which deliver higher powers in a compact enclosure, are becoming more popular. A particular mention is to be made of recent progress in optoelectronics that has given rise to compact, low-cost and very bright excitation sources such as semiconductor lasers and high-power light-emitting diodes. These sources mainly operate in the visible and NIR light and can thus be used with NIR luminescent lanthanide complexes.

The high peak powers delivered by sub-picosecond pulsed lasers such as Ti-sapphire enable multiphoton excitation of endogenous and extrinsic fluoro- and luminophores.. The development of robust and easy-to-use laser systems has made multiphoton microscopy a rapidly emerging technique for imaging [75].

Studies of lanthanide complexes under such multiphoton excitation are rare [24, 76–78], and further work is necessary to identify complexes that have a suitable response towards multiphoton excitation under biological conditions. We note that multiphoton-excited luminescence, which involves the *simultaneous* absorption of several photons, is different from the upconversion luminescence for lanthanide-doped phosphors described earlier in this chapter, which is based on the *sequential* absorption of photons.

Not surprisingly, multiphoton excitation of the luminescence of lanthanide complexes has mainly been applied to Eu^{3+} and Tb^{3+} complexes, yielding NIR excitation of visible luminescence. Multiphoton excitation of NIR luminescent complexes has been pioneered [79] and seems a viable option, giving rise to NIR excitation of NIR luminescence, of particular interest for luminescence detection in biological media, since both the NIR excitation light and the NIR emitted light propagate well in these materials.

Photoexcitation of luminescence is convenient and allows for flexibility in wavelengths and excitation regimes. Alternative excitation modes of luminescence excitation may be of interest, such as chemiluminescence [80, 81] or electrogenerated chemiluminescence (ECL, see Sect. 7.2) [82]. For example, electrodes may be readily integrated in current "lab-on-chip" microfluidic systems for biological detection, and both chemiluminescence [83] and ECL [84] have been demonstrated for the detection in microfluidics. Examples of chemical or electrochemical excitation of NIR lanthanide luminescence are rare, but the chemiluminescence of Pr^{3+} [85] and ECL [86] of Yb^{3+} have been reported.

4.3 *Detection of Near-Infrared Luminescence*

The first spectroscopic studies of ytterbium(III) luminescence in solution by Crosby and Kasha were achieved by projecting the dispersed spectrum onto NIR-sensitised photographic paper using long exposure times [4]. Nowadays, sensitive detectors

Table 4 Detectors that have been demonstrated to be sensitive enough to detect certain NIR luminescent lanthanide complexes

Detector	Detection mode	Wavelength range (nm)	Typ. time resolution
Ge photodiode with preamplifier (liquid nitrogen cooling)	Analogue	800–1,800	100 ns
InGaAs photodiode with preamplifier (liquid nitrogen cooling)	Analogue	800–1,700	10 ms
NIR PMT InP/InGaAsP (strong thermoelectrical cooling)	Photon counting	950–1,200 950–1,400	1 ns
NIR PMT InP/InGaAs (strong thermoelectrical cooling)	Photon counting	950–1,700	1 ns
CCD, Si, normal	Digitised analogue	400–1,050	40 ms
CCD, Si, backthinned	Digitised analogue	200–1,100	40 ms

with significant sensitivity at wavelengths longer than 700 nm are still relatively rare in standard spectroscopic and microscopic equipment. Over the years, however, much progress in detector technology has been made, and a range of detectors suitable for the detection of the weak NIR luminescence of lanthanide complexes in solution is available (Table 4). These detectors are often available as an option to fluorimetric equipment. Integration into microscopes or microplate readers is at present not commercially available.

Most detection systems (photodiodes, PMT) need strong cooling, either by liquid nitrogen (77 K) or multi-stage Peltier cooling (210 K), which makes these systems relatively bulky and more costly but very sensitive. The NIR photomultiplier modules offer photon counting sensitivity and time-resolution. On the lower-cost, more compact side, certain silicon-based detectors, particularly CCD arrays, which only need modest cooling, or can even be operated at room temperature, are capable of detecting the luminescence of Yb^{3+} at 980 nm (e.g. the spectra in Fig. 6 were taken using a scientific-grade front-illuminated CCD) and the shortest wavelength emission band of Nd^{3+} at 880 nm.

The challenge of detecting weak NIR (700–1,600 nm) luminescence in organic and biological media is not limited to luminescent lanthanide complexes. In this respect, it is interesting to mention the efforts currently made for the microscopic detection of singlet oxygen phosphorescence [87, 88]. Singlet oxygen is a highly reactive photochemical species, usually obtained through the energy transfer from excited triplet states of organic chromophores such as porphyrins. It is an intermediate in photodynamic therapies, which are studied as treatments for certain cancers, macular degeneration, acne and bacterial infections [89, 90]. Singlet oxygen

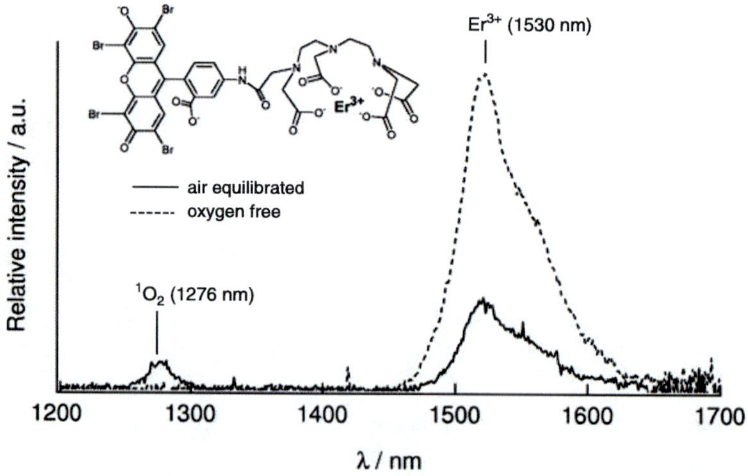

Fig. 8 The simultaneous generation of singlet oxygen phosphorescence and erbium(III) luminescence by an eosin-modified DTPA complex in buffered D2O (pD 8.3), demonstrating the comparable (very weak) luminescence intensities of erbium(III) luminescence and singlet oxygen emission. Upon deoxygenation, the erbium(III) luminescence yield increases, which shows the competition between oxygen and erbium(III) in the energy transfer from eosin

produces a weak phosphorescence emission at 1,276 nm, which has effective quantum yields on the order of 10^{-5}. Figure 8 shows that this emission is indeed comparable to the emission of certain NIR luminescent lanthanide complexes [18]. The efforts made in the instrumental development of microscopes that image the luminescence of singlet oxygen [88] are therefore of great interest for the imaging of NIR luminescent lanthanide complexes.

4.4 Sensing: Luminescent Probes Based on Near-Infrared Lanthanide Luminescence

Since lanthanide complexes can be designed to be sensitive to pH, temperature, oxygen or other analytes [91, 92] (see Chap. 5), they may become useful probes for live cell imaging. Luminescent probes have mainly been developed using europium(III) and terbium(III) as the emitting ions, but a few examples of modulation of the photophysical behaviour of NIR luminescent lanthanide complexes in response to external physicochemical parameters exist. For instance, the oxygen sensitivity of certain NIR luminescent lanthanide complexes [18] may be exploited for oxygen sensing.

Although not particularly designed to be a pH-sensitive luminescent probe, Beeby et al. described a phenanthridine-carrying water-soluble Yb^{3+} complex, in which the photosensitisation mechanism is switched as a function of pH [93]. A crown ether-modified neodymium(III) cryptand was proposed for barium ion detection [94], but as with many prototype cation sensors, the probe only works in acetonitrile. Moreover, the complex necessitates ultraviolet excitation, which removes one of the main advantages of using NIR luminescent lanthanide complexes.

The detection of DNA strands by NIR luminescent lanthanide complexes is possible through the modulation of Yb luminescence in Pd–porphyrin-containing complexes upon nucleic acid binding [48] or by the sensitivity of the intramolecular energy transfer towards the presence of DNA sensing in coumarin–rhodamine–modified ytterbium complexes [95].

4.5 Labelling, Bio-Imaging and Biosensing: Luminescent Labels Based on Near-Infrared Luminescent Lanthanide Ions

NIR labels based on lanthanide ions have been employed in proof-of-principle situations, which demonstrate their usability for certain applications. In all cases, specifically modified imaging and detection equipment needs to be used in order to accommodate NIR excitation and/or emission. The availability of NIR excitation sources and detectors in standard spectroscopic and microscopic equipment can be expected to grow in the coming years.

Visibly luminescent, NIR (980 nm) excitable Er/Yb-doped nanoparticles have been used for microscopic detection in DNA microarrays [65]. The complete absence of endogenous upconversion luminescence leads to very low backgrounds and enhanced contrast. An NIR luminescent ytterbium(III) complex based on the fluorescein–isothiocyanate motif was demonstrated to be an adequate luminescent label in a sandwich-type in vitro diagnostic test [27].

There are indications that NIR luminescent complexes can indeed be used for live cell imaging and in vivo detection. Ytterbium-porphyrins could be localised in mice by fibre-laser fluorimetry [15], whereas particular Yb^{3+} complexes can enter living cells and give detectable luminescence in cell suspensions [53, 96]. One of such complexes has been used as a dual-function probe, combining near-infrared luminescence for detection and imaging (although NIR imaging was not demonstrated) and singlet oxygen generation for photodynamic action on malignant cells [53].

For these biophotonic applications, questions such as cytotoxicity, in vitro and in vivo stability and specific targeting need to be addressed. The chemical similarity between lanthanide ions will help to translate know-how developed for visibly luminescent (Eu^{3+}, Tb^{3+}...) complexes into solutions for NIR luminescent complexes.

4.6 Comparison to Other Near-Infrared Luminescent Materials for Biophotonic Applications

It is interesting to compare NIR luminescent lanthanide complexes to other materials that can be and are being used in bioanalytical detection and imaging. In particular, "all-organic" NIR dyes such as indocyanine green already meet success in these applications. The NIR fluorescence quantum yields of organic NIR dyes can be on the order of 10% [97], although quantum yields are expected to be strongly dependent on the exact medium in which the dyes are used. The development of NIR fluorescent dyes is still ongoing [98]. As an interesting example of the application of this class of dyes, the protease activity of living cells has been investigated in fluorescence microscopy using a CCD camera [99].

The edge that Yb^{3+} complexes may have over NIR dyes for detection in biological, photonically complex samples lies in several aspects. One aspect is its spectrally concentrated emission, which allows for drastic optical filtering using narrow band pass filters. With current optical multilayer coating engineering, it is possible to obtain tailored filters with extremely high rejection ($OD > 6$) outside the pass band. Another aspect is its luminescence decay time, which may be as long as 10 μs, or longer, in biological samples, which may be used for time-gated detection.

In theory, the photostability of luminescent lanthanide complexes should be superior to the photostability of purely organic chromophores since excited lanthanide ions do not undergo photochemistry, and the energy transfer from the antenna chromophore to the lanthanide ion deactivates any "organic" excited states that would give rise to destructive photochemistry. Direct comparison of the photostability of NIR luminescent lanthanide complexes and organic fluorophores has not yet been carried out.

5 Conclusion

Clearly, there is still room for improvement in the photophysical performance of NIR luminescent lanthanide complexes. Without discarding the other NIR luminescent ions, it seems that ytterbium(III) at present is the most promising ion for bioanalytical applications, in view of its relatively good resistance to non-radiative deactivation and its emission being concentrated into one single emission band. The architectures of luminescent Yb^{3+} complexes may be further optimised to include powerful sensitising chromophores (high extinction coefficients and efficient energy transfer) and provide effective shielding against non-radiative deactivation (high quantum yields for the lanthanide luminescence step). In conclusion, NIR luminescent lanthanide still holds significant promise for bioanalytics and optical imaging, although demonstrations of their actual use in such applications are rare.

References

1. Taroni P, Pifferi A, Torricelli A, Comelli D, Cubeddu R (2003) Photochem Photobiol Sci 2:124–129
2. Valeur B (2008) Sons et lumière. Editions Belin, Paris, p 35
3. Abugo OO, Nair R, Lakowicz JR (2000) Anal Biochem 279:142–150
4. Crosby GA, Kasha M (1958) Spectrochim Acta 10:377–382
5. Heller A (1967) J Am Chem Soc 89:167–168
6. Werts MHV, Jukes RTF, Verhoeven JW (2002) Phys Chem Chem Phys 4:1542–1548
7. Werts MHV, Verhoeven JW, Hofstraat JW (2000) J Chem Soc Perkin Trans 2:433–439
8. Hasegawa Y, Wada Y, Yanagida S (2004) J Photochem Photobiol C 5:183–202
9. Beeby A, Dickins RS, Faulkner S, Parker D, Williams JAG (1997) Chem Commun 1401–1402
10. Davies GM, Adams H, Pope SJA, Faulkner S, Ward MD (2005) Photochem Photobiol Sci 4:829–834
11. Moore EG, Szigethy G, Xu J, Palsson LO, Beeby A, Raymond KN (2008) Angew Chem Int Ed 47:9500–9503
12. Zhang J, Badger PD, Geib SJ, Petoud S (2005) Angew Chem Int Ed 44:2508–2512
13. Hasegawa Y, Ohkubo T, Sogabe K, Kawamura Y, Wada Y, Nakashima N, Yanagida S (2000) Angew Chem Int Ed Engl 39:357–360
14. Gouterman M, Schumaker CD, Srivastava TS, Yonetani T (1976) Chem Phys Lett 40: 456–461
15. Gaiduk MI, Grigoryants VV, Mironov AF, Rumyantseva VD, Chissov VI, Sukhin GM (1990) J Photochem Photobiol B 7:15–20
16. Rusakova NV, Meshkova SB (1990) Zh Anal Khim 45:1914–1921
17. Rusakova NV, Meshkova SB, Venchikov VY, Pyatosin VE, Tsvirko MP (1992) J Appl Spectrosc 56:488–490
18. Werts MHV, Hofstraat JW, Geurts FAJ, Verhoeven JW (1997) Chem Phys Lett 276:196–201
19. Hasegawa Y, Kimura Y, Murakoshi K, Wada Y, Kim JH, Nakashima N, Yamanaka T, Yanagida S (1996) J Phys Chem 100:10201–10205
20. Beeby A, Faulkner S (1997) Chem Phys Lett 266:116–122
21. Ding X, Alford JM, Wright JC (1997) Chem Phys Lett 269:72–78
22. Korovin Y, Rusakova N (2004) J Alloys Compd 374:311–314

23. Li M, Selvin PR (1995) J Am Chem Soc 117:8132–8138
24. Werts MHV, Nerambourg N, Pélégry D, Le Grand Y, Blanchard-Desce M (2005) Photochem Photobiol Sci 4:531–538
25. Klink SI, Keizer H, van Veggel FCJM (2000) Angew Chem Int Ed 39:4319
26. Beeby A, Clarkson IM, Dickins RS, Faulkner S, Parker D, Royle L, Sousa AS, Williams JAG, Woods M (1999) J Chem Soc Perkin Trans 2:493–503
27. Werts MHV, Woudenberg RH, Emmerink PG, van Gassel R, Hofstraat JW, Verhoeven JW (2000) Angew Chem Int Ed 39:4542–4544
28. Bassett AP, Van Deun R, Nockemann P, Glover PB, Kariuki BM, Van Hecke K, Van Meervelt L, Pikramenou Z (2005) Inorg Chem 44:6140–6142
29. Glover PB, Bassett AP, Nockemann P, Kariuki BM, Van Deun R, Pikramenou Z (2007) Chem Eur J 13:6308–6320
30. Liu H, Yuan L, Yang X, Wang K (2003) Chem Biol Interact 146:27–37
31. Voloshin AI, Shavaleev NM, Kazakov VP (2001) J Lumin 93:199–204
32. Voloshin AI, Shavaleev NM, Kazakov VP (2000) J Lumin 91:49–58
33. Davies GM, Aarons RJ, Motson GR, Jeffery JC, Adams H, Faulkner S, Ward MD (2004) Dalton Trans 1136–1144
34. Sharma PK, Van Doorn AR, Staring AGJ (1994) J Lumin 62:219–225
35. Serra OA, Nassar EJ, Calefi PS, Rosa ILV (1998) J Alloys Compd 275:838–840
36. Zhang J, Petoud S (2008) Chem Eur J 14:1264–1272
37. Blasse G, Schipper W, Hamelink JJ (1991) Inorg Chim Acta 189:77–80
38. Azenha ME, Burrows HD, Fonseca SM, Ramos ML, Rovisco J, de Melo JS, Sobral AJFN, Kogej K (2008) New J Chem 32:1531–1535
39. Li WL, Mishima T, Adachi GY (1987) Inorg Chim Acta 131:287–291
40. Tapia MJ, Burrows HD (2002) Langmuir 18:1872–1876
41. Fivet V, Quinet P, Biemont E, Jorissen A, Yushchenko AV, Van Eck S (2007) Mon Not R Astron Soc 382:944
42. Fivet V, Quinet P, Biemont E, Jorissen A, Yushchenko AV, Van Eck S (2007) Mon Not R Astron Soc 380:771–780
43. Krause C et al (2003) ORNL Rev 36:3
44. Krupke WF (1972) IEEE J Quant Electron 8:725
45. Bailey MP, Rocks BF, Riley C (1984) Analyst (London) 109:1449–1450
46. Hebbink GA, Klink SI, Grave L, Alink PGBO, van Veggel FCJM (2002) Chem Phys Chem 3:1014–1018
47. Slooff LH, Polman A, Klink SI, Hebbink GA, Grave L, van Veggel FCJM, Reinhoudt DN, Hofstraat JW (2000) Opt Mater 14:101–107
48. Beeby A, Dickins RS, FitzGerald S, Govenlock LJ, Maupin CL, Parker D, Riehl J, Siligardi G, Williams J (2000) Chem Commun 1183–1184
49. Tsvirko M, Korovin Y, Rusakova N (2007) J Phys Conf Ser 79:012025
50. Zhuravlyov S, Rusakova N, Korovin Y (2008) J Alloys Compd 451:334–337
51. Seltzer MD, Fallis S, Hollins RA, Prokopuk N, Bui RN (2005) J Fluoresc 15:597–603
52. Chudinov AV, Rumyantseva VD, Lobanov AV, Chudinova GK, Stomakhin AA, Mironov AF (2004) Russ J Bioorganic Chem 30:89–93
53. Jiang FL, Poon CT, Wong WK, Koon HK, Mak NK, Choi CY, Kwong DWJ, Liu Y (2008) Chem Bio Chem 9:1034–1039
54. Turro NJ (1978) Modern molecular photochemistry. Benjamin/Cummings, Menlo Park, CA
55. Horrocks William DeW, Jr BJP, Smith WD, Supkowski RM (1997) J Am Chem Soc 119:5972–5973
56. Hofstraat JW, Oude Wolbers MP, Van Veggel FCJM, Reinhoudt DN, Werts MHV, Verhoeven JW (1998) J Fluoresc 8:301–308
57. Shi J, Zhang X, Neckers DC (1992) J Org Chem 57:4418–4421
58. Shi J, Zhang X, Neckers DC (1993) Tetrahedron Lett 34:6013–6016
59. Hebbink G, Grave L, Woldering L, Reinhoudt D, van Veggel F (2003) J Phys Chem A 107:2483–2491

60. Steemers FJ, Verboom W, Reinhoudt DN, Van der Tol EB, Verhoeven JW (1995) J Am Chem Soc 117:9408–9414
61. Werts MHV, Duin MA, Hofstraat JW, Verhoeven JW (1999) Chem Commun 799–800
62. Hebbink GA, Reinhoudt DN, van Veggel FCJM (2001) Eur J Org Chem 4101–4106
63. Eaton DF (1988) Pure Appl Chem 60:1107–1114
64. Beverloo HB, Van Schadewijk A, Van Gelderen-Boele S, Tanke HJ (1990) Cytometry 11:784–792
65. Van de Rijke F, Zijlmans H, Li S, Vail T, Raap AK, Niedbala RS, Tanke HJ (2001) Nat Biotechnol 19:273–276
66. Riwotzki K, Meyssamy H, Kornowski A, Haase M (2000) J Phys Chem B 104:2824–2828
67. Stouwdam JW, Hebbink GA, Huskens J, Van Veggel FCJM (2003) Chem Mater 15:4604–4616
68. Heer S, Kömpe K, Güdel HU, Haase M (2004) Adv Mater 16:2102–2105
69. Wei Y, Lu F, Zhang X, Chen D (2007) J Alloys Compd 427:333–340
70. Chengelis DA, Yingling AM, Badger PD, Shade CM, Petoud S (2005) J Am Chem Soc 127:16752–16753
71. Diamente PR, Burke RD, Van Veggel FCJM (2006) Langmuir 22:1782–1788
72. Lim SF, Riehn R, Ryu WS, Khanarian N, Tung C, Tank D, Austin RH (2006) Nano Lett 6:169–174
73. Driesen K, Van Deun R, Görller-Walrand C, Binnemans K (2004) Chem Mater 16:1531–1535
74. Sun L, Yu J, Zhang H, Meng Q, Ma E, Peng C, Yang K (2007) Microporous Mesoporous Mater 98:156–165
75. Zipfel WR, Williams RM, Webb WW (2003) Nat Biotechnol 21:1369–1377
76. Lakowicz JR, Piszczek G, Maliwal BP, Gryczynski I (2001) Chem Phys Chem 4:247–252
77. White GF, Litvinenko KL, Meech SR, Andrews DL, Thomson AJ (2004) Photochem Photobiol Sci 3:47–55
78. Fu LM, Wen XF, Ai XC, Sun Y, Wu YS, Zhang JP, Wang Y (2005) Angew Chem Int Ed 44:747–750
79. Piszczek G, Gryczynski I, Maliwal BP, Lakowicz JR (2002) J Fluoresc 12:15–17
80. Meijer EW, Wijnberg H (1982) J Chem Ed 59:1071–1072
81. Wildes PD, White EH (1971) J Am Chem Soc 93:6286–6288
82. Bard AJ (ed) (2004) Electrogenerated chemiluminescence. CRC, Boca Raton, FL
83. Liu BF, Ozaki M, Utsumi Y, Hattori T, Terabe S (2003) Anal Chem 75:36–41
84. Zhan W, Alvarez J, Crooks RM (2002) J Am Chem Soc 124:13265–13270
85. Kazakov VP, Voloshin AI, Shavaleev NM (1998) J Photochem Photobiol A 119:177–186
86. Ala-Kleme T, Haapakka K, Latva M (1999) Anal Chim Acta 395:205–211
87. Jimenez-Banzo A, Ragas X, Kapusta P, Nonell S (2008) Photochem Photobiol Sci 7: 1003–1010
88. Snyder JW, Zebger I, Gao Z, Poulsen L, Frederiksen PK, Skovsen E, McIlroy SP, Klinger M, Andersen LK, Ogilby PR (2004) Acc Chem Res 37:894–901
89. Bechet D, Couleaud P, Frochot C, Viriot ML, Guillemin F, Barberi-Heyob M (2008) Trends Biotechnol 26:612–621
90. Wilson BC, Patterson MS (2008) Phys Med Biol 53:R61–R109
91. Dossing A (2005) Eur J Inorg Chem 1425–1434
92. Parker D, Senanayake PK, Williams JAG (1998) J Chem Soc Perkin Trans 2:2129–2139
93. Beeby A, Faulkner S, Williams JAG (2002) J Chem Soc Dalton Trans 1918–1922
94. Coldwell JB, Felton CE, Harding LP, Moon R, Pope SJA, Rice CR (2006) Chem Commun 5048–5050
95. Bodi A, Borbas KE, Bruce JI (2007) Dalton Trans 4352–4358
96. Chauvin AS, Comby S, Song B, Vandevyver CDB, Bunzli JCG (2008) Chem Eur J 14:1726–1739
97. Shealy DB, Lipowska M, Lipowski J, Narayanan N, Sutter S, Strekowski L, Patonay G (1995) Anal Chem 67:247–251
98. Lin Y, Weissleder R, Tung CH (2002) Bioconjug Chem 13:605–610
99. Weissleder R, Tung CH, Mahmood U, Bogdanov A (1999) Nat Biotechnol 17:375–378

Lanthanide Assemblies and Polymetallic Complexes

Stephen Faulkner and Daniel Sykes

Abstract In this chapter, we discuss how luminescent lanthanides can be incorporated into complex architectures, and use the systems described to illustrate key aspects of the behaviour of lanthanide containing assemblies in the solid state and in solution.

Keywords Assemblies · d–f hybrids · Multi-metallic · Luminescence · Energy transfer

Contents

Abbreviations

BET	Back energy transfer
Bipy	2,2′-Bipyridine
Bpym	2,2′-Bipyrimidine

S. Faulkner (✉) and D. Sykes
Chemistry Research Laboratory, University of Oxford, Mansfield Road, Oxford OX1 3TA, UK
e-mail: Stephen.Faulkner@keble.ox.ac.uk

P. Hänninen and H. Härmä (eds.), *Lanthanide Luminescence: Photophysical, Analytical and Biological Aspects*, Springer Ser Fluoresc (2011) 7: 161–182, DOI 10.1007/4243_2010_4,

DTPA	Diethylenetriaminepentaacetic acid
Hat	Hexazatriphenylene
HHH	Head-to-head-to-head
HHT	Head-to-head-to-tail
IC	Internal conversion
ISC	Inter-system crossing
LMCT	Ligand-to-metal charge transfer
Ln	Lanthanide
MLCT	Metal-to-ligand charge transfer
NIR	Near infrared
PET	Photo-electron transfer
S_0	Singlet ground state
S_1	First singlet excited state
T_1	First triplet excited state

1 Introduction

More than 30 years after the term was first used, the idea of supramolecular chemistry has encompassed a multitude of molecules, extending from coordination chemistry into the realms of mechanical bonding. The common feature to all "supramolecular" structures is the idea of an array that comprises two or more molecular components that are held together by intermolecular forces. Such structures are, as we will see, effective tools for bringing together lanthanide ions and sensitizing chromophores; they have the added advantage that they facilitate synthesis and allow a degree of control of local geometry at the lanthanide site. The similarity of chemical behaviour between the lanthanides does, however, limit the scope of supramolecular assemblies – particularly when the preparation of well-defined heterometallic complexes is desirable. In such cases, much can be achieved through the use of kinetically stable complexes as building blocks, so we will also consider the use of such building blocks in the preparation of such system.

The reader will already have judged from the opening paragraph that the concept of lanthanide arrays is wide-ranging almost to the point of being nebulous. To keep matters in perspective, this chapter will focus on the preparation and luminescence properties of well-defined lanthanide containing systems in the solid state and in solution, using a range of examples to illustrate how complex architectures can be prepared and how such systems can be used to probe, exploit, and understand the nature of sensitized emission within the lanthanides.

For a lanthanide complex, the photophysical pathways that can lead to, or disrupt, sensitized luminescence from a lanthanide ion are shown in Fig. 1. Many of these have been discussed in detail in [1], and in a range of review articles [2], but it is worth reprising key points briefly here. For a typical sensitizing chromophore (e.g., an aromatic molecule or a transition metal complex), the first key step is

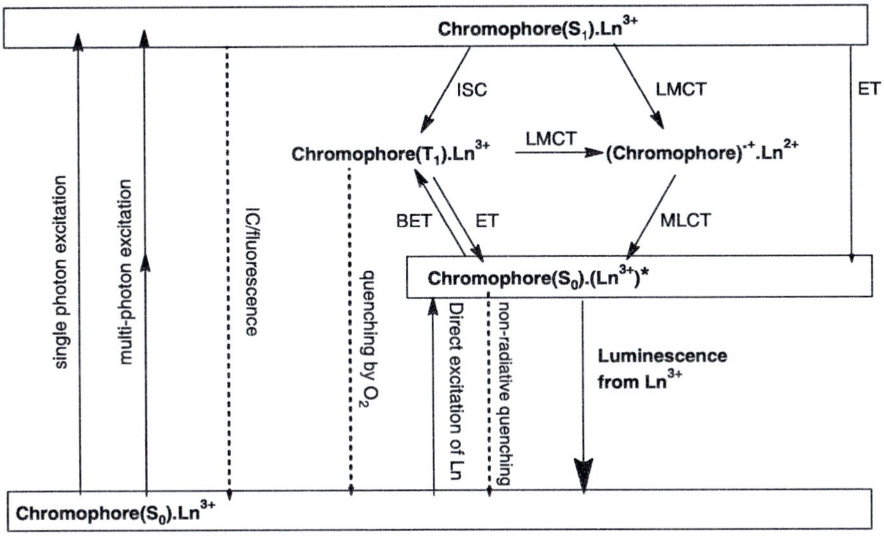

Fig. 1 Pathways for energy transfer and non-radiative deactivation in lanthanide complexes

absorption of a photon to generate an excited state – usually a singlet state. The most common pathway for energy transfer to a lanthanide ion then involves formation of a triplet state through inter-system crossing, which then populates the emissive state of the lanthanide by energy transfer (either via a Förster or a Dexter mechanism, depending on spectral overlap and array structure), though other pathways, including sequential charge transfer and direct energy transfer from the sensitizer singlet state, can also lead to formation of the lanthanide excited state. There are also a range of competing processes which can lead to non-radiative deactivation of chromophore-centred states before energy transfer to the lanthanide. In the case of ytterbium complexes, an alternative mechanism has also been observed [3, 4]; in this pathway, ytterbium-centred LMCT states can populate the $^2F_{5/2}$ state of Yb^{3+} through sequential double electron transfer. Nor is the chromophore triplet state alone responsible for energy transfer; there is considerable evidence that singlet-mediated energy transfer can also contribute significantly to population of the lanthanide excited state [5]. Such a pathway may also be responsible for the two-component risetime observed with a phenanthridinium appended lanthanide complex [6]; in Eu.**1**, the risetime of the emission (Fig. 2) at 595 nm consists of a component with the same lifetime as that of the triplet state and a second component corresponding to an effectively instantaneous rise within the envelope of the excitation pulse.

Once the lanthanide excited state has formed, luminescence is not an inevitable outcome. Non-radiative quenching of the excited state can occur through vibrational modes of the array, and particularly through vibrational quenching by X–H oscillators in the ligand backbone or in aqueous media. To optimize emission from the lanthanide it is necessary to minimize the non-radiative quenching pathways

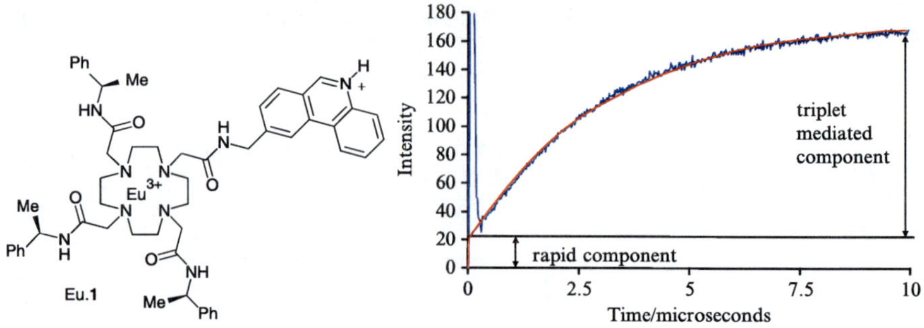

Fig. 2 Showing the rapid and triplet mediated components of the risetime of the europium-centred emission in Eu.**1**. The *blue line* shows experimental data, while the *red line* shows the fit obtained (reproduced with permission)

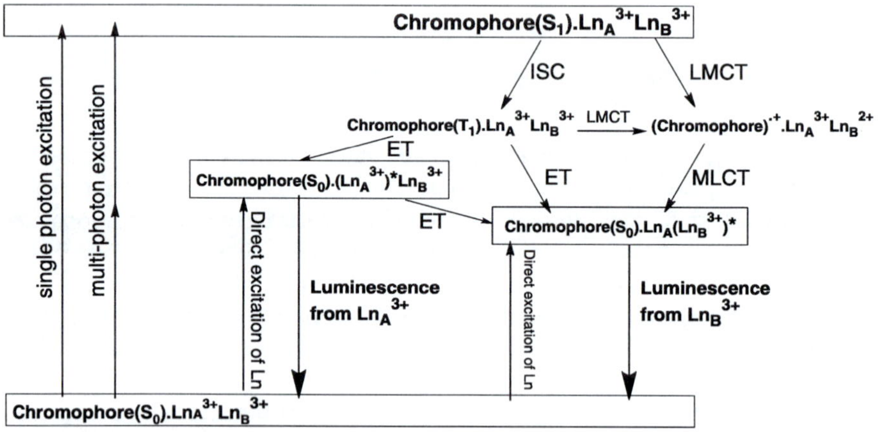

Fig. 3 Simplified scheme showing sensitized emission in polymetallic lanthanide complexes

available to the lanthanide emissive state (e.g., by excluding solvent from the inner coordination sphere) and maximizing the efficiency of energy transfer from the chromophore excited state.

In a polynuclear or polymetallic lanthanide complex, additional processes become feasible. Figure 3 is a simplified scheme that shows the energy transfer pathways available in a bimetallic lanthanide complex; singlet-mediated energy transfer and competitive non-radiative quenching pathways are not shown in the scheme to facilitate interpretation. Not only can both lanthanide ions be sensitized by the excited states of the chromophore, but energy transfer between lanthanide ions is also possible where appropriate spectral overlap exists between the two lanthanide ions.

In the case of chromophore sensitized emission, the efficiency of energy transfer to Ln_A will depend upon the separation between the lanthanide and the chromophore and upon the spectral overlap between the chromophore donor state and the lanthanide acceptor state. While the latter is effectively fixed for a given chromophore-lanthanide pair, the former can be varied by varying the chromophore separation. The rate of energy transfer governs the population of the excited state. For instance, Ala Kleme et al. [7] used a heterometallic assembly of the form $(HXTA)_2TbYb$ to show that the electron transfer mediated population of the ytterbium excited state dominates the photophysical behaviour of the system, quenching the terbium-centred luminescence.

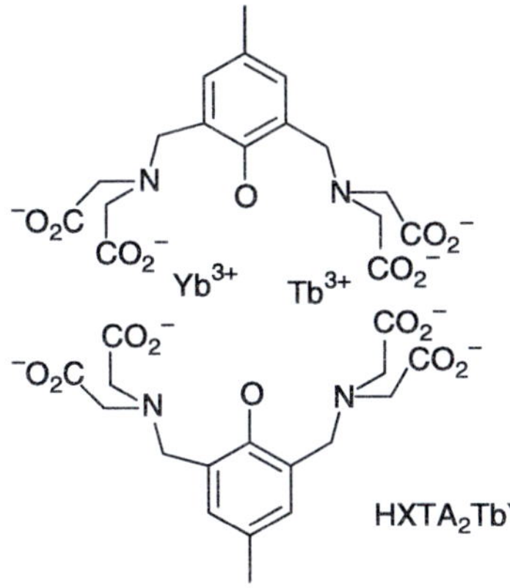

$HXTA_2TbYb$

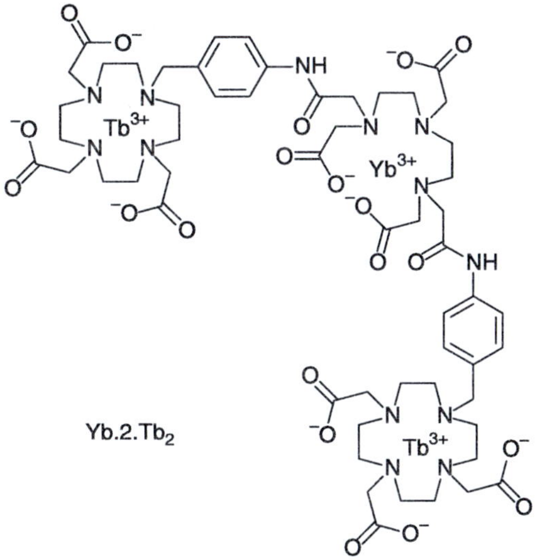

$Yb.2.Tb_2$

There is also the possibility of energy transfer between lanthanide ions, using the excited state of one lanthanide to facilitate energy transfer to a different lanthanide. We used Yb.**2**.Tb_2 to demonstrate energy transfer from terbium to ytterbium following direct excitation of the terbium $^5D_4$–$^7F_6$ transition [8].

As we will see from the examples on the following pages, a wide range of approaches have been developed to access lanthanide arrays. We will deal with them by using examples of a number of different classes of systems in the solid state and in solution to illustrate key aspects of the strategies used to prepare these complexes.

2 Lanthanide Arrays in the Solid State

The solid phase lends itself to the preparation of lanthanide arrays. Indeed, lanthanide doped yttrium aluminium garnets are well known as laser materials, while doped materials containing more than one kind of lanthanide have proved very effective at upconversion of energy – the sequential absorption of two photons giving rise to anti-Stokes emission [9]. In such systems, excited state absorption by the intermediate state gives rise to formation of a high energy emissive state. This is straightforward in cases where such a two step process can be mediated by the excited state manifold of a single ion. Among lanthanides Erbium is the most commonly used cation for upconversion applications, although several other ions also display this phenomenon. For instance, Güdel et al. have shown that upconversion can be observed in materials doped with Tm^{2+} [10]; although Tm^{2+} is an $4f^{13}$ ion in the ground state, it has a range of accessible excited 4f–5d states with significant excited state absorption cross-sections. Thus, NIR excitation results in luminescence in the visible part of the spectrum.

However, upconversion in heterometallic systems such processes must involve facile communication between lanthanide centres, mediated by super-exchange through bridging donor atoms. As such, dopant concentrations play a major role in determining the effectiveness of upconversion. By experimenting with different lanthanide ratios, van Veggel et al. [11] have prepared white light emissive thin films containing three types of upconverting nanoparticles, namely $La_{0.45}Yb_{0.5}Er_{0.05}F_3$, $La_{0.75}Yb_{0.2}Tm_{0.05}F_3$, and $Yb_{0.75}La_{0.2}Eu_{0.05}F_3$. Upon excitation at 980 nm (corresponding to the Yb^{3+} $^2F_{7/2} - {}^2F_{5/2}$ transition), the erbium doped system emits red and green light, the europium doped system emits red light, and the thulium doped system emits green light. While the lanthanide ratios for such systems have to be determined empirically, two things are clear: low concentrations of the emissive lanthanide ion are essential if self-quenching by proximate ions of the same type is to be avoided, while high concentrations of ytterbium optimize the likelihood of excited state absorption generating a high energy emissive state on a proximate lanthanide ion.

Coordination polymers offer a simple and effective way to incorporate chromophores and lanthanides into the same structure, particularly when the chromophore can be used as a bridging ligand. Such lanthanide containing coordination polymers are extended assemblies which assemble under kinetic control. The lability of lanthanide ions means that such systems tend to be unstable in solution, meaning that their optical properties must be determined in the solid phase using front surface techniques to observe emission from the sample and reject scatter. Ward's group have had particular success in this field, using luminescent complex anions with emissive MLCT states as sensitizers for lanthanide emission in the NIR. For instance, $[Ru(Bipy)(CN)_4]^{2-}$ [12, 13], $[Ru(Phen)(CN)_4]^{2-}$ [14], $[Ru(Bpym)(CN)_4]^{2-}$ [15], $\{[Ru(CN)_4]_2(\mu\text{-Bpym})\}^{2-}$ [15], $[Ru(Hat)(CN)_4]^{2-}$ [16], $\{[Ru(CN)_4]_3(\mu^3\text{-Hat})\}^{6-}$ [16], $\{[Ru(CN)_4]_2(\mu^2\text{-Hat})\}^{4-}$ [16], $[Ru(Hat)(CN)_4]^{2-}$ [16], $[Cr(CN)_6]^{3-}$ [17], and $[Co(CN)_6]^{3-}$ [17] have all been used as chromophores and building blocks in lanthanide containing arrays. Examples of these are shown in Fig. 4.

An alternative approach is illustrated by the work of Mazzanti and co-workers, who used ligand **3** as a building block to prepare 2D luminescent coordination polymers [18]; **3** possesses two pentadentate binding domains linked by a four atom chain that favours bridging between two lanthanide ions. These multidentate ligands also exclude solvent from the inner coordination sphere of the lanthanide ions, increasing the luminescence lifetime and intensity. This, thus, offers some important advantages over the use of simpler bridging domains, though it requires considerably more effort to be devoted to ligand design and synthesis.

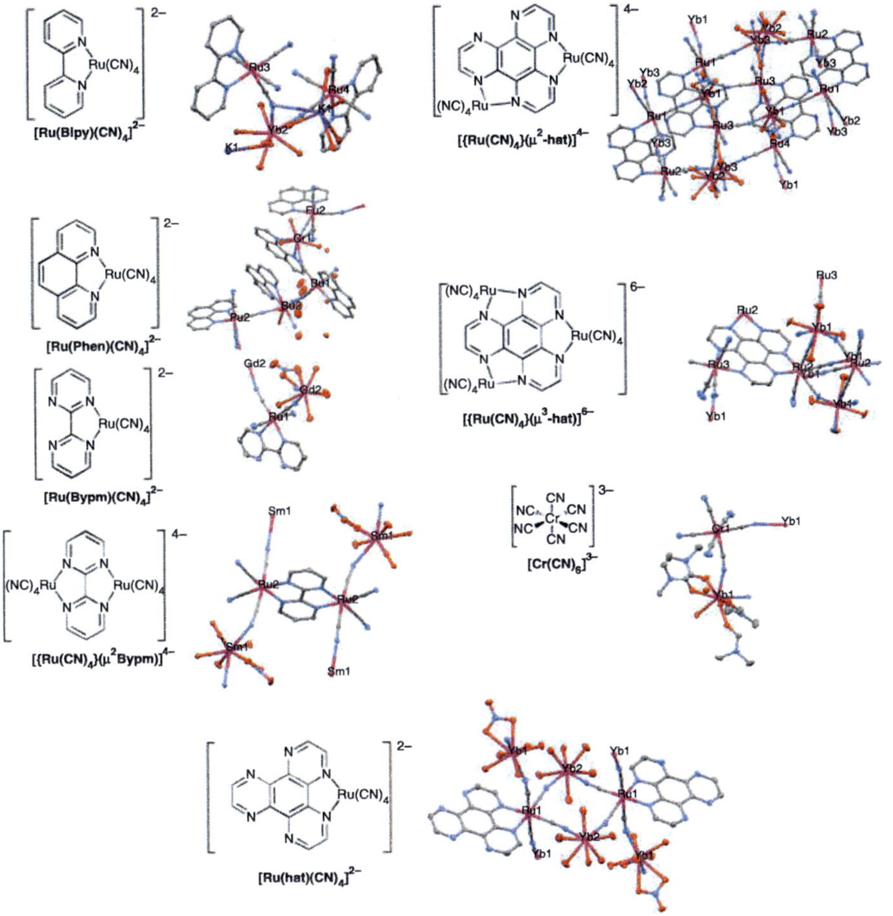

Fig. 4 Building blocks and drawings of repeat structures in luminescent coordination polymers derived from complex anions, taken from data in [12–17], with permission

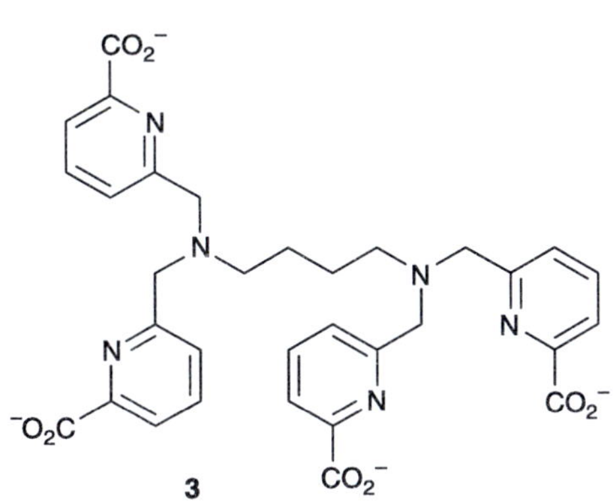

3

3 Self-Assembled Systems

Self-assembly has long been exploited in bioassay [19]; however, its use in the preparation of architectures based on small molecules is relatively recent, and it only now begins to achieve its potential. Self-assembly in solution requires more extensive ligand design and synthetic control.

3.1 Lanthanide Helicates

For the better part of two decades, the chemistry of self-assembled systems containing lanthanides has been dominated by helicate systems [20, 21]. Simple tridentate ligands such as 2,6-dipicolinate, terpyridyl and 6-(diaminocarbonyl)-picolinic acid form stable and luminescent triple helicate complexes in water [21–24]. Polymetallic chelates can be assembled by using related ligand structures that can bridge between two metal centres.

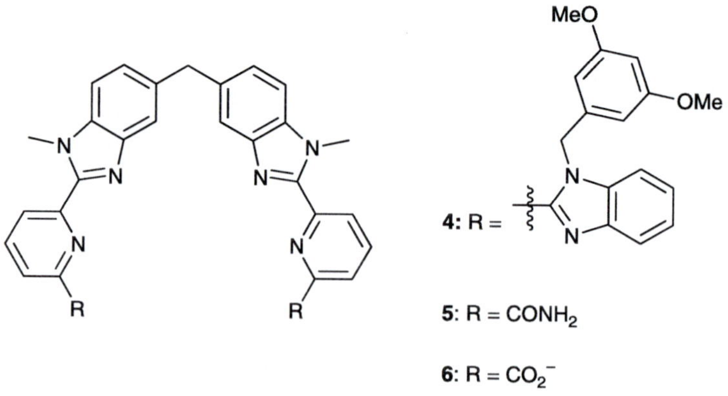

Ligands such as **4–6** contain two tridentate binding domains that must bind to different lanthanide ions, giving rise to binuclear helicates of the form Ln_2L_3. In the case of **4** [25], binuclear helicates of the form $(Ln_2.\mathbf{4}_3)^{6+}$ are only observed in acetonitrile, as the N_3 donor set does not bind to the lanthanide in protic media, where solvent molecules compete effectively for the lanthanide centre. Structural modification can change the nature of the assembly process; for instance, **5** forms two species $(Ln_2.\mathbf{5}_2)^{6+}$ and $(Ln_2.\mathbf{5}_3)^{6+}$ in solution, and the ratio of these can be varied by careful control of stoichiometry and concentration and by varying the nature of the solvent [26]. In the case of **6**, charge neutral complexes $(Ln_2.\mathbf{6}_3)$ form in all solvents, including water, assisted by the electrostatic attraction between the metal ions and the ligand fragments [27]. These principles apply across a wide range of structure types; for instance, the dicarboxylic acid **7** also forms robust helicates of the form $(Ln_2.\mathbf{7}_3)$ in aqueous solution [28]. All these systems exhibit lanthanide luminescence sensitized by the chromophores that make up the ligand skeleton.

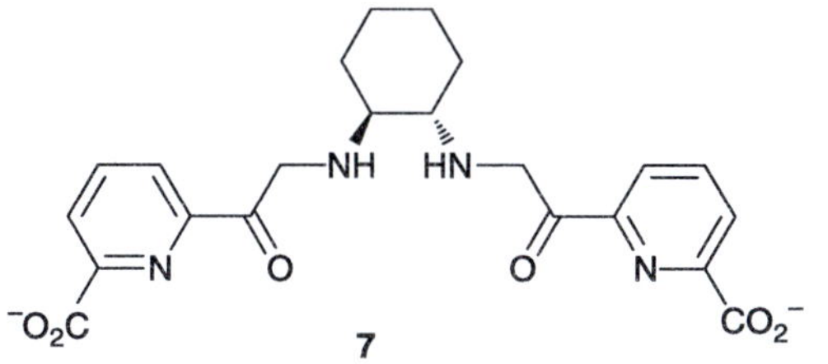

7

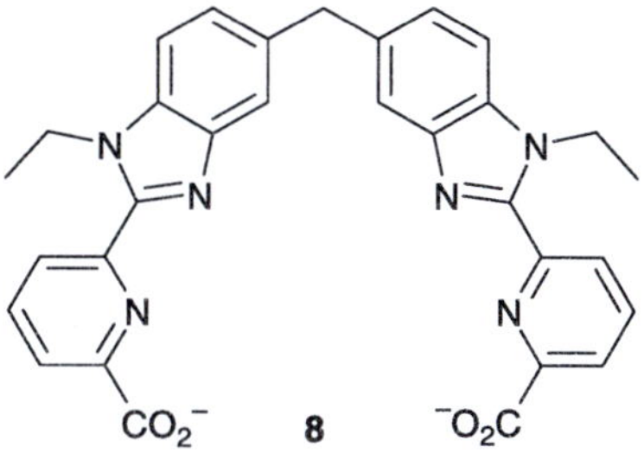

8

It is worth noting that the compartmental ligand structure enforces a separation between the two metal centres that is sufficient to prevent magnetic coupling between lanthanide ions. Gonçalves et al. carried out a detailed study on the mechanisms of energy transfer involved in sensitizing lanthanide luminescence in (Ln_2.**8**$_3$) (Ln = Sm, Yb) [5]. Upon helicate formation, the authors observed a red shift in the lowest energy π-π * absorption band of the ligand. Ligand fluorescence was effectively (though not totally) quenched as a consequence of helicate formation, while calculations implied that both triplet and singlet-mediated energy transfer processes contribute to energy transfer in (Sm_2.**8**$_3$), although triplet mediated energy transfer is by far the dominant pathway. In the case of the ytterbium containing system, poor spectral overlap between the chromophore triplet state and the metal centre mean that sequential charge transfer (vide supra) dominates. In the course of their study, the authors also attempted to observe upconversion processes within the helicate system without success, adding further evidence that magnetic communication between the two lanthanide centres in such systems is minimal.

Binuclear helicates have proved a very effective family of ligands; indeed, Bunzli and co-workers have developed a series of helicates for application in intracellular imaging, modifying the structure to ensure that complexes such as Ln_2.**9**$_3$ combine high binding constants with high solubility [29, 30]. While such systems exist in equilibrium, limiting their practicality in whole animals, they make excellent intracellular imaging agents.

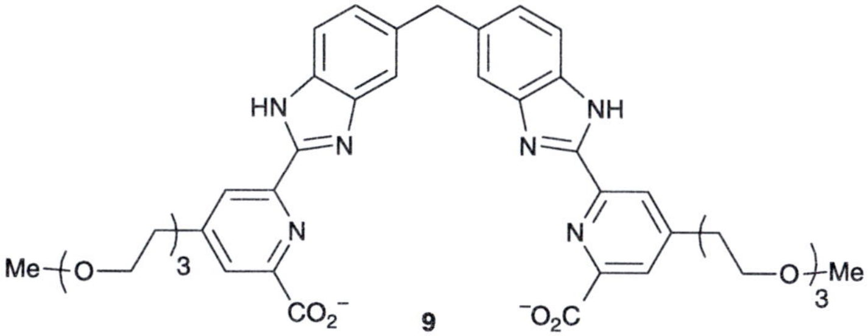

9

Nor are helicates restricted to binuclear complexes. Additional binding sites can be used to extend the system. For instance, Piguet and co-workers have used extended variants upon these compartmental structures to prepare ligands such as **10** and **11**, from the trinuclear and tetranuclear assemblies $(Ln_3.\mathbf{10}_3)^{9+}$ and $(Ln_4.\mathbf{11}_4)^{12+}$ [31, 32]. Such systems produce linear arrays of lanthanide ions, but trinuclear lanthanide helicates have also been prepared using simpler ligand structures [33]. Lama et al. prepared a helical trinuclear complex from **12** of the form $[Ln.\mathbf{12}_6(\mu_3\text{-}OH)(H_2O)_3]^{2+}$ [33], in which the chiral ligand induces chirality into the assembly (Fig. 5). In this structure, three ligands act as tridentate N_2O donors to one lanthanide ion, bridging to a second lanthanide through the remaining oxygen atom (these are shown in simplified form in blue in Fig. 5). The remaining three ligands form carboxylate bridges between two lanthanide centres.

Selective self-assembly of heterometallic helicates requires the use of unsymmetrical compartmental ligands. The preparation of heterobimetallic lanthanide complexes represents a particular challenge, given the ease with which lanthanide ions exchange ligands. This raises a new difficulty: a triple stranded helicate formed from unsymmetrical strands can exhibit isomerism as shown in Fig. 6 [34]. Since the system is under thermodynamic control, cooperative effects can govern the species that dominates at equilibrium. Thus two broadly similar ligands such as **13** and **14** can have very different properties. The head-to-head-to-head (HHH) arrangement predominates in helicates formed from **13,** while HHT isomers predominate in helicates formed from **14**. Under controlled conditions, heterometallic helicates of the form $(\mathbf{13}.Ln^a.Ln^b)^{6+}$ can be prepared; in cases where one lanthanide has a significantly larger radius than the other, heterometallic complexes dominate the speciation. The photophysical properties of $(\mathbf{13}.La.Eu)^{6+}$ and comparing them with those of $(\mathbf{13}.Eu_2)^{6+}$ are remarkably similar, and the lifetimes of the europium luminescence identical within experimental error. One surprising feature of the high-resolution spectra is the form of the $\Delta J = 0$ transition around 580 nm.

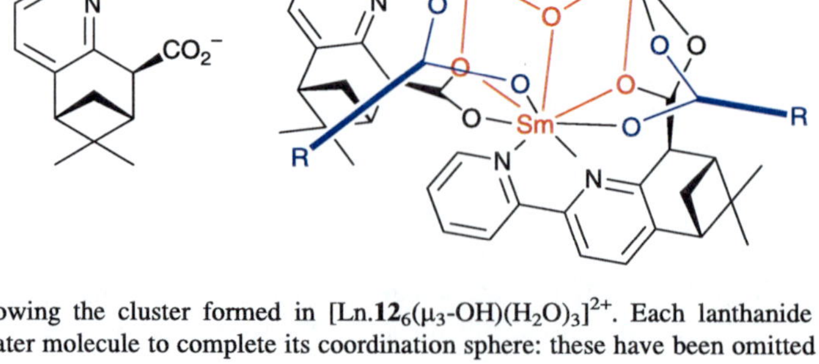

Fig. 5 Showing the cluster formed in $[Ln.\mathbf{12}_6(\mu_3\text{-}OH)(H_2O)_3]^{2+}$. Each lanthanide also has a capping water molecule to complete its coordination sphere: these have been omitted for clarity

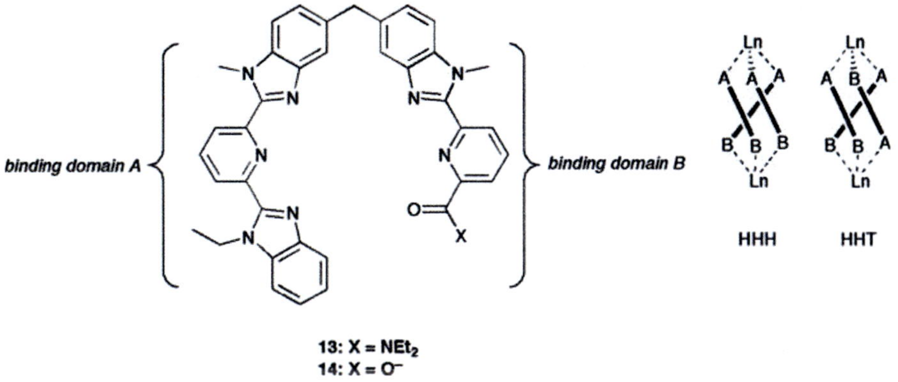

Fig. 6 Isomerism in helicates formed from unsymmetrically substituted compartmental ligands

The authors observed that both complexes (**13**.La.Eu)$^{6+}$ and (**13**.Eu$_2$)$^{6+}$ showed bands consistent with the same single species, suggesting that only one complex domain is luminescent, and allowing them to establish beyond doubt that the (**13**.La.Eu)$^{6+}$ contains a lanthanum ion in the bis-imidazolylpyridine binding domain (domain A in the figure above), since such domains do not result in emissive europium complexes (as a consequence of quenching of the europium excited state by low energy ligand-to-metal charge transfer states).

While the preparation of f-f′ heterometallic helicates remains challenging, d–f heterometallic complexes (often called d–f hybrid helicates) can be prepared with ease, since the d-block chromophore component imparts robustness to the array, effectively making it into a podand. Bunzli, Piguet and colleagues have had considerable success with luminescent helicates containing a chromium (III) centre and a luminescent lanthanide {[(**14**)$_3$.CrLn] [35, 36]}. In such systems, the large separation between the donor state and the lanthanide acceptor can be used to control the rate of energy transfer to the point at which energy transfer becomes rate determining.

3.2 *Self-Assembly of Systems Other Than Helicates*

Other approaches can be used to form polymetallic assemblies under thermodynamic control. One of the simplest is to exploit micellar structures that encapsulate lanthanide ions. For instance, during our early studies on lanthanide emission in the NIR, we used reverse micelles derived from sodium bis(2-ethylhexyl) sulphosuccinate to control the solvation of lanthanide ions [37]. While this approach gives great flexibility, micellar assemblies tend not to be well defined, meaning that a range of species exist in solution at any time.

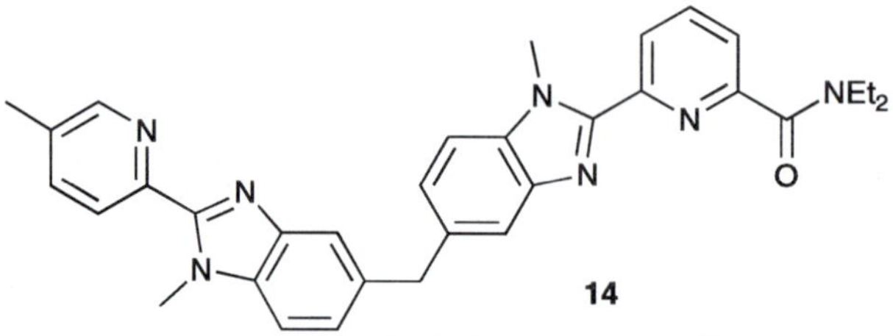

14

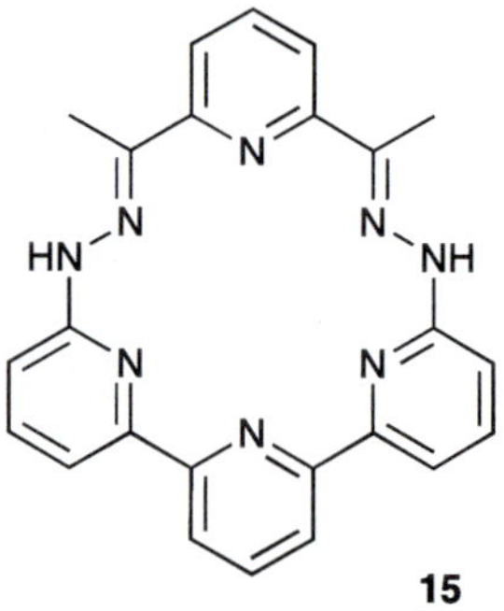

15

By contrast, coordinatively unsaturated lanthanide complexes exhibit much simpler speciation in solution. For instance, **15** acts as a hexadentate ligand to ytterbium in aqueous solution. At low pH, three bound solvent molecules give rise to effective non-radiative quenching and thus to weak, short-lived luminescence [38]. However, at high pH, an oxo-dimer is formed, giving rise to long lived luminescence as a consequence of the exclusion of O–H oscillators from the inner coordination sphere. This controlled hydrolysis can also be exploited in the formation of larger arrays. For instance effective routes to luminescent penta-metallic [39], hexa-metallic [40], and non-metallic lanthanide containing complexes [41] have been demonstrated with admirable synthetic control.

Nanoparticles also offer an effective way to access polynuclear complexes. Once again, the relatively diverse nature of the speciation in such systems means that there is relatively little control of the number of ions in the complex. However, unlike micelles, nanoparticles are robust on the timescale of imaging and assay. Gunnlaugsson et al. have made great use of this in the development of displacement assays in which a lanthanide is bound in a stable binding domain attached to the surface of a gold nanoparticle [42, 43]. In such systems, care needs to be taken to ensure sufficient separation between the nanoparticle surface and the luminescent centre and remove the possibility of quenching of the lanthanide excited state through interactions with the nanoparticle core [44].

3.3 Assembly of Arrays from Stable Building Blocks

The similarities in chemical behaviour across the lanthanides mean that kinetically stable lanthanide complexes offer a degree of control that cannot readily be achieved through selective self-assembly of small building blocks. Azamacrocycle complexes derived from cyclen amino-carboxylates are particularly suited to this, and ternary complexes can be used to study the photophysical behaviour of lanthanide complexes. We initially used such an approach to probe the effectiveness of chromophores as sensitizers for lanthanides without recourse to extensive synthesis (as illustrated in Fig. 7). This approach can be used with simple chromophores to probe the mechanism of energy transfer. For instance, tetrathiafulvalene

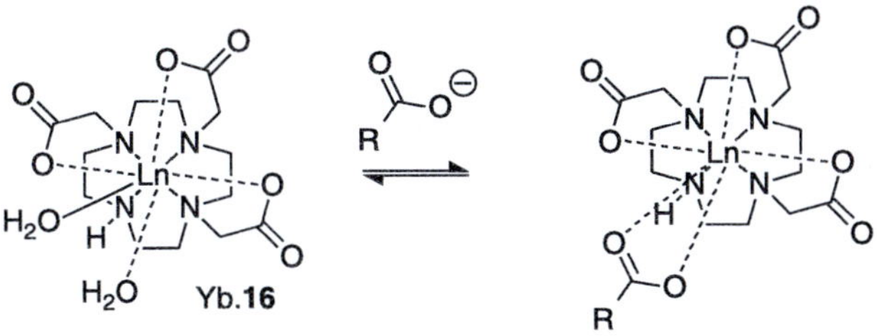

Fig. 7 Self-assembly of ternary complexes from carboxylates and Ln.DO3A complexes

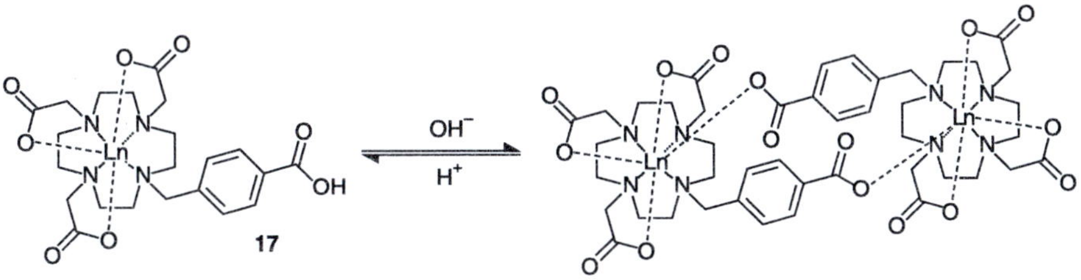

Fig. 8 Dimerization of **17** at high pH

carboxylic acid forms a ternary complex with Yb.DO3A (**16**) in which ytterbium luminescence is sensitized via charge transfer from the strongly electron donating tetrathiafulvalene ring system [45]. Similarly, carboxylic acid appended transition metal complexes, e.g., can be used to sensitize Ln.DO3A and probe the mechanisms of energy transfer [46].

Ligands derived from carboxylic acids can be displaced by phosphate and phosphonic acids owing to the stronger binding of the more highly charged species to the lanthanide centre [47].

Lanthanide complexes such as **17** also exhibit self-association (Fig. 8) to form binuclear complexes [48]; this is dependent upon pH as at high pH, the carboxylate ions coordinate to a second metal centre, acting as bridging ligands. However, at low pH the carboxylic acids are protonated and unable to coordinate to a lanthanide. These changes in the inner coordination sphere are all mirrored by changes in the luminescence lifetime of the lanthanide ion.

The approaches to heterometallic systems outlined above all rely upon coordination chemistry, whether to a metal ion or a nanoparticle surface. Supramolecular interactions can also play a role in assembling heterometallic systems. For instance, Sambrook et al. prepared a pseudo-rotaxane (**18**) templated around a chloride ion (Fig. 9) in which lanthanide ions are bound to the "thread" while a luminescent rhenium complex is bound to the macrocycle component [49]. Thus there is no direct link between the donor chromophore and the lanthanide ion, and Dexter exchange is not viable in the absence of bonds enforcing Förster energy transfer. This extreme of behaviour helps to rationalize the behaviour of more conventionally

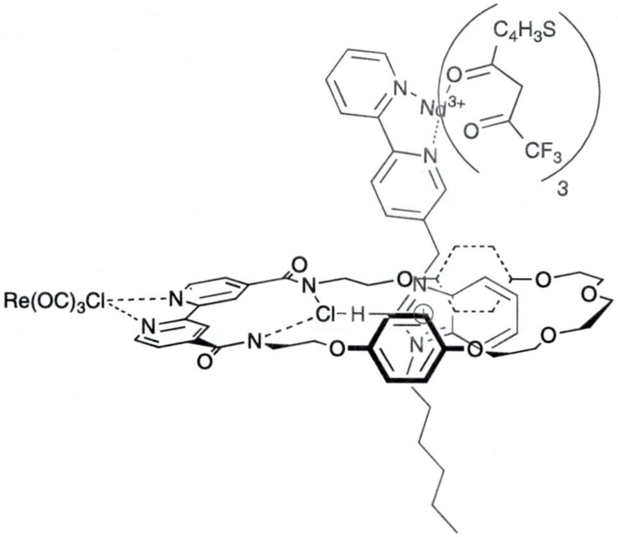

Fig. 9 The d–f hybrid pseudo-rotaxane, **18**

assembled systems, in which Dexter pathways have been shown to dominate wherever super-exchange mediated by ligand π electrons can assist energy transfer and Förster energy transfer dominates when saturated alkyl spacers separate the donor and the acceptor [50].

4 Robust Arrays Derived from Bridging and Multi-Compartmental Ligands

We have already discussed the difficulties inherent in preparing heterometallic lanthanide containing systems under thermodynamic control. However, such robust systems are essential if we are to understand the properties of lanthanides completely. A range of approaches have been developed that can be used to ensure that single species are formed, and we will now discuss these in turn.

The significant differences in coordination chemistry between the d-block and the lanthanides mean that the preparation of kinetically robust lanthanide complexes bearing d-block chromophores is relatively straightforward, and requires a ligand domain consistent with lanthanide binding. Early approaches to systems such as **19–21** relied upon the stability of a transition metal complex bearing additional binding sites, and were prepared by addition of the lanthanide as the final stage in the synthetic pathway [51–54]. This represents an effective approach, and it has been extended to the preparation of metallostars bearing six lanthanide ions around a ruthenium centre [55].

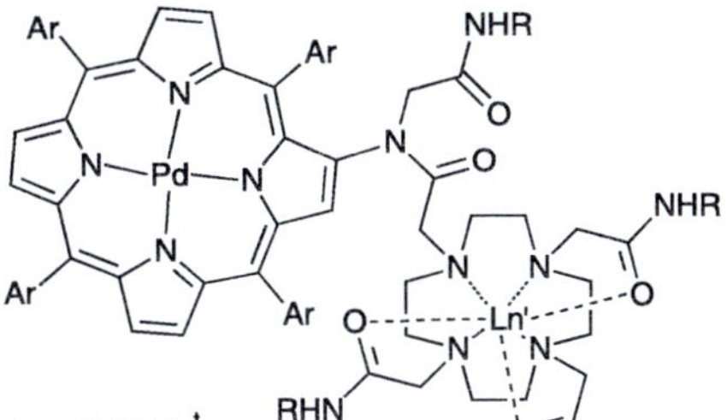

20 Ar = $C_6H_3Bu^t_2$
R = *R*1-phenyl ethyl

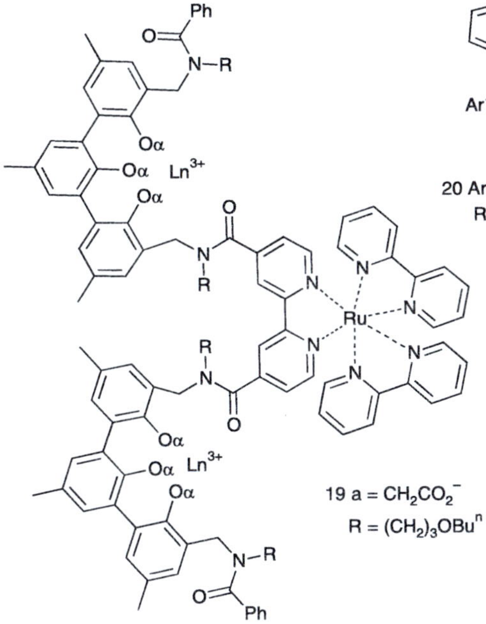

19 a = $CH_2CO_2^-$
R = $(CH_2)_3OBu^n$

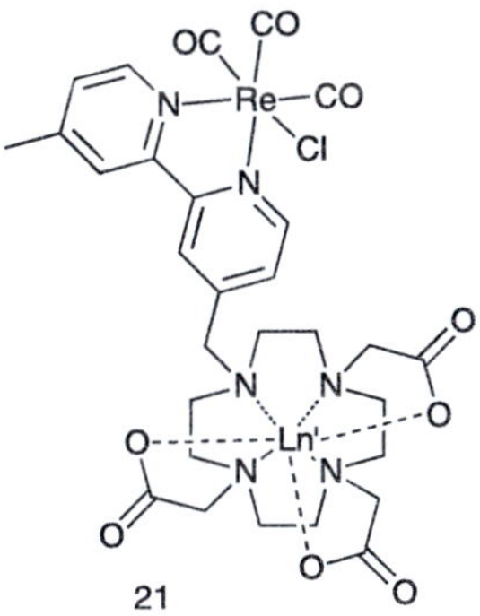

21

The photophysical properties of **20** are particularly interesting, in that the triplet energy of the porphyrin chromophore is comparable to the emissive state of Nd^{3+} and can be thermally repopulated at room temperature [52]. This leads to a dependence of the emission intensity and lifetimes upon the degree of sample aeration. In most cases, energy transfer from the chromophore to the lanthanide occurs via the ^{3}MLCT state of the chromophore, and normal behaviour is observed. However, the broad nature of the emission from luminescent MLCT states means that the tail of this emission is often superimposed upon the relatively weak luminescence from the lanthanide ion. In such situations the two signals can be resolved readily provided that their lifetimes differ significantly. This is not the case with many complexes related to $[Ru(bpy)_3]^{2+}$, which have luminescence lifetimes comparable with those of neodymium and ytterbium complexes.

The order of synthetic steps can be reversed if kinetically robust lanthanide complexes are used. For instance, Gunnlaugsson et al. showed that **23** is formed when **22** reacts with $[Ru(bpy)_2Cl_2]$ (Fig. 10) [56], and also showed that such an approach can be extended to the preparation of complexes with relatively labile transition metal ions such as Fe(II) [57].

Our own studies on the reaction between Ln.**24** and $Re(bpy)(CO)_3OTf$ revealed that the products [Ln.**24**.$Re(bpy)(CO)_3$] have very different photophysical properties to other rhenium containing df hybrids (such as **18** and **21**) [58]. While **18** and **21** both have short-lived luminescence from the rhenium MLCT state (<20 ns), the luminescence lifetimes of [Ln.**24**.$Re(bpy)(CO)_3$] are all longer than 200 ns.

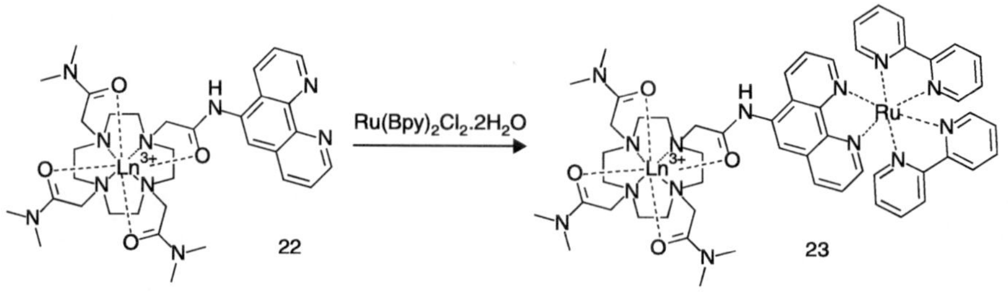

Fig. 10 Preparation of **23**

Indeed, they are sufficiently long to permit time-gating of the MLCT luminescence from background fluorescence. This dramatic change in lifetime is the consequence of excluding chloride from the rhenium coordination environment, and thus removing the possibility of non-radiative quenching of the MLCT through PET. It does, however, mean that the MLCT luminescence is difficult to separate from those of the lanthanide centred emission.

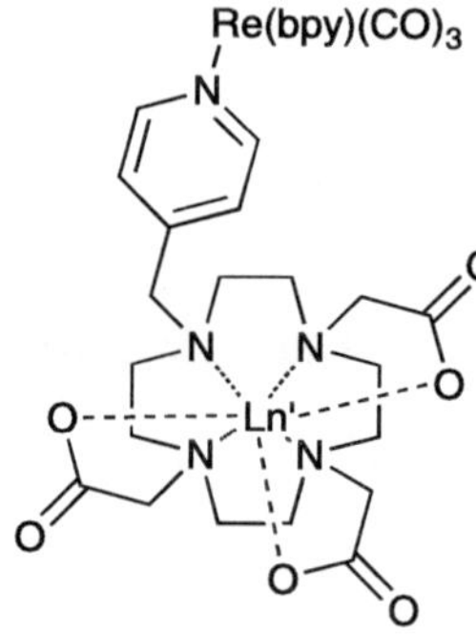

Ln.**24**.Re(bpy)(CO)$_3$

Controlled synthesis of heterometallic lanthanide complexes represents a bigger challenge, and much work has been devoted over the last few years to identifying ways in which these systems can be prepared effectively. Kinetically stable complexes are the key to all the approaches we will discuss here.

The synthesis of Yb.**2**.Tb$_2$ (Fig. 11) revolves around the stability of Tb.**25**, which is stable across a wide range of pH in the presence of competitor ligands [8]. Furthermore, Tb.**25** bears a pendent amine group which can react with a variety of organic molecules. Reaction with DTPA anhydride gives H$_3$.**2**.Tb$_2$, which contains an additional metal binding domain. Luminescence is a powerful tool for assessing whether the terbium ion remains in the original binding pocket or moves to this new site. Measuring luminescence lifetimes in H_2O and D_2O can give information about the solvation of the metal centre, since proximate OH oscillators will give rise to effective quenching of the excited state while the reduced Frank–Condon overlap with O–D oscillators will give rise to reduced quenching. The number of bound solvent molecules, q, can be calculated using the equation:

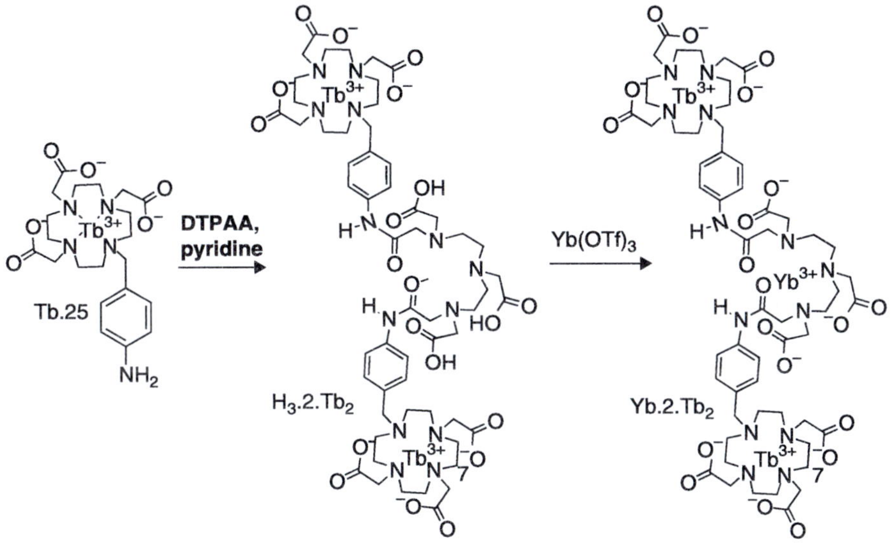

Fig. 11 Synthesis of Yb.**2**.Tb_2

$$q = A(k_{H_2O} - k_{D_2O} - B)$$

where A and B are constants for a given lanthanide and k_{H_2O} and k_{D_2O} are the observed rate constants for the decay of the luminescence in H_2O and D_2O, respectively [59]. In this case, q would be expected to decrease on moving from a heptadentate to an octadentate binding site. In fact, q increases marginally and the stability of the complex is borne by NMR studies. Addition of Yb^{3+} results in filling the third binding site and giving rise to the complex Yb.**2**.Tb_2 whose properties were discussed above in Sect. 1.

This approach to the synthesis of polymetallic complexes from stable metal containing building blocks has great scope. For instance, tetrametallic arrays can be prepared as shown in Fig. 12 [60]. Ln_2.**26** also contains an amine group that can be reacted further. The complex is sufficiently robust that it does not decompose under the highly acidic conditions associated with diazotization, but instead forms a diazonium salt that reacts with a second complex Ln'_2.**27** to form the tetrametallic complex Ln_2.**28**.Ln'_2. This not only ties the components together, but also incorporates a sensitizing chromophore which absorbs visible light.

Lanthanide complexes bearing pendent groups can now be derivatized in a large number of ways, including the use of click [61] and Ugi [62] chemistry to assemble the building blocks.

Tremblay and Sames used a slightly different approach in the preparation of heterometallic complexes [63]. In this, they prepared the ligand first, then added a metal which is complexed in the macrocycle cavity, which acts as a kinetic trap. Any residual metal ion bound in the second binding site can be removed owing to its kinetic instability. Thus, heterometallic complexes can be prepared.

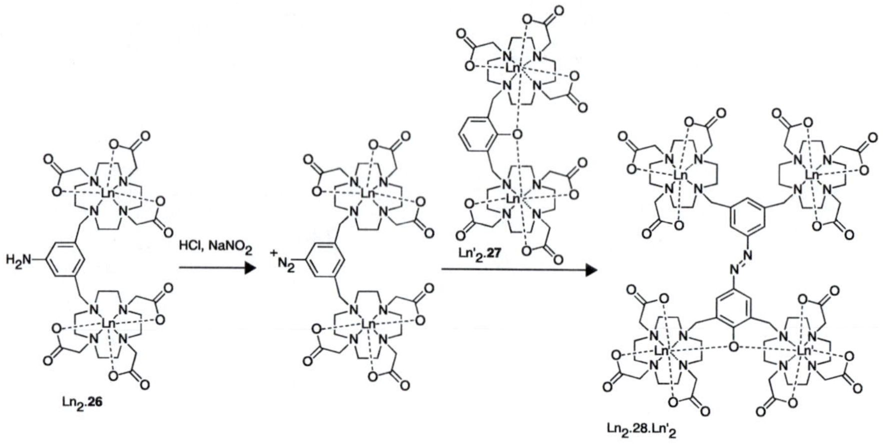

Fig. 12 Preparation of Ln_2.**28**.Ln′$_2$

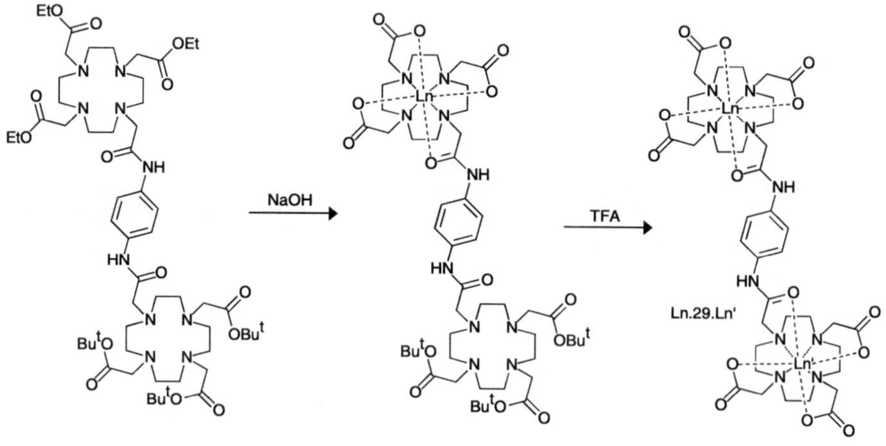

Fig. 13 Preparation of Ln.**29**.Ln′

Some molecules are amenable to neither of the above approaches. For instance, Ln.**29**.Ln′ cannot readily be prepared by coupling complexes together owing to its symmetrical nature. In such circumstances, the only facile way to prepare complexes is through orthogonal protection (and sequential deprotection) of the two binding sites as shown in Fig. 13 [64].

5 Conclusions and Perspectives

Lanthanide containing assemblies can be prepared in a vast variety of ways. Classical bridging ligand chemistry and self-assembly can combine to produce elegant architectures that can be used to probe the interaction between metal centres and which have useful properties and functions. This is particularly true where stable lanthanide complexes can be used as building blocks. In such circumstances it is reasonable to expect that the whole arsenal of organic chemical reactivity can be exploited in facilitating the synthesis of arrays. Much remains to be done spectroscopically. We need to understand the pathways that influence luminescence in such systems, and their dependence upon molecular structure. This is particularly true with d–f hybrid ytterbium complexes, where the pathways leading to emission still need to be established in detail.

References

1. Tanke HJ (2010) Imaging of lanthanide luminescence by time-resolved microscopy. Springer Ser Fluoresc. doi: 10.1007/4243_2010_2
2. Faulkner S, Natrajan LS, Perry WS, Sykes D (2009) Sensitised luminescence in lanthanide containing arrays and d-f hybrids. Dalton Trans 3890–3899
3. Horrocks WD, Bolender JP, Smith WP, Supkowski JM (1997) Photosensitized near infrared luminescence of ytterbium(III) in proteins and complexes occurs via an internal redox process. J Am Chem Soc 119:5972–5973
4. Beeby A, Faulkner S, Williams JAG (2002) pH dependence of the energy transfer mechanism in a phenanthridine-appended ytterbium complex. J Chem Soc Dalton Trans 1918–1922
5. Silva FRGE, Malta OL, Reinhard C, Güdel HU, Piguet C, Moser JE, Bünzli JCG (2002) Visible and near infrared luminescence of lanthanide-containing dimetallic triple-stranded helicates; modelling of the energy transfer processes in the Sm(III) and Yb(III) molecular edifices. J Phys Chem A 106:1670–1677
6. Beeby A, Faulkner S, Parker D, Williams JAG (2001) Sensitised luminescence from phenanthridine appended lanthanide complexes: analysis of triplet mediated energy transfer processes in terbium, europium and neodymium complexes. J Chem Soc Perkin Trans 2:1268
7. Ala Kleme T, Latva M, Haapakka K (2000) Radiative $^5D_4$-^7F_n dynamics of terbium in self-assembled dimeric chelates. Anal Chim Acta 403:161–171
8. Faulkner S, Pope SJA (2003) Lanthanide sensitized lanthanide luminescence. J Am Chem Soc 125:10526–10527
9. Suyver JF, Aebischer A, Biner D, Gerner P, Grimm J, Heer S, Kramer J, Reinhard C, Güdel HU (2005) Novel materials doped with trivalent lanthanides and transition metal ions showing near-infrared to visible photon upconversion. Opt Mater 27:1111–1130
10. Grimm J, Beurer E, Gerner P, Gudel HU (2007) Upconversion between 4f–5d excited states in Tm-doped $CsCaCl_3$, $CsCaBr_3$, and $CsCaI_3$. Chem Eur J 13:1152–1157
11. Sivakumar S, van Veggel FCJM, Raudsepp M (2005) Bright white light through up-conversion of a single NIR source from sol-gel-derived thin film made with Ln^{3+}-doped LaF_3 nanoparticles. J Am Chem Soc 127:12464–12465
12. Miller TA, Jeffery JC, Ward MD, Adams H, Pope SJA, Faulkner S (2004) Photoinduced Ru-Yb energy transfer and sensitised near-IR luminescence in a coordination polymer containing co-crystallised $[Ru(bipy)(CN)_4]^{(2-)}$ and Yb(III) units. Dalton Trans 1524–1526
13. Davies GM, Pope SJA, Adams H, Faulkner S, Ward MD (2005) Structural and photophysical properties of coordination networks combining [Ru(bipy)(CN)(4)](2-) anions and lanthanide (III) cations: Rates of photoinduced Ru-to-lanthanide energy transfer and sensitized near-infrared luminescence. Inorg Chem 44:4656–4665
14. Baca S, Adams H, Sykes D, Faulkner S, Ward MD (2007) Three component coordination networks based on $Ru(phen)(CN)_4]^{2-}$ anions, near-infrared luminescent lanthanide cations and

ancillary oligopyridine ligands; structures and photophysical properties. Dalton Trans 2419–2430

15. Herrera JM, Pope SJA, Adams H, Faulkner S, Ward MD (2006) Structural and photophysical properties of coordination networks combining [Ru(bpym)(CN)4]$^{2-}$ or [{Ru(CN)4–2(μ-bpym)]$^{4-}$ anions (bpym=2, $^{2'}$-bipyrimidine) with lanthanide(III) cations: Sensitized near-infrared luminescence from Yb(III), Nd(III), and Er(III) following Ru-to-lanthanide energy transfer. Inorg Chem 45:3895–3904
16. Herrera PJM, Pope SJA, Meijer AJHM, Easun TL, Adams H, Alsindi WZ, Sun XZ, George MW, Faulkner S, Ward MD (2007) Photophysical and structural properties of cyanoruthenate complexes of hexaazatriphenylene. J Am Chem Soc 129:11491
17. Lazarides T, Davies GM, Adams H, Sabatini C, Barigelletti F, Barbieri A, Pope SJA, Faulkner S, Ward MD (2007) Ligand-field excited states of hexacyanochromate and hexacyanocobaltate as sensitisers for near-infrared luminescence from Nd(III) and Yb(III) in cyanide-bridged d-f assemblies. Photochem Photobiol Sci 6:1152–1157
18. Marchal C, Filinchuk Y, Chen XY, Imbert D, Mazzanti M (2009) Lanthanide coordination polymer from a flexible multidentate linker. Chem Eur J 15:5273
19. Hemmilä I, Laitala V (2005) Progress in lanthanides as luminescent probes. J Fluoresc 15:529
20. Bunzli JCG, Piguet C (2002) Lanthanide-containing molecular and supramolecular polymetallic functional assemblies. Chem Rev 102:1897–1928
21. Piguet C, Bernardinelli G, Hopfgratner G (1997) Helicates as versatile supramolecular complexes. Chem Rev 97:2005–2062
22. Grenthe I (1961) Stability relationships among the rare earth dipicolinates. J Am Chem Soc 83:360–364
23. Mürner HR, Chassat E, Thummel RP, Bunzli JCG (2000) Strong enhancement of the lanthanide-centred luminescence in complexes with 4-alkylated 2,2′;6′,2″-terpyridines. J Chem Soc Dalton Trans 2809–2816
24. Renaud F, Piguet C, Bernardinelli G, Bunzli JCG, Hopfgartner G (1997) In search for mononuclear helical lanthanide building blocks with predetermined properties: triple-stranded helical complexes with N,N,N′,N′-tetraethylpyridine-2, 6-dicarboxamide. Chem Eur J 3:1646–1659
25. Piguet C, Bunzli J-CG, Bernardinelli G, Hopfgartner G, Williams AF (1993) Self-assembly and photophysical properties of lanthanide dinuclear triple-helical complexes. J Am Chem Soc 115:8197–8206
26. Martin N, Bunzli JCG, McKee V, Piguet C, Hopfgartner G (1998) Self-assembled dinuclear lanthanide helicates: substantial luminescence enhancement upon replacing terminal benzimidazole groups by carboxamide binding units. Inorg Chem 37:577–589
27. Elhabiri M, Scopelliti R, Bunzli JCG, Piguet C (1999) Lanthanide helicates self-assembled in water: a new class of highly stable and luminescent dimetallic carboxylates. J Am Chem Soc 121:10747–10762
28. Lessman JJ, Horrocks WD (2000) Supramolecular coordination chemistry in aqueous solution. Inorg Chem 39:3114–3124
29. Song B, Vandervyver CDB, Chauvin AS, Bünzli JCG (2008) Time-resolved luminescence microscopy of bimetallic lanthanide helicates in living cells. Org Biomol Chem 8:4125–4133
30. Deiters E, Song B, Chauvin AS, Vandevyver CDB, Gumy F, Bünzli JCG (2009) Luminescent bimetallic lanthanide bioprobes for cellular imaging with excitation in the visible-light range. Chem Eur J 15:885–900
31. Floquet S, Ouali N, Bocquet B, Bernardinelli G, Imbert D, Bünzli JCG, Hopfgartner G, Piguet C (2003) The first self-assembled trimetallic lanthanide helicates driven by positive cooperativity. Chem Eur J 9:1860–1875
32. Zeckert K, Hamacek J, Senegas JM, Dalla-Favera N, Floquet S, Bernardinelli G, Piguet C (2007) Predictions, synthetic strategy, and isolation of a linear tetrametallic triple-stranded lanthanide helicate. Angew Chem Int Ed 44:7954–7958

33. Lama M, Mamula O, Kottas GS, Rizzo F, de Cola L, Nakamura A, Kuroda R, Stoeckli-Evans H (2007) Lanthanide class of a trinuclear enantiopure helical architecture containing chiral ligands. Chem Eur J 13:7358–7373
34. André N, Jensen TB, Scopelliti R, Imbert D, Elhabiri M, Hopfgartner G, Piguet C, Bünzli JCG (2004) Supramolecular recognition of heteropairs of lanthanide ions: a step towards self-assembled bifunctional probes. Inorg Chem 43:515–529
35. Cantuel M, Gumy F, Bünzli JCG, Piguet C (2006) Encapsulation of labile trivalent lanthanides into a homobimetallic chromium(III) containing triple stranded helicate. Synthesis, characterization, and divergent intramolecular energy transfers. Dalton Trans 2647–2660
36. Imbert D, Cantuel M, Bünzli JCG, Bernardinelli G, Piguet C (2003) Extending lifetimes of lanthanide based near-infrared emitters in the millisecond range through Cr(III) sensitization in discrete bimetallic edifices. J Am Chem Soc 124:15698–15699
37. Beeby A, Clarkson IM, Eastoe J, Faulkner S, Warne B (1997) Lanthanide containing reversed micelles: a structural and luminescence study. Langmuir 13:5816–5819
38. Hall J, Hamer R, Aime S, Botta M, Faulkner S, Parker D, DeSousa AS (1998) Relaxometric and luminescence behaviour of triaquahexaazamacrocyclic complexes, the gadolinium complex displaying high relaxivity with a pronounced pH dependence. New J Chem 22:627–631
39. Thompson MK, Vuchkov M, Kahwa IA (2001) Cooperative process governing formation of small pentanuclear Ln nanoclusters and energy transport within and between them. Inorg Chem 40:4332–4341
40. Chen XY, Bretonnière Y, Pecaut D, Imbert D, Bünzli JCG, Mazzanti M (2007) Selective self-assembly of hexameric homo- and heteropolymetallic lanthanide wheels: synthesis, structure, and photophysical properties. Inorg Chem 46:625–637
41. Comby S, Scopelliti R, Imbert D, Charbonniere L, Ziessel R, Bünzli JCG (2006) Dual emission from nonalanthanide clusters. Inorg Chem 45:3158–3160
42. Massue J, Quinn SJ, Gunnlaugsson T (2008) Lanthanide luminescent displacement assays. J Am Chem Soc 130:6900–6901
43. Bonnet CS, Massue J, Quinn SJ, Gunnlaugsson T (2009) Lanthanide luminescent gold nanoparticles: pH-driven self-assembly formation between Eu(III)-conjugated AuNps and sensitising β-diketonate antenna in water. Org Biomol Chem 7:3074–3078
44. Lewis DJ, Day TM, MacPherson JV, Pikramenou Z (2006) Luminescent nanobeads: attachment of surface reactive Eu(III) complexes to gold nanoparticles. Chem Commun 1433–1435
45. Faulkner S, Burton-Pye BP, Khan T, Martin LR, Wray SD, Skabara PJ (2002) Interaction between tetrathiafulvalene carboxylic acid and ytterbium DO3A: solution state self-assembly of a ternary complex which is luminescent in the NIR. Chem Commun 1668–1669
46. Pope SJA, Coe BJ, Faulkner S, Bichenkova EV, Yu X, Douglas KT (2004) Self-assembly of hetero-bimetallic d-f hybrid complexes: sensitization of lanthanide luminescence by d-block metal-to-ligand charge-transfer excited states. J Am Chem Soc 126:9490–9491
47. Pope SJA, Burton-Pye BP, Khan, T, Berridge R, Skabara PJ, Faulkner S (2006) Self-assembly of luminescent ternary complexes between seven-coordinate lanthanide (III) complexes and chromophore bearing carboxylates and phosphonates. Dalton Trans:2907–2912
48. Faulkner S, Burton-Pye BP (2005) pH dependent self-assembly of dimetallic lanthanide complexes. Chem Commun 259–261
49. Sambrook MR, Curiel D, Hayes EJ, Beer PD, Pope SJA, Faulkner S (2006) Sensitised near infrared emission from lanthanides via anion-templated assembly of d-f heteronuclear [2] pseudorotaxanes. New J Chem 30:1133–1136
50. Lazarides T, Sykes D, Faulkner S, Barbieri A, Ward MD (2008) On the mechanism of d-f energy-transfer in Ru(II)/Ln(III) and Os(II)/Ln(III) dyads: Dexter energy-transfer over a distance of 20 Å. Chem Eur J 14:9389–9399
51. Klink SI, Keiser H, van Veggel FCJM (2000) Organo-d-metal complexes as new class of photosensitizers for near-infrared lanthanide emission. Angew Chem Int Ed 39:4319–4321

52. Beeby A, Dickins RS, Fitzgerald S, Govenlock LJ, Maupin CL, Parker D, Riehl JP, Siligardi J, Williams JAG (2000) Porphyrin sensitization of circularly polarized near-IR lanthanide luminescence: enhanced emission with nucleic acid binding. Chem Commun 1183–1184
53. Pope SJA, Coe BJ, Faulkner S (2004) Re(I) sensitised near-infrared lanthanide luminescence from a hetero-trinuclear Re_2Ln array. Chem Commun 1550–1551
54. Pope SJA, Coe BJ, Faulkner S, Laye RH (2005) Metal-to-ligand charge-transfer sensitisation of near-infrared emitting lanthanides in trimetallic arrays M_2Ln. Dalton Trans:1482–1490
55. Moriggi L, Aebischer A, Cannizzo C, Sour A, Borel A, Bünzli JCG (2009) A ruthenium metallostar: synthesis, sensitised luminescence, and ^1H relaxivity. Dalton Trans:2088–2095
56. Senechal-David K, Pope SJA, Quinn S, Faulkner S, Gunnlaugsson T (2006) Sensitized near-infrared lanthanide luminescence from Nd(III)- and Yb(III)-based cyclen-ruthenium coordination conjugates. Inorg Chem 45:10040–10042
57. Senechal-David K, Leonard JP, Plush SE, Gunnlaugsson T (2006) Supramolecular self-assembly of mixed f-d metal ion conjugates. Org Lett 7:2727–2730
58. Koullourou T, Natrajan LS, Bhavsar H, Pope SJA, Feng J, Narvainen J, Shaw R, Scales E, Kauppinen R, Kenwright AM, Faulkner S (2008) Synthesis and spectroscopic properties of a prototype single molecule dual imaging agent comprising a heterobimetallic rhenium gadolinium complex. J Am Chem Soc 130:2178–2179
59. Beeby A, Clarkson IM, Dickins RS, Faulkner S, Parker D, de Sousa AS, Williams JAG (1999) Non-radiative deactivation of the excited states of europium, terbium and ytterbium complexes by proximate energy-matched OH, NH and CH oscillators: an improved luminescence method for establishing solution hydration states. J Chem Soc Perkin Trans 2:493–503
60. Placidi MP, Villaraza AJL, Natrajan LS, Sykes D, Kenwright AM, Faulkner S (2009) Synthesis and spectroscopic properties of azo-dye derivatives of polymetallic lanthanide complexes: using diazotization to link metal complexes together. J Am Chem Soc 131: 9916–9917
61. Jauregui M, Perry WS, Allain C, Vidler LR, Willis MC, Kenwright AM, Stasiuk G, Lowe MP, Faulkner S (2009) Changing the local coordination environment in mono- and bi-nuclear lanthanide complexes through "click" chemistry. Dalton Trans 6283–6285
62. Tei L, Gugliotta G, Avedano S, Giovenzana GB, Botta M (2009) Application of the Ugi four-component reaction to the synthesis of ditopic bifunctional chelating agents. Org Biomol Chem 7:4406–4414
63. Tremblay MS, Sames D (2006) Luminescent heterometallic bis lanthanide complexes via selective sequential metallation. Chem Commun 4116–4118
64. Natrajan LS, Villaraza AJL, Kenwright AM, Faulkner S (2009) Controlled preparation of a heterometallic lanthanide complex containing different lanthanides in symmetrical binding pockets. Chem Commun 6020–6022

Lanthanide Luminescence in Solids

Peter A. Tanner

Abstract A tutorial introduction to the spectra of lanthanide ions is given. The chapter begins with a brief comparison of luminescence of lanthanide ions (Ln^{3+}) in the solid, liquid, and gas phases. Then a description of the importance of nonradiative versus radiative processes is made. The various types of transition of lanthanide ions encountered are then introduced: $4f^N$–$4f^N$; $4f^N$–$4f^{N-1}5d$; charge transfer; host band-to-band; and defect site or impurity transitions. Reference is briefly made to the spectra of dipositive ions. The luminescence of lanthanide ions in non-lanthanide hosts is examined, and the importance of locating the relative positions of Ln^{3+} energy levels relative to the host band gap is stressed. Some of the various upconversion phenomena, including second harmonic generation, two-photon absorption, ground state/excited state absorption, energy transfer upconversion, and photon avalanche, are then described with reference to Ln^{3+} and Ln^{3+}-TM^{n+} (transition metal) systems. The experimental distinctions of which particular process is operative in a given system are explained by considering the power, concentration, and lifetime dependences, and the upconversion excitation spectrum. Some applications of lanthanide luminescence are briefly reviewed and a mention of the luminescence in nanomaterials is also included.

Keywords Applications of luminescence · Charge transfer · Divalent lanthanide · Electronic transitions · Lanthanide luminescence · Nonradiative processes · Upconversion

P.A. Tanner
Department of Biology and Chemistry, City University of Hong Kong, Tat Chee Avenue, Kowloon, Hong Kong S.A.R., People's Republic of China
e-mail: bhtan@cityu.edu.hk

P. Hänninen and H. Härmä (eds.), *Lanthanide Luminescence: Photophysical, Analytical and Biological Aspects*, Springer Ser Fluoresc (2011) 7: 183–234, DOI 10.1007/4243_2010_6,

Contents

Abbreviations

BAM	$BaMgAl_{10}O_{17}:Eu^{2+}$
CB	Conduction band
CRT	Cathode ray tube
CT	Charge transfer
ED	Electric dipole
ETU	Energy transfer upconversion
FED	Field emission display
GS	Ground state
GSA/ESA	Ground state absorption/excited state absorption
irrep	Irreducible representation
LED	Light-emitting diode
OLED	Organic light-emitting diode
PA	Photon avalanche

PDP	Plasma display panel
SHG	Second harmonic generation
TM	Transition metal
TPA	Two-photon absorption
VB	Valence band
VFPD	Volatile flat panel display
VUV	Vacuum ultraviolet
YAG	$Y_3Al_5O_{12}$
YAH	Hexagonal-$YAlO_3$
YAM	$Y_4Al_2O_9$
YAP	Orthorhombic-$YAlO_3$

1 Scope of the Chapter and Comparison with Vapor and Solution Spectra

The luminescence of lanthanide ions in solids has been the subject of several books or chapters [1–12] and has an immense scope. Since I was asked to write this review within a short time, I have failed to give a comprehensive survey but have given a qualitative overview of several areas with some references to more quantitative treatments. Topics such as lanthanide luminescence of laser materials [4, 8, 13, 14], sensors [15], hybrid materials [16], organolanthanide, and coordination compounds [17] are missing. So what is here in October 2009? – basically, a description of the luminescence spectra of lanthanide ions in the solid state and some applications of phosphor materials.

The luminescence of tripositive lanthanide ions exhibits very different spectral features in the gaseous, liquid, and solid states. The investigation of the spectra of gaseous ions was performed, for example, by Dieke's group around 50 years ago, using photographic detection with a high resolution vacuum ultraviolet (VUV) spectrometer. One publication concerns the rather simple case of tripositive praseodymium [Pr(IV)] [18]. In the gas phase, line widths are Doppler-broadened and could be measured to 0.005 cm^{-1} or better for good lines [18] with the resolution of nuclear hyperfine structure. In the gaseous state, there is no crystal band gap so the emission spectra extend to much higher energy (actually measured up to 122.6 nm; 81,540 cm^{-1} in [18]) and concern only the free-ion electric dipole (ED) allowed transitions: 6s–6p (230–270 nm), 4f–5d (150–220 nm), and 5d–6p (120–150 nm). The spectral lines are very sharp so that energy levels could be measured to two decimal places (e.g., 4f5d $^3H_4$ at 61170.95 cm^{-1}). Note that in the absence of crystal field splitting, multiplet terms are split only by spin–orbit coupling. A comprehensively derived energy level scheme for the free-ion Pr(IV) was given by Crosswhite et al. and since the intraconfigurational transitions $4f^2$–$4f^2$ are forbidden, the relevant energies of the 13 multiplet terms of $4f^2$ were inferred from 4f5d to $4f^2$ transitions. A simultaneous determination and publication by Sugar [19] gave these energies at exactly the same values to within 0.2 cm^{-1}. The gaseous free-ion multiplet term energy levels are higher than for Pr^{3+} in the solid state and this is attributed to the nephelauxetic effect, in which the interelectronic repulsion between the 4f electrons is decreased because of the penetration of the Pr^{3+} ion by the ligand electrons.

The gaseous ion spectra contrast markedly with the spectra of Ln^{3+} in solution, where fewer multiplet terms luminescence (due to coordination by ligands with high frequency vibrations as discussed subsequently) and the spectral bandwidths are broad (~10 nm). The high energy cut-off in the spectra is determined by the absorption of the solvent (which was subtracted out in the classic absorption study of Carnall et al. of aquo complexes [20]), and the spectra do not extend to shorter wavelengths. This limits investigations of Ln^{3+} to $4f^N$ and sometimes part of the $4f^{N-1}5d$ levels. There are further differences from gas phase spectra, as exemplified, for example, in the room temperature emission spectrum of Gd^{3+} in dilute acid solution [21], where vibronic structure due to H_2O bend and stretch vibrations is observed. The derived vibrational energies from this emission spectrum do not parallel the H_2O vibrational energies measured in the infrared or Raman spectra since they correspond to the vibrations of nearest-neighbor ligands and not to the water molecules of the bulk medium. Just as for solid state spectra, broad bands due to the transfer of charge between the metal ion and ligands are present in the spectra of lanthanide ions in solution. Naturally, there are no analogous features for gas phase spectra.

It will become evident from the descriptions of solid state spectra given below that we are referring to an intermediate scenario between the two extremes of gaseous and solution state spectra.

2 Luminescence and Nonradiative Processes

The lanthanides are long known as the lighting elements. Luminescence occurs from a lanthanide energy level if the gap to the next lower energy level is spanned by five (sometimes, even by four) or more phonons provided that the luminescent level is suitably populated and that it is not nonradiatively deactivated. Looking at the $4f^N$ energy levels of tripositive lanthanide ions (Ln^{3+}) from recent calculations (Fig. 2 in [22] and [23]), many gaps are observed. Particularly, wide energy gaps occur below the well-known luminescent levels of Tb^{3+} ($^5D_4$), Eu^{3+} ($^5D_0$), Yb^{3+} ($^2F_{5/2}$), but also many other levels of these and other Ln^{3+} are luminescent. This luminescence is characterized by long lifetimes (even up to several seconds) and sharp spectral lines (i.e., color purity). The natural lifetime, τ, comprises radiative, τ_r, and nonradiative, τ_{nr}, components:

$$1/\tau = 1/\tau_r + 1/\tau_{nr} \tag{1}$$

For efficient luminescence of Ln^{3+}, the inter-ionic nonradiative rate is decreased by diluting the ion into a transparent host lattice. Thus, fast migration between ions, for which the excitation finally ends up at defect or killer sites, and energy transfer between ions [24] are both minimized.

The intra-ionic nonradiative decay rate of Ln^{3+} by multiphonon relaxation in ionic crystals or glasses has been expressed as a product of "electronic" and "lattice" parts, where the latter has been shown to be proportional to the infrared absorption coefficient [25], although it is useful to distinguish between promoting and accepting modes [26]. The modes which drive the nonradiative transition are called the promoting modes, where each promoting mode consumes only one phonon. The remaining energy difference between the two energy levels is taken up by multiple quanta of accepting modes. There are selection rules operative upon the "electronic" part of the nonradiative rate, which, for example, forbid to first order of the relaxation between $J = 1$ and $J = 0$ multiplets (e.g., $^5D_1$ to $^5D_0$ in Eu^{3+}). The empirical observation above concerning the maximum phonon energy is consistent with the energy gap law, which describes the nonradiative rate of Ln^{3+} in a particular host lattice. This has been expressed in many forms [11, 26, 27] such as [28]:

$$W_{NR}(T = 0K) = \beta_{el} \exp[-\alpha(\Delta E - 2h\nu_{max})] \quad (2)$$

where the prefactor β_{el} (s^{-1}) is considered to contain the electronic coupling element; ΔE (cm^{-1}) is the energy gap between the two levels, $h\nu_{max}$ is the maximum phonon energy (cm^{-1}), and α (cm), which depends upon the Huang–Rhys parameter, S, is constant for the particular host lattice. This latter parameter, S, measures the difference in the electron-lattice coupling between the two states, due to displacement $Q_0(b)$-$Q_0(a)$, of the potential energy curve along the configuration coordinate diagram (Fig. 1), and is defined, with M as an effective ionic mass, as:

$$S = (2\pi^2 M\nu/h)[Q_0(b) - Q_0(a)]^2 = E_{dis}/h\nu \quad (3)$$

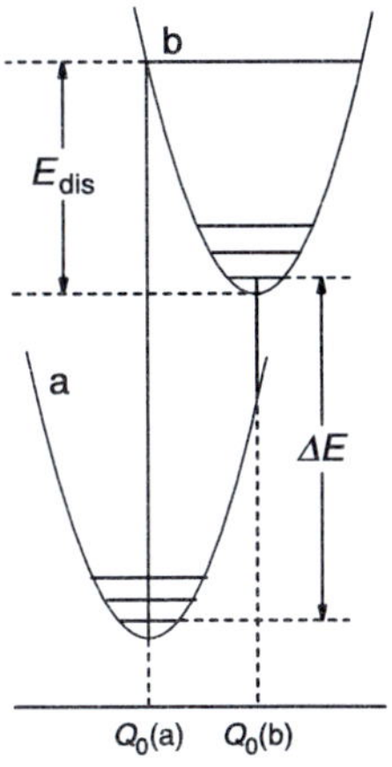

Fig. 1 Configuration coordinate diagram showing the potential energy curves for two electronic states a and b. The ordinate is energy and the abscissa is the configuration coordinate, which in one dimension can be thought of as a change in bond distance. Equilibrium (minimum) values of potential energy are identified by the 0 subscript

The maximum phonon energy is therefore critical in determining the nonradiative rate, and rare earth coordination to ligands with high vibrational energy, such as OH_2, NH_2, is usually avoided if slow nonradiative decay is required. For example, Orlovskii et al. [29] examined the decay kinetics of the excited $^4I_{9/2}$ initial level of Er^{3+} (at ~12,000 cm^{-1}) doped at 0.5 at.% in fluorite crystals MF_2 (M = Ca, Pb) at 77 K. The measured lifetime in CaF_2 was 14.3 μs, whereas it was 102 μs in PbF_2. The effective longitudinal optical phonon frequencies of these hosts are 474 cm^{-1} and 337 cm^{-1}, respectively, so that the $^4I_{9/2}$–$^4I_{11/2}$ energy gap is bridged by five phonons in the case of CaF_2, and seven phonons for PbF_2.

The expression for the temperature dependence of the multiphonon relaxation rate has been given ([2], p 256), and it simplifies for low temperatures and small Huang–Rhys parameter, to:

$$\begin{aligned} W(T) &= W(0)\left\{1+[\exp(\mathrm{h}\nu/k\mathrm{T}) - 1]^{-1}\right\}^p \\ &= W(0) \times \{1/[1 - \exp(-\mathrm{h}\nu/k\mathrm{T})]^p\} \end{aligned} \tag{4}$$

where the mean thermal occupancy of the vibrational mode (i.e., number of phonons per mode):

$$m = [\exp(\mathrm{h}\nu/k\mathrm{T}) - 1]^{-1} \tag{5}$$

and p is the order of the multiphonon process (ΔE/hν).

The above discussion has focused upon multiphonon decay (i.e., ΔE is greater than the highest energy phonon). When crystal field energy levels are closer together, nonradiative relaxation from the upper to lower level can occur by direct phonon emission, or by Raman or Orbach processes ([2], pp 228–234).

In some cases, there are large differences in multiphonon relaxation rates among crystal field levels of a multiplet term, and the relaxation rate between the crystal field levels does not follow Boltzmann thermalization. In the Eu^{3+}-doped vanadate crystal, the $^5D_1$–$^5D_0$ relaxation is a 2-phonon process due to the high energy V-O stretch vibration. In the 2 K excitation spectra of Eu^{3+} emission for a YVO_4:Eu^{3+} crystal, the relative intensity of the two $^5D_1$ crystal field levels in the excitation spectrum changes by a factor of 2 depending upon whether $^5D_1$ or $^5D_0$ emission is monitored [30].

3 Types of Luminescent Transitions

The luminescence of lanthanide ions in solids is characterized by several types of electronic transition, which differ markedly in spectral intensity and linewidth. Some of these are now illustrated.

3.1 *Intraconfigurational 4f^N → 4f^N Transitions*

As discussed in [22], the spherical symmetry of Ln^{3+} is destroyed when these ions are situated in solids, so that a $^{2S+1}L_J$ multiplet term level can be split up to $2J + 1$ crystal field levels for a non-Kramers ion. Due to the parity selection rule for pure electronic transitions in solids, the $4f^N(i) \rightarrow 4f^N(f)$ transition between states i and f is ED forbidden to first order. Parity describes the inversion behavior of the wavefunction of an electronic orbital, so that s,d... orbitals have even parity whereas p,f... orbitals are odd. The spectral feature representing the pure electronic transition is termed the electronic origin or the zero phonon line. An ED transition requires a change in orbital parity because the transition dipole operator (μ_e) is odd, and the overall parity for the nonzero integral involving the Einstein coefficient of spontaneous emission, A(ED):

$$\mathrm{A(ED)} \sim |\langle \mathrm{i}|\mu_e|\mathrm{f}\rangle|^2 \tag{6}$$

must be even. On the other hand, when Ln^{3+} is situated at a site in a solid without inversion symmetry, the interaction between d and f orbitals has no parity, and they can mix to some extent, so that the transition becomes $x[4f^N(i)] + y[4f^{N-1}5d(i)] \rightarrow t[4f^N(f)] + u[4f^{N-1}5d(f)]$, where although $x >> y$ and $t >> u$, there is now a pathway to make the transition become ED allowed. These transitions are termed forced dipole transitions and their strength is about 10^{-3} to 10^{-4} of a fully allowed ED transition. The most important (multiplet–multiplet) selection rules and intensities for $^{2S+1}L_J \rightarrow {}^{2S'+1}L'_{J'}$ transitions may be derived from Judd–Ofelt theory [22] including $\Delta J = 2, 4, 6$, so for example, the $^5D_0 \rightarrow {}^7F_2$ transition of Eu^{3+} is much stronger than $^5D_0 \rightarrow {}^7F_3$. Naturally, to enable these transitions, $\Delta S = 0$, so mixing of septet and quintet spin wavefunctions is also required. The Judd–Ofelt Ω_i $(i = 2, 4, 6)$ parameters are readily obtained from the luminescence spectra of lanthanide ions, as has been demonstrated for $YOCl{:}Gd^{3+}$ [31].

Each multiplet term is split into crystal field levels when Ln^{3+} is situated in a crystal host, and the assignment of irreducible representations of the site symmetry point group to these energy levels is described by Tsukerblat [32]. The crystal field energy levels can be calculated by using a parametrized Hamiltonian:

$$\begin{aligned} H(4\mathrm{f^N}) &= E_{\mathrm{avg}} + \sum_{k=2,4,6} F^k \mathbf{f}_k(\mathrm{f}) + \zeta_{4\mathrm{f}} \mathrm{A_{so}}(\mathrm{f}) \\ &+ \sum_{k=2-4,6-8} T^k \mathbf{t}_k(\mathrm{f}) + \alpha \mathbf{L}^2(\mathrm{f}) + \beta \mathbf{G}[G_2(\mathrm{f})] + \gamma \mathbf{G}[G_7(\mathrm{f})] \\ &+ \sum_{k=0,2,4} M^k \mathbf{m}_k(\mathrm{f}) + \sum_{k=2,4,6} P^k \mathbf{p}_k(\mathrm{f}) + \sum_{kq} B^k_q \mathbf{C}^{(k)}_q(\mathrm{f}) \end{aligned} \tag{7}$$

where the italic and bold letters represent parameters and operators, respectively. The parameter E_{avg} in $H(4f^N)$ serves to shift all the levels so that the energy of the lowest level of $4f^N$ is zero; the second and third terms in $H(4f^N)$ are the two strongest interactions for the $4f^N$ configuration, i.e., the Coulomb and spin–orbit interactions; the last term in $H(4f^N)$ is the crystal field interaction experienced by 4f electrons; and the other terms are effective interactions to describe various effects due to configuration interaction, as described in [33]. More accurate calculations specifically include the structure of interacting configurations [34].

The orbital selection rules for transitions between individual crystal field levels are displayed in terms of the point group irreps of the initial and final state wavefunctions, and for a forced ED transition between $4f^N\ \Gamma_i \rightarrow 4f^N\ \Gamma_f$:

$$\Gamma_i \times \Gamma_o \times \Gamma_f \text{ contains } \Gamma_1 \tag{8}$$

where Γ_o is the irrep corresponding to the ED operator (which transforms as a Cartesian vector) and Γ_1 is the totally symmetric irrep of the relevant point group which corresponds to the site symmetry of Ln^{3+}. A typical emission spectrum is shown in Fig. 2b for Nd^{3+} at the D_2 site in $Y_3Al_5O_{12}$ (YAG:Nd^{3+}) together with the energy

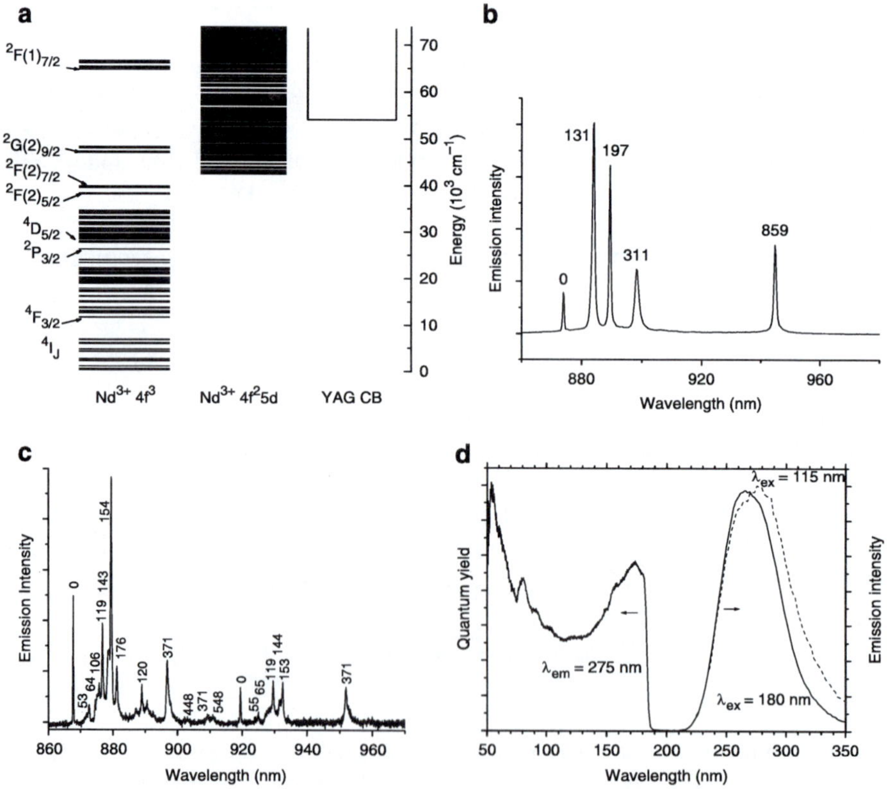

Fig. 2 (continued)

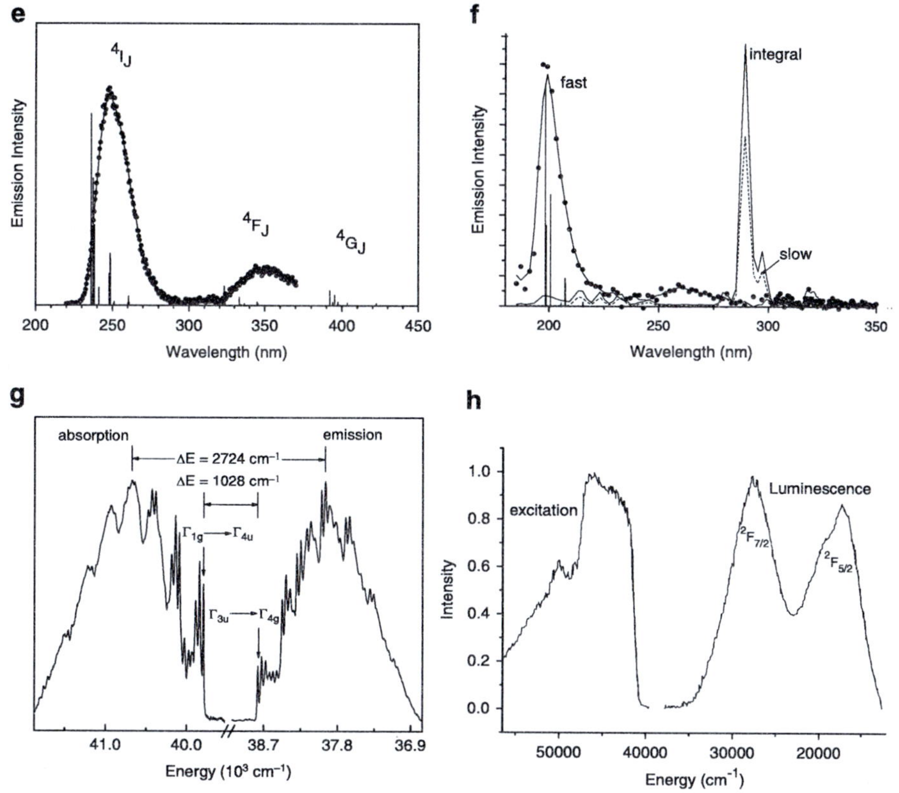

Fig. 2 Types of transitions observed in lanthanide ion spectra. (**a**) Calculated $4f^3$ and $4f^25d$ energy level schemes of the ion Nd^{3+} (doped at 0.5 at.%) at the D_2 site in $Y_3Al_5O_{12}$ (YAG:Nd^{3+}). The CB energy is shown on the *right hand side*. (**b**) $^4F_{3/2} \rightarrow {}^4I_{9/2}$ emission spectrum of YAG:Nd^{3+} at 10 K. The zero phonon line marked 0 is at 11,442 cm^{-1} and the displacement energies of the other $^4I_{9/2}$ crystal field levels are marked in cm^{-1}. (**c**) $^4F_{3/2} \rightarrow {}^4I_{9/2}$ emission spectra of Nd^{3+} in the Cs_2NaYF_6 host at 9 K. The two magnetic dipole zero phonon lines are marked by 0 and some of the derived vibrational energies are marked. (**d**) Emission spectra excited at 180 nm (*solid curve*) and 115 nm (*dashed curve*) and excitation spectrum of 275 nm emission of pure YAG at 8.5 K. (**e**) Luminescence spectrum of 192 nm excited fast emission component from YAG:Nd^{3+} at 10 K. The *vertical bars* indicate the calculated $4f^25d \rightarrow 4f^3$ emission line positions and intensities. The terminal $4f^3$ multiplets are marked. (**f**) Time-resolved 180 nm excited emission spectra of Cs_2NaYF_6:Nd^{3+} at 10 K. The fast (time window 40 ns), slow (time window 127 ns: also delayed by 40 ns after the excitation pulse) components and the integral spectra are shown. The *vertical bars* are the calculated $4f^25d \rightarrow 4f^3$ emission intensities. (**g**) Lowest energy $4f^2 \rightarrow 4f5d$ absorption and highest energy $4f5d \rightarrow 4f^2$ emission bands of Pr^{3+} in the Cs_2NaYCl_6 host. Note the abscissa scale break. (**h**) Excitation spectrum and charge transfer emission of Y_2O_3:Yb^{3+}. The terminal multiplets are marked (adapted from [35]). Note the change of abscissa scale from wavelength (nm) to energy (cm^{-1} units) in (**g, h**)

level scheme of this $4f^3$ ion in the host (Fig. 2a, left hand side). The emission spectrum thus comprises sharp zero phonon lines representing transitions between the initial and final crystal field levels. In this case, for the $^4F_{3/2} \rightarrow {}^4I_{9/2}$ transition of this $4f^3$ Kramers ion, there are $J + 1/2 = 5$ terminal levels. Equation (8) holds for all of these and the terminal state energies (in cm^{-1}) are indicated in the figure. Transition intensity theories enable simulations of crystal field electronic transitions in emission and absorption spectra, and the reader is referred to some relevant literature [36–41].

The occupation of many different types of site and environment, such as in glasses, leads to inhomogeneous broadening of spectral features. This broadening occurs to a smaller extent when Ln^{3+} occupies just several sites in a crystal (such as in β-$NaYF_4$:Ln^{3+}) or ceramic [42] material.

When the Ln^{3+} ion is situated at a centrosymmetric site (i.e., with an inversion center), the pure electronic transitions between $4f^N$ levels are ED forbidden [10]. Magnetic dipole transitions (which are up to 10^6 times weaker than ED transitions) may then be allowed between states of the same parity in the solid if (8) is satisfied, since the magnetic dipole operator, Γ_o, is of even parity. The only way to destroy the centrosymmetry of Ln^{3+} and permit an ED transition between two electronic states is by motions of odd (ungerade) vibrations so that the electronic spectra of Ln^{3+} at an inversion center of a crystal are vibronic (vibrational-electronic) in nature. The transition selection rules then become:

$$\Gamma_i \times \Gamma_o \times \Gamma_f \text{ contains, } \Gamma_v \times \Gamma_{v'} \tag{9}$$

where v and v′ refer to the particular vibration involved in the initial and final states, and at low temperatures since no vibrational quanta are excited in the initial state, $\Gamma_v = \Gamma_1$. Figure 2c shows the 10 K emission spectrum of Cs_2NaYF_6:Nd^{3+}, where the dopant ion Nd^{3+} is situated at an octahedral symmetry site. Note the sharp magnetic dipole allowed zero phonon lines (labeled 0) and the dominant vibrational structure, for which the derived energies (in cm^{-1}) are marked. The vibronic transitions may be distinguished from pure electronic transitions by their (almost) symmetrical arrangement about the zero phonon line and the Boltzmann temperature dependence of anti-Stokes features. Calculations of vibronic intensities have been performed for systems where the vibrations of the lanthanide ion with the nearest-neighbor ligands are of most importance (as for $LnCl_6^{3-}$ [43–45]).

3.2 *Band-to-Band Transition*

The transition from the valence band (VB) to the conduction band (CB) of the host crystal overlaps the $4f^N$–$4f^N$ transitions for low band gap hosts, such as oxides or chlorides. In this case, luminescence does not occur from the overlapping and higher $4f^N$ levels, although efficient host–guest ion energy transfer can give strong luminescence from lower $4f^N$ levels upon band-to-band excitation. The host excitonic luminescence is characterized in the case of YAG by a lifetime in the microsecond range, and nonelementary overlapping, broad emission bands with large Stokes shift, as shown on the right hand side of Fig. 2d. Note the band wavelength shift when the excitation wavelength is changed. The excitation spectrum of the emission is shown on the left hand side of the figure.

Much effort has been directed to the utilization of the intense band-to-band transition in semiconductors, especially in the nano-regime, in order to sensitize lanthanide activators. However, complications due to size and charge discrepancies arise when Ln^{3+} is doped in materials such as TiO_2 and ZnO so that discrete lattice sites are not occupied by the guest. For the latter host, there have been conflicting reports of host–guest energy transfer and usually, upon excitation into the CB, the Ln^{3+} $4f^N$–$4f^N$ emission, such as for Eu^{3+}, is weak and superimposed upon a broad background [46–50]. The use of Li^+ co-doping has been employed in order to elucidate the nature of the Eu^{3+} sites in ZnO [51, 52].

3.3 Interconfigurational $4f^{N-1}5d \rightarrow 4f^N$ Transitions of Ln^{3+}

The $4f^{N-1}5d$ configuration is at higher energy than the $4f^N$ configuration, and a simple phenomenological rationalization of the energy of the lowest level has been made by Dorenbos [53]. By knowing the energy of the lowest $4f^{N-1}5d$ level of a lanthanide ion in one host, $E_{4f\text{-}5d}$(Ln1, host), the lowest $4f^{N-1}5d$ energy level can approximately be calculated for another lanthanide ion, Ln2, in the same host:

$$E_{4f-5d}(\mathrm{Ln2, host}) = E_{4f-5d}(\mathrm{Ln1, host}) + \Delta E_{4f-5d}(\mathrm{Ln2, Ln1}) \tag{10}$$

where the values of $\Delta E_{4f\text{-}5d}$(Ln2,Ln1) are tabulated in [53] and are assumed to be host independent. So for example, knowing the lowest 5d level for Ce^{3+} in the Cs_2NaYCl_6 host lattice (28,193 cm^{-1} [54]) and $\Delta E_{4f\text{-}5d}$(Ln2, Ln1) $= 12{,}240 \pm 750\ cm^{-1}$ [53], the predicted value for Pr^{3+} in Cs_2NaYCl_6 is $40{,}433 \pm 750\ cm^{-1}$, which compares with the experimental value for the lowest 4f5d level of 39,017 cm^{-1} [55].

The standard phenomenological crystal field Hamiltonian H(4f) and its extension have been employed to model the energy levels of the $4f^{N-1}5d$ configuration [56]:

$$H = H(4f^{N-1}) + H(5d) + H_{\mathrm{int}}(4f, 5d) \tag{11}$$

where $H(4f^{N-1})$ (as above), H(5d), and H_{int}(4f, 5d) describe the interactions experienced by or between the 4f electrons; experienced by the 5d electron; and the interaction between 4f and 5d electrons, respectively, as shown below:

$$\begin{aligned} H(5\mathrm{d}) &= \zeta_{5\mathrm{d}}\mathbf{A}_{\mathrm{so}}(\mathrm{d}) + \sum_{kq} B_q^k(\mathrm{d})\mathbf{C}_q^{(k)}(\mathrm{d}) \\ H_{\mathrm{int}}(4\mathrm{f}, 5\mathrm{d}) &= E_{\mathrm{exc}} + \sum_{k=2,4} F^k(\mathrm{fd})\mathbf{f}_k(\mathrm{fd}) + \sum_{k=1,3,5} G^k(\mathrm{fd})\mathbf{g}_k(\mathrm{fd}) \end{aligned} \tag{12}$$

where (f), (d), and (fd) are used to show that the operators are interactions for 4f electrons, the 5d electron, and interactions between 4f and 5d electrons, respectively. The two terms in H(5d) are the spin–orbit and the strong crystal field interactions experienced by the 5d electron. The H_{int}(4f,5d) parameter E_{exc} describes the separation between the barycenters of the $4f^N$ and $4f^{N-1}5d$ levels; and the second and third terms are the direct and exchange Coulomb interaction between 4f and 5d electrons. This model has been employed together with intensity calculations to simulate the energy levels and spectra of interconfigurational transitions. Most studies have utilized fluoride hosts where the band gaps are high so that $4f^{N-1}5d$ levels can be probed by excitation spectra [57, 58].

An example of an interconfigurational emission spectrum is shown in Fig. 2e for the highest energy 5d → 4f transitions of YAG:Nd^{3+}, with the simulated zero phonon line intensities shown as vertical bars. The presence of broad features is the characteristic of 5d → 4f emission transitions. Even for Nd^{3+} doped into Cs_2NaYF_6 (Fig. 2f), the features are similar and show that the transition from the lowest 5d level to the ground multiplet $^4I_{9/2}$ is much stronger than ^4I_J ($2J$ = 11, 13, 15). This spectrum (Fig. 2f) shows that time resolution can easily separate the inter and intraconfigurational transitions. The breadth of features in Fig. 2e, f is typical of 5d → 4f emission spectra and is due to unresolved vibrational progressions, in addition to overlapping electronic transitions. The 4f5d → $4f^2$ emission spectrum of Cs_2NaYCl_6:Pr^{3+} is, however, well resolved and permits a clearer understanding of the nature of the type of vibronic interaction [55]. The highest energy emission bands are shown in Fig. 2g.

This spectrum of Cs_2NaYCl_6:Pr^{3+} elucidates two points. First, although the 4f–5d transition is ED allowed, the site group selection rules in (8) may forbid transitions between certain crystal field levels. This is the case for Pr^{3+} at the O_h symmetry site, so that transitions from the lowest 4f5d level, Γ_{3u}, are only allowed to terminal Γ_{4g} and Γ_{5g} $4f^2$ levels. The transition to the electronic ground state $^3H_4$ Γ_{1g} is therefore forbidden. By contrast, the absorption transitions from the $4f^2$ Γ_{1g} electronic ground state are only allowed to terminal 4f5d Γ_{4u} states. The selection rules therefore account for the gap of 1,028 cm^{-1} between the absorption and emission bands of Pr^{3+} in Cs_2NaYCl_6 (Fig. 2g). Each transition is marked by vibrational progressions upon the zero phonon lines, with the most intense progressions occurring for the totally symmetric Pr-Cl stretch vibration. The progressions occur because the excitation from the $4f^2$ to the 4f5d configuration is a Franck–Condon transition (i.e., vertical in the configuration coordinate diagram, Fig. 1) and since there is a bond length change between the two configurations, the most intense transitions correspond to terminal vibrationally excited states. It turns out that the bond Pr-Cl distance is slightly shorter in the lower 4f5d states than in $4f^2$ [59]. The separation of the most intense band in the absorption spectrum from that in emission is normally termed the Stokes shift, which can be related to the electron–phonon coupling strength, via the Huang–Rhys parameter ([2], p 199). Note that this reasoning only applies when the respective bands correspond to the same transition (i.e., with coincident zero phonon lines in absorption and emission), and is not applicable here.

The energy levels of the $4f^{N-1}5d$ configuration can be envisaged under a simple model where only the dominant terms in the Hamiltonian are considered. The Hamiltonian in (11) can be written as [60]:

$$H = H_{\mathrm{Coul}}(\mathrm{ff}) + H_{\mathrm{cf}}(\mathrm{d}) + \lambda_{\eta_f S_f L_f S}\mathbf{S}\cdot\mathbf{L}_{\mathrm{f}} - J_{ex}\mathbf{S}_{\mathrm{f}}\cdot\mathbf{s}_{\mathrm{d}} \qquad (13)$$

where the first two terms, i.e., the Coulomb interaction between 4f electrons and the crystal field interaction for the 5d electron, are the strongest interactions. The next two terms of (13) are the f-electron spin–orbit interaction and the f–d Coulomb interaction which have been simplified within the subspace $(4f^{N-1}\eta_f S_f L_f, {}^2\Gamma_d)$. In this scheme, the smaller interactions such as the d-electron spin–orbit coupling and the f-electron crystal field have been neglected. This model has been applied to the interpretation of the d–f spectra of several systems. For example, in the $4f^2 5d \rightarrow 4f^3$ transitions of Nd^{3+} in YPO_4 crystal each of the transitions from the lowest 5d state to ${}^4I$, ${}^4F$, ${}^4G$ was modeled to comprise two prominent peaks with the transitions to ${}^4I$ most intense [61].

Luminescence is not observed from 5d levels for some lanthanide ions (Sm^{3+}, Eu^{3+}, Tb^{3+}) because fast nonradiative relaxation occurs from the lowest 5d level to the ladder of 4f levels below. A different scenario occurs for the quenching of $5d \rightarrow 4f$ emission of Ce^{3+}. Often, the 5d level is located within the CB of the host so that immediate transfer and delocalization of the Ce^{3+} 5d electron occurs to the host lattice CB upon excitation from the 4f ground state. In some cases for Ce^{3+}, the 5d emission can be turned on or off by the application of pressure [62]. The application of high pressure to Lu_2O_3:Ce^{3+} influences the energy of the 4f and 5d levels of Ce^{3+} as well as the CB edge of Lu_2O_3. However, the major outcome is a lowering of the lowest 5d level with respect to the CB so that luminescence can then occur.

For the second half of the lanthanide series, there is the possibility that the 5d electron spin is paired or opposed to the resultant spin of the $4f^{N-1}$ core, giving high and low spin levels, respectively. By Hund's rule, the high spin level is lowest so that the emission transition from this state to the $4f^N$ electronic ground state is spin-forbidden. Thus, the emission lifetime in these cases (such as for Er^{3+}, Tm^{3+}) is in the microsecond – and not nanosecond – regime [57].

3.4 Charge Transfer Transitions

The valence band (VB) of an oxide or halide crystal mostly comprises p-orbital character. The charge transfer (CT) transition of a lanthanide ion in the host corresponds to ligand to Ln^{3+} metal ion ground state (GS) transfer:

$$Ln^{3+}(\mathrm{GS}) + \mathrm{VB}(e^-) \rightarrow Ln^{2+}(\mathrm{vibr.ex.}) + \mathrm{VB}(\mathrm{hole}) \qquad (14)$$

where the dipositive ion is in a highly vibrationally excited state. In the case of $Y_2O_3:Yb^{3+}$, an ab initio calculation has shown that the description of the CT state involves the hole localized on the 2p orbital of just one oxygen ion surrounding Yb^{3+}, rather than being delocalized over all oxygen ligands [35]. The analogous transition in emission is:

$$\mathrm{Ln^{2+}(GS) + VB(hole) \rightarrow Ln^{3+}(vibr.ex.) + VB(e^-)} \tag{15}$$

so that a large energy separation occurs between the emission and absorption bands, both of which are usually very broad. The CT energy for a given system can be calculated roughly by knowing the optical electronegativities of the relevant ions [63], or by the refined spin-pairing theory [64], but more recently the model of Dorenbos [65] has been employed:

$$E_{\mathrm{CT}}(\mathrm{Ln2, host}) = E_{\mathrm{CT}}(\mathrm{Ln1, host}) + \Delta E_{\mathrm{CT}}(\mathrm{Ln2, Ln1}) \tag{16}$$

It is assumed here that the CT energy represents the energy of the ion above the top of the VB. So that knowing the CT energy of one lanthanide ion $E_{\mathrm{CT}}(\mathrm{Ln1, host})$, such as Eu^{3+} in a particular host, and using the tabulated values of $\Delta E_{\mathrm{CT}}(\mathrm{Ln2, Ln1})$ [65], under several assumptions [66], the value for any other lanthanide ion in the same host can be estimated. An alternative calculation scheme has been given by Nakazawa and Shiga [67].

The lowest energy CT transitions occur for Eu^{3+} and Yb^{3+} and the CT emission of Yb^{3+} has been studied in a variety of hosts [68–70]. An example is shown in Fig. 2h for the CT transitions of $Y_2O_3:Yb^{3+}$. The emission comprises two broad bands, corresponding to transitions to the $^2F_{5/2}$ ground state and $^2F_{7/2}$ excited state, separated by $\sim$10,000 cm^{-1}, with the highest energy band peaking at 195–228 nm in phosphates $MPO_4:Yb^{3+}$ (M = Sc, Y, La, Lu); 250 nm in $LiScO_2:Yb^{3+}$; 368 nm in $M_2O_3:Yb^{3+}$ (M = Sc, Y); 390 nm and 439 nm in $M_2O_2S:Yb^{3+}$, M = Y, La, respectively; and 181 nm in $LiYF_4:Yb^{3+}$ [68]. Thus, the position of CT luminescence shifts to longer wavelengths with increasing covalency of the host lattice and with increasing size of the cation site [68]. An alternative description of the dependence of CT energy upon crystal structure and composition in terms of an environmental factor has been given by Zhang's group [71, 72].

Attention has been directed to the quenching of CT emission, particularly with respect to the configuration coordinate diagram [68, 73]. For a given hosts, the quenching temperatures of Eu^{3+} emission are much higher than for Yb^{3+} emission. The temperature dependence of the Yb^{3+} CT emission in YAG (where the highest energy band has a maximum at 334 nm) and orthorhombic-$YAlO_3$ (YAP: maximum at 360 nm) follows a very different behavior from that of the Yb^{3+} infrared emission [73]. The CT emission lifetime at 7 K is about 0.1 μs for Yb^{3+} doped into these hosts, compared with the value $\sim$1 ns at room temperature [70]. The CT luminescence intensity roughly halves from 7 to 120 K. The decrease has been modeled by the consideration of three processes: (i) photoionization of the CT state involving the escape of a hole to the VB, (ii) radiative and (iii) nonradiative energy transfer to 4f levels [73]. Another factor which may quench the CT emission is the overlap of this emission with the absorption bands of intrinsic defects [70]. The role of CT states themselves in the quenching of 4f–4f luminescence of lanthanide ions has also been discussed [74] and modeled [75]. In the case of $LuCl_3:Pr^{3+}$ [76], excitation below, or into the band gap, gives 4f5d $\rightarrow$ $4f^2$ or $4f^2$ $\rightarrow$ $4f^2$ emission, respectively. The role, (iii), of the CT state in populating $4f^2$ levels has been suggested.

3.5 *Transitions Due to Defect Sites and Impurities*

Since crystals are imperfect, lanthanide ions can reside at minority sites or in different phases; in crystal interstices; or coordinated to ions such as OH^- at the surface. These situations have different repercussions upon the emission spectra, giving broad features for the second and third cases together with quenching of emission for the third. The compound YAG is an example of the first scenario since $Y_4Al_2O_9$ (YAM), YAP, and hexagonal-$YAlO_3$ (YAH) phases may coexist [77]. A sensitive test for the purity of the product is to use excitation lines which strongly excite one phase, but not another. This is the case for 488 nm excitation (which strongly excites $YAG:Er^{3+}$, but not $YAP:Er^{3+}$), and 457 nm (which does the reverse). Figure 3a, b shows a pure sample of $YAG:Er^{3+}$ under these two excitation lines, and although the intensity scale differs by a factor of about 60, there are no apparent differences between the spectra. On the other hand, the impure sample 3(c) shows fairly strong $YAG:Er^{3+}$ emission under 488 nm excitation, although some other additional, weaker bands are evident. Under 457 nm excitation of the same sample (Fig. 3d), a totally different spectrum is observed, corresponding to $YAP:Er^{3+}$ impurity.

In concentrated materials, satellite bands appear next to zero phonon lines due to the consequence of interactions between neighboring ions [78–80]. Very often, energy transfer occurs between the different types of sites so that even if the excitation is not directed into one particular site, its spectrum appears [81].

It is also important to recognize the characteristic luminescence of each lanthanide ion so that the presence of unintentional impurities in a crystal can be recognized. This is generally easy because the transitions to different lower multiplet levels provide a distinctive fingerprint. However, some reports have attributed unusual emission bands from trace lanthanide ion impurities to exotic phenomena.

3.6 *Other Transitions*

Finally, some $4f^N$ transitions with oscillator strengths of 10^{-10} or less are briefly mentioned. Pure electric quadrupole transitions have been reported in conjunction with hypersensitive transitions [82]. Not only two-center vibronic transitions (such as involving the electronic excitation of Yb^{3+} and the intra-vibrational excitation of

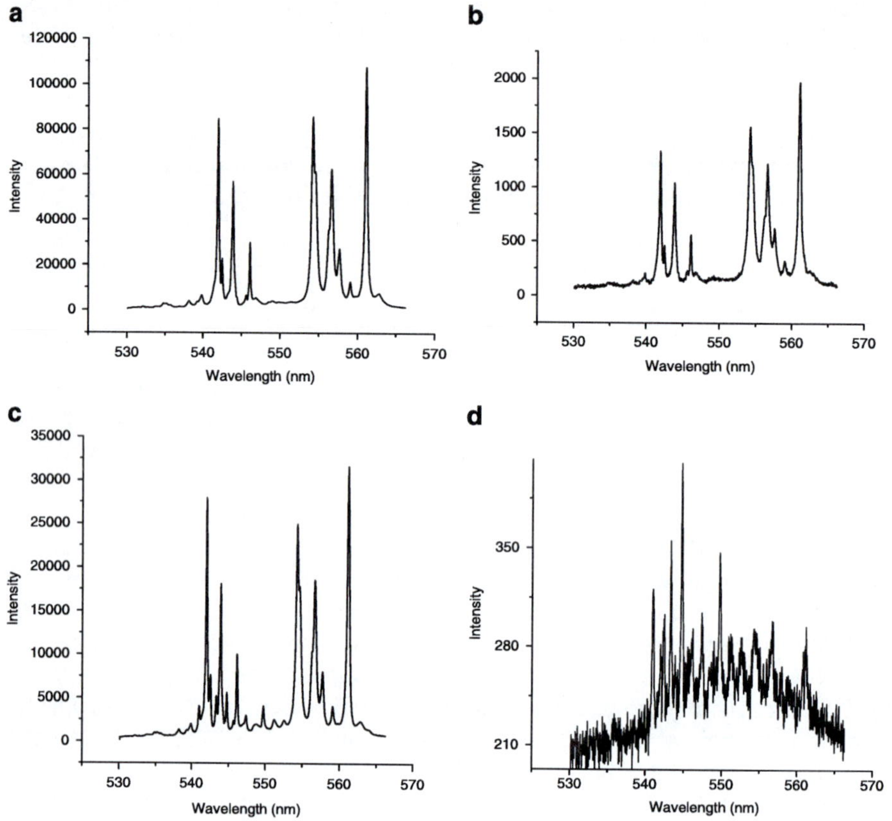

Fig. 3 Effects of impurity phases upon the $^2H_{11/2}/^4S_{3/2} \rightarrow {}^4I_{15/2}$ emission spectrum of YAG:Er^{3+}. Note the intensity scales. (**a, c**) 488 nm and (**b, d**) 457 nm excited luminescence spectra of Er^{3+}-doped samples of (**a, b**) pure and (**c, d**) contaminated YAG samples (adapted from [77])

an OH_2 ligand) have been observed and analyzed but also "dimer" pure electronic excitations, involving, for example, Yb^{3+} clusters, have also been observed and simulated. The simple model for the former systems stressed the importance of infrared active modes, although contributions from Raman modes in the vibronic sidebands also appear to be important [83]. The model for "dimer" excitations has largely been based upon the Dexter formalism so that short bond distances were considered to play a major role, although it now appears that the relationship between covalency and transition probability is important [84].

3.7 *Luminescence from the Divalent State*

The luminescence of the divalent lanthanide ions Eu^{2+}, Tm^{2+}, Sm^{2+}, Yb^{2+} has been reviewed by Rubio O [85]. Generally, the d $\rightarrow$ f emission transitions occur at lower energy than for trivalent lanthanides. Hence, divalent europium-activated phosphors find diverse applications, some of which are summarized in [86]. Notably, $BaMgAl_{10}O_{17}:Eu^{2+}$ (BAM) and $Sr_5(PO_4)_3Cl:Eu^{2+}$ are efficient tricolor lamp phosphors. In these hosts, and others such as $SrCl_2$ [87], $4f^65d \rightarrow 4f^7$ emission of Eu^{2+} is observed [88]. The structure comprises only the transition to the lowest multiplet term of $4f^7$, $^8S_{7/2}$, since the next highest term, $^6P_{7/2}$, is at ~30,000 cm^{-1} to high energy. At room temperature, a broad band between 390 and 440 nm is observed for $SrCl_2:Eu^{2+}$, but this becomes structured at low temperatures. The 10 K spectrum is shown in Fig. 4 and reveals a dominant progression in the Eu-Cl stretch. The location of the lowest $4f^65d$ level with respect to $4f^7$ $^6P_{7/2}$ can be modified by weakening the crystal field experienced by Eu^{2+}. Thus, $^6P_{7/2} \rightarrow {}^8S_{7/2}$ intraconfigurational emission with millisecond lifetime has been reported in a group of compounds such as $SrAlF_5$ [88], and more recently, $^6I_{7/2} \rightarrow {}^8S_{7/2}$ emission was observed in $KMgF_3:Eu^{2+}$ below 25 K, where Eu^{2+} occupies the large K^+ site with 12-coordination [89]. The preference for $4f^7$ emission at low temperature disappears when $4f^65d$ levels are thermally populated at higher temperatures.

Although the competition between intra and interconfigurational emission is not uncommon, being also observed for tripositive lanthanide ions such as Pr^{3+}, and Gd^{3+} (mentioned subsequently), the emission from lower and higher $4f^{N-1}5d$ levels is less common. Grimm and Güdel investigated the rich emission spectrum of $CsCaBr_3:Tm^{2+}$ at 10 K and observed sharp $4f^{13} \rightarrow 4f^{13}$ emission at 8,796 cm^{-1}; broad $4f^{12}5d \rightarrow 4f^{13}$ emission bands from the first (high spin at 12,240 cm^{-1}, and low spin at 13,640 cm^{-1}) $^3H_6$, t_{2g} 5d states to the ground and first excited $4f^7$ states; as well as emission bands from the next higher 5d state, $^3F_4$, t_{2g} [90].

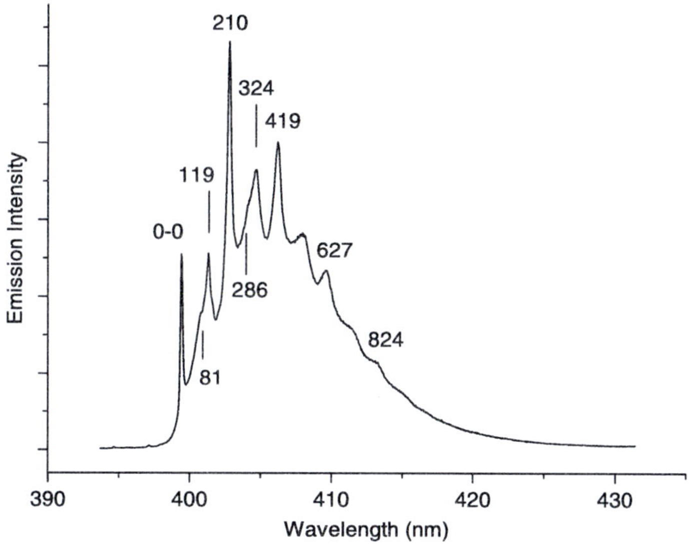

Fig. 4 355 nm excited emission spectrum of $SrCl_2:Eu^{2+}$ at 10 K. The major vibrational energy displacements from the zero phonon line (0–0) are marked in cm^{-1}

4 Location of Lanthanide Ion Energy Levels Relative to Those of the Host

Studies of the photoconductivity of doped lanthanide systems by conventional [91, 92] or novel [93] techniques have provided valuable information concerning the relative locations of host and lanthanide guest energy levels. In one of the experiments of Yen and co-workers [91, 92], the impurity Er^{3+} energy levels were determined relative to the host Gd_2O_3 band gap, 5.2 eV (Fig. 5). Under the action of 488 nm Ar^+ laser excitation (lower thick vertical arrow), Er^{3+} was excited to the $^4F_{7/2}$ state, from which it underwent rapid nonradiative decay to the metastable $^4S_{3/2}$ multiplet. The energy ΔE in Fig. 5 (corresponding to the upper thick vertical arrow) was adjusted by using radiation from a xenon lamp passed through a monochromator. Measurement of the wavelength at the onset of photoconductivity enabled ΔE to be determined as 3.6 eV and hence for the complete energy diagram to be constructed. Photoconductivity measurements, together with conventional spectral information from the transitions of divalent and trivalent ions, and thermoluminescence data [65, 94] or resonant photoemission spectra [95] have thus enabled a more complete rationalization of the energy level schemes of both the host and guest ions. It is recognized that the relative location of the lowest 4f state within this scheme controls such aspects as carrier trapping, luminescence quantum efficiency, and valence stability [65].

As an illustration, the locations of energy levels of Ln^{3+} in the series $Cs_2NaLnCl_6$ are depicted in Fig. 6. From (16) (using Eu^{3+}) and (10) (using Ce^{3+}), the CT energy for one Ln^{3+} and the location of the lowest 5d → 4f transition of another, are the only data required to plot the variation of these quantities for the entire series. The band gaps for $Cs_2NaLnCl_6$ can be estimated from (diffuse) reflectance spectra, or more roughly from excitation spectra [96]. The final

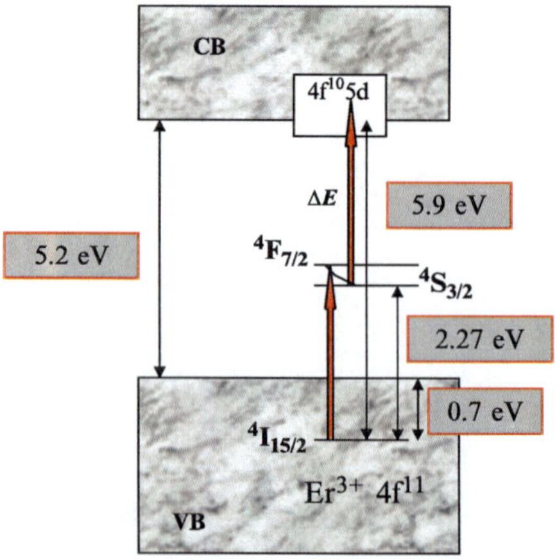

Fig. 5 Energy diagram of the impurity ion Er^{3+} in Gd_2O_3, illustrating the two-step photoconductivity experiment. The ordinate is energy in eV (adapted from [92])

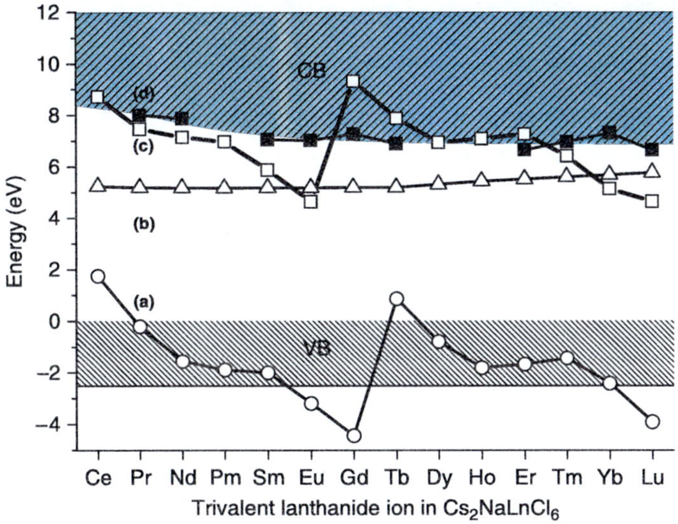

Fig. 6 Energy diagram for $Cs_2NaLnCl_6$ relative to the top of the valence band. The band gaps are smoothed and interpolated in curve (d). Curve (a) represents the relative position of the Ln^{3+} $4f^N$ ground state; (b) the relative position of the lowest Ln^{3+} $4f^{N-1}5d$ state; and (c) the relative position of the Ln^{2+} $4f^N$ vibrationally excited ground state

information that is required is the location of the 4f ground state relative to the top of the VB, which is estimated from that of Lu^{3+} in $Cs_2NaLuCl_6$ (−3.9 eV) [97], with the widths of the VB and CB about 3 and 8 eV, respectively [98]. As Dorenbos has pointed out, the energy levels in Fig. 6 should not be interpreted as due to single electron states, but rather as total energy states [65]. Thus, for example, in this figure, the fact that the Ce^{3+} ground state is located at 1.8 eV above the top of the VB means that it requires 1.8 eV less energy to remove an electron from the ground state of Ce^{3+} than from the top of the VB; or that a hole in the VB can be trapped by Ce^{3+} with 1.8 eV binding energy; or that Ce^{4+} is more stable by 1.8 eV than a free hole at the top of the VB. The Ln^{n+} ($n = 2,3$) energies are therefore referenced with respect to the VB and CB.

Energy level diagrams for Ln^{3+} have also been constructed by taking advantage of the energy of the intervalence CT band in the ultraviolet spectral region. It has been observed that the luminescence of some tripositive lanthanide ions which are more susceptible to oxidation (Tb^{3+}, Pr^{3+}) may be quenched by electron transfer to an acceptor group such as VO_4^{3-} [99]. Thus, whereas VO_4^{3-} is an exceptional antenna for the $^5D_0$ luminescence of Eu^{3+}, the luminescence from the $^3P_0$ multiplet of Pr^{3+} is quenched in YAG: V^{5+}, Pr^{3+} [100]. Krumpel et al. [101] have positioned the ground state energy levels of Pr^{3+} and Tb^{3+} in vanadate hosts based on the intervalence electron transfer to the CB.

5 Luminescence in Non-Lanthanide Hosts

There are several reasons to dope Ln^{3+} in host lattices which comprise s-block or transition metal ions. Naturally, for lighting purposes, a host lattice with Ca, say, instead of Y, would be cost-effective. In other applications, the thermal stability of materials such as Al_2O_3 would be advantageous. Upon doping, it is necessary to determine the location of the lanthanide guest in the host lattice.

The rationale for the use of Eu^{3+} as a probe of local site symmetry by using polarized electronic absorption spectroscopy has been given by Görller–Walrand and Binnemans ([33], p 230), and [22]. By counting the number of (polarized) bands in each multiplet–multiplet transition, some deductions can be made about site symmetry. Caution must be adopted when counting the number of spectral features in each transition because (i) some bands may overlap or be too weak to be observed; (ii) bands may correspond to vibronic structure and not to pure electronic transitions; and (iii) the scheme applies to ED and not magnetic dipole transitions. A similar scheme has been given by Jia [102] for the luminescence transitions of Eu^{3+} and it is displayed in Fig. 7. In this case, the magnetic dipole intensity of the $^5D_0 \rightarrow {}^7F_1$ transition has been included. An example of the application of the deductions, such as in this scheme, is now given.

Some studies have reported the replacement of Al^{3+} by Er^{3+} in α-Al_2O_3. This has particular relevance to optical communications using the transition at 1.6 μm. Our studies have shown that this replacement does not occur for sol-gel methods and that instead erbium perovskite or garnet phases are formed when Er^{3+} is doped into α-Al_2O_3 at atom percent levels [103]. However, with the use of combustion synthesis [104] it can be demonstrated that some of the Eu^{3+} ions occupied a C_{3v} symmetry site in α-Al_2O_3, which is consistent with the replacement of Al^{3+}. The 10 K emission spectrum under 471.6 nm excitation, in the region from 570 to 770 nm, is shown in Fig. 8. Other bands are observed in the spectra excited by different excitation lines and some of these are still present in Fig. 8 and are starred. The assignments in terms of C_{3v} symmetry are included in the figure.

As mentioned above, it is an ongoing quest to effectively dope lanthanide ions into semiconductors and utilize the band gap absorption to sensitize efficient emission from Ln^{3+}. In most cases, doping Ln^{3+} into the host lattice brings charge compensation problems. Alarcon et al. put forward a qualitative model [105], without taking into account ionic size differences, to describe the effects upon Eu^{3+} luminescence in such cases. Doping Eu^{3+} into a M^{2+} site brings an effective positive charge so that the electronic ground state lowers in energy more than the CT state, and the luminescence efficiency upon excitation into the CT state (q_{CT}) decreases. A "stiff" environment around Eu^{3+}, such as with highly charged M^{n+} in the second coordination sphere so that angle Eu^{3+}– O^{2-}– M^{n+} is 180°, can alleviate this decrease. The model predicts a high q_{CT} when Eu^{3+} situates at a M^{4+} site, such as Zr^{4+}. Certainly, these authors found q_{CT} was higher for MgO:Eu^{3+} than CaO:Eu^{3+}.

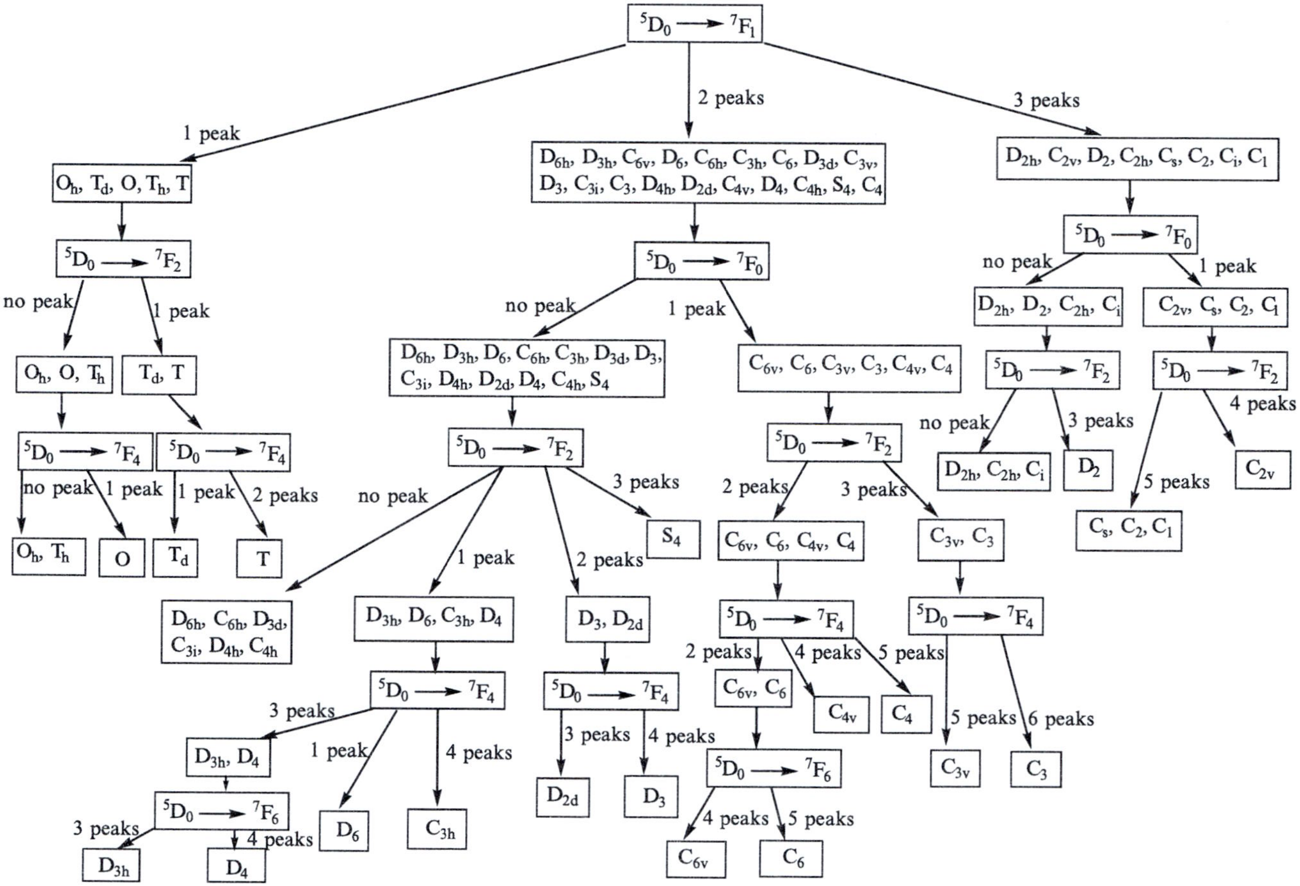

Fig. 7 Use of $^5D_0 \rightarrow {}^7F_J$ Eu^{3+} luminescence as a probe of site symmetry in crystals. The transitions $^5D_0 \rightarrow {}^7F_3$, $^7F_5$ are forbidden under Judd selection rules and are usually very weak so they are not included. The scheme refers to forced electric dipole-allowed transitions except for the case of the $^5D_0 \rightarrow {}^7F_1$ transition, which is allowed by the magnetic dipole mechanism (adapted from [102])

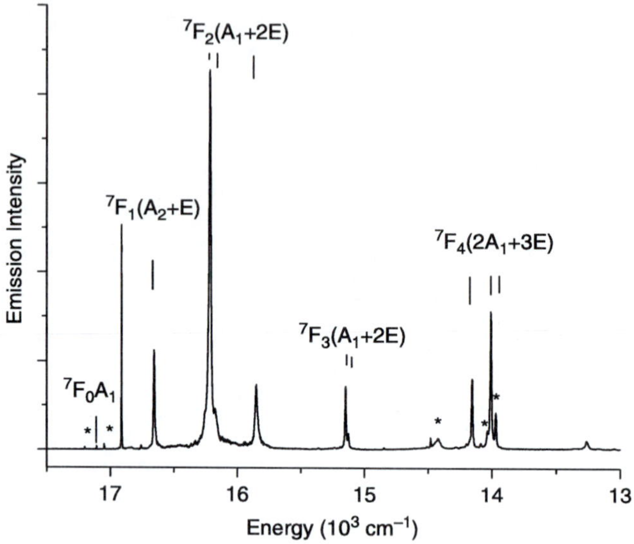

Fig. 8 10 K spectra of a sample of α-Al_2O_3:Eu^{3+} (1 at.%) prepared by combustion synthesis under 471.6 nm excitation. The initial luminescence state is $^5D_0$. Assignments are given in terms of C_{3v} site symmetry of Eu^{3+}. The *starred bands* correspond to other phases (adapted from [104])

6 Upconversion Luminescence in Solids

Upconversion luminescence from lanthanide ions, especially from infrared to visible, has been increasingly studied during recent years due to the availability of inexpensive high power infrared laser diodes. Thus, new applications have been forthcoming, particularly for nanomaterials, such as pH sensing [106], security devices [107] and cell-imaging [108, 109]. Some reviews have described the upconversion phenomena in detail [110–114], and a tutorial review has been given for lanthanide-doped upconversion nanocrystals [115]. The ion Yb^{3+}, with an absorption band at ~970 nm is a well-studied donor ion, often in conjunction with Ho^{3+} or Er^{3+} to produce green and red emissions, or together with Tm^{3+} for violet upconversion. These ions fulfill the necessity of two or more metastable states for the upconversion process to occur. The choice of host lattice is important to minimize unwanted nonradiative processes and the change from oxide or fluoride lattices to chloride, bromide, or iodide hosts can often lead to new upconversion, cross-relaxation, and luminescence processes [116]. Particular use has been made of hosts such as $Cs_3Ln_2X_9$ (X = Cl, Br, I) or Ba_2LnCl_7 where the Ln^{3+}–Ln^{3+} distance is small, and of β-$NaYF_4$, where the presence of several sites produces broad spectral features so that spectral overlaps are greater. Herein a broad overview of upconversion is given together with some recent developments. The more fundamental aspects of the upconversion are considered elsewhere, as well as more general reviews of energy transfer phenomena (for example [2, 5, 117, 118]).

In resonant energy transfer, the spectral overlap between the donor emission and acceptor absorption is of major importance.

6.1 Second Harmonic Generation

Figure 9 displays some processes where radiation of higher frequency is generated from that of lower frequency. Second (or multiple) harmonic generation (SHG) produces harmonics of the incident frequency via materials with large first hyperpolarizability, β, and has not been well investigated for lanthanide materials. The phenomenon is readily distinguished from other processes by changing the incident wavelength and checking for its harmonic. It is not exclusive of other processes, and the production of SHG and third harmonic generation simultaneously with intraconfigurational 4f emission upon multiphoton excitation has been reported [119]. Within a two-level model of the ground (g) and excited (e) states, β depends upon the transition dipole moment from g to e, the difference between the dipole moments of the excited and ground states and the optical band gap. Evidence has been provided that the 4f electrons contribute directly to β from its general increase across the lanthanide ion series in hydrated sodium tris(dipicolinato) lanthanates from La to Yb [120], and in terpyridyl-like complexes [121]. In other studies, a decrease in SHG intensities has been observed along the series $K_2Ln(NO_3)_5{\cdot}2H_2O$ (Ln = Pr, Nd, Sm) although the SHG intensities were larger than for KDP [122]. By contrast, Bogani and co-workers [123] remarked that the understanding of the origin of SHG properties requires the investigation of the building blocks (crystal

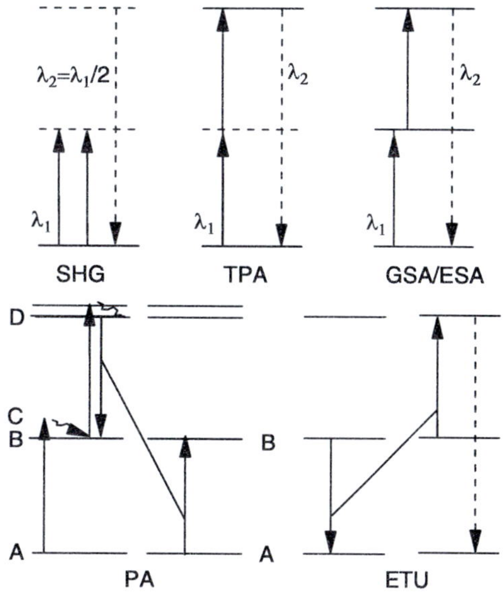

Fig. 9 Some upconversion processes: *SHG* second harmonic generation, *TPA* two-photon absorption, *GSA/ESA* ground state absorption/excited state absorption, *PA* photon avalanche, *ETU* energy transfer upconversion. Refer to the text for explanation

symmetry and geometrical arrangements of subunits in the unit cell) and concluded that for a family of rare earth based single chain magnets the main contribution to nonlinear processes arises from intermolecular π-stacking interactions.

Figure 10 (lower) shows the variation, as a function of the number of unpaired electrons, of the relative SHG intensity in the solid state, measured using 800 nm excitation, of a series of fifteen dipolar polymeric lanthanide complexes of trans-

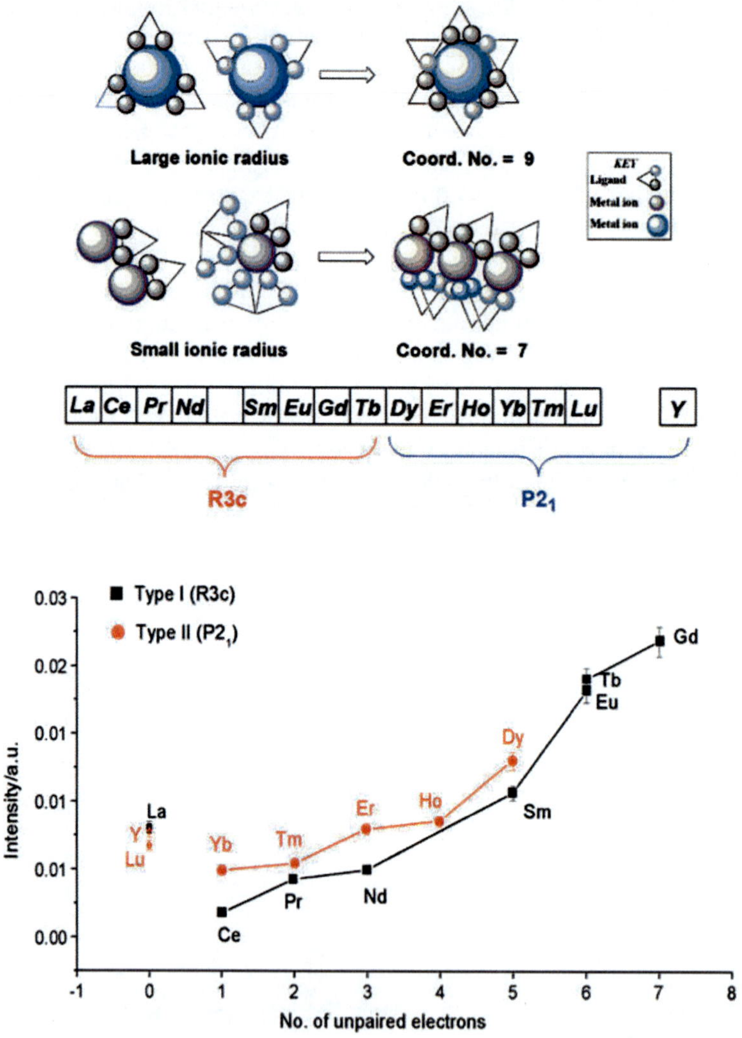

Fig. 10 Series of 15 dipolar polymeric lanthanide complexes of trans-cinnamic acid, $[Ln(C_9H_7O_2)_3]_n$. *Above*: Schematic of structural properties of the complexes formed with different coordination numbers of 9 and 7 in the non-centrosymmetric space groups R3c and $P2_1$, respectively; *Below*: The variation, as a function of the number of unpaired electrons, of the relative SHG intensity in the solid state, measured using 800 nm excitation (adapted from [124])

cinnamic acid, $[Ln(C_9H_7O_2)_3]_n$ [124]. There is a change in crystal structure and coordination geometry from 9 (Type I) to 7 (Type II), for the members of the second half of the lanthanide series as also shown in Fig. 10 (upper). The SHG relative intensity is a maximum near the center of the series, showing a very different trend from those mentioned above. Clearly, the diverse trends in SHG properties mentioned present unanswered questions.

6.2 *Two-Photon Absorption, Ground State/Excited State Absorption, and Energy Transfer Upconversion*

Two-photon absorption (TPA) and ground state absorption followed by excited state absorption (GSA/ESA) (Fig. 9) are both single ion processes occurring within the laser pulse and differ in that the intermediate state is virtual in the former case and metastable in the latter. TPA differs from SHG in that the former is system dependent whereas the latter is laser dependent. Denning and co-workers have exploited the window between the infrared and visible spectral regions of Tb^{3+} and Eu^{3+} in order to investigate the higher energy level schemes of these ions in elpasolite crystals, using linear polarization and measurements in a magnetic field to resolve ambiguities (for example, [125, 126]). Present calculations provide a reasonable understanding and rationalization of two-photon spectral intensities [127].

The simplest energy transfer upconversion (ETU) involves a cross-relaxation process between two ions. The processes ESA and ETU can be distinguished by energy dependence measurements, such as by recording the excitation spectrum of the luminescent state. For example, the mechanism of upconversion luminescence of Tm^{3+} in TmP_5O_{14} was investigated by Chen et al. [128]. Pumping the $^3F_2$ multiplet (15,153 cm^{-1}) by a 659 nm laser was observed to lead the luminescence $^1D_2 \rightarrow {}^3F_4$ at 450 nm. The intensity of the $^1D_2$ emission, I_{em}, varied as:

$$I_{em} \propto {P_{las}}^2 \tag{17}$$

where P_{las} is the laser power. Thus, a log–log plot of I_{em} versus P_{las} gave the slope of 2, as expected for a two-photon process either involving GSA/ESA or ETU [128]. The excitation spectrum of the $^1D_2$ emission was measured and found to involve bands not only associated with the ground state excitation to $^3F_2$, i.e., $^3H_6 \rightarrow {}^3F_2$, but also with additional features. These were assigned to the ESA transition $^3H_4 \rightarrow {}^1D_2$. Thus, the upconversion was found to be due to an ESA of a single ion comprising two consecutive absorption steps: excitation to $^3F_2$, followed by nonradiative decay to $^3H_4$ (12,674 cm^{-1}), and then excitation $^3H_4 \rightarrow {}^1D_2$. Alternatively, if the upconversion had been due to ETU between two ions, such as:

$$^3F_2(\text{ion A}) + {}^3F_2(\text{ion B}) \rightarrow {}^1D_2(\text{ion A}) + {}^3H_6(\text{ion B}) \tag{18}$$

the peaks of excitation spectrum and that of the absorption spectrum $^3H_6 \rightarrow {}^3F_2$ would overlap, and the line profile of excitation peaks would be approximately the square of that of the absorption peaks. The ETU and ESA mechanisms can also be distinguished by their concentration dependences, since the luminescence intensities would vary quadratically or be independent of the concentration of dopant luminescent ion in the host lattice, respectively. Another method to differentiate these two processes concerns the lifetime decay curve measurements when monitoring the upconversion level. If the upconversion process is due to ESA of a single ion, the line profile of the decay curve should be a single-exponential curve. However, if the upconversion is due to ETU between two ions, the population of the final (luminescent) state requires a cross-relaxation between two intermediate states, so an initial rise may be detected if the intermediate level has a sufficiently long lifetime. Usually, ETU dominates over GSA/ESA at high dopant ion concentrations. The two or three step upconversion mechanism for polycrystalline $NaYF_4$: Er^{3+} (20 at.%) from the 1.52 μm $^4I_{13/2}$ level thus involves ETU to $^4I_{9/2}$ and $^4S_{3/2}$, respectively. This material was attached to the rear of a bifacial silicon solar cell for enhanced near infrared solar cell response [129].

The GSA/ESA and ETU processes are not exclusive and GSA/ESA upconversion can be enhanced by the co-participation of a cross-relaxation, as reported by Goldner et al. [130] for $LiYF_4$:Nd^{3+}. Also, completely different mechanisms can follow the same excitation regime for different systems. For example, for $LiNbO_3$: MgO:Tm^{3+}, pulsed red laser excitation gives $^3H_6 \rightarrow {}^3F_2$ absorption but in the same laser pulse ESA $^3H_4 \rightarrow {}^1D_2$ occurs, giving violet emission [131]. However, a looping mechanism occurs for $Gd_3Ga_5O_{12}$ where the long-lived $^3F_4$ level (~5,000 cm^{-1}) is fed by cross-relaxation [131]. Looping mechanisms are not uncommon in upconversion processes. Another example [132] has been reported for the near infrared to red and green upconversion emission from silica sol-gel films made with $La_{0.45}Yb_{0.5}Er_{0.05}F_3$ nanoparticles. In this case, the temporal evolution of the green or red Er^{3+} emission under 980 nm laser diode excitation exhibit millisecond rise-times. The mechanism of this process has been postulated to involve sequential energy transfers from Yb^{3+} to excite Er^{3+} to the $^4F_{7/2}$ multiplet. Multiphonon relaxation to $^2H_{11/2}$ then occurs. Then, there is a step which initiates the feedback loop, involving a cross-relaxation between $Er^{3+}(^2H_{11/2})$ and the ground state $Yb(^2F_{7/2})$ to give $Er^{3+}(^4I_{11/2})$ and $Yb^{3+}(^2F_{5/2})$. The Yb^{3+} excited state then populates $Er^{3+}(^4I_{11/2})$ of another ion, which is the reservoir level for upconversion. The loop is reset by energy transfer from $Yb^{3+}(^2F_{5/2})$ to $Er^{3+}(^4I_{11/2})$ to regenerate $Er^{3+}(^2H_{11/2})$ (refer to Fig. 7 in [132]). However, several problems may be identified with this proposed mechanism. First, no explanation is given for the maximum log I_{em} versus log P_{las} slope of 8, other than "it represents a feedback mechanism". Second, the proposed cross-relaxations are not compatible at room temperature with the energy level schemes of LaF_3:Er^{3+} and LaF_3:Yb^{3+} (which are expected to be similar in the nano and bulk systems). For example, relaxation for Er^{3+} from $^2H_{11/2}$ to $^4S_{3/2}$ ($\Delta E = 678$ cm^{-1}) is rapid at room temperature and the energy difference from $^4S_{3/2}$ to $^4I_{11/2}$ is 8,162 cm^{-1}, compared with the energy of the lowest Yb^{3+} $^2F_{5/2}$ level (10,260 cm^{-1}). A detailed analysis is out of the scope of this review although it is noted that the high laser power densities promote thermal emission with a supralinear dependence upon laser intensity. The heat generated enables thermal occupation of excited states, such as those of $^2H_{11/2}$, and enables new cross-relaxation pathways.

6.3 *Photon Avalanche*

The final process shown in Fig. 9 is photon avalanche (PA), which presents some unusual characteristics [133, 134] in that nonresonant GSA leads to strong upconverted emission. In Fig. 9, C is not an electronic energy level. The absorption transition A → C is weak since it is due to such reasons as phonon sideband or defect absorption. Level B is then populated by nonradiative decay. The laser pump frequency matches the ESA B → D. Then, this can be followed by emission from D, or the cross-relaxation process:

$$\mathrm{D(ion\,1) + A(ion\,2) \rightarrow B(ion\,1) + B(ion\,2) + phonon(s)} \tag{19}$$

The pump photons are therefore only required to be resonant with B → D. One characteristic of PA is a power threshold above which the emission intensity increases by orders of magnitude. The Ln^{3+} ion concentration needs to be sufficiently high for efficient ETU. Whereas many examples of the PA phenomenon exist in the literature, only one study has been made for elpasolite systems, for $Cs_2NaGdCl_6:Tm^{3+}$, where the upconverted emission is due to the $^1G_4 \rightarrow {}^3H_6$ transition at ~480 nm [135]. The situation is rather more complex than in Fig. 9 because several other processes can occur, which lead also to emission from $^1D_2$. A quadratic emission $^1G_4$ intensity–excitation power dependence was obtained at low excitation intensities for samples of $Cs_2NaGdCl_6$ doped between 6 and 15 mol % Tm^{3+}. However, a dramatic increase of the emission intensity appears above the excitation threshold value, ~9 kW cm^{-2}. The slope of the log I_{em} versus log P_{las} increases to 6 for the 10 mol% Tm^{3+}-doped sample. The time dependence of the upconverted emission exhibits different behavior at different excitation powers. At the threshold excitation power, the upconversion emission has an almost linear risetime which is followed by a further slower rise over several seconds. At high excitation powers, the establishment of the stationary state is quicker, and the $^3F_4 \rightarrow {}^1G_4$ ESA decreases the transmitted laser light by several percent.

Most of the upconversion processes have been restricted to $4f^N \rightarrow 4f^N$ transitions, employing hosts such as β-$NaYF_4$ [136] and rare earth sesquioxides [137], which have relatively low phonon energies, together with tripositive lanthanide ions. Figure 11 shows a schematic upconversion process using 810 nm excitation, involving GSA/ESA and ETU, for $CsCaI_3:Tm^{2+}$, where the highest energy phonon energy is only ~170 cm^{-1} [138]. The initial excitation populates the lowest $4f^{12}5d$ level, ($^3H_6$, t_{2g}), from which (i) d → f emission, (ii) cross-relaxation with a neighbor, or (iii) ESA can occur to populate a higher subset of $4f^{12}5d$ energy levels, ($^3F_4$, t_{2g}),

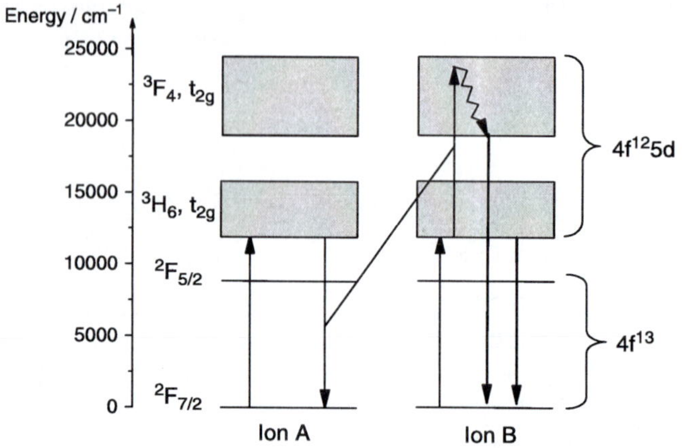

Fig. 11 Schematic upconversion process using 810 nm excitation, involving GSA/ESA and ETU, for $CsCaI_3:Tm^{2+}$. *Thick downward arrows* represent emission, which occurs from both $4f^{12}5d$ levels

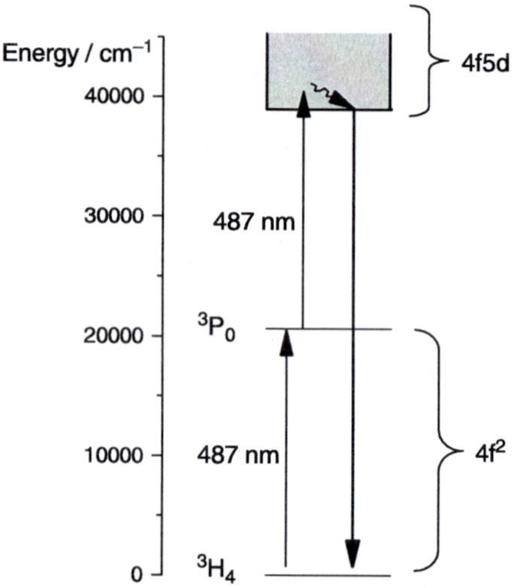

Fig. 12 Schematic of ESA upconversion to the fd configuration of Cs_2NaYCl_6:Pr^{3+}. Intermediate levels are omitted

which leads to higher energy d → f emission. The improvement in this upconversion process rests in the much stronger f → d, compared with f → f emission. The emission from the $4f^15d^1$ configuration of Pr^{3+} has also been investigated by upconversion (Fig. 12) in $Cs_2NaYCl_6:Pr^{3+}$. In this case, the metastable $^3P_0$ level, with millisecond lifetime, was pumped and ESA populated the d-electron levels. Broad, intense emission was observed from the lowest d-level at ~40,000 cm^{-1} [139]. The 4f5d → $4f^2$ emission between 30,000 and 44,000 cm^{-1} was also observed in $YAlO_3:Pr^{3+}$ using ESA from $^3P_0$ or $^1D_2$ levels.

6.4 *Many Body Processes*

Some energy transfer mechanisms involving more than two ions have been highlighted [140], although three-body processes are normally orders of magnitude weaker than two-body ones. One well-studied process is cooperative upconversion between Yb^{3+} and Yb^{3+}, to sensitize Tb^{3+} emission by infrared excitation, since there is a close match of energies between the $^5D_4$ Tb^{3+} levels and the combined $^2F_{5/2}$ energies of two Yb^{3+} ions. Experiments to study the weak Tb^{3+} emission are complicated by the cooperative luminescence involving the emission of the Yb^{3+} dimer system (this luminescence has been employed as a signature of ytterbium ion clustering in glasses [141]), which also produces emission bands near 500 nm [142]. Also, the presence of trace impurities of Er^{3+} leads to green and red emissions from dominant Yb^{3+}–Er^{3+} upconversion. Under near infrared laser excitation, the Tb^{3+} $^5D_4$ luminescence was measured in the range from 10 K to 300 K and gained intensity by three orders of magnitude from the lower to the upper temperature in the case of $Cs_3Tb_2Br_9$:Yb^{3+} (1 at.%), where the shortest Tb-Yb distance in the $TbYbBr_9{}^{3-}$ unit is 389 pm [143]. The dominant upconversion mechanism above 100 K was found to be the three-body process involving sensitization of Tb^{3+} by two Yb^{3+} ions. The same mechanism was cited to account for the Tb^{3+} emission intensity increase in Tb^{3+}, Yb^{3+} co-doped tellurite glass under anti-Stokes excitation [144]. However, it was reported [143] that at 10 K in $Cs_3Tb_2Br_9$:Yb^{3+} the Yb^{3+}-Tb^{3+} upconversion follows a different mechanism, associated with an exchange-coupled Yb^{3+}– Tb^{3+} dimer, i.e., a two-body process where ESA leads to Tb^{3+} emission. Excitation into the dimer levels around 12,000–14,500 cm^{-1}, where neither Yb^{3+} nor Tb^{3+} have single ion levels, leads to Yb^{3+} emission.

A clear example of three-body upconversion is the $^5D_0 \rightarrow {}^7F_J$ Eu^{3+} emission in Y_2O_3:Eu^{3+}, Yb^{3+} upon 970 nm laser diode excitation into the $^2F_{5/2}$ multiplet of Yb^{3+} [145]. The schematic energy level diagram is shown in Fig. 13a, where the lowest $^2F_{5/2}$ and $^5D_1$ energies are (in cm^{-1}): 10,225 and 18,937, respectively. The two-photon nature of the process is confirmed by the emission intensity–laser power plot in Fig. 13b. It is observed that the Eu^{3+} emission intensity increases considerably with temperature in the range from 10 K up to room temperature (Fig. 13c) and this was accounted for by two simulations. One of these simulations was based upon the thermalization of the $^7F_1$ levels of Eu^{3+}, whereas the alternative simulation focused on the dependence of upconversion rate upon temperature [145].

6.5 *Transition Metal-Lanthanide Upconversion Systems*

Güdel and coworkers have exploited an alternative upconversion approach by employing lanthanide–transition metal (TM) donor–acceptor systems. The selection of the host lattice can optimize the luminescence properties of the transition metal ion, since the spectra of these ions (by contrast to Ln^{3+}) are sensitive to ligand coordination and geometry. Many Ln^{3+}–TM^{n+} systems were investigated

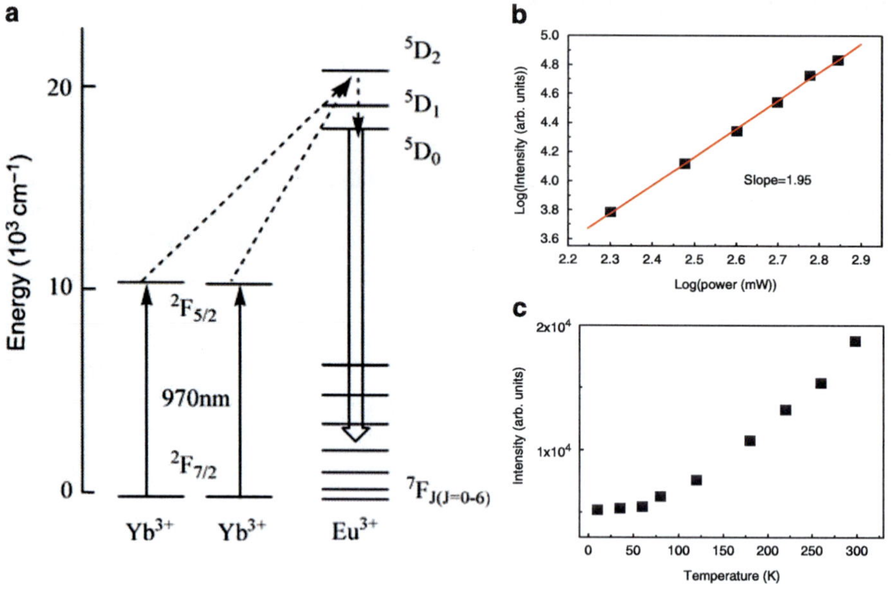

Fig. 13 (**a**) Schematic diagram of Yb^{3+}–Eu^{3+} cooperative upconversion mechanism; (**b**) log–log dependence of $^5D_0 \rightarrow {}^7F_2$ transition intensity upon excitation power; (**c**) temperature dependence of the integrated emission intensity at 610 nm for the sample Y_2O_3:1 at.% Eu^{3+}, 2 at.% Yb^{3+}

by this group [111–113] and two strategies have been employed. The first strategy involves absorption by Ln^{3+} (invariably Yb^{3+} for near infrared excitation) to give broad band visible emission from the TM ion. The second involves broadband absorption of near infrared radiation by the TM ion with upconversion to Ln^{3+} to give sharp line emission. This case has the potential application of converting "waste light" of an incandescent lamp into visible radiation [113], and it is essential that the TM sensitizer does not have absorption bands in the region of the upconverted emission. Several examples of the first strategy [111–113] are briefly discussed now.

In $CsMnBr_3$:Yb^{3+} (Fig. 14a), Yb^{3+} ions are incorporated into Mn^{2+} sites in a Yb^{3+}–vacancy–Yb^{3+} arrangement. Excitation into $^2F_{5/2}$ gives a broad red emission band due to the $^4T_1 \rightarrow {}^6A_1$ transition. The upconversion process was interpreted as due to an exchange-coupled dimer process. The energy level scheme and mechanism are shown in Fig. 15a. The ratio of the Mn^{2+} visible emission intensity to the Yb^{3+} near infrared emission was taken as a measure of the efficiency of the process and as a measure of the coupling strength between Mn^{2+} and Yb^{3+} in the dimer. In fact, the ratio was measured as 0.05% for $CsMnBr_3$, 4.2% for $MnCl_2$:Yb^{3+}, and 28% for Rb_2MnCl_4:Yb^{3+}, where the Yb-Cl-Mn angle changes from 74.8° to 92.8° and 180°, in the face, edge, and corner sharing bridging geometries, respectively [146–148]. This increase agrees with the remarks of Alarcon et al. [105] above. Figure 15b shows that tetrahedrally (rather than octahedrally) coordinated Mn^{2+} provides a smaller ligand crystal field strength, so that the Mn^{2+} levels are elevated (whereas the Yb^{3+} levels are insignificantly shifted). The situation in Fig. 15b would

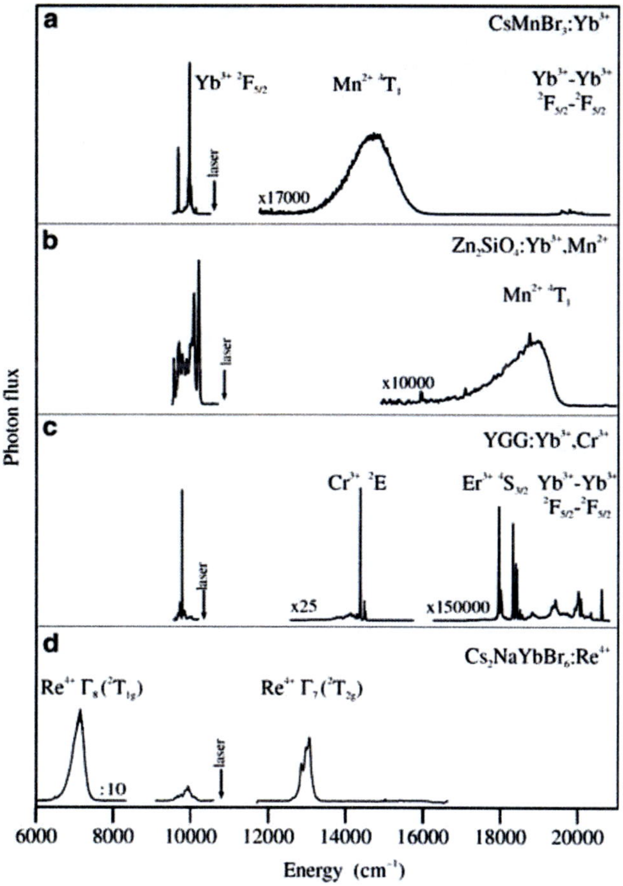

Fig. 14 15 K emission spectra of Yb^{3+}–TM systems, with laser excitation as shown into Yb^{3+} absorption bands. (**a**) $CsMnBr_3$:Yb^{3+}; (**b**) Zn_2SiO_4:Yb^{3+}, Mn^{2+}; (**c**) $Y_3Ga_5O_{12}$:Yb^{3+}, Cr^{3+}; (**d**) $Cs_2NaYbBr_6$:Re^{4+} (adapted from [113])

therefore enhance the upconversion emission, relative to 15a, because nonradiative decay from ($^4T_1$, $^2F_{7/2}$) to ($^6A_1$, $^2F_{5/2}$) is slower. Such tetrahedral coordination of Mn^{2+} occurs in Zn_2SiO_4:Tb^{3+}, Mn^{2+} (Fig. 14b), where the upconversion emission is stronger and situated at higher energy. By contrast, the upconversion mechanism for $Y_3Ga_5O_{12}$:Yb^{3+}, Cr^{3+} (1 at.%, 2 at.%, respectively; Fig. 15c) is a three-body process, just as for Y_2O_3:Yb^{3+}, Eu^{3+} described above. The emission spectrum is shown in Fig. 14c. Figure 14d shows the energy transfer from Yb^{3+} to Re^{4+} in the elpasolite lattice, where after Yb^{3+} excitation, 99% of the emitted photons originate from Re^{4+} [113]. The two-photon mechanism is shown in Fig. 14d.

We have observed that the site symmetry of an ion of inappropriate size in the elpasolite lattice can differ greatly from octahedral symmetry at moderate dopant concentrations. For example, it is suspected that this occurred in the system

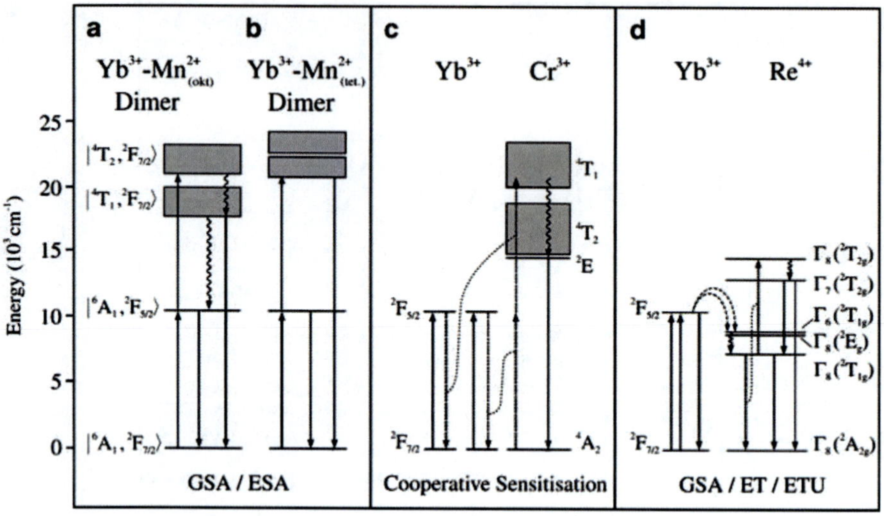

Fig. 15 Energy level diagrams showing upconversion mechanisms for systems in Fig. 14. (**a**) $CsMnBr_3:Yb^{3+}$; (**b**) $Zn_2SiO_4:Yb^{3+}$, Mn^{2+}; (**c**) $Y_3Ga_5O_{12}:Yb^{3+}$, Cr^{3+}; (**d**) $Cs_2NaYbBr_6:Re^{4+}$. Refer to the text (adapted from [113])

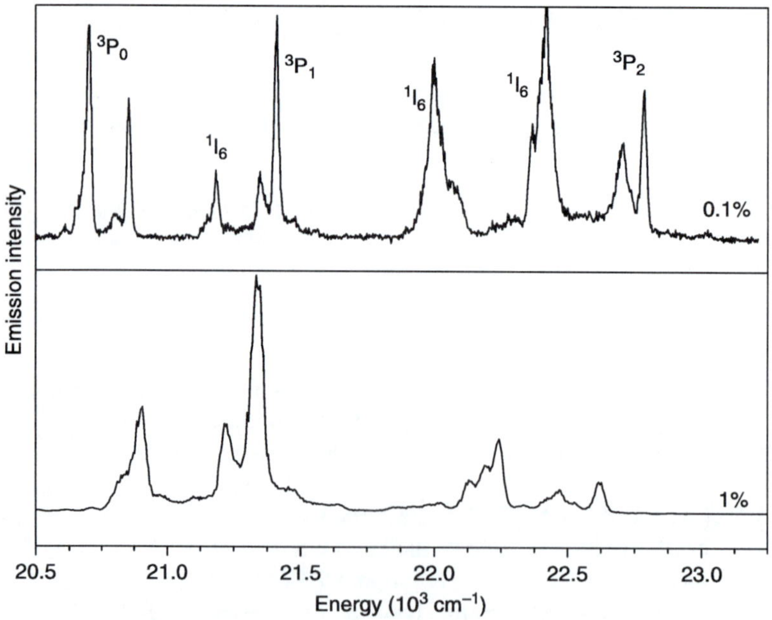

Fig. 16 Comparison of survey 10 K $^3P_0 \rightarrow {}^3F_2$ excitation spectra of 0.1 at.% and 1 at.% Pr^{3+} doped into $Cs_2NaScCl_6$ showing the effects of concentration increase. The terminal multiplet terms are marked for the 0.1 at.% sample

$Cs_2NaScCl_6:V^{3+}$, Pr^{3+} which was employed for ETU from Pr^{3+} to V^{3+} [113]. The 10 K excitation spectra of $Cs_2NaScCl_6:Pr^{3+}$ are shown for 0.1 at.% and 1 at.% dopant concentrations in Fig. 16. The ionic radii of Pr^{3+} and Sc^{3+} are 99 pm and 75 pm, respectively, whereas the spectrum in Fig. 16a can be assigned with O_h site occupation by Pr^{3+}, several other sites are occupied by Pr^{3+} at the higher concentration [149].

7 Some Applications of Luminescent Lanthanide Ions

7.1 *Fluorescent Lamps*

The long-term use of lanthanide ions in solids has been as phosphors in fluorescent lamps. These are available in different color temperatures such as cool white (2,700 K), warm white (4,000 K), and daylight (6,000 K). The phosphors which are coated on the inside of the lamp vary according to the color required. The criteria for a phosphor are stringent: it must be chemically stable, retain its properties at the operating lamp temperature, be synthesized as spherical particles of the order 10 μm, be non-toxic, have a strong ultraviolet absorption and high quantum efficiency of emission. Thus, europium doped strontium fluoroborate is employed in a backlight bulb, and $Y_2O_3:Eu^{3+}$ (red) together with $LaPO_4:Tb^{3+}$, Ce^{3+} (green, blue) are used for cool white. A 20–40 μm phosphor layer is coated inside the lamp in an atmosphere of a noble gas (400 Pa) and mercury (0.8 Pa: an amount of about 12 mg). Compact fluorescent lamps have lower power consumption than traditional fluorescent lamps, greater radiation stability, and they can be produced in a variety of shapes and sizes, with reduction of diameter to 1 cm. Supposedly, they have longer lifetimes if they are not switched on and off repeatedly. However, they still contain mercury (2–6 mg) and since this element has a high vapor pressure and it is toxic, the disposal of compact fluorescent lamps also leads to contamination of the environment. Another disadvantage of their use, in addition to higher cost, is that they do not give immediate high brightness. A barium cathode at ~1,200 K emits electrons, which produce a plasma in the mixture of mercury and neon so that mercury emission occurs at the predominant wavelengths 185 nm (12%) and 254 nm (85%). Each lanthanide ion emits a fairly pure color and their summation gives white light. Typical phosphors which are employed are $(La,Ce)PO_4:Tb^{3+}$ and $(Ce,Gd)MgB_5O_{10}:Tb^{3+}$.

7.2 *White Light*

A method of white light illumination uses the electroluminescence of semiconductor light-emitting diodes (LEDs), either singly or by the combination of red, green, and blue emitting diodes. However, white organic LEDs suffer from relatively short operational lifetimes [150]. Alternatively, the combination of a blue LED with phosphor(s) can be employed, but due to initial cost, these have not yet penetrated the general illumination market [151]. One such white light source is a blue GaN laser diode encapsulated with a YAG:Ce^{3+} yellow-emitting phosphor coating. Due to the absence of green light, the color rendering index is not good as for multiple phosphors and the color changes with input power. An alternative coating for the blue LED is the $Ba_3MgSi_2O_8$:Eu^{2+}, Mn^{2+} phosphor [152].

With multiple phosphors, under ultraviolet excitation, the resulting color addition of red, green, and blue emissions in appropriate intensities produces white light. The primary color emissions need to be controllable, and the device must be efficient, cost-effective, and durable. One such method to produce white light uses downconversion in core-shell materials. This was achieved by employing Y_2O_3:Eu^{3+} core–$LnPO_4$ (Ln = Ce, Tb) shell micron-size particles [153]; by the use of nanoparticles comprising LaF_3 doped with Eu^{3+} as core; with Tb^{3+} as first shell; with Tm^{3+} as second shell; and then with LaF_3 passive coating [154]; or by core-shell $CaWO_4$ microspheres co-doped with Na^+ and Ln^{3+} (Ln = Tb, Sm, Dy, Eu) [155]. The use of several lanthanide ions for white light generation can be avoided if use is made of CT emission in conjunction with the emission of one lanthanide ion. A rather expensive example of this is the sensitization of a Eu^{3+} in an organometallic d–f complex with iridium to produce "almost" white light [156]. Alternatively, the substitution of a lanthanide ion at two sites with very different crystal field strengths in a phosphor can produce blue and yellow emission. This was exploited for the ED allowed $4f^65d \rightarrow 4f^7$ transition of Eu^{2+} in $Mg_{0.1}Sr_{1.9}SiO_4$:Eu^{2+} where this ion occupies two different Sr^{2+} sites [157].

Upconversion, rather than downconversion, provides another method of white light generation by lanthanide ions. In most cases, color addition of the emission from several lanthanide ions has been utilized to give white light. As an alternative to core-shell technology, ceramic materials provide an easier and cheaper option. Transparent glass ceramic embedding β-YF_3:Tm^{3+}, Er^{3+}, Yb^{3+} nanocrystals has been utilized to generate white light using 976 nm Ti:sapphire laser excitation [158]. White light was also generated by upconversion using SiO_2 and ZrO_2 sol-gel thin films doped with different combinations of lanthanide ions: Yb^{3+}, Tm^{3+}, Yb^{3+}, Eu^{3+} and Yb^{3+}, Er^{3+} [159].

Current interest is centered upon the search for a single phosphor which can emit white light by photoluminescence. It has been observed that under vacuum, powders of lanthanide oxides have broad emission bands which give a supralinear power dependence [160]. Figure 17a shows the build-up of this emission for Tm_2O_3 using various 808 nm laser diode powers. The intensity rise lasts for several seconds and it is shorter for higher powers. The effect, over several orders of magnitude, of pressure change upon the emission intensity of Tm_2O_3 is shown in Fig. 17b. Figure 17c shows the color ranges for various oxide systems. White light can be achieved using Tm_2O_3, and the color can be modified by co-doping. The contributions of thermal and multiphoton processes to these upconversions are under investigation [160].

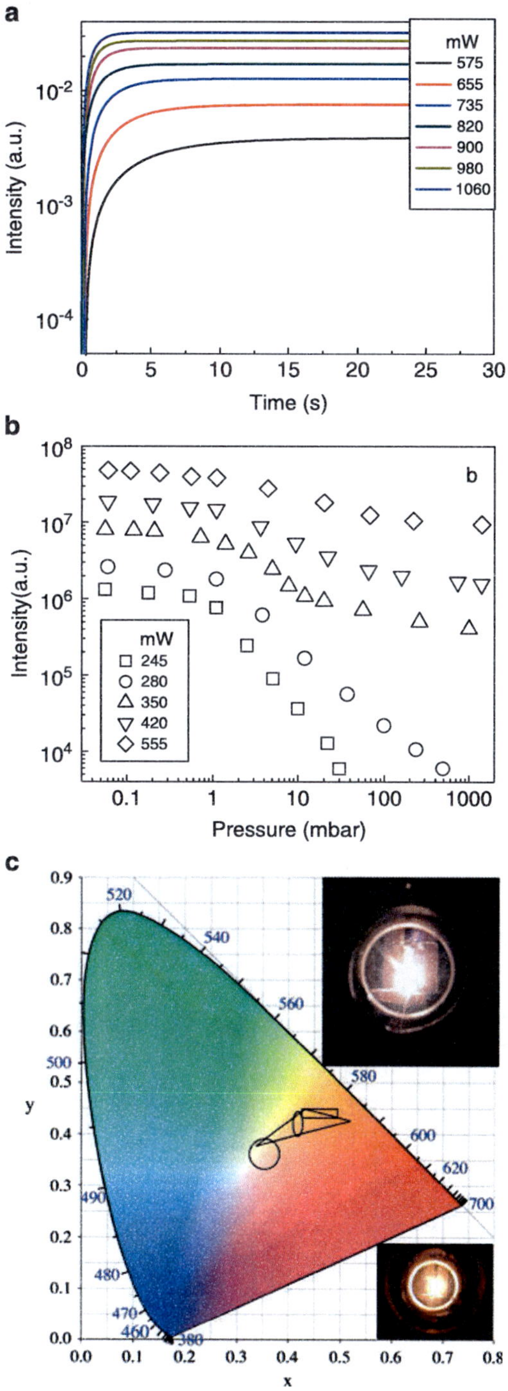

Fig. 17 (**a**) Time dependence of the upconversion emission intensity for Tm_2O_3 excited by a 808 nm laser diode in vacuum; (**b**) effect of ambient pressure on the upconversion intensity of Tm_2O_3 at various incident laser powers; (**c**) calculated *color* coordinate ranges for upconversion emission of powders under different power conditions between 400 and 1,000 mW. The *circle* represents the coordinate range for Tm_2O_3 (808 nm excitation), the *triangle* region for Nd_2O_3; the *ellipse* region for Sm_2O_3 and the *rectangle* for CeO_2 (all under 975 nm excitation). The *insets* are photographs of the upconversion emission of Tm_2O_3 under 808 nm excitation (*top*) and Nd_2O_3 under 975 nm excitation (*bottom*) powder samples in the groove of a copper holder behind the sapphire window of the closed chamber

7.3 *Quantum Cutting*

The trend to more environmentally friendly lighting has led to research into other forms of lighting, such as using light-emitting diodes and non-toxic discharge gases. Xenon emits lines at 147 nm and 172 nm and with a continuum at the longer wavelength region. Since the mercury discharge has a higher light output in terms of lumens per watt than the xenon discharge, and phosphors in mercury fluorescent lamps are more than 90% efficient, the efficiency of phosphors in xenon lamps must exceed 100% to be competitive. Thus, Wegh et al. [161] suggested various quantum cutting schemes, which employ between one to three ions (Fig. 18) where one ultraviolet photon is transformed into two visible photons whose summation can (in principle) produce white light. One ion (A) has a sufficiently high energy level (30,000–50,000 cm^{-1}). Scheme 1, which is shown with some simplifications, involves one ion only which sequentially emits two visible photons, such as for $^1S_0 \rightarrow {}^1I_6$ and $^3P_0 \rightarrow {}^3F_2$, for Pr^{3+}. The trick is to match the complementary emission colors to obtain white light, which is not the case for Pr^{3+}. The system $LiYF_4$:Gd^{3+} exhibits quantum cutting upon excitation into the ^6G_J levels around 200 nm (see below), but an efficient quantum cutter for white light based solely on Gd^{3+} is not possible [162]. Schemes 2–4 in Fig. 18 involve two ions. In Scheme 2, a cross-relaxation occurs from the high energy level of A to excite the type B ion, B1. Then, energy transfer occurs from the lower energy level of A to another B ion, B2. However, B1 and B2 emit the same color. In Schemes 3 and 4, different colors can be emitted by ions A and B. A variant of scheme 2 was proposed by Wegh et al., involving excitation of $LiGdF_4$:Eu^{3+} (0.5 at.%) at 202 nm into the ^6G_J levels. The cross-relaxation of Gd^{3+} from ^6G_J to ^6P_J excites a neighbor Eu^{3+} to $^5D_0$. Then, energy transfer from Gd^{3+} ^6P_J can occur to an excited state of Eu^{3+}, followed by nonradiative relaxation to ^5D_J levels from which emission occurs. The problems with this scheme are the weak absorption of Gd^{3+} and the loss of energy by nonradiative processes.

The quantum cutter K_2GdF_5:Eu^{3+}, Pr^{3+} has a quantum efficiency of 138% under 210 nm excitation into the 4f5d band of Pr^{3+}. The cross-relaxation:

$$^1S_0 + {}^7F_6 \rightarrow {}^3P_J + {}^5D_3 \quad (20)$$

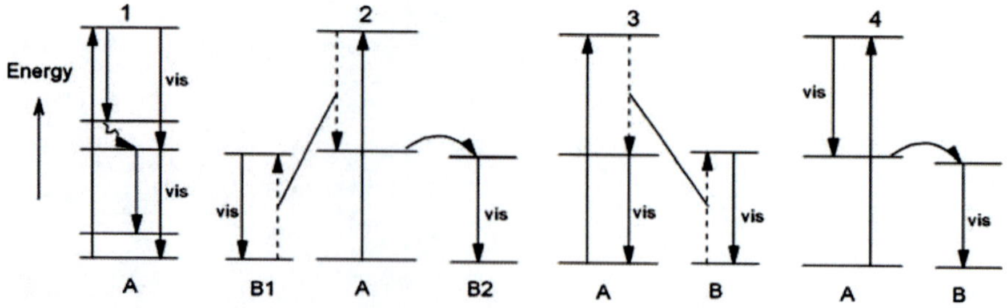

Fig. 18 Various quantum cutting schemes. Ions A and B are lanthanide ions. B1, B2 represent two B ions. *Upward arrows* represent excitation; *dotted arrows* represent cross-relaxation processes; *thick downward arrows* represent emission (adapted from [161])

enables energy transfer to the Eu^{3+} ^5D_J levels, giving predominantly red emission. However, in this system, the energy transfer from ^3P_J to ^5D_J is also able to feed Eu^{3+} [163].

The $^1S_0 \rightarrow {}^3P_1 + {}^1I_6, {}^1D_2$ emission of Pr^{3+} is too high in energy to be a useful component of white light. Energy transfer processes were investigated in $PbWO_4$: Pr^{3+} to convert this emission into visible light [164]. The cross-relaxation:

$$Pr^{3+}(^1S_0) + WO_4{}^{2-}(GS) \rightarrow Pr^{3+}(^1D_2) + WO_4{}^{2-}(exc.) \tag{21}$$

is then followed by relaxation of the excited tungstate ions:

$$Pr^{3+}(^3H_4) + WO_4{}^{2-}(exc.) \rightarrow Pr^{3+}(^3P_2, {}^1I_6, -{}^3P_0) + WO_4{}^{2-}(GS) \tag{22}$$

where the dash indicates subsequent nonradiative decay. Thus, tungstate is an effective catalyst to give emission in the region 490–650 nm from $^3P_0$. This type of scheme is a novel fifth alternative to quantum cutting (not depicted in Fig. 18). A study of the factors which determine how much emission is forthcoming from the second quantum cutting step ($^3P_0$, $^1D_2 \rightarrow {}^{2S+1}L_J$) has been made by Vink et al. [165].

The lanthanide ion Er^{3+} has many luminescent levels and rather than quantum cutting, it is photon cascade downconversion that occurs upon high energy excitation [166, 167].

7.4 Volatile Flat Panel Displays

7.4.1 Field emission displays

Volatile flat panel displays (VFPDs) include field emission displays (FEDs) which have thin panels, wide viewing angle, quick response, high brightness and contrast ratio, light mass, and low power consumption. By contrast to a cathode ray tube (CRT), the single electron gun is replaced by individual nanoscopic electron guns, situated at ~1 mm from the phosphor in vacuum. FEDs operate at lower excitation voltages ($\leq$5 kV) and higher current densities (10–100 μA cm^{-2}) than CRTs. Thus, the cathodoluminescent phosphors are required to have a high efficiency at low voltage, high resistance to current saturation as well as chromaticity and long-lasting properties. Thus, rare earth doped sesquioxide phosphors, which are more stable and environmentally friendly than sulfide phosphors, have great potential for application in FEDs.

7.4.2 Organic Light-Emitting Diodes

Another type of VFPD is an organic light-emitting diode (OLED). In this case, no external source is required since an emissive electroluminescent layer is printed in rows and columns on a flat carrier. In order to achieve sharp, rather than broad emission bands, the incorporation of a lanthanide ion with a π-conjugated ligand such as the β-diketonato ligand has been employed as the emitter [168]. Sensitization of Ln^{3+} (with most attention being directed to Eu^{3+} and Tb^{3+}) occurs via the ligand triplet state. In fact, the combination of these two lanthanide ions in a binuclear complex has generated white light in a triple-layer OLED [169]. More recently, white electroluminescence has been obtained from a single europium complex, where the ligand emits a broad band at ~500 nm and the $^5D_0 \rightarrow {}^7F_2$ Eu^{3+} emission is intense at ~610 nm [170].

7.4.3 Plasma Display Panels

A plasma display panel (PDP) is another type of VFPD which is common to large-area television screens. These slim devices produce more accurate color reproduction than liquid crystal displays. In a PDP, a mixture of noble gases (90% Ne, 10% Xe or 69% He, 27% Ne, 4% Xe) under a pressure of 350–600 torr is contained in many cells between glass screens. Each pixel comprises three elementary discharge cells, one for each primary color. The phosphors, coated on glass, are excited by radiation emitted by the noble gas discharge. Certainly, the phosphors employed in xenon lamps require more stringent thermal properties so that high-band gap materials with strong covalent linkages such as P-O, B-O, Si-O, and Al-O are employed. For example, although BAM shows efficient blue emission under VUV excitation, it has a strong thermal degradation. Red phosphors $(Y,Gd)BO_3:Eu^{3+}$ are commercialized and have been compared to the other commercial red phosphors $Y_2O_3:Eu^{3+}$ and $Y(V,P)O_4:Eu^{3+}$ by Jüstel et al. [171], specifically for their response to photonic excitation in the VUV range, with in view their application in PDPs. Green phosphors $(Y,Gd)BO_3:Tb^{3+}$ have also been considered for PDP applications.

It is advantageous to improve the efficacy of the phosphor mixture in a condition which is as near as possible to that of the application. "The plasma set-up" was developed to enable the determination of the relative intensities of emission of phosphors under the appropriate final PDP application conditions [172]. In this equipment, a hermetically sealed chamber is filled with a gas mixture under a certain pressure Fig. 19a. Parallel and planar electrodes, at an adjustable distance, allow the initiation of a discharge in the gas mixture. The emitted VUV photons excite the phosphor sample and the luminescence spectrum can then be acquired and recorded using a monochromator and detector. Variable pressures between 50 and 170 torr can be employed. At low pressure, the VUV photons are emitted by the resonance states of xenon (147 nm), but with increase of pressure the first continuum (150 nm) and second continuum (173 nm) increase in intensity too. Figure 19b shows a typical emission spectrum of $YBO_3:Eu^{3+}$ obtained with this set-up and the comparison of integrated emission intensities of Y_2O_3, YBO_3, and $Y_{1-x}Gd_xBO_3$ doped with 5 at.% Eu^{3+} in the 470–750 nm spectral region is displayed in Fig. 19c. It is evident that the sesquioxide has similar emission intensity to the borates at low gas pressures, but is superior at high pressure.

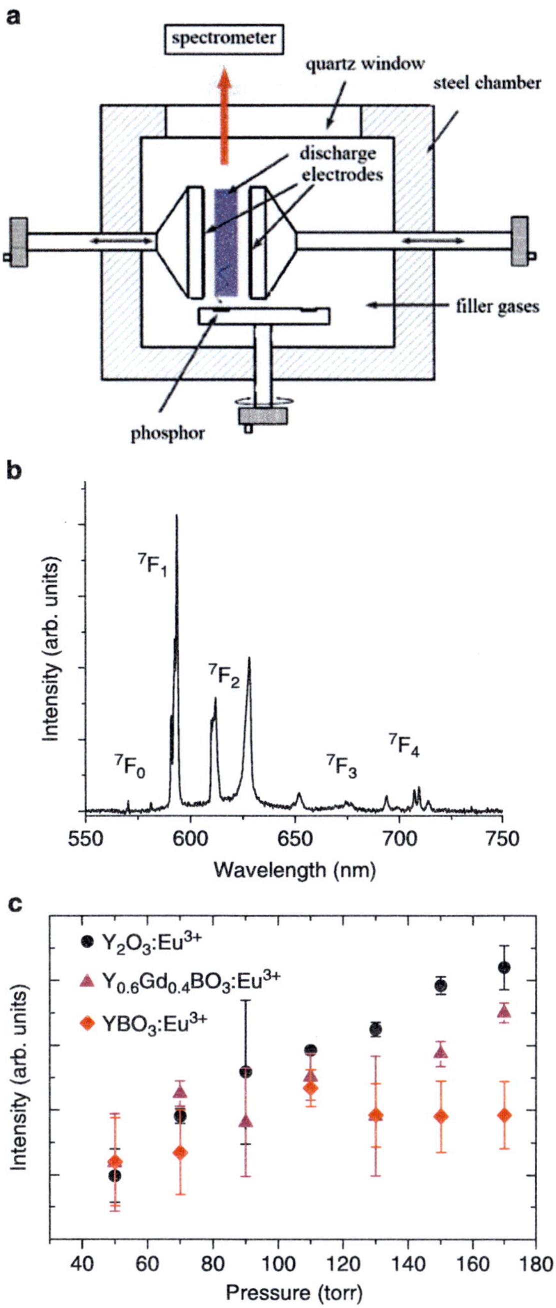

Fig. 19 (**a**) The plasma set-up for the determination of emission intensities of phosphors under different gas pressures; (**b**) room temperature emission spectrum of $YBO_3:Eu^{3+}$ (5 at.%) under plasma excitation; (**c**) integrated 298 K emission intensities versus pressure in the Ne-Xe 50–50% plasma chamber for samples containing 5 at.% Eu^{3+}. The *error bars* indicate standard deviations of 5 measurements

7.5 *Scintillators*

Lanthanide ions also find application in high energy radiation detectors such as scintillators. A higher energy application of rare earth doped inorganic crystals is γ-ray detection, with applications in measurement systems and medical diagnostic imaging. Optimum scintillator requirements involve fast response (10–100 ns), high light yield, proportional response, and high density and high atomic number. Naturally, the scintillator emission, which is excited by γ-rays, must match the wavelength response of the detection system (a photomultiplier tube or silicon diode). The fast 5d → 4f emission of Ce^{3+} (with lifetime in the range from 10–70 ns) makes it an ideal candidate for incorporation in appropriate hosts for such applications. The preference has been for halide hosts, such as $LuI_3:Ce^{3+}$, with a light yield of 95,000 photons per MeV [173], and for Ce^{3+} doped into elpasolite hosts such as Cs_2LiYBr_6 [174].

8 Emission from High-Lying Energy $4f^N$ Levels

Photon cascade occurs in neat Gd^{3+} compounds, such as $GdBaB_9O_{16}$ under 202 nm excitation [175]. Emission occurs from ^6G_J (near 50,000 cm^{-1}) to ^6I_J and ^6P_J, giving infrared and red radiation, respectively. This is followed by nonradiative decay to the lowest excited $4f^7$ term, ^6P_J, from which ultraviolet emission occurs to the electronic ground state $^8S_{7/2}$. It was concluded, however, that although quantum cutting occurs for several lanthanide ions, the efficiency in the useful visible spectral region is not above 100% for excitation at $< 50{,}000$ cm^{-1} [176]. In conjunction with the easier access to synchrotron radiation, this prompted the investigation of emission from even higher $4f^N$ energy levels in suitable high-band gap hosts. In the case of Er^{3+} in fluoride hosts, emission was observed from $^2F(2)_{5/2}$ (~63,300 cm^{-1}) by using synchrotron excitation [176]. In $LaF_3:Er^{3+}$, this level is located just below the lowest $4f^{10}5d$ state, but in $LiYF_4:Er^{3+}$, it lies in-between the two lowest d-electron states. In the latter case, the slow nonradiative relaxation from $^2F(2)_{5/2}$ was attributed to its spin-forbidden nature.

A glance at Fig. 2 in [22] shows that for many Ln^{3+} the gaps between energy levels get narrower at higher energies, and (not shown) the overlap of $4f^N$ and $4f^{N-1}5d$ levels occurs. Therefore, a careful optimization of the host lattice is required in some cases to achieve intraconfigurational luminescence. The highest $4f^2$ multiplet term of Pr^{3+} is $^1S_0$ and this can be situated above or below the $4f^15d^1$ energy levels. A high coordination number of Pr^{3+} results in a weak crystal field so that the 4f5d level is at high energy. In the latter case, such as for $SrAl_{12}O_{19}$: Pr^{3+}, emission from $^1S_0$ can be observed at below 46,500 cm^{-1} [177]. By using high pressure to increase the crystal field experienced by Pr^{3+} in the YAG host lattice, the 4f5d → $4f^2$ emission was shifted to lower energy, linearly with pressure [178].

9 Solar Energy Conversion

Silicon solar cells suffer from about 50% losses because the solar spectrum does not match well with silicon absorption. Photons with wavelengths longer than the band gap (for multicrystalline silicon, *m*Si, $E_g = 1.1$ eV) are wasted, whereas shorter wavelengths are absorbed but the excess energy is lost due to thermalization of the electrons. The optimum wavelength for absorption is 1,100 nm. Effort has therefore been directed to tailor incident radiation upon the cell by using a front panel so that the optimum response is achieved. It therefore entails that both upconversion and downconversion processes are required to be applied to modify the energies of incident photons.

Several quantum cutting regimes have been utilized with most of them involving Gd^{3+} and Pr^{3+} since these ions have high energy levels from which the initial photon is released. However, these levels lie too high for effective absorption of solar energy. The emission from Yb^{3+} is just above the band gap of crystalline Si. Hence, the downconversion in which one blue photon (~483 nm) is absorbed by Tb^{3+} and used to sensitize two Yb^{3+} ions for $^2F_{5/2} \rightarrow {}^2F_{7/2}$ emission near 1 μm has been investigated [179]. Three types of downconversion mechanism: cooperative, accretive, and phonon-assisted were considered in a Monte Carlo simulation, and the first one was considered to be dominant. A further mechanism which involves the CT state Yb^{2+}-Tb^{4+} was also suggested by Yuan et al. [180]. The ultraviolet–near infrared downconversion couple Er^{3+}–Yb^{3+} was found to be inefficient in $NaYF_4$:Er^{3+}, Yb^{3+} [181] because fast nonradiative decay occurs from the Er^{3+} $^4F_{7/2}$ level to $^4S_{3/2}$, instead of the energy transfer:

$$Er^{3+}\left({}^4F_{7/2}\right) + Yb^{3+}\left({}^4I_{15/2}\right) \rightarrow Er^{3+}\left({}^4I_{11/2}\right) + Yb^{3+}\left({}^2F_{5/2}\right) \tag{23}$$

Mechanisms to utilize the "wasted" infrared radiation at wavelengths longer than the band gap have also been considered and mainly involve upconversion from the Er^{3+} $^4I_{13/2}$ level, as mentioned above for $NaYF_4$:Er^{3+}.

10 Luminescence of Lanthanide Nanomaterials

Interest has been generated in the luminescence processes of nanomaterials, particularly for biolabels and high-definition imaging. The morphology of particles can be tailored into various shapes and protective shells can be superimposed. Since this field has been extensively reviewed recently [182–184], brief comments are given here for only four topics.

First, the change in spontaneous emission lifetime with refractive index of the surrounding medium has been modeled by several theories as well as being investigated by experimental work, which ideally requires the measurement of emission lifetimes over a wide range of refractive index without other changes in the system [183]. A summary of the theories has been given by Duan and Reid [185], and the basic problem is that the local microscopic electric field at the position of the luminescent ion differs from the macroscopic field. These theories apply to ED allowed transitions and are not appropriate for $4f^N$–$4f^N$ transitions, although they have been applied to the forced dipole luminescence of Eu^{3+} in various media. It has more recently been pointed out that since the linear sizes of nanoparticles are considerably smaller than light wavelengths, the expression for the rate of spontaneous emission (A_{nano}) also needs to be modified from the bulk expression (A_{bulk}) [186, 187]. It turns out that the ratio A_{nano}/A_{bulk} can be estimated without recourse to a specific local-field model.

The Yb^{3+}, Er^{3+} co-doped systems are the most well-studied upconversion systems, particularly in Y_2O_3 and β-$NaYF_4$ crystals, to give red, green, and violet emission. The relative intensities of these emissions depend upon several factors, including temperature and dopant ion concentration [188]. Recently, the intensity ratio of green to red emission (G/R) was reported to increase with decreasing particle size of single nanocrystals, using moderate laser powers [189]. In other studies [190], the red emission was enhanced relative to green for smaller nanoparticles. This is expected from the much smaller energy gap below $^4S_{3/2}$ compared with that below $^4F_{9/2}$ and the faster nonradiative decay from $^2H_{11/2}$, $^4S_{3/2}$ due to the occurrence of more surface imperfections, such as dangling OH_2 bonds in very small particles. Hence, the former result [189] was surprising. It does, however, show the difficulties and problems in severe chromatic aberration when interfacing an AFM with a laser spectrometer, because it appears that this gives the apparent reduction in red light intensity, relative to green light. This conclusion may be inferred from the different intensity–power dependence of red and green emission. Both the red and green upconversions mostly correspond to two-photon processes, and the laser powers employed [189] were not sufficiently high as to give anomalous slopes differing markedly from 2 [191]. However, as deduced from Fig. 3a of [189], the slope of the $\log(I_{em})$ versus $\log(P_{las})$ plot is 1.8 for green emission and 0.5 for red.

Quantum confinement exhibits novel spectroscopic effects in semiconductors and quantum dots, where the carriers are pinched into a dimension smaller than the Bohr exciton radius. As Zych has pointed out [192], the situation is different in insulators where electronic states are strongly localized. Nevertheless, there have been reports in the literature of quantum confinement effects of Ln^{3+} doped into sesquioxides, as evidenced by excitation or emission spectra. The spectroscopic effects are very minor so that other causes due to impurities, phase changes, or physical effects such as scattering, certainly play an important role. In particular, combustion syntheses produce variable spectroscopic results, particularly when non-stoichiometric ratios are employed [193]. Zych has ruled out quantum confinement effects for insulators doped with lanthanide ions, at least down to 6 nm.

Another anomalous effect which has been investigated and reported for lanthanide ions is the persistence of hot bands in spectra at very low temperatures. This anomalous thermalization results from the inability of nonradiative relaxation to occur for an energy level because the mechanism is restricted to the direct (one phonon) process and the phonon energy is too large for very small particle size.

Although this effect clearly occurs for small energy level gaps of 10–20 cm^{-1} [194], suggestions that it also occurs for gaps of 100s of cm^{-1} are to be questioned. In these cases, particularly for nanotubes, the actual temperature of the system may deviate considerably from that of a sensor, or bulk material, due to the very poor thermal conductivity. The slow nonradiative relaxation rate between $J = 1$ and $J = 0$ levels also appears to play a role [195], and it would be instructive to investigate other systems without this restriction and with single Ln^{3+} sites.

Acknowledgments Financial support for this work under the City University Strategic Research Grant 7002462 is gratefully acknowledged. Thanks are given to Dr Guohua Jia for helping to draw the figures and for constructively reading the manuscript, as well as to Prof. Xueyuan Chen for comments on the manuscript. Prof. Chen has also drawn attention to a comprehensive new review concerning quantum cutting [196].

References

1. Blasse G, Grabmeier BC (1994) Luminescent materials. Springer-Verlag Telos, Berlin
2. Henderson B, Imbusch GF (1989) Optical spectroscopy of inorganic solids. Clarendon Press, Oxford
3. Yen WM, Shionoya S, Yamamoto H (2007) The phosphor handbook. CRC Press, Boca Raton, FL
4. Luo Z, Huang Y, Chen X (2006) Spectroscopy of solid-state laser and luminescent materials. Nova Publishers, New York
5. Holstein T, Lyo SK, Orbach R (1981) Excitation transfer in disordered systems. In: Yen WM, Selzer PM (eds) Laser spectroscopy of solids. Springer-Verlag, Berlin, pp 39–82
6. Garcia-Solé J, Bausá LE, Jaque D (2005) Optical spectroscopy of inorganic solids. Wiley, New York
7. Kaplyanskii AA, Macfarlane RM (1987) Spectroscopy of solids containing rare earth ions. North-Holland, Amsterdam
8. Hendersen B, Bartram RH (2000) Crystal-field engineering of solid-state laser materials. Cambridge University Press, Cambridge
9. Hüfner S (1978) Optical spectra of transparent rare earth compounds. Academic, New York
10. Tanner PA (2004) Spectra, energy levels and energy transfer in high symmetry lanthanide compounds. Top Curr Chem 241:167–278
11. Riseberg LA, Weber MJ (1976) Relaxation phenomena in rare earth luminescence. In: Wolf E (ed) Progress in optics, vol XIV, Chap. 3. North-Holland Publishers, Amsterdam, pp 90–159
12. Liu G, Jacquier B (eds) (2005) Spectroscopic properties of rare earths in optical materials. Springer Series in Material Science, vol 83. Springer-Verlag, Berlin
13. Kaminskii AA (1981) Laser crystals. Springer-Verlag, Berlin
14. Scheps R (1996) Upconversion laser processes. Prog Quantum Electron 20:271–358
15. dos Sanotos CMG, Harte AJ, Quinn SJ, Gunnlaugsson T (2008) Recent developments in the field of supramolecular lanthanide sensors and self-assemblies. Coord Chem Rev 252:2512–2527
16. Carlos LD, Ferreira RAS, de Zea BV, Ribeiro SJL (2008) Lanthanide-containing light-emitting organic-inorganic hybrids: a bet on the future. Adv Mater 20:1–26
17. Leonard JP, Nolan CB, Stomeo F, Gunnlaugsson T (2007) Photochemistry and photophysics of coordination compounds: lanthanides. Top Curr Chem 281:1–43

18. Crosswhite HM, Dieke GH, Carter WJ (1965) Free-ion and crystalline spectra of Pr^{3+} (Pr(IV)). J Chem Phys 43:2047–2054
19. Sugar J (1965) Energy levels of Pr^{3+} in the vapor state. Phys Rev Lett 14:731–732
20. Carnall WT, Fields PR, Rajnak K (1968) Electronic energy levels of the trivalent lanthanide aquo ions I. Pr^{3+}, Nd^{3+}, Pm^{3+}, Sm^{3+}, Dy^{3+}, Ho^{3+}, Er^{3+}, and Tm^{3+}. J Chem Phys 49:4424–4442
21. Tanner PA, Wang J (2008) Energy transfer from Gd^{3+} to Tb^{3+} in solution. Chem Phys Lett 455:335–338
22. Bünzli J.-C.G (2010) Basics of lanthanide photophysics. Springer Ser Fluoresc. doi:10.1007/4243_2010_3
23. Piejzel PS, Meijerink A, Wegh RT, Reid MF, Burdick GW (2005) A complete $4f^n$ energy level diagram for all trivalent ions. J Solid State Chem 178:448–453
24. Malta OL (2008) Mechanisms of non-radiative energy transfer involving lanthanide ions revisited. J Non-Cryst Solids 354:4770–4776
25. Pukhov KK, Pelle F, Heber J (2003) Multiphonon relaxation of excited rare-earth ions in ionic matrices. Mol Phys 101:1001–1006
26. Schuurmans MFH, van Dijk JMF (1984) On radiative and non-radiative decay times in the weak coupling limit. Physica 123B:131–155
27. Delsart C, Pelletier-Allard N (1973) Probabilities for radiative and nonradiative decay of Pr^{3+} ion in $LaAlO_3$. J Phys C Solid State Phys 6:1277–1291
28. De Mello DC, Meijerink A, Blasse G (1995) Non-radiative relaxation processes of the Pr^{3+} ion in solids. J Phys Chem Solids 56:673–685
29. Orlovskii YuV, Basiev TT, Pukhov KK, Vorobiev IN, Papashvili AG, Pelle F, Osiko VV (2001) Multiphonon relaxation of mid-IR transitions of rare earth ions in the crystals with fluorite structure. J Lumin 94–95:791–795
30. Reed ED, Moos HW (1973) Nonthermalization and large variation in multiphonon relaxation rate among rare-earth-ion Stark levels. Phys Rev B 8:988–992
31. Sytsma J, Imbusch GF, Blasse G (1989) The spectroscopy of Gd^{3+} in yttrium oxychloride: Judd-Ofelt parameters from emission data. J Chem Phys 91:1456–1461 (erratum: ibid 92:3249)
32. Tsukerblat BS (1994) Group theory in chemistry and spectroscopy. Academic, New York
33. Görller-Walrand C, Binnemans K (1996) Rationalization of crystal-field parametrization. In: Gschneidner KA Jr, Eyring L (eds) Handbook on the physics and chemistry of rare earths, vol 23, Chap. 155. Elsevier BV, Amsterdam, pp 121–283
34. Tanner PA, Mak CSK, Faucher MD (2001) Configuration interaction of Pr^{3+} in $PrCl_6^{3-}$. J Chem Phys 114:10860–10871
35. Krasikov DN, Scherbinin AV, Vasil'ev AN, Kamenskikh IA, Mikhailin VV (2008) Model of Y_2O_3 – Yb charge-transfer luminescence based upon ab initio cluster calculations. J Lumin 128:1748–1752
36. Gruber JB, Hills ME, Allik TH, Jayasankar CK, Quagliano JR, Richardson FS (1990) Comparative analysis of Nd^{3+} ($4f^3$) energy levels in four garnet hosts. Phys Rev B 41:7999–8012
37. Burdick GW, Jayasankar CK, Richardson FS, Reid MF (1994) Energy-level and line-strength analysis of optical transitions between Stark levels in Nd^{3+}:$Y_3Al_5O_{12}$. Phys Rev B 50:16309–16325
38. Burdick GW, Richardson FS, Reid MF, Kooy HJ (1995) Direct calculation of lanthanide optical transition intensities Nd^{3+}:YAG. J Alloys Compd 225:115–119
39. Malta OL, Couto dos Santos MA, Thompson LC, Ito NK (1996) Intensity parameters of 4f–4f transitions in the Eu(dipivaloylmethanate)$_3$ 1, 10-phenanthroline complex. J Lumin 69:77–84
40. Malta OL, Ribeiro SJL, Faucher M, Porcher P (1991) Theoretical intensities of 4f-4f transitions between Stark levels of the Eu^{3+} ion in crystals. J Phys Chem Solids 52:587–593
41. Görller-Walrand C, Binnemans K (1998) Spectral intensities of f-f transitions. In: Gschneidner KA Jr, Eyring L (eds) Handbook on the physics and chemistry of rare earths, vol 25, Chap. 167. Elsevier BV, Amsterdam, pp 101–264

42. Gruber JB, Sardar DK, Yow RM, Allik TH, Zandi B (2004) Energy-level structure and spectral analysis of Nd^{3+} ($4f^3$) in polycrystalline ceramic garnet $Y_3Al_5O_{12}$. J Appl Phys 96:3050–3056
43. Faulkner TR, Richardson FS (1978) Vibronic coupling model for the intensities of f-f transitions in octahedral lanthanide (III) complexes. Mol Phys 35:1141–1161
44. Acevedo R, Tanner PA, Meruane T, Poblete V (1996) Vibronic intensities in the absorption spectra of Yb^{3+}. Phys Rev B 54:3976–3988
45. Malta OL (1995) The theory of vibronic transitions in rare earth compounds. J Phys Chem Solids 56:1053–1062
46. Ishizumi A, Kanemitsu Y (2005) Structural and luminescence properties of Eu-doped ZnO nanorods fabricated by a microemulsion method. Appl Phys Lett 86:253106-1–253106-3
47. Tanner PA, Yu L (2008) Photoluminescence of $ZnO:Eu^{3+}$ nanoflowers. J Nanosci Nanotechnol 8:1307–1311
48. Cheng B-M, Yu L, Duan C-K, Wang H, Tanner PA (2008) Vacuum ultraviolet and visible spectra of $ZnO:Eu^{3+}$ prepared by combustion synthesis. J Phys Condens Matter 20:345231-1–345231-4
49. Armelao L, Heigl F, Jürgensen A, Blyth RIR, Regier T, Zhou X-T, Sham TK (2007) X-ray excited optical luminescence studies of ZnO and Eu-doped nanostructures. J Phys Chem C 111:10194–10200
50. Hayashi Y, Narahara H, Uchida T, Noguchi T, Ibuki S (1995) Photoluminescence of Eu-doped ZnO phosphors. Jpn J Appl Phys 34:1878–1882
51. Liu Y, Luo W, Li R, Chen X (2007) Spectroscopic evidence of the multiple-site structure of Eu^{3+} ions incorporated in ZnO nanocrystals. Opt Lett 32:566–568
52. Liu Y, Luo W, Li R, Liu G, Antonio MR, Chen X (2008) Optical spectroscopy of Eu^{3+} doped ZnO nanocrystals. J Phys Chem C 112:686–694
53. Dorenbos P (2000) Predictability of 5d level positions of the triply ionized lanthanides in halogenides and chalcogenides. J Lumin 87–89:970–972
54. Tanner PA, Mak CSK, Edelstein NM, Liu G, Huang J, Seijo L, Barandiaran Z (2003) Absorption and emission of Ce^{3+} in elpasolite lattices. J Am Chem Soc 125:13225–13233
55. Tanner PA, Mak CSK, Faucher MD, Kwok WM, Phillips DL, Mikhailik V (2003) 4f–5d transitions of Pr^{3+} in elpasolite lattices. Phys Rev B 67:115102-1–115102-19
56. Reid MF, van Pieterson L, Wegh RT, Meijerink A (2000) Spectroscopy and calculations for $4f^N \rightarrow 4f^{N-1}5d$ transitions of lanthanide ions in $LiYF_4$. Phys Rev B 62:14744–14749
57. Peijzel PS, Vergeer P, Meijerink A, Reid MF, Boatner LA, Burdick GW (2005) $4f^{n-1}5d \rightarrow 4f^n$ emission of Ce^{3+}, Pr^{3+}, Nd^{3+}, Er^{3+}, and Tm^{3+} in $LiYF_4$ and YPO_4. Phys Rev B 71:045116-1–045116-9
58. Tanner PA, Duan C-K, Makhov VN, Kirm M, Khaidukov NM (2009) Emission spectra of lanthanide ions in hexafluoroelpasolite lattices excited by synchrotron radiation. Opt Mater 31:1729–1734
59. Barandiarán Z, Edelstein NM, Ordejón B, Ruipérez F, Seijo L (2005) Bond lengths of $4f^1$ and $5d^1$ states of Ce^{3+} hexahalides. J Solid State Chem 178:464–469
60. Duan CK, Jiang S (2010) Study of the d-f emission, and f-d ground-state and excited-state absorption of Nd^{3+} in crystals using the simple model. Spectr Lett doi:10.1080/00387010.2010.487022
61. Ning LX, Duan CK, Xia SD, Reid MF, Tanner PA (2004) A model analysis of $4f^N$ - $4f^{N-1}5d$ transitions of rare-earth ions in crystals. J Alloys Compd 366:34–40
62. Shen Y, Gatch DB, Rodróguez Mendoza UR, Cunningham G, Meltzer RS, Yen WM, Bray KL (2002) Pressure-induced dark-to-bright transition in $Lu_2O_3:Ce^{3+}$. Phys Rev B 65:212103-1–212103-4
63. Jørgensen CK (1962) Electron transfer spectra of lanthanide complexes. Mol Phys 5: 271–277
64. Krupa JC (2005) High energy optical absorption in f-compounds. J Solid State Chem 178:483–488

65. Dorenbos P (2009) Lanthanide charge transfer energies and related luminescence, charge carrier trapping, and redox phenomena. J Alloys Compd 488:568–573
66. Dorenbos P (2004) Locating lanthanide impurity levels in the forbidden band of host crystals. J Lumin 108:301–305
67. Nakazawa E, Shiga F (2003) Lowest 4f-to-5d and charge transfer transitions of rare- earth ions in $LaPO_4$ and related host lattices. Jpn J Appl Phys 42:1642–1647
68. van Pieterson L, Heeroma M, de Heer E, Meijerink A (2000) Charge transfer luminescence of Yb^{3+}. J Lumin 91:177–193
69. Legendziewicz J, Guzik M, Szuszkiewicz W (2008) Charge transfer and f-f emission of trivalent ytterbium observed in double phosphates $M^IM^{III}(PO_4)_2$ (M^I = Na, Rb; M^{III} = Lu, Y). J Alloys Compd 451:165–171
70. Nikl A, Yoshikawa A, Fukuda T (2004) Charge transfer luminescence in Yb^{3+}-containing compounds. Opt Mater 26:545–549
71. Li L, Zhang S (2006) Dependence of charge transfer energy on crystal structure and composition in Eu^{3+}-doped compounds. J Phys Chem B 110:21438–21443
72. Li L, Zhou S, Zhang S (2009) Relationship between charge transfer energies of Yb^{3+} and Sm^{3+} and crystal environmental factor. J Lumin 129:187–191
73. Kamenskikh I, Dujardin C, Garnier N, Guerassimova N, Ledoux G, Mikhailin V, Pedrini C, Petrosyan A, Vasil'ev A (2005) Temperature dependence of the charge transfer and f-f luminescence of Yb^{3+} in garnets and YAP. J Phys Condens Matter 17:5587–5594
74. Blasse G (1976) The influence of charge-transfer and Rhydberg states on the luminescence properties of lanthanides and actinides. Struct Bond 26:43–79
75. Faustino WM, Malta OL, de Sá GF (2006) Theoretical modeling of thermally activated luminescence quenching through charge transfer states in lanthanide complexes. Chem Phys Lett 429:595–599
76. Srivastava AM, Happek U, Schmidt P (2008) Luminescence of $LuCl_3:Pr^{3+}$ under interconfigurational ($4f^2 \rightarrow 4f^15d^1$) and band gap excitations. Opt Mater 31:213–217
77. Tanner PA, Law P-T, Fu L (2003) Preformed sol-gel synthesis and characterization of lanthanide ion-doped yttria-alumina materials. Phys Status Solidi C 199:403–415
78. Lupei V, Lupei A, Tiseanu C, Georgescu S, Stoicescu C, Nanau PM (1995) High-resolution optical spectroscopy of YAG:Nd: a test for structural and distribution models. Phys Rev B 51:8–17
79. Merkle LD, Powell RC (1979) Energy transfer among Nd^{3+} ions in garnet crystals. Phys Rev B 20:75–84
80. Guillot-Noël O, Mehta V, Viana B, Gourier D (2000) Evidence of ferromagnetically coupled Nd^{3+} ion pairs in weakly doped $Nd:LiYF_4$ and $Nd:YVO_4$ crystals as revealed by high resolution optical and EPR spectroscopies. Phys Rev B 61:15338–15346
81. Lupei V, Lupei A, Ikesue A (2010) Multicenter structure and dynamical processes in the rare earth doped garnet and sesquioxide laser crystals and ceramics. Spectr Lett doi:10.1080/00387010903329409
82. Faucher MD, Tanner PA (2006) Energy levels and hypersensitivity of samarium(III) in the elpasolite $Cs_2NaSmCl_6$. J Phys Condens Matter 18:8503–8522
83. Tanner PA, Mak CSK, Mak TCW (2000) Synthesis, structure and spectroscopy of rare earth hypophosphites: 3. Ytterbium hypophosphite. Polyhedron 19:863–870
84. Ishii T (2005) First-principles calculations for the cooperative transitions of Yb^{3+} dimer clusters in $Y_3Al_5O_{12}$ and Y_2O_3 crystals. J Chem Phys 122:024705-1–024705-6
85. Rubio O (1991) Doubly-valent rare-earth ions in halide crystals. J Phys Chem Solids 52:101–174
86. Gahane DH, Kokode NS, Muthal PL, Dhopte SM, Moharil SV (2009) Luminescence of some Eu^{2+} activated bromides. J Alloys Compd 484:660–664
87. Pan Z, Ning L, Cheng B-M, Tanner PA (2006) Absorption, excitation and emission spectra of $SrCl_2:Eu^{2+}$. Chem Phys Lett 428:78–82

88. Hewes RA, Hoffman MF (1970) $4f^7$–$4f^7$ emission from Eu^{2+} in the system MF_2 – AlF_3. J Lumin 3:261–280
89. Ellens A, Meijerink A, Blasse G (1994) The first observation of $^6I \rightarrow {}^8S$ emission from Eu^{2+} in $KMgF_3$. J Lumin 60&61:70–73
90. Grimm J, Güdel HU (2005) Five different types of spontaneous emission simultaneously observed in Tm^{2+} doped $CsCaBr_3$. Chem Phys Lett 404:40–43
91. Jia D, Meltzer RS, Yen WM (2002) Location of the ground state of Er^{3+} in doped Y_2O_3 from two-step photoconductivity. Phys Rev B 65:235116-1–235116-5
92. Yen WM, Jia S, Lu L, Meltzer RS (2003) Two-step photoconductivity: determination of impurity levels of Er^{3+} relative to the host bands of Gd_2O_3. J Lumin 102–103:333–337
93. Joubert M-F, Kazanskii SA, Guyot Y, Gâcon J-C, Pédrini C (2004) Microwave study of photoconductivity induced by laser pulses in rare-earth-doped dielectric crystals. Phys Rev B 69:165217-1–165217-13
94. Grimm J, Fleniken J, Krämer K, Biner D, Happek U, Güdel HU (2007) On the determination of photoionization thresholds of Ce^{3+} doped Cs_3LuCl_6, $Cs_2LiLuCl_6$ and Cs_2LiYCl_6 by thermoluminescence. J Lumin 122–123:325–328
95. Thiel CW, Cruguel H, Sun Y, Lapeyre GJ, Macfarlane RM, Equall RW, Cone RL (2001) Systematics of 4f electron energies relative to host bands by resonant photoemission of rare earth doped optical materials. J Lumin 94–95:1–6
96. Tanner PA, Duan C-K, Cheng B-M (2010) Excitation and emission spectra of $Cs_2NaLnCl_6$ crystals using synchrotron radiation. Spectr Lett (in press)
97. Bessière A, Dorenbos P, van Eijk CWE, Krämer KW, Güdel HU, Galtayries A (2006) Scintillation and anomalous emission in elpasolite $Cs_2LiLuCl_6$:Ce^{3+}. J Lumin 117:187–198
98. Rodnyi PA, Mikhailik VB, Stryganyuk GB, Voloshinovskii AS, van Eijk CWE, Zimmerer GF (2000) Luminescence properties of Ce-doped $Cs_2LiLaCl_6$ crystals. J Lumin 86:161–166
99. Boutinaud P, Putaj P, Mahiou R, Cavalli E, Speghini A, Bettinelli M (2007) Quenching of lanthanide emission by intervalence charge transfer in crystals containing closed shell transition metal ions. Spectr Lett 40:209–220
100. Jia G-H, Tanner PA, Cheng B-M (2009) Contrasting emission behaviors of YAG:V^{5+} co-doped with Pr^{3+} or Eu^{3+}. Chem Phys Lett 474:97–100
101. Krumpel AH, van der Kolk CE, Boutinaud P, Bettinelli M, Dorenbos P (2009) Lanthanide 4f-level location in AVO_4:Ln^{3+} (A = La, Gd, Lu) crystals. J Phys Condens Matter 21:115503-1–115503-8
102. Jia G (2009) Spectroscopy and energy transfer in metal ion doped complexes. PhD Thesis, City University of Hong Kong, p 21
103. Tanner PA, Wong KL, Liang Y (2004) Multiple phase production on doping Er^{3+} into α-Al_2O_3. Chem Phys Lett 399:15–19
104. Tanner PA, Pan Z, Rakov N, Maciel GS (2006) Luminescence of Eu^{3+} in α-Al_2O_3 powders. J Alloys Compd 424:347–349
105. Alarcon J, van der Voort D, Blasse G (1992) Efficient Eu^{3+} luminescence in non-lanthanide host lattices. Mat Res Bull 27:467–472
106. Sun LN, Peng HS, Stich MIJ, Achatz D, Wolfbeis OF (2009) pH sensor based upon upconverting luminescent lanthanide nanorods. Chem Commun 33:5000–5002
107. Kim WJ, Nyk M, Prasad PN (2009) Color coded multilayer photopatterned microstructures using lanthanide(III) ion co-doped $NaYF_4$ nanoparticles with upconversion luminescence for possible applications in security. Nanotechnology 20:185301-1–185301-7
108. Wong K-L, Law G-L, Murphy MB, Tanner PA, Wong W-T, Lam PK-S, Lam MH-W (2008) Functionalized europium nanorods for in vitro imaging. Inorg Chem 47:5190–5196
109. Chatteriee DK, Rufalhah AJ, Zhang Y (2008) Upconversion fluorescence imaging of cells and small animals using lanthanide-doped nanocrystals. Biomaterials 29:937–943
110. Pollnau M, Gamelin DR, Lüthi SR, Güdel HU (2000) Power dependence of upconversion luminescence in lanthanide and transition metal systems. Phys Rev B 61:3337–3346

111. Gamelin DR, Güdel HU (2001) Upconversion processes in transition metal and rare earth metal systems. Top Curr Chem 214:1–56
112. Gamelin DR, Güdel HU (2000) Design of luminescent inorganic materials: new photophysical processes studied by optical spectroscopy. Acc Chem Res 33:235–242
113. Suyver JF, Aebischer A, Biner D, Gerner P, Grimm J, Heer S, Krämer K, Reinhard C, Güdel HU (2005) Novel materials doped with trivalent lanthanides and transition metal ions showing near-infrared to visible photon upconversion. Opt Mater 27:1111–1130
114. Auzel F (2004) Upconversion and anti-Stokes processes with f and d ions in solids. Chem Rev 104:139–173
115. Wang F, Liu X (2009) Recent advances in the chemistry of lanthanide-doped upconversion nanocrystals. Chem Soc Rev 38:976–989
116. Güdel HU, Pollnau M (2000) Near-infrared to visible photon upconversion processes in lanthanide doped chloride, bromide and iodide lattices. J Alloys Compd 303–304:307–315
117. Kushida T (1973) Energy transfer and cooperative optical transitions in rare-earth doped inorganic materials. J Phys Soc Jpn 34:1318–1337
118. Collins JM, Di Bartolo B (1996) Temperature and energy gap dependence of energy transfer between rare-earth ions in solids. J Lumin 69:335–341
119. Law G-L, Wong K-L, Kwok W-M, Wong W-T (2007) Tanner P A (2007) Terbium luminescence sensitized through three-photon excitation in a self-assembled unlinked antenna. J Phys Chem B 111:10858–10861
120. Tancrez N, Feuvrie C, Ledoux I, Zyss J, Toupet L, Le Bozec H, Maury O (2005) Lanthanide complexes for second order nonlinear optics: evidence for the direct contribution of f electrons to the quadratic hyperpolarizability. J Am Chem Soc 127:13474–13475
121. Sénêchal-David K, Hemeryck A, Tancrez N, Toupet L, AG WJ, Ledoux I, Zyss J, Boucekkine A, Guêgan J-P, Le Bozec H, Maury O (2006) Synthesis, structural studies, theoretical calculations, and linear and nonlinear optical properties of terpyridyl lanthanide complexes: new evidence for the contribution of f electrons to the NLO activity. J Am Chem Soc 128:12243–12255
122. Dong W, Zhang H, Su Q, Lin Y, Wang S, Zhu C (1999) Crystal growth, structure, and properties of new nonlinear optical materials: $K_2Ln(NO_3)_5{\cdot}2H_2O$ (Ln = La, Ce, Pr, Nd, Sm). J Solid State Chem 148:302–307
123. Bogani L, Cavigli L, Bernot K, Sessoli R, Gurioli M, Gatteschi D (2006) Evidence of intermolecular π-stacking enhancement of second-harmonic generation of a family of single chain magnets. J Mater Chem 16:2587–2592
124. Law GL, Wong KL, Lap ST, Lau KK, Tanner PA, Wong WT (2010) Nonlinear optical activity in dipolar organic-lanthanide complexes. J Mater Chem 20:4074–4079
125. Thorne JRG, Jones M, McCaw CS, Murdoch KM, Denning RG, Khaidukov NM (1999) Two-photon spectroscopy of europium(III) elpasolites. J Phys Condens Matter 11:7851–7866
126. McCaw CS, Murdoch KM, Denning RG (2003) Energy levels of terbium(III) in the elpasolite $Cs_2NaTbBr_6$. I. Luminescence and two-photon spectroscopy. Mol Phys 101:427–438
127. Wang D, Ning L, Xia S, Tanner PA (2003) Theoretical analysis of the two-photon absorption spectrum of Tb^{3+} in $Cs_2NaTbCl_6$. J Phys Condens Matter 15:2681–2691
128. Chen XB, Zhang GY, Mao YH, Hou YB, Feng Y, Hao Z (1996) Research on the upconversion luminescence of Tm^{3+} ion in crystal and amorphous pentaphosphate materials. J Lumin 69:151–160
129. Shalav A, Richards BS, Trupke T, Krämer KW, Güdel HU (2005) Application of $NaYF_4$:Er^{3+} up-converting phosphors for enhanced near-infrared silicon solar cell response. Appl Phys Lett 86:013505-1–013505-3
130. Goldner P, Pellé F (1996) Photon avalanche fluorescence and lasers. Opt Mater 5:239–249
131. Brenier A, Garapon C, Madej C, Pédrini C, Boulon G (1994) Excited state absorption in Tm^{3+} doped $LiNbO_3$ and $Gd_3Ga_5O_{12}$ and looping mechanism. J Lumin 62:147–156

132. Sivakumar S, van Veggel FCJM, Stanley-May P (2007) Near infrared (NIR) to red and green up-conversion emission from silica sol-gel thin films made with $La_{0.45}Yb_{0.50}Er_{0.05}F_3$ nanoparticles, hetero-looping-enhanced energy transfer (Hetero-LEET): a new upconversion process. J Am Chem Soc 129:620–625
133. Joubert MF, Guy S, Jacquier B, Linarès C (1994) The photon-avalanche effect: review model and application. Opt Mater 4:43–49
134. Joubert M-F (1999) Photon avalanche upconversion in rare earth laser materials. Opt Mater 11:181–203
135. Joubert MF, Guy S, Cuerq S, Tanner PA (1997) Room-temperature blue upconverted luminescence via photon avalanche pumping in $Cs_2NaGdCl_6$:Tm. J Lumin 75:287–293
136. Yang Y, Tu LP, Zhao JW, Sun YJ, Kong XG, Zhang H (2009) Upconversion luminescence of beta-$NaYF_4$:Yb^{3+}, Er^{3+} @beta $NaYF_4$ core-shell nanoparticles: excitation power, density and surface dependence. J Phys Chem C 113:7164–7169
137. Chen GY, Liu Y, Zhang ZG, Aghahadi B, Somesfalean G, Sun Q, Wang FP (2007) Chem Phys Lett 448:127–131
138. Grimm J, Beurer E, Gerner P, Güdel HU (2007) Upconversion between 4f-5d excited states in Tm^{2+}-doped $CsCaCl_3$, $CsCaBr_3$, and $CsCaI_3$. Chem. Eur J 13:1152–1157
139. Laroche M, Bettinelli M, Girard S, Moncorgé R (1999) Chem Phys Lett 311:167–172
140. Chua M, Tanner PA (1996) Three-body energy transfer processes of lanthanide ions. J Lumin 66 & 67:203–207
141. Schaudel B, Goldner P, Prassas M, Auzel F (2000) Cooperative luminescence as a probe of clustering in Yb^{3+} doped glasses. J Alloys Compd 300–301:443–449
142. Dou CG, Yang QH, Hu XM, Xu J (2008) Cooperative up-conversion luminescence of ytterbium doped yttrium lanthanum oxide transparent ceramic. Opt Commun 281:692–695
143. Salley GM, Valiente R, Güdel HU (2003) Cooperative Yb^{3+}-Tb^{3+} dimer excitations and upconversion in $Cs_3Tb_2Br_9$:Yb^{3+}. Phys Rev B 67:134111-1–134111-9
144. Vermelho MVD, dos Santos PV, de Araújo MT, Gouveia-Neto AS, Cassanjes FC, Ribeiro SJL, Messaddeq Y (2003) Thermally enhanced cooperative energy-transfer frequency upconversion in terbium and ytterbium doped tellurite glass. J Lumin 102–103:762–767
145. Wang H, Duan C-K, Tanner PA (2008) Visible upconversion luminescence from Y_2O_3:Eu^{3+}, Yb^{3+}. J Phys Chem C 112:16651–16654
146. Reinhard C, Gerner P, Valiente R, Wenger OS, Güdel HU (2001) Upconversion phenomena in the Yb^{3+} doped transition metal compounds Rb_2MnCl_4 and $CsMnBr_3$. J Lumin 94–95:331–335
147. Valiente R, Wenger O, Güdel HU (2000) New photon upconversion processes in Yb^{3+} doped $CsMnCl_3$ and $RbMnCl_3$. Chem Phys Lett 320:639–644
148. Valiente R, Wenger O, Güdel HU (2002) Upconversion luminescence in Yb^{3+} doped $CsMnCl_3$: spectroscopy, dynamics and mechanisms. J Chem Phys 116:5196–5204
149. Zhou X-J, Tanner PA, Faucher MD (2007) Luminescence of $Cs_2NaScCl_6$:Pr^{3+}: effects of changing the lattice parameter. Spectr Lett 40:349–366
150. Sun Y, Giebink NC, Kanno H, Ma B, Thompson ME, Forrest SR (2006) Management of singlet and triplet excitons for efficient white organic light-emitting devices. Nature 440:908–912
151. Steele RV (2007) The story of a new light source. Nat Photonics 1:25–26
152. Kim JS, Jeon PE, Choi JC, Park HL, Mho SI, Kim GC (2004) Warm-white-light emitting diode utilizing a single-phase full-color $Ba_3MgSi_2O_8$:Eu^{2+}, Mn^{2+} phosphor. Appl Phys Lett 84:2931–2933
153. Li L, Tao J, Pan H, Chen H, Wu X, Zhu F, Xu X, Tang R (2008) Colour tuning of core-shell fluorescent materials. J Mater Chem 18:5363–5367
154. DiMaio JR, Kokuoz B, Ballato J (2006) White light emissions through down-conversion of rare-earth doped LaF_3 nanoparticles. Opt Express 14:11412–11417
155. Su Y, Li L, Li G (2009) Generation of tunable lights in core-shell $CaWO_4$ microspheres via co-doping with Na^+ and Ln^{3+} (Ln = Tb, Sm, Dy, Eu). J Mater Chem 19:2316–2322

156. Coppo P, Duati M, Kozhevnikov VN, Hofstraat JW, De Cola L (2005) White-light emission from an assembly comprising luminescent iridium and europium complexes. Angew Chem Int Ed 44:1806–1810
157. He H, Fu R, Song X, Wang D, Chen J (2008) White light-emitting $Mg_{0.1}Sr_{1.9}SiO_4:Eu^{2+}$ phosphors. J Lumin 128:489–493
158. Chen D, Wang Y, Zheng K, Guo T, Yu Y, Huang P (2007) Bright upconversion white light emission in transparent glass ceramic embedding $Tm^{3+}/Er^{3+}/Yb^{3+}$:β-YF_3 nanocrystals. Appl Phys Lett 91:251903-1–251903-3
159. Sivakumar S, Boyer J-C, Bovero E, van Veggel FCJM (2009) Up-conversion of 980 nm light into white light from sol-gel derived thin film made with new combinations of $LaF_3:Ln^{3+}$ nanoparticles. J Mater Chem 19:2392–2399
160. Wang J, Tanner PA (2010) Upconversion for white light generation by a single compound. J Am Chem Soc 132:947–949
161. Wegh RT, Donker H, Oskam KD, Meijerink A (1999) Visible quantum cutting in $LiGdF_4$: Eu^{3+} through downconversion. Science 283:663–666
162. Wegh RT, Donker H, Meijerink A, Lamminmäki HJ (1997) Vacuum ultraviolet spectroscopy and quantum cutting for Gd^{3+} in $LiYF_4$. Phys Rev B 56:13841–13848
163. Lee T-J, Luo L-Y, Cheng B-M, Diau W-G, Chen T-M (2008) Appl Phys Lett 92:081106-1–081106-3
164. Xiong F, Lin Y, Chen Y, Luo Z, Ma E, Huang Y (2006) Visible quantum cutting in $Pr^{3+}:PbWO_4$ crystal through energy transfer between the host and Pr^{3+} ions. Chem Phys Lett 429:410–414
165. Vink AP, Dorenbos P, van Eijk CWE (2003) Observation of the photon cascade emission process under $4f^15d^1$ and host excitation in several Pr^{3+}-doped materials. J Solid State Chem 171:308–312
166. Zhou X, Tanner PA, Duan CK, Cheng BM (2007) Downconversion in $Cs_2NaErCl_6$. Chem Phys Lett 442:302–306
167. Peijzel PS, Meijerink A (2005) Visible photon cascade emission from the high energy levels of Er^{3+}. Chem Phys Lett 401:241–245
168. Kido J, Okamoto Y (2002) Organolanthanide metal complexes for electroluminescent materials. Chem Rev 102:2357–2368
169. Quirino WG, Legnani C, Cremona M, Lima PP, Junior SA, Malta OL (2006) White OLED using β-diketones rare earth binuclear complex as emitting layer. Thin Solid films 494:23–27
170. Law G-L, Wong KL, Tam H-L, Cheah K-W, Wong W-T (2009) White OLED with a single component europium complex. Inorg Chem 48:10492–10494
171. Jüstel T, Krupa JC, Wiechert DU (2001) VUV spectroscopy of luminescent materials for plasma display panels and Xe discharge lamps. J Lumin 93:179–189
172. Dexpert-Ghys J, Mauricot R, Caillier B, Guillot P, Beaudette T, Jia G, Tanner PA, Cheng BM (2010) VUV excitation of YBO_3 and $(Y,Gd)BO_3$ phosphors doped with Eu^{3+} or Tb^{3+}: comparison of efficiencies and effect of site-selectivity. J Phys Chem C 114:6681–6689
173. Krämer KW, Dorenbos P, Güdel HU, van Eijk CWE (2006) Development and characterization of highly efficient new cerium doped rare earth halide scintillator materials. J Mater Chem 16:2773–2780
174. Bessiere A, Dorenbos P, van Eijk CWE, Kråmer K, Güdel HU (2004) New thermal neutron scintillators: $Cs_2LiYCl_6:Ce^{3+}$ and $Cs_2LiYBr_6:Ce^{3+}$. IEEE Trans Nucl Sci 51:2970–2972
175. Yang Z, Lin JH, Su MZ, Tao Y, Wang W (2000) Photon cascade luminescence of Gd^{3+} in $GdBaB_9O_{16}$. J Alloys Compd 308:94–97
176. Wegh RT, van Loef EVD, Burdick GW, Meijerink A (2003) Luminescence spectroscopy of high-energy $4f^{11}$ levels of Er^{3+} in fluorides. Mol Phys 101:1047–1056
177. Huang S, Lu L, Jia W, Wang X-J, Yen WM, Srivastava AM, Setlur AA (2001) The spectral properties of the $^1S_0$ state in $SrAl_{12}O_{19}$:Pr. Chem Phys Lett 348:11–16
178. Meltzer RS, Zheng H, Wang JW, Yen WM, Grinberg M (2005) Pressure dependence of the $4f^15d^1 \rightarrow 4f^2$ emission of Pr^{3+}:YAG using excited state absorption. Phys Stat Sol C 2:284–288

179. Vergeer P, Vlugt TJH, Kox MHF, den Hertog MI, van der Eerden JPJM, Meijerink A (2005) Quantum cutting by cooperative energy transfer in $Yb_xY_{1-x}PO_4:Tb^{3+}$. Phys Rev B 71:014119-1–014119-11
180. Yuan J-L, Zeng X-Y, Zhao J-T, Zhang Z-J, Chen H-H, Yang X-X (2008) Energy transfer mechanisms in Tb^{3+}, Yb^{3+} codoped Y_2O_3 downconversion phosphor. J Phys D Appl Phys 41:105406-1–105406-6
181. Aarts L, van der Ende BM, Meijerink A (2009) Downconversion for solar cells in $NaYF_4$:Er, Yb. J Appl Phys 106:023522-1–023522-6
182. Liu G, Chen XY (2007) Spectroscopic properties of lanthanides in nanomaterials. In: Gschneidner KA, Bünzli J-CG, Pecharsky VK (eds) Handbook of physics and chemistry of rare earths, vol 37, Chap. 233. Elsevier BV, Amsterdam
183. Song H, Tanner PA (2010) Luminescence properties of rare earth doped nanophosphors. In: Chen W (ed) Doped nanomaterials and nanodevices, Chap. 4. American Scientific Publishers, Stevenson Ranch, CA
184. Tanner PA (2005) Synthesis and luminescence of nano-insulators doped with lanthanide ions. J Nanosci Nanotechnol 5:1455–1464
185. Duan CK, Reid MF (2007) Macroscopic models for the radiative relaxation lifetime of luminescent centers embedded in surrounding media. Spectr Lett 40:237–246
186. Pukhov KK, Basiev TT, Orlovskii YV (2008) Spontaneous emission in dielectric nanoparticles. JETP Lett 88:12–18
187. Basiev TT, Orlovskii YV, Pukhov KK (2008) Spontaneous and induced emission in dielectric nanoparticles. Nanotechnol Russ 3:551–559
188. Vetrone F, Boyer J-C, Capobianco JA, Speghini A, Bettinelli M (2004) Significance of Yb^{3+} concentration on the upconversion mechanisms in codoped $Y_2O_3:Er^{3+}$, Yb^{3+} nanocrystals. J Appl Phys 96:661–667
189. Schietinger S, Menezes L de S, Lauritzen B, Benson O (2009) Observation of size dependence in multicolor upconversion in single Yb^{3+}, Er^{3+} codoped $NaYF_4$ nanocrystals. Nano Lett 9:2477–2481
190. Pires AM, Serra OA, Davalos MR (2005) Morphological and luminescent studies on nanosized Er, Yb-yttrium oxide up-converter prepared from different precursors. J Lumin 113:174–182
191. Qu Y, Kong X, Sun Y, Zeng Q, Zhang H (2009) Effect of excitation power density on the upconversion luminescence of $LaF_3:Yb^{3+}$, Er^{3+} nanocrystals. J Alloys Compd 485:493–496
192. Zych E, Wójtowicz M, Kepiński L, Małecka M (2008) Size effects in the low temperature spectroscopy of Lu_2O_3 nanopowders. Opt Mater 31:241–246
193. Tanner PA, Fu L, Cheng B-M (2009) Spectral band shifts in the electronic spectra of rare earth sesquioxide nanomaterials doped with europium. J Phys Chem C 113:10773–10779
194. Liu GK, Zhang HZ, Chen XY (2002) Restricted phonon relaxation and anomalous thermalization of rare earth ions in nanocrystals. Nano Lett 2:535–539
195. Macedo AG, Ferreira RAS, Ananias D, Reis MS, Amaral VS, Carlos LD, Rocha J (2010) The effects of phonon confinement on anomalous thermalization, energy transfer, and upconversion in Ln^{3+}-doped Gd_2O_3 nanotubes. Adv Func Mater 20:624–634
196. Zhang QY, Huang XY (2010) Recent progress in quantum cutting phosphors. Progr Mater Sci 55:353–427

Luminescent Chemical and Physical Sensors Based on Lanthanide Complexes

Corinna Spangler and Michael Schäferling

Abstract Luminescent lanthanide complexes (LLCs) are versatile probes for the determination of a variety of analytes such as metal ions, anions, hydrogen peroxide, ATP, pH, or oxygen. Antenna chromophores have been introduced as ligands for sensitizing lanthanide emission. The ligand system or the lanthanide ion can act as receptors for reversible binding of analytes. Thus, the response of luminescence emission can occur via ligand or metal-centered processes, which will both be discussed. LLCs, particularly Eu^{3+} and Tb^{3+} complexes, have been incorporated in solid state matrices, e.g., polymer hydrogels or silica to obtain optical sensor layers. A selection of such sensitive materials will be outlined, including sensors for the analytically relevant parameters oxygen, pH, hydrogen peroxide, humidity, and temperature.

Keywords Luminescence lifetime · Luminescent lanthanide complex · Luminescent probe · Optical sensor

Contents

C. Spangler and M. Schäferling (✉)
Institute of Analytical Chemistry, Chemo- and Biosensors, University of Regensburg, Regensburg, Germany
e-mail: michael.schaeferling@chemie.uni-regensburg.de

P. Hänninen and H. Härmä (eds.), *Lanthanide Luminescence: Photophysical, Analytical and Biological Aspects*, Springer Ser Fluoresc (2011) 7: 235–262, DOI 10.1007/4243_2010_1,

Abbreviations

ADP	Adenosine-5′-diphosphate
AMP	Adenosine-5′-monophosphate
ATP	Adenosine-5′-triphosphate
BSA	Bovine serum albumin
CCD	Charge-coupled device
DNA	Desoxyribonucleic acid
EuTc	Europium(III) tetracycline
GTP	Guanosine-5′-triphosphate
HSA	Human serum albumin
IUPAC	International Union of Pure and Applied Chemistry
LED	Light-emitting diode
LLC	Luminescent lanthanide complex
NADP	Nicotinamide adenine dinucleotide phosphate
PSP	Pressure sensitive paint
Ras	Rat sarcoma
Tc	Tetracycline
TSP	Temperature sensitive paint
vol	Volume

1 Introduction

Complexes of the lanthanide ions europium(III) and terbium(III) are widely used luminescent probes and stains for biomolecular systems. Their applications have already been discussed 1982 in a pioneering review by Richardson [1]. Lanthanide complexes exhibit coordination numbers from six to twelve, with eight and nine being most common. Primary hydration numbers for Eu^{3+} and Tb^{3+} in water are also between eight and nine. According to the *Pearson* scheme, lanthanide cations can be classified as "hard" acids. The preference for binding donor atoms is, therefore, in the order O > N > S. Ligand coordination occurs predominantly via ionic interactions leading to a strong preference for donor groups with negatively charged oxygen (hard bases). Water molecules also can act as strong ligands for lanthanides. In aqueous solution, these can usually only be replaced by other hard donor groups.

The electronic shielding of the f-orbitals of lanthanides is quite weak. Therefore, they are contracted and insulated from the environment. The interaction with ligand orbitals is very weak and the f → f transitions result in very sharp peaks with quite long-lived luminescence lifetimes [1–4]. The luminescence spectra of Eu^{3+} complexes, which exhibit a $4f^6$ electronic configuration, are dominated by emission bands corresponding to the $^5D_0 \rightarrow {}^7F_J$ transitions. The strongest intensities are observed for $^5D_0 \rightarrow {}^7F_1$ and $^5D_0 \rightarrow {}^7F_2$ transitions (Fig. 1). Particularly, the latter one with its very strong and sharp emission line around 615 nm is the basis for the application of europium complexes as luminescent probes and labels. The hypersensitivity of this transition is due to its electric dipole character, and the radiative transition probability is very sensitive to the nature of the ligand environment. The same is the case for the $^5D_4 \rightarrow {}^7F_5$ transition of Tb^{3+} compounds ($4f^8$ electronic configuration) centered at 543 nm. Thus, the emission intensity responds to chemical (or biochemical) analytes that can interfere with these transitions. Their fluorimetric determination is the basis for the use of luminescent lanthanide ligand complexes (LLCs) in chemical sensors (Table 1).

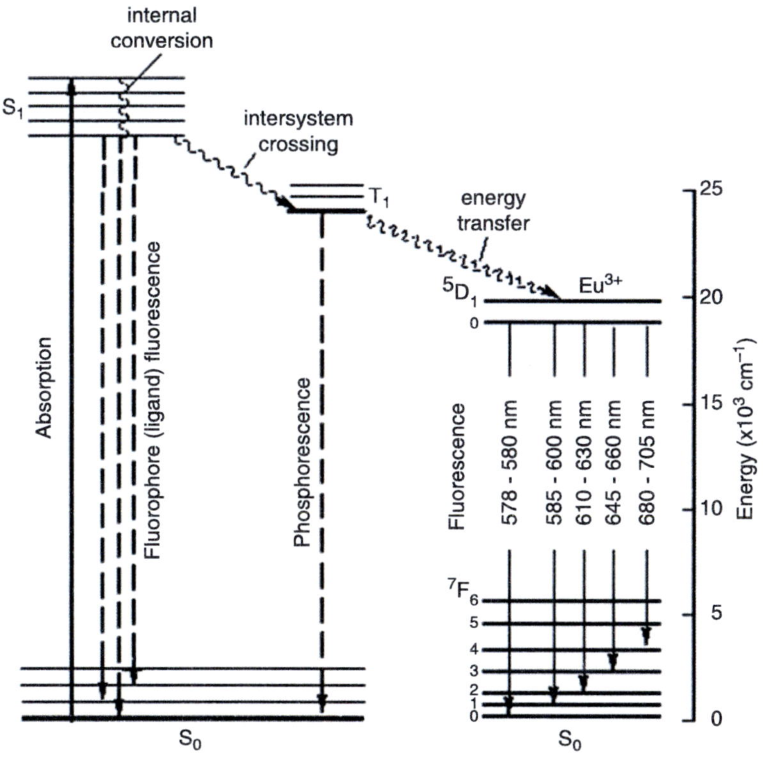

Fig. 1 Fluorescence emission mechanism of Eu^{3+} complexes sensitized by antenna chromophores. S_0, S_1, and T_1 are singlet ground state, singlet excited state, and triplet state, respectively

Table 1 Characteristics of emission spectra of Tb^{3+} and Eu^{3+} and their complexes in aqueous solution adapted from [1]

Transition	Emission (nm)	Relative intensity	Characteristics and sensitivity to ligand environment
Terbium (III)			
$^5D_4 \rightarrow {}^7F_6$	485–500	Medium-strong	Moderate sensitivity
$\rightarrow {}^7F_5$	540–555	Strongest	Hypersensitivity, best probe transition
$\rightarrow {}^7F_4$	580–595	Medium	Moderate sensitivity
$\rightarrow {}^7F_3$	615–625	Medium-weak	Fine-structure at high resolution
$\rightarrow {}^7F_2$	645–655	Weak	Moderate sensitivity
Europium (III)			
$^5D_0 \rightarrow {}^7F_0$	578–580	Weak	Single sharp line
$\rightarrow {}^7F_1$	585–600	Strong	Magnetic dipole, sharp and structured
$\rightarrow {}^7F_2$	610–630	Strongest	Hypersensitivity, best probe transition
$\rightarrow {}^7F_3$	645–660	Weak	Weak intensity and sensitivity
$\rightarrow {}^7F_4$	680–705	Weak	Intensity and structuring moderate sensitive

The direct electronic excitation of lanthanide ions is very inefficient because of their low absorption coefficients and the occurrence of nonradiative deactivation processes mediated by solvent molecules, particularly by water. Therefore, sensitizing ligands are applied. These sensitizers are often termed as "antenna" chromophores. By using antenna chromophores like acridone or diaryl ketones, the excitation wavelength for europium complexes, which is usually <370 nm, can be shifted to the visible region [5–7].

These chromophores have a small energy gap between the lowest singlet excited state (S_1) and the triplet state (T_1). The common sensitization mechanism for luminescent europium complexes includes a triplet pathway after intersystem crossing of the electronically excited states of the ligand. The transfer of the energy absorbed by the ligand to the Eu^{3+} center takes place from the triplet excited state (T_1) of the ligand (Fig. 1). In particular cases, singlet sensitization pathways also have been observed, leading to an even extended excitation window to longer wavelengths [8]. However, the energy transfer from ligand to Eu^{3+} occurs only if the ligand donor state is located at wavenumbers in a range from ~17,200 to 25,000 cm^{-1} [1]. Similar effects can be obtained with complexes of Tb^{3+}, whose excitation window can be shifted due to ligand energy transfer up to wavelengths of ~ 340 nm. If the triplet state of the donor is in the frequency range between 20,500 and 40,000 cm^{-1}, indirect excitation of Tb^{3+} leads to relaxation via the $^5D_4$ emitting level. The $4f^n$ energy levels of all trivalent lanthanide ions have been calculated by Peijzel et al. [9]. First, $4f^n$ energy level diagrams for trivalent lanthanide ions in different matrices have been presented by Dieke [10]. These diagrams were found to be very useful because the energy levels in different crystals are almost constant.

The following chapter will be limited to Eu^{3+} and Tb^{3+} complexes, as other lanthanides are much less frequently used as molecular probes or in chemical sensors. There are only very rare examples of probes containing Yb^{3+} or Nd^{3+}, all emitting in the near infrared [11, 12].

At this point it is also helpful to emphasize the commonly acknowledged definition of a chemical sensor, the so-called *Cambridge definition* [13]:

> Chemical sensors are miniaturized devices that can deliver real time and on-line information on the presence of specific compounds or ions in even complex samples.

Variant definitions can be found in literature. The IUPAC defines the main constituents of a chemical sensor [14]:

> Chemical sensors usually contain two basic components connected in series: a chemical (molecular) recognition system (receptor) and a physicochemical transducer.

Accordingly, this review is focused on actual sensors, and does not cover the multitude of molecular probes or indicators that have been designed for applications in solution. A notable number of research articles and also reviews on LLCs have been published that term responsive luminescent complexes misleadingly as "(molecular) sensor". These cannot be considered in the main part of this overview. Nevertheless, since several of these molecular probes can be incorporated into optical sensors as the receptor part, the next section will summarize a selection of LLC structures that respond to different types of analytes and discuss the basic processes involved.

2 Classification of Luminescent Lanthanide Probes

The hypersensitivity of certain emission bands turns LLCs into promising candidates as probes for analytes such as anions, pH, oxygen, nucleic acids, DNA, proteins, cofactors and coenzymes. The number of probes that have been reported in literature is rather large. Therefore, this chapter is confined to complexes with the capability to be applied in sensor devices. Foremost, they have to be responsive in aqueous solution at a pH range from 6 to 9. Furthermore, this overview does not cover lanthanide systems in which the analyte itself actuates as sensitizer for the lanthanide ion. This principle can be used for the determination of antibiotics in aqueous solution [15–17].

Generally, lanthanide complexes can be divided into two subgroups. The first comprises antennae that satisfy a high degree of coordination and rigidity. These ligands bear macrocyclic or polydentate moieties that form chelate complexes with the lanthanide ion. Such ligands are often based on cyclen, cryptand, crown ether, or diethylenetriaminepentaacetic acid (DTPA) structures. They can constitute polycyclic structures or polychelates of one or more lanthanide ions even in combination with transition metals such as Pd^{2+}, Pt^{2+}, Ru^{2+}, and Ir^{2+} [2, 11, 18–21]. In case of these complexes, the luminescence response is usually based on interactions of the analyte with the antenna ligand system, which bears a corresponding receptor unit. Moreover, intercalators for DNA have been designed on the basis of such mixed f–d metal complexes [11, 12].

The second group of lanthanide complexes includes ligands such as tetracyclines that interact less strongly with the respective lanthanide ions. In these cases, the analyte acts as additional ligand for the lanthanide center. The response occurs due to the exchange of more or less quenching ligands. The overall structure of these complexes is often not determined and best sensitivities are sometimes obtained by combining an odd ratio of lanthanide ion to ligand, usually with an excess of metal ions. In a general perspective, the probes of the first type are based on an intramolecular energy transfer between sensitizer and lanthanide ion. Modulation of luminescence emission occurs via ligand-centered processes. Probes of the second type depend on an intermolecular energy transfer. Their luminescence is actuated by metal-centered interactions.

2.1 *Complexes with Macrocyclic and Polydentate Ligands*

A multitude of lanthanide complexes comprises ligands based on 1,4,7,10-tetra-azacyclododecane (cyclen) substructures (Scheme 1). Cyclen-based ligands occupy eight or even nine coordination sites of the lanthanide ion and, therefore, shield solvent molecules effectively from its coordination sphere. Examples of such ligands are depicted in Table 2. The europium complex **1** was designed by Pope et al. to probe Zn^{2+} ions in aqueous solution at pH 7.4 [22]. The receptor for Zn^{2+} is based on a dipicoyl moiety. The shape of the lanthanide emission changes upon binding of Zn^{2+}.

Chelate **2** was found to respond to pH changes from alkaline to acidic medium by an increasing luminescence emission due to protonation of the quinoline nitrogen [23]. Further probes reported by Gunnlaugsson et al. are represented by the complexes **3** and **4** [24]. The crown ether derived receptor **3** undergoes a large enhancement of its luminescence in neutral aqueous solution upon addition of Na^+, whereas **4** responds to K^+. The phenanthroline ligand **5** responds to changes in pH in the range from pH 5.5 to 12.5 [25, 26]. Furthermore, it can act as probe for Cu^{2+} at pH 7.4. However, higher concentrations of Fe^{2+} and Co^{2+} interfere with the determination of copper ions.

Complex **6** is one of the rare halide probes that is applicable in neutral aqueous solution [27]. The quenching effect on **6** follows the order $I^- > Br^- > Cl^-$.

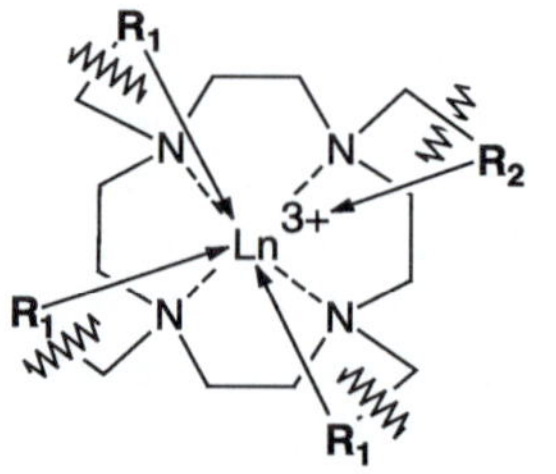

Scheme 1 Basic structure of macrocyclic cyclen ligands

Table 2 Cyclen-based lanthanide complexes applied as molecular probes according to scheme 1

Ln^{3+}/**R_1**	Antenna **R_2**	Analyte/refs
Eu^{3+} 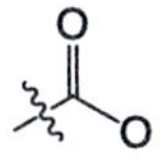 **1**	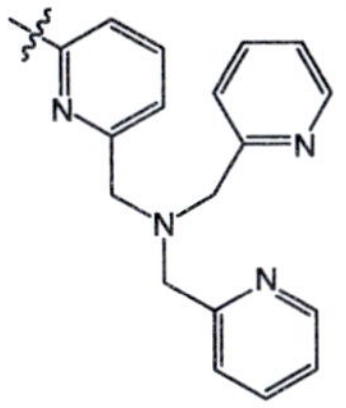	Zn^{2+} [22]
Eu^{3+} 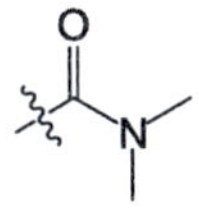 **2**	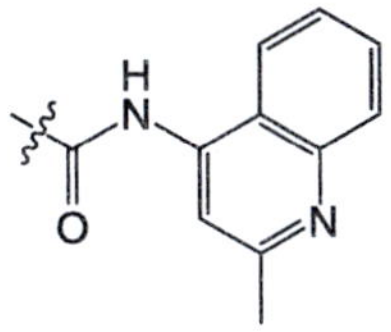	pH [23]
Tb^{3+} 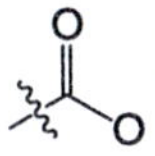 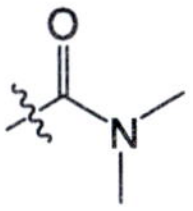 **3** **4**	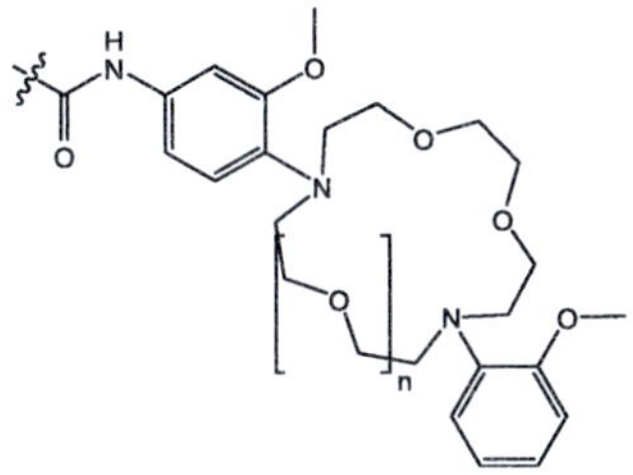	Na^+, K^+ [24]
Eu^{3+} 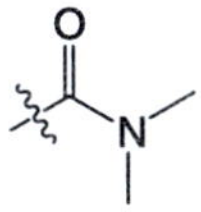 **5**	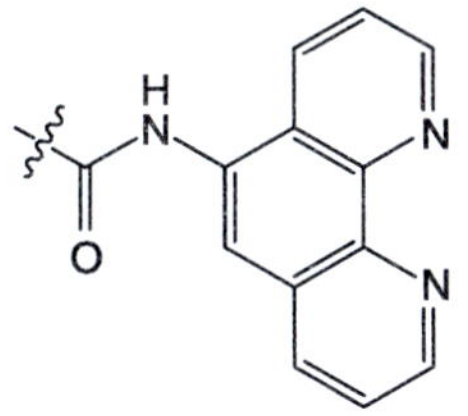	pH, Cu^{2+}, Fe^{2+}, Co^{2+} [25, 26]
Eu^{3+}, Tb^{3+} 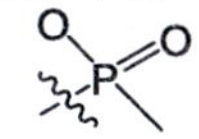 **6** **7**	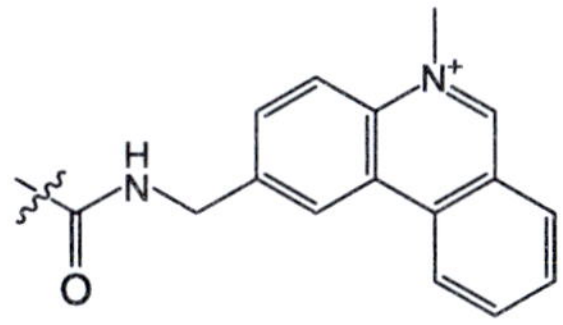	I^-, Br^-, Cl^-, O_2, pH [25, 27]
Eu^{3+} 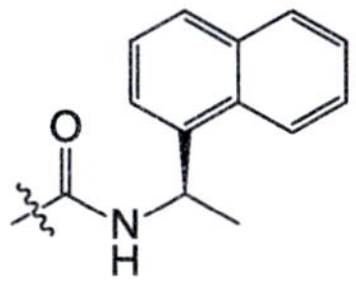 **8**	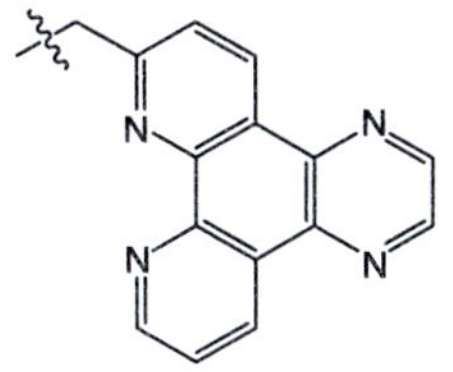	DNA, poly(dGdC) [28]

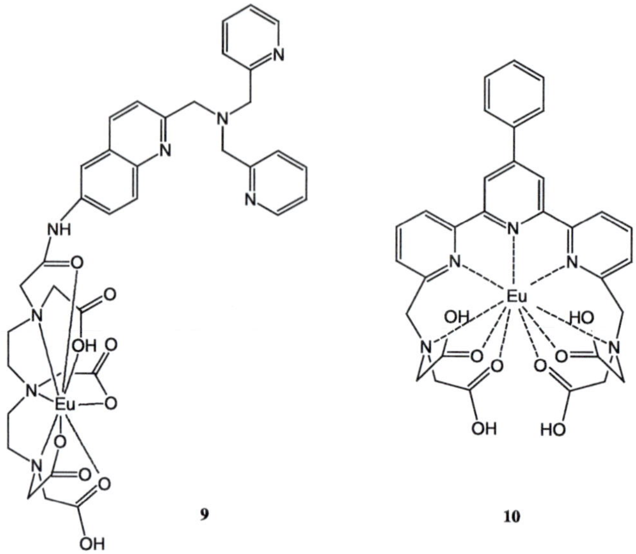

9 10

Scheme 2 Lanthanide complexes with noncyclic polydentate ligands

It is further reported to respond to pH with a pK_b value of 10.6, whereat the fluorescence is switched off in basic environment. The analogous terbium complex **7** can be used as oxygen-sensitive probe. The principle of detection is based on the quenching of aromatic triplet excited states by oxygen. The response of **7** to pO_2 is independent of pH between pH 2 and 9 and is only marginally interfered by the presence of anions. The cationic nine-coordinate Eu^{3+} complex **8** comprises naphthyl groups and a tetraazatriphenylene moiety as the sensitizing chromophore. The lanthanide emission increases up to 3.5-fold upon binding of poly (dGdC) or DNA [28].

Furthermore, lanthanides form stable complexes with polydentate chelators like DTPA, which exhibit a noncyclic structure. Two structures are depicted in Scheme 2. The following examples are only representatives for the variety of analyte molecules that can be determined by these kind of lanthanide complexes. Structure **9** employs a quinolyl ligand both as chromophore and acceptor for Zn^{2+}. The emission of the europium ion is strongly enhanced upon binding of Zn^{2+} and showing distinct selectivity over other biologically relevant metal cations in aqueous solution at neutral pH [29]. The luminescence of the Eu^{3+} chelate **10** is efficiently quenched by Cu^{2+} ions in aqueous medium [30]. The presence of Fe^{3+}, Co^{2+}, Ni^{2+}, Cr^{3+}, and Mn^{2+} interferes with the determination of Cu^{2+}, although to a relatively small extent, whereas the ions Zn^{2+}, Cd^{2+}, Hg^{2+}, and Pb^{2+} do not interact with probe **10**.

In contrast to the receptors shown above, chelating structures with polydentate ligands can also be applied for determination of analytes that act as secondary ligands for the lanthanide center, e.g., hydrogen carbonate, acetate, lactate, citrate, or amino acids [31–33], and form thereby ternary complexes. Heptadentate ligands based on cyclen leave two binding sites unallocated, which can be occupied by additional small ligands and cause a luminescence response. Ligand exchange dynamics and resulting changes in vibrational energy transfer processes have been studied in detail by Parker et al. [34]. Those mechanisms are also involved in case of the probes discussed in the next section.

2.2 *Complexes with Non-Chelating Ligands*

In addition to the highly coordinating ligands discussed in Sect. 2.1, several antenna chromophores have been described that form more or less stable complexes with lanthanide ions. Their chemical structures are usually not revealed. The same is true for the mechanisms that lead to fluorescence quenching or enhancement after coordination of certain analytes [35].

Europium(III) tetracycline **11** (EuTc) was first reported by Hirschy et al. as 1:1 complex with high energy transfer efficiency from tetracycline to Eu^{3+} [36]. Later, it was discovered that the luminescence intensity of this complex is strongly increased in presence of hydrogen peroxide [37]. Accordingly, the group of Wolfbeis et al. showed that EuTc in a 3:1 stoichiometry can be applied as a very sensitive probe for the determination of hydrogen peroxide in aqueous solution. The complex shows a 15-fold increase in emission intensity in presence of H_2O_2 [38]. This is the first reported Eu^{3+} probe that can be excited at wavelengths >400 nm. It can be used to detect hydrogen peroxide in river water [39] or to monitor the activity of enzymes that convert or produce hydrogen peroxide such as glucose oxidase, catalases, or peroxidases [40–42]. Due to its quenching effect, the determination of copper ions also can be achieved by the EuTc hydrogen peroxide complex [43].

EuTc in a 3:1 stoichiometry responds to phosphate by a strong fluorescence enhancement [44], whereas its luminescence in a 1:1 stoichiometry is quenched by phosphate [45] and also by nucleoside phosphates such as ATP, ADP, or GTP [46]. Both effects can be used to apply EuTc as a probe for the determination of the activities of alkaline phosphatase [45] and creatine kinase [46]. Furthermore, EuTc was reported to form complexes with citrate and other species of the citrate cycle in cellular metabolism in a 1:1:2 stoichiometry of Eu^{3+}/Tc/citrate. Their coordination induces a strong luminescence increase [47].

Tetracycline belongs to a group of antibiotics which includes oxytetracycline, doxycycline, chlortetracycline and many more (see Table 3). After the initial reports of EuTc as probe for hydrogen peroxide a great variety of probes based on tetracycline derivatives was investigated. In 2007, Bel'tyukova et al. reported on various Eu^{3+} tetracycline derived complexes responding to hydrogen peroxide [48]. These include tetracycline itself, oxytetracycline **12**, chlortetracycline **13**, rolitetracycline **14**, methacycline **15**, meclocycline **16**, doxycycline **17**, and demeclocycline **18**.

Table 3 Luminescent probes based on europium(III) tetracycline complexes: ligand structures and corresponding analytes

Ligand basic structure	Tetracyclines	Analytes/refs
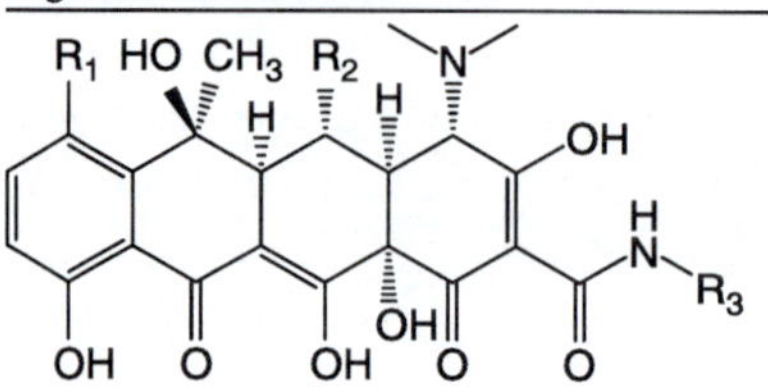	Tetracycline **11** $R_1 = R_2 = R_3 = H$	H_2O_2 [38–42] P_i, ATP, Citrate [44–47]
	Oxytetracycline **12** $R_1 = H$; $R_2 = OH$; $R_3 = H$	H_2O_2 [48] ATP [49]
	Chlortetracycline **13** $R_1 = Cl$, $R_2 = R_3 = H$	H_2O_2 [48]
	Rolitetracycline **14** $R_1 = R_2 = H$; $R_3 =$ 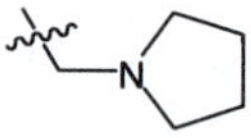	H_2O_2 [48]
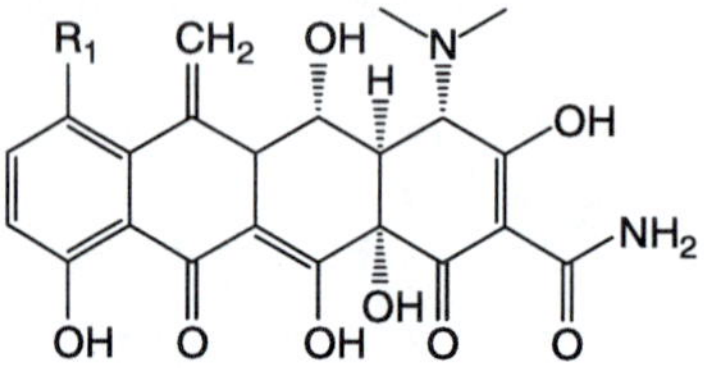	Methacycline **15** $R_1 = H$	H_2O_2 [48] Lysozyme [49]
	Meclocycline **16** $R_1 = Cl$	H_2O_2 [48]
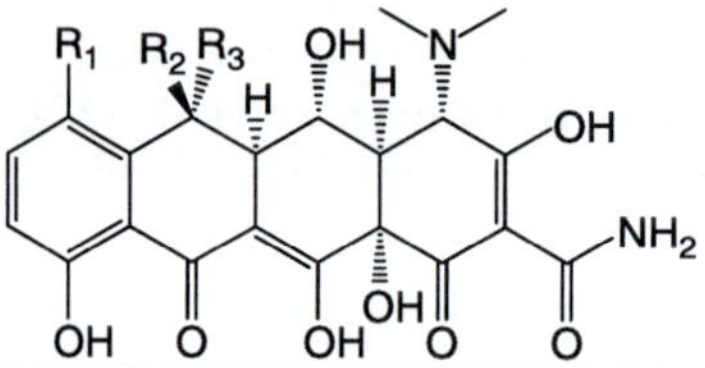	Doxycycline **17** $R_1 = R_2 = H$; $R_3 = CH_3$	ATP [51] Heparin [52] NADP [53]
	Demeclocycline **18** $R_1 = Cl$, $R_2 = OH$, $R_3 = H$	H_2O_2 [48]

The europium(III) methacycline complex **19** can be used for spectrofluorimetric determination of lysozyme [49]. The oxytetracycline and doxycycline complexes are both sensitive to ATP in that their luminescence emission is increased in presence of ATP [50, 51]. In addition, complex **17** can be used to determine heparin and NADP in aqueous solution [52, 53]. The detection of these biomolecules occurs by means of an enhanced fluorescence emission. Bile acid quenches the luminescence of Eu^{3+} doxycycline in a concentration dependent manner [54].

Further europium complexes responding to inorganic phosphate or ATP are depicted in scheme 3. All three ligands **19–21** are used in 1:1 stoichiometry with Eu^{3+}. Inorganic phosphate causes an increase of luminescence emission in case of complex **19**, whereas **20** is quenched [55, 56]. On the other hand, the presence of ATP, ADP, or AMP induces a strong increase in luminescence of complex **21** [57].

Apart from the Eu^{3+} complexes specified above, several Tb^{3+} complexes stand out because of their immunities towards interferences of ubiquitous metal ions like Mg^{2+}, Ca^{2+}, or Mn^{2+}, and of proteins such as BSA. Terbium(III) norfloxacin **22** is

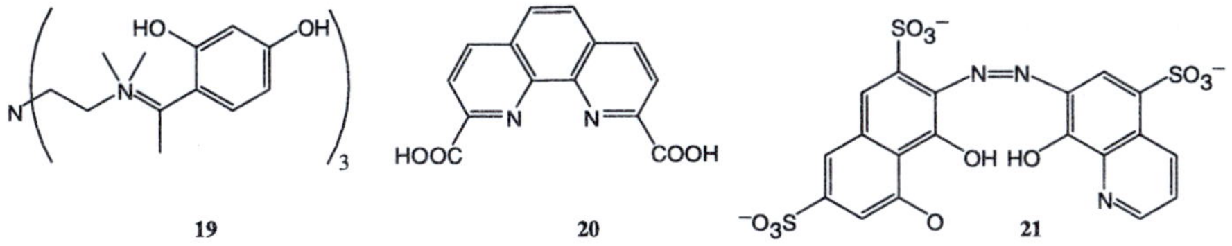

19 20 21

Scheme 3 Sensitizing ligands forming phosphate-sensitive europium complexes

one prominent example. With a stoichiometry of 4.5:1 of Tb^{3+} to norfloxacin its luminescence responds to several biologically relevant analytes. Miao et al. reported that ATP enhances the luminescence of Tb^{3+} norfloxacin [58]. The same is observed in case of NADP [59], DNA [60], heparin [61], GTP, and GDP [62]. On the other hand, the emission is quenched by phosphate [59], pyrophosphate [63], and lecithin [64]. These responses can be exploited to monitor enzyme activities and was applied to screening assays for enzyme inhibitors, e.g., for adenylyl cyclases [63], and for the determination of GTPase activity of Ras proteins [62].

Norfloxacin belongs to a group of antibiotics referred to as fluoroquinolones. Other members of the fluoroquinolone group have also been applied to sensitize lanthanide luminescence (Table 4). A further interesting example of a fluoroquinolone as sensitizer for terbium based probes is ciprofloxacin **23**. Tb^{3+} and ciprofloxacin were applied in a 10:1 stoichiometry to determine ATP [65], heparin [66], coenzyme A [67], and lecithin [68]. A lomefloxacin complex **24** can be alternatively adopted to probe heparin [69].

The luminescence of complexes **22**, **23** and **25** is enhanced by interaction with human serum albumin [70]. Tb^{3+} prulifloxacin **26** can be applied for the determination of ATP in pharmaceutical samples [71]. It is worthwhile to remind that all these complexes can act as reversible indicators for the specified analytes, as the luminescence response is based on the exchange of water ligands.

3 Lanthanide Complex Based Sensors

Optical chemical or physical sensors have found widespread applications in industrial process control, environmental monitoring, biomedical research and medical diagnostics. The analytes mainly targeted are oxygen, pH, CO_2, H_2O_2, water, metal ions, anions, glucose, and temperature. The luminescence response of the indicator to a particular analyte should be specific and reversible. These are the premises to make it applicable as a receptor unit of a sensor device. It is obvious that an ideal sensor should response reversibly and highly selective to the analyte of interest. The main features of sensors are their fast response, the capability of continuous monitoring of complex samples, and the recording of changing analyte concentrations over time.

In order to integrate luminescent probes into sensor devices as chemical receptors, they have to be immobilized on a solid support and connected to the transducer

Table 4 Luminescent probes based on terbium(III) fluoroquinolones: ligand structures and corresponding analytes

Ligand basic structure	Fluoroquinolone	Analytes/refs
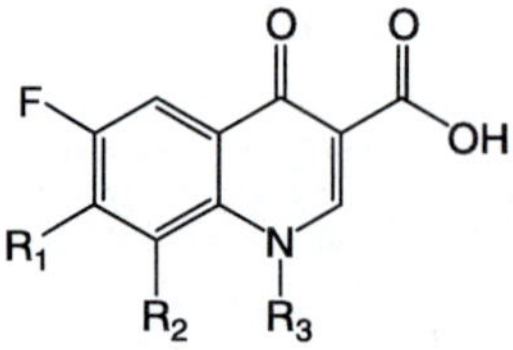	Norfloxacin **22** R_1 = 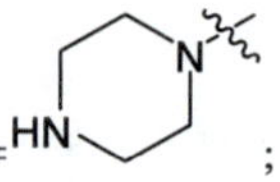 ; R_2 = H; R_3 = 	ATP [58, 62] NADP [59] DNA [60] Heparin [61] GTP, GDP [62]
	Ciprofloxacin **23** R_1 = 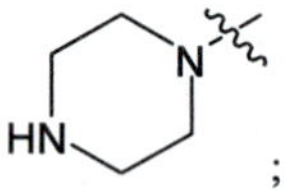 ; R_2 = H; R_3 = 	ATP [65] Heparin [66] Coenzyme A [67] lecithin [68] HSA [70]
	Lomefloxacin **24** R_1 = 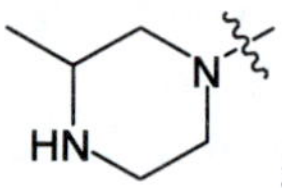 ; R_2 = F; R_3 = 	Heparin [69]
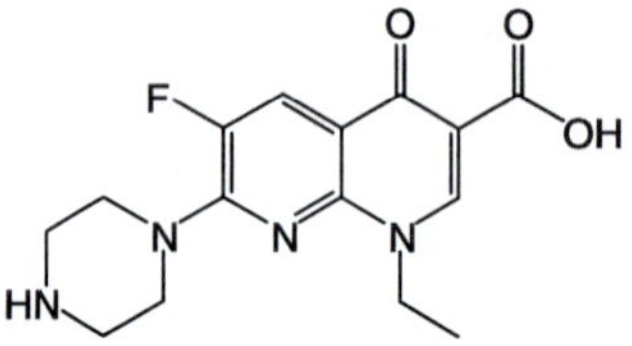	Enoxacin **25**	HSA [70]
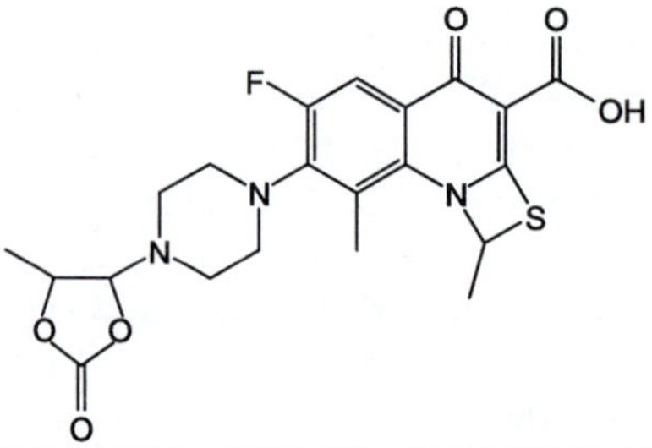	Prulifloxacin **26**	ATP [71]

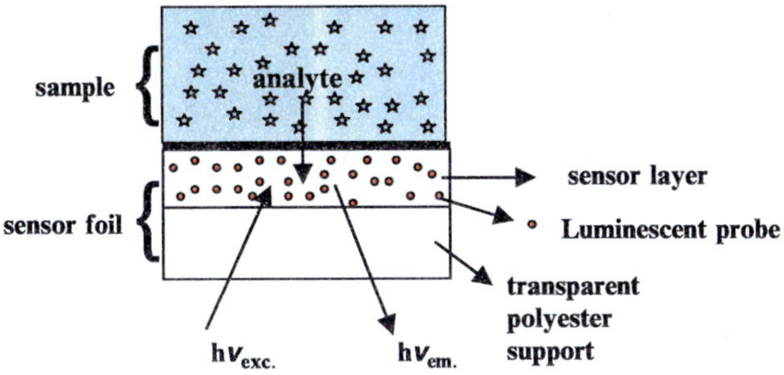

Fig. 2 Schematic of a luminescent optical sensor layer. It consists of an indicator doped solid matrix which is spread onto a transparent support, the bottom of a microwell plate, or on the tip of an optical waveguide. The matrix should provide a highly selective analyte permeability. The indicators (probes) interact with the respective analyte and their luminescence emission (or lifetime) reflects the (local) analyte concentration

part. Usually, dyes can be incorporated in polymer matrices, silica, ormosil (organically modified siloxanes), or attached to nano- or microparticles. The resulting sensor layers can be used to coat fiber-optic microsensors, microwell plates, glass, or plastic supports (Fig. 2). They can be integrated in microfluidic systems or applied as sensor paints for larger objects to study transport processes on surfaces [72–74].

Photoluminescent sensors require a light source for excitation of the probe molecules and the assembly of a photodetector. Immobilization or incorporartion into polymer matrices does usually not affect the spectral characteristics or the lifetime of a luminescent probe. But its encapsulation can significantly change the response characteristics to a given analyte [75]. Ideally, the sensitivity and selectivity of the luminescent probe can be increased by choosing an appropriate matrix. The luminescence decay of lanthanide complexes is distinguished by its long lifetime in the range from microseconds to milliseconds. Thus, changes of luminescence lifetimes can easily be recorded and provide a useful quantity for the sensor read out, in addition to intensity-based measurements [76]. What makes europium or terbium complexes also attractive as sensor elements is the large shift between excitation and emission wavelengths due to the ligand intersystem crossing and the ligand to metal energy transfer.

In principle, any of the LLCs shown in the last section may be incorporated in polymer matrices or other solid supports and integrated in sensor systems. Basic approaches have been discussed by Pandya et al. [77]. This chapter will highlight some examples of different sensor types classified with respect to the corresponding analyte.

3.1 *Oxygen Sensors*

The determination of oxygen partial pressure (pO_2) is a major task in chemical and biotechnological process control, pharmaceutical screening, medical chemistry and diagnosis (e.g., blood gas analysis), and environmental analysis (e.g., cellular respirometry) [78, 79]. Typical luminescent oxygen-sensitive probes applied in sensors consist of transition metal–ligand complexes, particularly of Ru^{2+}, Os^{2+}, and Ir^{3+}, or platinum and palladium porphyrins [80, 81]. The response to oxygen is caused by dynamic quenching processes and can be described in solid state sensor layers by a modified *Stern–Volmer* equation according to the *two-site model* [82]:

$$\frac{\tau}{\tau_0} = \frac{I}{I_0} = \frac{f_1}{1 + k_{SV}^{(1)}[O_2]} + \frac{f_2}{1 + k_{SV}^{(2)}[O_2]} \quad (1)$$

Deviations from the linear *Stern–Volmer* relation derive from different micro-environments in polymer matrices [83], with f_1 and f_2 as the emissive fraction of the probes in the different environment, k_{SV} as the Stern–Volmer quenching constants of the different components, and $[O_2]$ as the oxygen partial pressure. Assuming a constant proportion of oxygen in atmospheric air, the amount of oxygen quenching can be used for air pressure determination. Hence, luminescence quenching by oxygen is the basis for barometric sensors and *pressure sensitive paints* (PSPs) used in fluid mechanics and aerodynamic measurements [84, 85]. The latter example outlines the importance of optical oxygen sensors for imaging applications. These can not be approached by conventional oxygen or pressure sensors (*Clark* oxygen electrode or pressure taps, respectively). Polymers with high oxygen permeability and mechanical stability are suited materials to immobilize the oxygen sensitive dyes on a solid support. These include siloxanes, silicones, organic glassy polymers (e.g., polystyrene), fluoropolymers, or cellulose derivatives such as ethyl cellulose [84].

Unlike many other luminescent metal–ligand complexes, LLCs usually display only low sensitivities towards oxygen. Neither the emission from the excited state of the lanthanide ion nor the energy transfer from the triplet state of the antenna chromophore to the metal center is significantly affected by oxygen quenching. This is particularly the case for Eu^{3+} complexes. Some of that species are completely insensitive to oxygen (and also to temperature), so that they can be used as inert reference dye in oxygen sensors or PSPs [86]. Yet, some Tb^{3+} complexes that are sensitive to oxygen have been reported. These include complexes with salicylaldehyde derived Schiff base or thenoyltrifluoroacetone ligands. They can be incorporated in poly(methyl methacrylate) to form oxygen-sensitive layers. Also in this case, a non oxygen-quenchable Eu(III) complex is used as internal reference. The matrix, containing indicator and reference, is disposed at the distal tip of a single optical fiber [87]. The resulting fiber-optic sensor is operated with a time-resolved fluorescence detection scheme by means of a pulsed light source and a time-gated photodetector. The determined *Stern–Volmer* constant (K_{SV}) of 9×10^{-3} mm Hg^{-1} (37°C, 546 nm) in a linear range from 0 to 180 mm Hg is comparable to oxygen sensors containing palladium(II) or platinum(II) porphyrin complexes.

In a different approach, a tris(acetylacetonato)-1,10-phenanthroline terbium(III) complex $[Tb(acac)_3phen]$ was immobilized on a thin layer alumina plate [88]. Figure 3 shows the quenching of the terbium $^5D_4 \rightarrow {}^7F_J$ transitions (compare

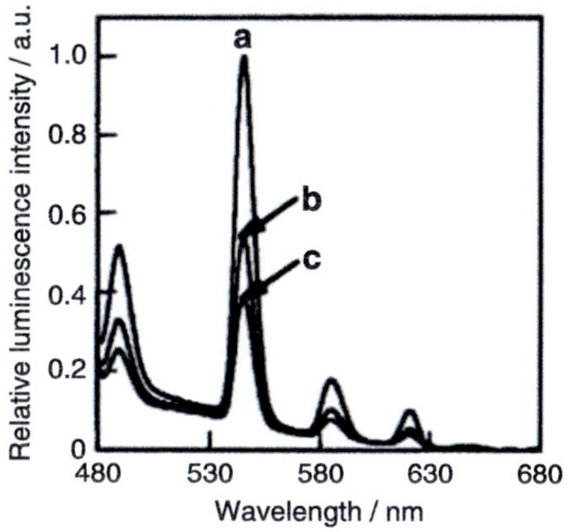

Fig. 3 Luminescence spectrum of a thin film of Tb(acac)$_3$phen on alumina in presence of 0% (**a**), 20% (**b**), and 100% (**c**) [O_2] at atmospheric pressure; λ_{exc} = 350 nm [88]

Table 1) by oxygen. Though, the *Stern–Volmer* plot shows only a weak quenching of luminescence intensity. Thus, this oxygen sensor can provide only a moderate sensitivity. A selection of oxygen-sensitive Tb^{3+} complexes with macrocyclic ligands based on a 1,4,7,10-tetraazacyclododecane (cyclen) structure (see, e.g., complex **7** in Sect. 2.1) has been presented by Parker [25].

3.2 *pH Sensors*

Optical pH sensors have found widespread applications in environmental analysis, biomedical research, medical diagnosis and industrial process control. Fluorescent pH sensitive dyes like 8-hydroxypyrene-1,3,6-trisulfonic acid or fluorescein derivatives are typically dual wavelength probes that have different absorption and/or emission maxima in the protonated and basic form. These can be immobilized on the tip of a waveguide by sol-gel techniques or by embedding in polymeric hydrogels [89]. Disposable microplates (e.g., a 96-well format) with integrated fluorescent pH sensors can be used for screening of enzyme reactions or monitoring the proliferation of cells [79].

Fiber optic pH sensors have distinct advantages over pH electrodes. They are small, not interfered by electromagnetic fields and have remote sensing capability. They can be used in extreme environments, such as deep-water analysis, chemical reactors, or wastewater. Moreover, they can be intrinsically referenced due to the dual wavelength measurement capability and do not require a reference electrode [90]. Optical pH sensors also pave the way for imaging applications, whereas pH electrodes only enable punctiform pH measurements. Sensors for pH determination are also of high significance in environmental and marine research because they provide the basis for CO_2 sensors.

A smart method to introduce pH sensitivity to LLCs is to conjugate a sensitizer chromophore to a cyclen ligand system which can be protonated. Parker has shown that phenanthridine derivatives **7** are suitable donors [25] (see Sect. 2.1, Table 1).

Protonation of the aromatic phenanthridine ring lowers the excited singlet and triplet states. A highly selective excitation of the protonated ligand is possible by selecting an appropriate excitation wavelength. This leads to pH dependent efficiency of the energy transfer to the lanthanide and, finally, to a pH dependent emission from the sensitized lanthanide.

This basic principle was extended by Gunnlaugsson et al. to the preparation of pH sensors [23, 91]. They incorporated the pH sensitive Eu^{3+} complex **2** (Table 2) in a water-permeable hydrogel matrix consisting of poly(methyl methacrylate-*co*-2-hydroxyethyl methacrylate). The cyclen ligand was conjugated to a quinoline ring. Deprotonation of the quinoline nitrogen by increasing the pH from 3 to 11.5 leads to a deactivation of the $^5D_0 \rightarrow {}^7F_J$ transitions. Thus, a decrease in lanthanide emission can be recorded as a function of pH if the quinoline is excited at 330 nm (selective excitation of the protonated species). In principle, the most intense transitions to $^7F_1$ (593 nm), $^7F_2$ (615 nm), and $^7F_4$ (683/701 nm) can all be considered to monitor pH changes (Fig. 4). The sensor provides a dynamic range from pH 4 to 8 with a pK_a of around 6. The sensor films were evaluated by means of confocal laser-scanning microscopy.

Lobnik et al. have used DTPA as ligand for Eu^{3+} (Scheme 4). This chelate was linked to 4-trifluoromethylcarbostyril acting as antenna system **27** [92]. The

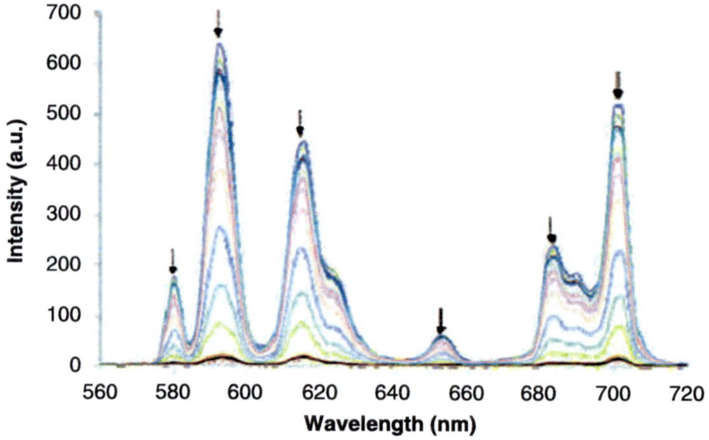

Fig. 4 Decreasing Eu(III) emission of complex **2** as a function of increasing pH (pH 3–11.5) in solution, showing the deactivation of the $^5D_0 \rightarrow {}^7F_J$ transitions [91]

Scheme 4 Eu(III)–trifluoromethylcarbostyril–DTPA complex **27** as luminophore for optical pH sensors

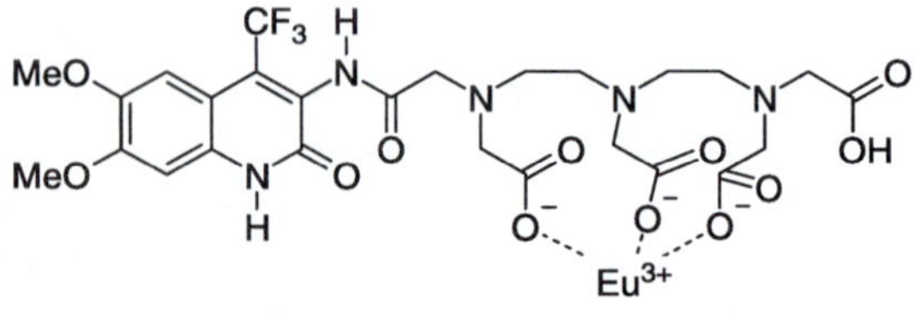

complex was co-immobilized with the non-fluorescent pH indicator bromothymol-blue in an ormosil matrix. The indicator provides a pH dependent absorption of the 615-nm main band of europium luminescence as this emission peak coincides with the absorption of the basic form of the indicator. This sol-gel sensor layer can reversibly monitor pH with an excitation wavelength of 370 nm in the range from 4.5 to 10.7, with a pK_a around 7.3.

3.3 Sensors for Hydrogen Peroxide

Hydrogen peroxide as one of the reactive oxygen species can cause substantial damage to living cells, and also to DNA. It is used in industry for bleaching, cleaning, and disinfection. Therefore, it is also released to the environment. Methods for determination or continuous sensing of H_2O_2 are of high interest. Another demand in analytical assays and sensors for H_2O_2 results from the fact that it is produced or consumed by many enzymes such as oxidases, catalases, or peroxidases. Hence, numerous tests for the determination of H_2O_2 have been developed, particularly amperometric electrochemical methods. A large number of optical assays do also exist. Both chemiluminescent and electrochemiluminescent methods have very low limits of detection [93, 94], but work at pH above 8.0 only. In addition, these assays are usually irreversible. Other optical assays are based on the oxidative power of H_2O_2 that can convert a non-fluorescent substrate (such as Amplex Red) into a fluorescent product (resorufin), usually in the presence of the enzyme horseradish peroxidase. But this assay is prone to interferences [95].

As outlined in Sect. 2.2, the luminescence of europium complexes in aqueous solution can be enhanced by addition of an excess of H_2O_2. In this case, a reversible ligand exchange of highly quenching H_2O by H_2O_2 occurs. H_2O_2 ligands also act as "hard" electron donor groups but are weaker OH oscillators. As a consequence, the rate of radiationless deactivation of the excited lanthanide ion by vibrational energy transfer is reduced. The removal of quenching water OH oscillators is accompanied by an increase of intensity and lifetime of the lanthanide emission (Fig. 5).

Europium(III) tetracycline **11** (Eu/Tc 3:1) was immobilized in a matrix of polyacrylonitrile-*co*-polyacrylamide for the fabrication of sensors for hydrogen peroxide [96, 97]. As outlined, this probe undergoes a 15-fold increase of the luminescence intensity of the strong $^5D_0 \rightarrow {}^7F_2$ emission band at 615 nm and an increase of its average lifetime from 30 to 60 μs [38]. A fast read out of sensor arrays can be achieved by means of fluorescence lifetime imaging (FLIM). This method requires pulsed LEDs as light source (excitation at 405 nm) and a CCD camera with mechanical shutter. The lifetimes can be determined by time gated detection schemes such as *Rapid Lifetime Determination* (RLD) [98]. The sensor has a dynamic range from 0.5 to 50 ppm with a limit of detection of 0.50 ppm of H_2O_2 in aqueous solutions of pH 7 [97].

A general problem of this type of lanthanide probes is their lack of selectivity, as their luminescence is interfered by a series of other chelating agents like citrate,

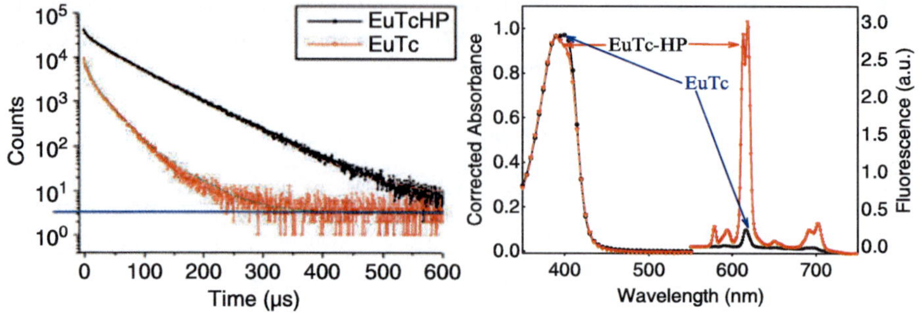

Fig. 5 *Left*: Luminescence decay profiles of the 615 nm emission line of europium(III) tetracycline **11** in absence (EuTc) and presence of H_2O_2 (EuTcHP) as recorded by time-correlated single photon counting. *Right*: absorption and emission spectra of EuTc at pH 6.9 in presence and absence of H_2O_2 (λ_{exc} = 405 nm), displaying the $^5D_0 \rightarrow {}^7F_J$ transitions (see. Fig. 1)

phosphate, and acetates. Also biophosphates such as ATP or NADP and biopolymers such as DNA or different proteins induce luminescence responses of lanthanide complexes of the tetracycline family. Nevertheless, this kind of optical sensors can be employed for the determination of enzymatically produced (or consumed) H_2O_2 and thus be useful for the screening of enzyme activities.

This H_2O_2 transducer can also be used for the fabrication of glucose sensors. In this case, the enzyme glucose oxidase is coadsorbed in the polymer film. A H_2O_2 sensitive transducer has distinct advantages over the commonly used oxygen or pH transducers in glucose sensors. It is reversible, measures at a zero background concentration of H_2O_2, and can also be applied to high throughput screening and imaging applications. With this sensor, glucose concentrations can be monitored in a dynamic range from 1 to 150 μmol L^{-1} at a limit of detection of 0.5 μmol L^{-1} [99]. Background fluorescence from the sample or the optical set up is efficiently eliminated by time-resolved measurements, collecting the lanthanide luminescence after delay times of 20–30 μs after excitation.

3.4 Humidity Sensors

The measurement of humidity is a significant task in many industrial fields, particularly in the pharmaceutical, food, and electronic industry. Even small amounts of water can cause damaging effects on moisture-sensitive products such as the deterioration of food or drugs and the corrosion of electronic components. Unwanted exposure to humidity can occur during processing, transport, or storage. A large variety of probes have been utilized in luminescent humidity sensors, including rhodamine [100], perylene [101], Ru(II), and Os(II) complexes [102–104], or Pt- and Pd-porphyrins [105]. The transition metal complexes are prefered because they enable luminescence lifetime-based sensing modes.

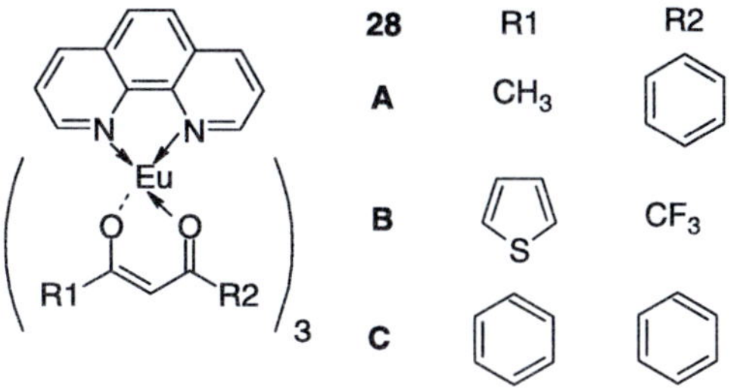

Scheme 5 Eu(III) β-diketonate complexes tested as luminescent probes for humidity sensing

LLCs are promising candidates as probes for humidity sensors due to the distinct quenching effect of water molecules which can reversibly coordinate to lanthanide ions. The decrease in lifetime of the $^5D_0 \rightarrow {}^7F_2$ transition of europium(III) perchlorate was used for the determination of small amounts of water in DMF and DMSO [106]. Wang and Li have presented luminescent nanospheres for the determination of small amounts of water (0.05–3.0 vol%) in ethanol [107]. They coated silica nanoparticles with a thin layer of a salicylic acid-La^{3+}/Tb^{3+} coordination compound. The green fluorescence peaking at 549 nm corresponds to the $^5D_4 \rightarrow {}^7F_5$ transition of Tb^{3+} and is strongly quenched by trace amounts of water. The nanoparticles can be excited at wavelengths around 350 nm.

It would be worthwhile to follow up and optimize the capability of LLCs entrapped in polymer matrices for the indication of humidity in the gas phase. This approach seems promising since it was shown that the luminescence of $EuCl_3$ included in a dehydrated Nafion membrane is quenched in the course of adsorption of water vapor [108]. A series of β-diketonate complexes of Eu^{3+} (**28 A–C** in Scheme 5) embedded in Nafion membranes were also evaluated for their responses to relative humidity [109]. There still remains some analytical work to be done because astonishingly it depends on the type of β-diketonate ligand whether the luminescence is suppressed or enhanced in presence of water.

Knall et al. have tested a series of europium complexes with antenna chromophores in different polymer matrices in terms of sensitivity, response time, and dynamic range for the sensing of water vapor. A copolymer containing a Eu^{3+} chelate that can be spin-cast as thin film on glass slides was developed [110]. Scheme 6 shows the chemical structure of the polymerizable precursor **29**. This complex can be copolymerized with norbornene-2,3-dicarboxylic acid dimethyl ester by means of ring opening metathesis polymerization. The sensor spots respond to water vapor by reversible luminescence quenching, which can be analyzed by means of phase sensitive measurements of luminescence lifetime [110].

3.5 Sensors for Copper Ions

The dynamic quenching of lanthanide luminescence by heavy metal ions has been described in Chap. 2. Particularly, copper ions induce a strong decrease in

Scheme 6 Europium complex as monomer for the preparation of humidity sensitive luminescent copolymers

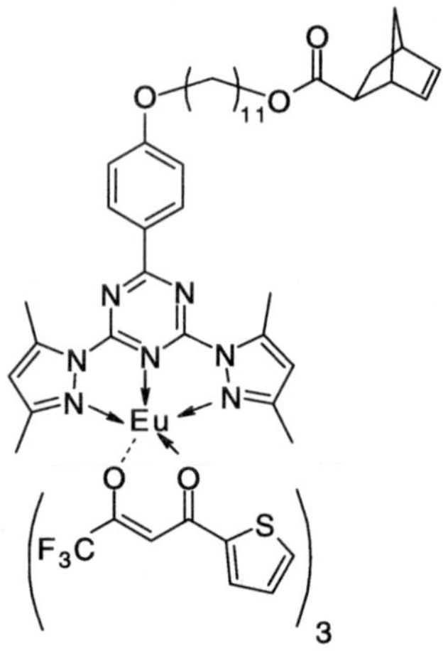

29

luminescence intensity and lifetime of the 615-nm emission line of europium(III) complexes. The response can be described by a *Stern–Volmer* equation. The determination of Cu^{2+} in drinking water and also in environmental waters such as river water and industrial wastewater is an important task due to its toxicity on living organisms. The legal maximum limit for copper ions in drinking water in Europe is 2 mg L^{-1}.

The luminescence intensity and decay time of $EuCl_3$ in carboxymethyl cellulose membranes is decreased in presence of heavy metal ions like Cu^+ or Cu^{2+}, but also Cr^{3+} and Fe^{3+} exert a distinct quenching effect [111]. It is not likely that a sensor with high specificity can be prepared on the basis of LLCs that is free of interferences from other metal ions. However, an adequate choice of the ligand system may help to improve the selectivity of the response. Another approach uses a sol-gel technique to embed a complex of Eu^{3+} and silanized 2,6-pyridine-dicarboxylic acid as antenna in a silica network. This luminescent material can sense copper ion concentrations in water down to 50 µg L^{-1}, but the sensor was not evaluated with respect to interferences of other metal ions or in environmental samples [112].

3.6 Sensors for Small Organic Compounds

The responses of LLCs to organic (bio)molecules capable of coordinating to the lanthanide as additional ligand have been outlined in Chap. 2.2. Examples have been shown for ATP or NADP. The plurality of these interactions indicates the problems of sensors including lanthanide complexes as receptors for the specific recognition of a certain analyte. The same is the case for free lanthanide ions that have been presented for the luminescent determination of pharmaceutically active compounds such as antibiotics [15–17]. Nevertheless, Molina-Diaz et al. have described a sensor for flow-injection analysis (FIA) that can be applied to the determination of p-aminobenzoic acid in pharmaceutical preparations [113]. In this work, Tb^{3+} was immobilized on a Sephadex anion exchange gel. This method provides a limit of detection of 0.06 μg mL^{-1}. It is prone to interfering species but their presence is limited in certain sample matrices such as pharmaceutical products. Thus, this could be a valuable and straightforward approach for the quality control of drugs by means of FIA systems.

3.7 Temperature Sensors

All luminescent indicators applied in optical sensors display a more or less strong sensitivity towards temperature. The Boltzmann distribution is one factor for this phenomenon because it governs the populations of the different vibrational levels of the electronic states involved. This evokes the demand for temperature sensitive probes for two purposes: (a) measurement or imaging of temperature by means of optical sensors and (b) compensation of temperature effects on optical sensors.

Sensing temperature by itself is of high importance in industrial process control, marine research, geochemical studies, biotechnology, and in diagnostic or therapeutic applications. The monitoring of hyperthermal cancer therapy is one prominent example which is discussed as an application for luminescent temperature sensing and imaging. Temperature sensitive paints (TSPs) are useful tools for the study of flows on surfaces, e.g., in cryogenic wind tunnels, or of combustion processes and heat fluxes, e.g., in turbochargers and turbines [84]. Whoever has worked with LLCs, be it in form of molecular probes or labels, may have noticed the high temperature sensitivity of many of these complexes. Particularly, the luminescence of sensitized Eu(III) β-diketonate complexes is highly temperature-dependent. Therefore, these are promising candidates as temperature sensitive probes for optical sensors.

The ubiquitous temperature effects on luminescent sensors can be referenced and compensated with dual luminophore preparations. Several approaches of dual optical sensors have been demonstrated, that can be used for simultaneous determination of oxygen and temperature, or CO_2 and temperature, respectively [114, 115]. Luminescent temperature indicators have also been employed as reference components in PSPs (see Sect. 3.1). These have found widespread application in fluid mechanics and aerodynamic wind tunnel tests. The real-time imaging of dynamic flow processes on model surfaces are of high significance for aerospace and car industry. To avoid interferences or energy transfer between the oxygen and temperature sensitive dyes, these can be incorporated into different types of polymer microparticles [116].

An ideal polymer binder for temperature sensitive layers should eliminate interferences (e.g., by oxygen quenching) and amplify the temperature dependency. Dual pressure- and temperature-sensitive paints have found increasing interest in the past years. The response of the luminescence emission (or lifetime) of the two indicators can be separated either spectrally by appropriate optical filters [117] or by advanced time-resolved imaging methods [116].

The main temperature-dependent nonradiative relaxation mechanisms of the $^5D_0$ level of Eu(III) chelates are discussed by Berry et al. [118]. These are (1) multi-phonon emission – in this pathway the energy difference between the electronic states is converted to vibrational energy in the ligand or in the matrix; (2) energy transfer from the Eu(III) $4f^n$ states to electronic states of the ligand or to fitting electronic states of other species in the environment; and (3) crossover from the excited $^5D_0$ state to a ligand to metal charge transfer state with lower energy. The latter radiationless pathway causes the unusually strong temperature dependence of the lifetime of the $^5D_0$ around room temperature.

Applied in sensors, the complex is usually immobilized in solid polymer matrices. Hence, the first two mechanisms will also play a significant role. Additionally, at higher temperatures the triplet excited state of the ligand can also be deactivated leading to a less efficient energy transfer to the lanthanide ion. The rate constants w of the crossover processes involved can be described approximatively by an *Arrhenius*-type equation, where the barrier height is expressed by the activation energy E_a [118, 119]:

$$w = A \exp\left(\frac{-E_a}{kT}\right) \quad (2)$$

where A is the pre-exponential factor, k the Boltzmann constant, and T the absolute temperature. Accordingly, the lifetime τ of the electronic excited $^5D_0$ state can be expressed by

$$\frac{1}{\tau} = k_0 + k_1 \exp\left(-\frac{\Delta E}{RT}\right) \quad (3)$$

where k_0 is the temperature independent decay rate for the deactivation of the excited state, k_1 the pre-exponential factor, and ΔE the energy gap between emitting level and the deactivating excited state level [120].

Examples of europium complexes **30–32** that have been applied in temperature sensors or dual pressure- and temperature-sensitive paints are listed in Table 5. The respective ligand structures are shown in Scheme 7. The temperature dependency is quantified as average luminescence intensity temperature coefficient I [%/°C]. Usually, it is determined in a temperature range from 1 to 40 or 50°C. These examples exceed the intensity temperature coefficients of other established temperature sensitive probes such as ruthenium(II)-tris-(1,10-phenanthroline) [121]. Generally, the lifetime temperature coefficients are significantly lower. This indicates that thermal quenching of the triplet state of the antenna chromophore plays an important role. Due to the narrow emission band of europium complexes at 615 nm even triple sensors for temperature, oxygen, and pH are achievable [122].

Thermographic phosphors are other materials applicable to temperature sensing or to the imaging of temperature distributions on surfaces. They consist of inorganic

Table 5 Temperature sensors based on europium chelates excitable at wavelengths >400 nm

Indicator	Polymer matrix	λ_{exc} (nm)	*I* (%/°C)	Ref.
1,10-phenanthroline-tris(3-(3-phenanthryl)-1-(9-phenanthryl) propane-1,3-dione) europium **EuD2**	FIB	400	−4.4 −1.3[a] (5–45°C)	[117]
Eu(III)-tris(thenoyltrifluoroacetonato)-(2-(4-diethylamino-phenyl)-4,6-bis (3,5-dimethylpyrazol-1-yl)-1,3,5-triazine) **Eu(tta)$_3$(dpbt)** [123]	PBTS or PVC microbeads in polyurethane hydrogel	405	−1.8 (1–25 °C) −0.7[a] (1–40°C)	[124] [116]
Eu(III)-tris(thenoyltrifluoroacetonato)-4-(4,6-di(1H-indazol-1-yl)-1,3,5-triazin-2-yl)-*N,N*-diethylbenzenamine **Eu(tta)$_3$DEADIT**	PViCl-PAN	425	−1.3[a] (1–50°C)	[125]

FIB Poly(hexafluoroisopropyl methacrylate-*co*-hepta-fluoro-*n*-butyl methacrylate), *PBTS* Poly (4-*t*-butyl styrene), *PVC* Poly(vinyl chloride), *PViCl-PAN* Poly(vinylidene chloride-*co*-acrylonitrile)
[a]Luminescence lifetime response

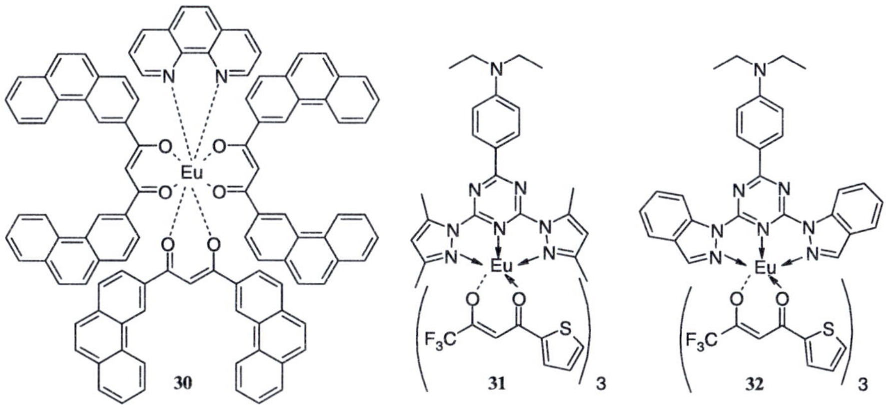

Scheme 7 Temperature sensitive europium probes: EuD2 **30**, Eu(tta)$_3$(dpbt) **31**, and Eu(tta)$_3$ DEADIT **32**

ceramic compounds and are very robust in terms of thermal stability. Typical examples are La_2O_2S or Y_2O_3 doped with Eu^{3+} and YAG doped with Tb^{3+} or Dy^{3+} [126]. These lanthanide-doped materials represent an own class of sensors that cannot be covered within a review on lanthanide complexes. The photophysical principles of luminescence thermometry by means of thermographic phosphors have been reviewed by Allison and Gillies [127]. In case of europium phosphors, the excited ^5D_J electronic states can be deactivated by a thermal promotion to a nonemitting charge transfer state followed by nonradiative relaxation. The temperature sensitive ^5D_J emission lines span a broad range of the visible spectrum from ~470 to 640 nm.

4 Conclusion and Outlook

A multitude of LLCs have been introduced as luminescent probes in the past years. They are able to respond reversibly to a broad spectrum of analytes such as pH, anions (e.g. phosphate, citrate, or hydrogen carbonate), metal ions, H_2O_2, ATP, or biopolymers. The response mechanisms can be classified in two different categories: (1) interaction of the analyte with the complex ligand which serves as antenna chromphore for the lanthanide or (2) reversible replacement of small ligands (e.g., water molecules) in ternary complexes. Most of the lanthanide probes, particularly those discussed in Sect. 2.1, require rather shortwave excitation in a range between 340 and 380 nm. This is disadvantageous because usually all biological matter displays strong intrinsic fluorescence when exposed to UV light. Furthermore, many biological samples strongly absorb in this wave-length range, and the resulting inner filter effect can inhibit the optical excitation of the lanthanide complex.

Another problem of using lanthanide complexes as luminescent probes is their susceptibility to interferences. This diminishes the selectivity of the response to a certain analyte and impedes analytical applications in complex samples. The incorporation of LLCs in solid state matrices leads to responsive layers that can be integrated in optical devices for continuous sensing.

Examples of such chemosensors have been outlined including oxygen (or air pressure), pH, humidity, hydrogen peroxide, copper, and temperature sensors. However, the encapsulation of lanthanide probes in solid state matrices or polymer nano- or microparticles can contribute to reduce the impact of interfering agents. Many polymers provide a selective permeability and block certain species. A number of copolymers, particularly block copolymers, are available that on the one hand are well suited for the immobilization of lanthanide complexes, and on the other hand hinder the diffusion of certain interfering species into the layer.

Despite the drawbacks discussed above, lanthanide complexes have two unique properties which make them ideally suited as luminescent probes: their long-wavelength emission and long luminescence lifetimes. By application of time gated measurements, background fluorescence, and autofluorescence of samples can be excluded improving the signal-to-noise ratios of luminescent sensors. Microwell plate assays can be simplified by the implementation of luminescent sensors. These functionalized plates can be used for high throughput screening or parallel analysis of a multitude of samples.

Overall, the implementation of lanthanide probes in chemical sensor technology is still in its initial stage. Up to now they have not found their way into commercialized sensor systems. Particularly, with respect to pO_2 and pH sensors, it cannot be foreseen that LLCs may displace established fluorescent indicators. Sensors for small molecules such as hydrogen peroxide, phosphate, or ATP can be useful in enzymatic assays in which the conversion of the substrate has to be monitored. In this case, the selectivity is provided by the enzyme involved. A concrete example is presented by means of a glucose sensor based on immobilized glucose oxidase and a hydrogen peroxide sensitive transducer. Other potential fields of application are the determination of enzyme activities or the screening of enzyme inhibitors. Particularly, enzymes that convert ATP or GTP (such as kinases, adenylyl cyclases, ATPases or GTPases), or release phosphate anions (phosphatases) are of high pharmaceutical interest as drug targets. Many suitable lanthanide probes have been developed in the past years but it is evident that possible applications are limited to assays in which the specificity of the probe is not crucial and the amount of interferences can be minimized. The sample matrices must not be too complex.

Temperature sensors including Eu^{3+} complexes or thermographic phosphors are mainly used for research purposes. Their luminescence emission exhibits high sensitivity to thermal quenching and in case of inorganic lanthanide phosphors high thermal stabilities. Remote sensors can be processed for imaging of temperature distributions on surfaces. This paved the way for dual pressure and temperature sensitive paints that are frequently used in aerodynamic and fluid mechanic research. Presumably, the field of thermometry is the most attractive and promising one for a commercial implementation of lanthanide-based sensors.

Acknowledgment We would like to thank Prof. Otto S. Wolfbeis, Dr. Meng Wu, and Dr. Zhihong Lin for their support and the Deutsche Forschungsgemeinschaft (DFG) for funding of our research.

References

1. Richardson FS (1982) Chem Rev 82:541
2. Dos Santos CMG, Harte AJ, Quinn SJ, Gunnlaugson T (2008) Coord Chem Rev 252:2512
3. Georges J (1993) Analyst 118:1481
4. Gunnlaugson T, Stomeo F (2007) Org Biomel Chem 5:1999
5. Werts MHV, Duin MA, Hofstraat JW, Verhoeven JW (1999) Chem Commun 799
6. Dadabhoy A, Faulkner S, Sammes PG (2000) J Chem Soc Perkin Trans 2:2359
7. Beeby A, Bushby LM, Maffeo D, Williams JAG (2000) J Chem Soc Perkin Trans 2:1281
8. Yang C, Fu LM, Wang Y, Zhang JP, Wong WT, Ai XC, Qiao YF, Zou BS, Gui LL (2004) Angew Chemie 116:5120; Angew Chemie Int Ed 43:5010
9. Peijzel PS, Meijerink A, Wegh RT, Reid MF, Burdick GW (2005) J Solid State Chem 178:448
10. Dieke GH (1968) Spectra and energy levels of rare earth ions in crystals. Interscience Publishers, New York
11. Glover PB, Ashton PR, Childs LJ, Rodger A, Kercher M, Williams RM, DeCola L, Pikramenou Z (2003) J Am Chem Soc 125:9918
12. Beeby A, Dickins RS, FitzGerald S, Govenlock LJ, Parker D, Williams JAG, Maupin CL, Riehl JP, Siligardi G (2000) Chem Commun 1183
13. Cammann K, Guibault EA, Hall H, Kellner R, Wolfbesi OS (1996) The cambridge definition of chemical sensors. In: Proceedings of the Cambridge Workshop on Chemical Sensors and Biosensors. Cambridge University Press, New York
14. Thevenot DR, Tóth K, Durst RA, Wilson GS (1999) Pure Appl Chem 71:2333
15. Rodriguez-diaz RC, Fernandez-Romero JM, Aguilar-Cabellos MP, Gomez-Hens A (2006) J Argric Food Chem 54:9670–9676
16. Hernandez-Arteseros JA, Compano R, Prat MD (1998) Analyst 123:2729–2732
17. Arnaud N, Georges J (2001) Analyst 126:694–697
18. Li C, Law GL, Wong WT (2004) Org Lett 6:4841

19. Tremblay MS, Sames D (2006) Chem Commun 4116
20. Senechal-David K, Pope SJA, Quinn S, Faulkner S, Gunnlaugsson T (2006) Inorg Chem 45:10040
21. Leonard JP, Nolan CB, Stomeo F, Gunnlaugsson T (2007) Top Curr Chem 281:1
22. Pope SJA, Laye RH (2006) J Chem Soc Dalton Trans 3108
23. Gunnlaugsson T, McCoy CP, Stomeo F (2004) Tetrahedron Lett 45:8403
24. Gunnlaugsson T, Leonard JP (2003) Chem Commun 2424
25. Parker D (2000) Coord Chem Rev 205:109
26. Gunnlaugsson T, Leonard JP, Senecal K, Harte AJ (2003) J Am Chem Soc 125:12062
27. Parker D, Senanayake PK, Williams JAG (1998) J Chem Soc Perkin Trans 2:2129
28. Bobba G, Bretonniere Y, Frias JC, Parker D (2003) Org Biomol Chem 1:1870
29. Hanaoka K, Kikuchi K, Kojima H, Urano Y, Nagano T (2004) J Am Chem Soc 126:12470
30. Kessler MA (1998) Anal Chim Acta 364:125
31. Dickins RS, Aime S, Batsanov AS, Beeby A, Botta M, Bruce JI, Howard JAK, Love CS, Parker D, Peacock RD, Puschmann H (2002) J Am Chem Soc 124:12697
32. Bretonniere Y, Cann MJ, Parker D, Slater R (2004) Org Biomol Chem 2:1624
33. Parker D, Yu J (2005) Chem Commun 3141
34. Parker D, Dickins RS, Puschmann H, Crossland C, Howard JAK (2002) Chem Rev 102:1977
35. Spangler CM, Spangler C, Schäferling M (2008) Ann NY Acad Sci 1130:138
36. Hirschy LM, Van Geel TF, Winefordner JD, Kelly RN, Schulman SG (1984) Anal Chim Acta 166:207
37. Rakicioglu Y, Perrin JH, Schulman SG (1999) J Pharm Biomed Anal 20:397
38. Wolfbeis OS, Dürkop A, Wu M, Lin Z (2002) Angew Chem 114:4681, Angew Chem Int Ed 41:4495
39. Lei W, Dürkop A, Lin Z, Wu M, Wolfbeis OS (2003) Microchim Acta 143:269
40. Wu M, Lin Z, Schäferling M, Dürkop A, Wolfbeis OS (2005) Anal Biochem 340:66
41. Wu M, Lin Z, Wolfbeis OS (2003) Anal Biochem 320:129
42. Lin Z, Wu M, Wolfbeis OS, Schäferling M (2006) Chem Eur J 12:2730
43. Cano-Raya C, Fernandez Ramos MD, Capitan Vallvey LF, Wolfbeis OS, Schäferling M (2005) Appl Spectrosc 59:1209
44. Schrenkhammer P, Rosnizeck IC, Dürkop A, Wolfbeis OS, Schäferling M (2008) J Biomol Screen 13:9
45. Duerkop A, Turel M, Lobnik A, Wolfbeis OS (2006) Anal Chim Acta 555:292
46. Schäferling M, Wolfbeis OS (2007) Chem Eur J 13:4342
47. Lin Z, Wu M, Schäferling M, Wolfbeis OS (2004) Angew Chem 116:1767, Angew Chem Int Ed 43:1735
48. Bel´tyukova SV, Vityukova EO, Egorova AV (2007) J Appl Spectrosc 74:344
49. Chionqiu J, Li L (2004) Anal Chim Acta 511:11
50. Hou F, Miao Y, Jiang C (2005) Spectrochim Acta A 61:2891
51. Hou F, Wang X, Jiang C (2005) Anal Sci 21:231
52. Li J, Liu J, Zhu X, Peng Q, Jiang C (2005) J Lumin 113:305
53. Peng Q, Hou F, Jiang C (2005) Spectrochim Acta A 65:62
54. Wang T, Wang X, Jiang C (2007) J Clin Lab Anal 21:207
55. Yang T, Qin W (2007) Microchim Acta 157:55
56. Coates J, Gay E, Sammes PG (1997) Dyes Pigm 34:195
57. Li SH, Yuan WT, Xu JG (2004) Anal Biochem 331:235
58. Miao Y, Liu J, Hou F, Jiang C (2006) J Lumin 116:67
59. Wang Y, Liu J, Jiang C (2005) Anal Sci 21:709
60. Tong C, Hu Z, Liu W (2005) J Agric Food Chem 53:6207
61. Wang H, Wang Y, Jiang C (2005) Anal Lett 38:167
62. Spangler C, Spangler CM, Spoerner M, Schäferling M (2009) Anal Bioanal Chem 394:989

63. Spangler CM, Spangler C, Göttle M, Shen Y, Tang W-J, Seifert R, Schäferling M (2008) Anal Biochem 381:86
64. Bian W, Jiang C (2006) Clin Chim Acta 368:144
65. Huo F, Liu J, Jiang C (2005) Anal Lett 38:281
66. Miao Y, Hou F, Jiang C (2005) Anal Sci 21:1207
67. Li J, Ge X, Jiang C (2007) Anal Bioanal Chem 387:2083
68. Bian W, Jiang C (2006) Anal Bioanal Chem 385:861
69. Wei W, Wang H, Jiang C (2006) Spectrochim Acta A 63:241
70. Wang Y, Feng L, Jiang C (2004) Spectrochim Acta A 61:2909
71. Yu F, Li L, Chen F (2008) Anal Chim Acta 610:257
72. Wolfbeis OS (2008) Adv Mater 20:3759
73. Borisov SM, Wolfbeis OS (2008) Chem Rev 108:423
74. Wolfbeis OS (2008) Anal Chem 80:4269
75. Draxler S, Lippitsch ME, Klimant I, Kraus H, Wolfbeis OS (1995) J Phys Chem 99:3162
76. Schäferling M (2005) Luminescence lifetime-based imaging of sensor arrays for high-throughput screening applications. In: Orellana G, Moreno-Bondi MC (eds) Frontiers in chemical sensors, Springer series on chemical sensors and biosensors 3. Springer, Berlin Heidelberg, pp 45–92
77. Pandya S, Yu J, Parker D (2006) Dalton Trans 2757
78. Wolfbeis OS, Weidgans BM (2006) Fiber optic chemical sensors and biosensors: a view back. In: Baldini F, Chester AN, Homola J, Martellucci S (eds) Optical chemical sensors. Springer, Dordrecht, pp 17–44
79. Arain S, John GT, Krause C, Gerlach J, Wolfbeis OS, Klimant I (2006) Sens Actuators B 113:639
80. Mingoarranz FJ, Moreno-Bondi MC, Garcia-Fresnadillo D, de Dios C, Orellana G (1995) Microchim Acta 121:107
81. Amao Y (2003) Microchim Acta 143:1
82. Carraway ER, Demas JN, DeGraff BA, Bacon JR (1991) Anal Chem 63:337
83. Draxler S, Lippitsch ME (1996) Anal Chem 68:735
84. Stich MIJ, Wolfbeis OS (2008) Fluorescence sensing and imaging using pressure-sensitive paints and temperature-sensitive paints. In: Resch-Genger U (ed) Springer series on fluorescence 5: Standardization and quality assurance in fluorescence measurements I. Springer, Berlin Heidelberg, pp 429–461
85. Liu T, Sullivan JP (2005) Pressure and temperature sensitive paints. Springer, Berlin Heidelberg
86. Klein C, Engler RH, Henne U, Sachs WE (2005) Exp Fluids 39:475
87. Hauenstein BL, Picerno R, Brittain HG, Nestor JR (1989) Luminescent oxygen sensor based on a lanthanide complex. US Patent 4861727
88. Amao Y, Okura I, Miyashita T (2000) Chem Lett 1286
89. Lin J (2000) Trends Anal Chem 19:541
90. Lee ST, Gin J, Nampoori VPM, Vallabhan CPG, Unnikrishnan NV, Radhakrishnan P (2001) J Opt A: Pure Appl Opt 3:355
91. McCoy CP, Stomeo F, Plush SE, Gunnlaugsson T (2006) Chem Mater 18:4336
92. Lobnik A, Majcen N, Niederreiter K, Uray G (2001) Sens Actuators B 74:200
93. Leca B, Blum LJ (2000) Analyst 125:789
94. Fähnrich KA, Prawda M, Guilbault GG (2001) Talanta 54:531
95. Votyakova TV, Reynolds IJ (2004) Arch Biochem Biophys 431:138
96. Wolfbeis OS, Schäferling M, Dürkop A (2003) Microchim Acta 143:221
97. Schäferling M, Wu M, Enderlein J, Bauer H, Wolfbeis OS (2003) Appl Spec 57:1386
98. Woods RJ, Scypinski S, Cline Love LJ, Ashworth HA (1984) Anal Chem 56:1395
99. Schäferling M, Wu M, Wolfbeis OS (2004) J Fluoresc 14:561
100. Choi MM, Tse OL (2000) Anal Chim Acta 423:229
101. Posch HE, Wolfbeis OS (1988) Sens Actuators 15:77

102. Bedoya M, Díez MT, Moreno-Bondi MC, Orellana G (2006) Sens Actuators B 113:573
103. McGaughey O, Ros-Lis JV, Guckian A, McEvoy AK, McDonagh C, MacCraith BD (2006) Anal Chim Acta 570:15
104. Chang Q, Murtaza Z, Lakowicz JR, Rao G (1997) Anal Chim Acta 350:97
105. Papkovsky DB, Ponomarev GV, Chernov SF, Ovchinnikov AN, Kurochkin IN (1994) Sens Actuators B 22:57
106. Lis S, Choppin GR (1991) Anal Chem 63:2542
107. Wang L, Li Y (2007) Small 3:1218
108. Petushkov AA, Shilov SM, Pak VN (2006) J Luminescence 116:127
109. Petushkov AA, Shilov SM, Puzik MV, Pak VN (2007) Russian J Phys Chem A 81:612
110. Knall AC, Pein A, Noormofidi N, Stelzer F, Slugovc C (2007) Polymer Preprints 48:575
111. Arakawa T, Akamine M (2003) Sens. Actuators B 91:252–255
112. Barja BC, Aramendia PF (2008) Photochem Photobiol Sci 7:1391
113. Ortega-Algar S, Ramos-Martos N, Molina-Diaz A (2008) Anal Bioanal Chem 391:715
114. Borisov SM, Vasylevska AS, Krause C, Wolfbeis OS (2006) Adv Funct Mater 16:1536
115. Stich MIJ, Borisov SM, Henne U, Schäferling M (2009) Sens Actuators B 139:204
116. Stich MIJ, Nagl S, Wolfbeis OS, Henne U, Schäferling M (2008) Adv Funct Mater 18:1399
117. Zelelow B, Khalil GE, Phelan G, Carlson B, Gouterman M, Callis JB, Dalton LR (2003) Sens Actuators B 96:304
118. Berry MT, May PS, Xu H (1996) J Phys Chem 100:9216
119. Mott NF (1938) Proc R Soc London Ser A 167:384
120. Coyle LM, Gouterman M (1999) Sens Actuators B 61:92
121. Liebsch G, Klimant I, Wolfbeis OS (1999) Adv Mater 11:1296
122. Stich MIJ, Schäferling M, Wolfbeis OS (2009) Adv Mater 21:2216
123. Fu LM, Wen XF, Ai XC, Sun Y, Wu YS, Zhang JP, Wang Y (2005) Angew Chem 117:757, Angew Chem Int Ed 44:747
124. Borisov SM, Wolfbeis OS (2006) Anal Chem 78:5094
125. Borisov SM, Klimant I (2008) J Fluoresc 18:581
126. Feist JP, Heyes AL, Seefeldt S (2003) Meas Sci Technol 14:N17
127. Allison SW, Gillies GT (1997) Rev Sci Instrum 68:2615

Time-Domain Measurements

Ari Kuusisto and Pekka Hänninen

Abstract The unique properties of lanthanides as analytical luminophores may only be efficiently exploited when the measuring instrument has been especially designed for this particular purpose. Nowadays, many commercial plate fluorometers can be equipped for time-resolved measurements. However, not all of them perform to the limits of the luminophores: time-gating alone does not constitute an optimally sensitive measuring device but other factors such as the performance of optical components need also to be considered. The scope of this chapter is to underline these special, practical design requirements and solutions rather than give generic outlines on how fluorometers need to be built.

Keywords Fluorescence lifetime · Fluorometer · Optical filters · Optical materials · Photon counting · Spectrofluorometer · Time-resolved fluorometry · Time-gating

Contents

A. Kuusisto
Solar Simulator Finland Ltd, Keskiläntie 5, 20660 Littoinen, Finland

P. Hänninen (✉)
Department of Cell Biology and Anatomy, University of Turku, Tykistökatu 6A, 20520, Turku, Finland
e-mail: pekka.hanninen@utu.fi

P. Hänninen and H. Härmä (eds.), *Lanthanide Luminescence: Photophysical, Analytical and Biological Aspects*, Springer Ser Fluoresc (2011) 7: 263–278, DOI 10.1007/4243_2011_18,

1 Introduction

High measurement sensitivity, the special analytical advantage of lanthanide-based labels, stems mostly from the long lifetime of the excited state and consequently the sensitive time-resolved (gated) measurement of lanthanide luminescence. The measurement of this delayed luminescence will be in the central role in representing the instrumental aspects in this chapter. Other important contributing factors to the high measurement sensitivity are the large Stokes shift between the excitation and emission and the extremely narrow emission band.

The time-resolved measurements may be split into two categories: time-domain and frequency-domain methods. In the frequency-domain methods, the excitation is carried out with a periodically modulated light, and the signal is measured as amplitude and phase by a "lock-in" detection system. These frequency-domain methods are discussed in Chap. 6.2. In contrast to modulation, a train of light pulses is used for excitation in the time-domain method, and a detection (integration) window is gated with these pulses. The detection window is then optimized in two ways – integration width and delay from the gating pulse. Beyond intensity, the time-domain method offers the possibility of recording the emission decay spectrum of the label. The decay spectrum may be used for gaining additional information about the sample, i.e., the environment of the used label.

Different variations of time-resolved luminescence assays were patented in 1982 by Wieder [1], in 1983 by Soini and Hemmilä [2], and in 1999 by Diamandis [3]. The Finnish company Wallac first commercialized the principle and introduced an assay reader for dissociation-enhanced lanthanide fluorescence immunoassay (DELFIA) in the beginning of the 1980s. The first (1982) DELFIA-based bioassay for diagnostic market was for Rubella antibodies, and it was the first sensitive nonradioisotope immunoassay marking the beginning of a new era [4].

2 Time-Resolved Measurement of Luminescence Intensity

With pulsed excitation, the emission from each luminescent component of the sample decays with its characteristic rate instantly after the excitation pulse. With the combination of time-gated detection and spectral separation, especially the longer lifetime compounds can be well separated from those with shorter luminescence lifetimes. Lanthanides have luminescence lifetimes from a few microseconds to milliseconds, while most of the compounds causing background luminescence in organic samples have luminescence lifetimes under 10 ns [1]. Separating the emission with orders of magnitude different luminescence lifetimes is relatively easy and can be performed with high contrast – in comparison to assays using conventional "fast" fluorochromes, the separation improvement and consequent sensitivity increase can be as much as 4–5 orders of magnitude.

Spectrofluorometers with continuous unmodulated excitation can be used to study the spectral properties of lanthanide emission as of any label, but time-resolving capabilities are required for sensitive detection and for studying the temporal properties of luminescence emission. Often time-resolved spectrofluorometers have pulsed excitation and advanced time-resolved detection system capable of resolving under nanosecond luminescence decay times. These have usually complex design and are optimized for flexibility making them expensive and clumsy for routine use.

The simplest and usually from the luminescence collection point of view, the most efficient fluorometer setup can be built using transmitted light principle. For example, the excitation light is focused to the sample using a lens or a lens group and collected from the opposite side by collector lens (group) to the detector on the same axis. Unfortunately, this setup for normal fluorescence is usually suboptimal due to the high background from the transmitted excitation light that cannot completely be blocked even with the best of optical filters. With long lifetime lanthanide labels the transmitted light principle, however, has proved to be an efficient method for sensitive and yet simple instrumentation – the gating of the measurement window suppresses efficiently the background created by the on-axis illumination source. A design advantage of this system is the separation of illumination and emission optics – i.e., the optics for focusing the illumination source and collecting the (delayed) luminescence may be designed separately and their materials optimized for low background signal.

To reduce the background caused by the transmitted illumination light, the detection may be build in an angle with respect to the excitation. This type of setup retains the advantage of separate optics for excitation and emission but offers a much better signal-to-background ratio for short lifetime luminescence measurements. Detection at an angle, however, becomes impractical when a plate of samples needs to be measured. A practical and currently the most common solution for even time-resolved instrumentation is to illuminate and detect through the same optics from a single side of the sample. This *epi-fluorescence* setup offers a better signal-to-background ratio as compared to transmitted light setup with an additional advantage of enabling measurements of nontranslucent samples. The optics of this arrangement, however, become more complicated since the used components, i.e., lenses, filters and dichroic mirrors need to be optimized for both excitation and emission. For optimized lanthanide measurements, this task is nontrivial since many otherwise useful materials cannot be used due to their intrinsic lanthanide emission band luminescence – especially europium, but also other "rare-earths" can often be found in trace quantities in different glass materials. For this particular reason, multilabel time-resolved instruments often have, e.g., separately optimized dichroic mirrors for each lanthanide label used, although one could expect that a single dichroic would do due to the large Stokes shift of these labels.

In a time-resolved measurement, the excitation is performed by a pulsed light source and the detector is optimally kept inactive during the excitation pulse and activated after a time delay. During the excitation pulse, the emission intensity is at its maximum and, thus, significant amount of emitted photons are lost during the time delay. Since long luminescence lifetime of the label also leads to low momentary emission, the sensitivity of the whole measurement instrument becomes critical. While in a standard fluorescence measurement the sensitivity is usually limited by other factors than the instrument itself, in a time-resolved measurement the instrument design becomes critical for maximal sensitivity. As with all instrumentation, there are tradeoffs in time-resolved instrument design especially between speed, instrumental complexity, and sensitivity of detection: use of shorter lifetime labels can increase the speed of recording at the expense of increased complexity and reduced sensitivity. Some part of this "reduced sensitivity" is due to the fact that most TRF instruments today are based on the use of Xe flash lamps. These by definition are suboptimal, and the more modern semiconductor-based illumination sources may in time prove that practicality of the TRF-technique can be improved using currently nonexistent or less common shorter lifetime labels.

Europium and terbium are the most frequently used lanthanides in bioassays. The labels based on Eu or Tb have normally luminescence lifetimes in the range of 500 μs to 1.5 ms. Samarium- and dysprosium-based labels are also commercially available. They have significantly shorter luminescence lifetimes and lower quantum efficiency, making them less optimal for current instrumentation.

3 Excitation

The spectrum of the excitation light is selected to match the maximal excitation efficiency of the used label with minimum excitation of other luminescent substances in samples, i.e., for maximum signal-to-background ratio. The common light sources usually emit wider spectrum than is required by the excitation and the optimum part of the spectrum has to be selected with a high contrast. Special care must be taken to eliminate an intense light-source emission at the emission wavelength of the used label.

After the excitation pulse, the intensity of luminescence emission decays with the rate that is characteristic to the used label. The photons emitted during the delay time, the time between the pulse and activation of the detector, are lost and, thus, a short delay time is preferred in the detection. The background luminescence decays typically very rapidly (within tens of nanoseconds) and long delay times are not required. However, many used light sources exhibit a switch-off delays and "afterglow" that prevent the use of short delay times. This switch-off time is an important characteristic of a light source and may limit the performance of a time-resolved fluorometer significantly.

3.1 Xenon Flash Lamps

In early days of time-resolved measurements, mechanical choppers and continuous light sources were used. Due to impracticality, these setups were replaced by Xenon flash lamps. Xe flash lamps can have average powers in excess of 1 kW and the energy of a single pulse can be several joules. Although Xe flash lamps emit a continuous spectrum, they also have strong UV lines between 200 and 300 nm (Fig. 1). Since the efficiency of a Xe flash lamp can be nearly 50%, they produce less heat with the same optical emission than alternative lamps used in fluorometry. High-power Xe flash lamps are only seldom used in fluorometers because they cannot normally reach sufficiently high repetition rates. Most commercial fluorometers use compact short arc Xe flash lamps with less than 100 W average power and at least 1 kHz repetition rates.

Xe flash lamps have long operating lifetime compared to other UV-light sources. Hamamatsu Photonics Inc. specifies the operating lifetime for their Xe flash lamp to be 10^9 pulses [5] converting to over 2,700 h of continuous running time with 1 kHz flash rate. Manufacturing technology for Xe flash lamps is mature and the cost is moderate (€500–€1,000).

Xe flash lamps require trigger voltages of several kilovolts, and the discharge current during the light pulse can be in the range of several hundred amperes per microsecond. This causes strong electromagnetic disturbance in measurement circuits and good shielding is required. Fortunately, no detection occurs during the excitation pulse in time-resolved fluorescence measurement, which makes shielding against the disturbances somewhat easier. Another difficulty with Xe lamps is

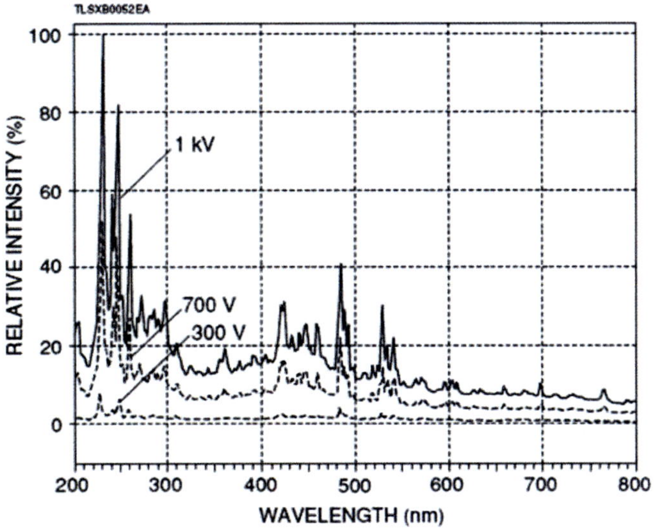

Fig. 1 Typical spectrum of a Xe flash lamp as a function of wavelength and operating voltage (*courtesy*: Hamamatsu Photonics)

the afterglow. Duration of this toward red shifting afterglow is characteristically in the range of several tens of microseconds. This prevents the use of short delay times reducing the efficiency of emission signal collection. The most commonly used lanthanide labels have sufficiently long luminescence lifetime to overcome these difficulties, making Xe flash lamps the most frequently used light source in commercial time-resolved fluorometers today. However, the development of new labels with shorter luminescence lifetimes raises the requirement for faster switch-off-time. New light sources, like lasers and light-emitting diodes (LEDs), are already challenging Xe flash lamps.

3.2 Nitrogen Laser

A nitrogen laser emits light pulses at 337 nm, matching well the excitation spectrum of most lanthanide complexes. A light pulse from a nitrogen laser may contain up to a couple hundred millijoules of energy, and the pulse width can be from a few hundred picoseconds up to 30 ns, but is typically in the range of 6–8 ns. The repetition rate of a typical nitrogen laser can be several kilohertz, provided that adequate gas flow and cooling are supplied. In practice, this is usually the case only with the larger models – the smaller nitrogen lasers have relatively low average power, yielding either significantly lower repetition rate or pulse energy. Most of the commercially available compact nitrogen lasers with adequate pulse energy for luminescence applications have repetition rates less than 100 Hz, which prolongs the measurement time. The greatest advantage of the nitrogen laser is the short pulse and relatively high pulse energy. The drawbacks are the low repetition rate, high cost, and large size.

Nitrogen lasers are specified to operate at least 20–60 million pulses within a service period (e.g., PFI-Optilas, Newport, LTB).This means that after 1,500 h of measuring time (note this model has maximum flash rate of 20 Hz), the laser needs service.

3.3 Diode-Pumped Solid-State Laser

Diode-pumped solid-state (DPSS) lasers have over recent years become available as replacement for different ion and flash lamp pumped lasers. They offer many of the similar characteristics as, e.g., nitrogen laser with the exception of considerable longer operating lifetime. Compared to nitrogen lasers, they are yet higher priced for similar performance but have almost totally replaced the use of, e.g., Ar:Ion lasers. Their primary wavelength lies in the near infrared (NIR) region at around 1 μm wavelength. For visible or UV wavelengths, their operation wavelength is generated by special frequency doubling or tripling optics. They are compact and can produce high powers with the higher efficiency than most other lasers. The most common DPSS lasers emit green (532 nm), blue (473 nm), and yellow (593.5 nm) light, but many other wavelengths are available including wavelengths suitable for lanthanide excitation. Most of them are continuous wave models, but pulsed models with pulse energies as high as 120 J are available.

A special construction of DPSS lasers base on the use of an integrated structure combining the pump laser and a microchip crystal into a small mass-producible package. These microchip lasers have a thin and usually monolithic laser crystal, which is coated with reflective coating on both surfaces. The reflective surfaces form the laser cavity, and the pump laser is often directly coupled to this crystal. These lasers can be very compact since the thickness of the crystal is often 1 mm or less. UV wavelengths, most often 355 and 266 nm, can be achieved by efficient intracavity frequency conversion [8]. Pulse rates of more than 1 kHz are possible, and pulse energies are now approaching microjoule levels. As microchip laser cost is reducing continuously in time, it has become a good choice for time-critical applications where large pulse energy is not needed.

3.4 Diode Lasers

Laser diodes have undergone a considerable development over the past two decades driven by the telecom industry. They are usually compact in size, electrically simple, and can be pulsed with a high repetition rate. They also have a long operating lifetime, and the emitted light can be focused into a small spot or fed into an optical fiber. Unfortunately, still today most diode lasers emit at the NIR or visible wavelength range being unsuitable for the lanthanide excitation. However, diode lasers emitting in the UV range can be found and their power today is typically around 150 mW at 404 nm and 15 mW at 375 nm (Power Technology Inc.). For lower wavelengths, it is even more difficult to build diode lasers, but, e.g., Hamamatsu has been reported to have a diode laser producing 3 mW at 336 nm in the pulsed mode [9].

Diode lasers at UV range are still quite expensive and the power is limited. Thus, they are seldom used in commercial fluorometers. The development of these devices is intense, and more affordable prices with good performance figures can be expected in the near future.

3.5 Fiber Lasers

Similar to DPSS, diode lasers pump most practical fiber lasers. In a fiber laser, a usually rare-earth doped fiber forms at the same time the gain region and (part of) the cavity. Since the fiber can readily be spliced, different types of fibers can be put together to form a monolithic light source. The primary wavelength of these lasers lies in the NIR, but they can readily be frequency converted to UV and visible wavelengths. Over the recent years, the development of these lasers has been tremendous, and it is expected that they become the ultimate choice for many areas since they present a practically optimal light source for many applications. Robustness, high optical quality, compact size, high power, and compatibility for mass production yield high expectations for these lasers.

3.6 *Light-Emitting Diodes*

LEDs have been applied as light sources since late 1980s. The technology is relatively mature and LEDs have been used in many instruments and setups. Eom has reviewed the use of LEDs in various instrumentation showing the wide applicability of LEDs for visual excitation [10]. However, LEDs are presently seldom applied for UV excitation, as high power LEDs are mostly available for visual light emission and only a limited number of UV-LEDs are available. The most powerful UV-LEDs emitting at 365 and 375 nm (e.g., Nichia I-LED) have an output power of 200–400 mW in the continuous wave mode, and in the pulsed mode the power can be doubled if adequate cooling is applied (NichiaNCSU033A specifications). Wavelengths below 365 nm are available, but the output powers are normally under 1 mW. In a special construction, several of these deep-UV-LEDs have been built together to form a LED lamp giving over 100 mW power in 25 ns pulses (Sensor electric technology Inc.).

The expected operating lifetime for standard LEDs in the visible range is over 50,000 h. Unfortunately, this is not currently the case with the UV-LEDs. In the case of Nichia I-LED, the operating lifetime under normal conditions has been specified to be around 500 h in the continuous wave operation mode (NichiaNCSU033A specifications). In addition, the operating lifetime of a LED depends substantially on the operating temperature, and the operating lifetime can be significantly increased at low operating temperature.

In contrast to the large pulse-to-pulse variation of a Xe flash lamp, the energy of a pulsed LED is very stable. This might make the correction circuitry of the excitation intensity unnecessary, which would lower the complexity and cost of the instrumentation even more.

Small footprint is an important advantage for a laboratory instrument since tabletop space in the laboratories and bedside clinics is usually limited and several instruments has to be placed there side by side. The small mechanical size and low cost of LED itself (e.g., Nichia LED NCSU033B emitting more than 300 mW at 365 nm costs presently €100 or less) enable building of smaller and lower cost instruments. This also makes it possible to use multiple LED light sources in one instrument adding flexibility without added size and cost. In LED, the light normally originate from an area of only millimeter or two wide and propagates in a relatively narrow angle compared to the lamps that normally emit in all directions.

Designing good collection optics for a LED is often easier than for a lamp, lowering the cost of the optics. These altogether give LED a significant cost-advantage over the most other UV-light sources.

4 Optical Requirements

The optical setup for the time-resolved luminescence measurement is in principle simple and includes the following components:

- Light source
- Lenses or lens groups for excitation and emission light
- Dichroic mirrors for steering the light beams
- Excitation and emission filters for selectively blocking and passing bands of wavelengths
- Detectors for emission and measuring the light pulse

The lanthanides are usually excited around their absorption maxima of 340 nm, but the light source (Xe flash) emits strongly between 200 and 300 nm. This UV light is absorbed by the normal glass optics, reducing the excitation light intensity at the excitation band and generating a strong luminescence (autofluorescence) background. Even if the optics had quite good transmission in UV (for instance quartz lenses), the remaining impurities in lens material and the lens coatings may cause photoluminescence emission. Often these impurities include some lanthanides and so the characteristics of this background emission can be very similar to the emission from the lanthanide labels making the optical design challenging.

4.1 Optical Material Requirements

Broadband ultraviolet excitation light source causes easily difficulties with the choice of materials. UV is heavily absorbed by most of the optical materials. This naturally lowers transmission of the optical components, and also very easily causes unwanted luminescence emission. Suitable materials are often more expensive and harder to machine than the standard optical glasses. Mostly, the other demands for the optics are not very strict, and a single lens can replace the objective lens. This can be manufactured out of fused silica at relatively low cost. Fused silica is highly transparent at UV wavelengths above 300 nm and is usually quite pure with only minute traces of background creating luminescent materials.

Good UV transmission is critical for all the components along the excitation light path. The components located before the excitation filter are normally causing less luminescence background because of the excitation filter that only allows the excitation wavelength band to pass – however, the excitation filter itself may also emit unwanted luminescence. Furthermore, the components after the excitation filter are at risk since they are illuminated by the excitation source – low background luminescence (autofluorescence) materials should be used.

Simple lenses can be easily made also for UV, but more complicated lenses required for, e.g., higher magnifications of microfluorometry are challenging. In these cases, the limited number of available glass materials makes the design more difficult and may lead to expensive solutions or lower performance. The objective lens in the epi-illumination geometry is an especially critical component for the performance of the instrument. Its design must take into account both the excitation and the emission wavelengths as well as the requirement for low autofluorescence background. Since the lanthanides have large Stokes shifts, color aberration is high and the compensation can be hard, especially if several different excitation and emission bands are used.

Intense ultraviolet light may also cause damage to some components. This can change the performance of a component significantly over time. It is important to know the UV sensitivity of the components along the excitation light path.

4.2 *Reflection and Scattering*

Reflection and scattering occur on the surfaces of every optical component causing loss of light and diverted light rays. This leads to reduced optical efficiency, but can be normally compensated quite easily. Unfortunately, the excitation light reflected or scattered by the surface of a component may hit other luminescent material and cause a significant rise of the background level - such background may be hard to compensate for. Surface reflections can be reduced with good antireflection coating, and lower scattering occurs if the surface quality of the optical component is good. Some scattering is still caused by imperfections inside the component material. Even with the best available antireflection coatings and surface quality, some excitation light is diverted away from the correct path and can cause increase in the background level. To minimize this, it is important to use non-luminescent material in all parts around the excitation light path and use design features that limit the dimensions of the light path to cut and "trap" the stray light.

4.3 *Spectral Selection*

Filters and dichroic mirrors are the most critical components in fluorometers since the signal-to-background ratio is mostly dictated by the performance of these components. Especially luminescence from a dichroic mirror may easily cause high background, but also the luminescence properties of the excitation filter are important. The emission filter is less critical for the luminescence background because it is normally exposed only to minute fractions of the excitation light, but it needs to block efficiently all possible background outside of the detection luminescence band. In case of conventional luminescent labels, the steepness of the band edges of the filter are often very important because the excitation and emission bands of the label often overlap each other. The large Stokes shift of lanthanide emission makes the filter design easier, but the steepness of the transmission band at the emission filter is still important because the emission bands of lanthanides can be very narrow and wide transmission band of the filter only allows unwanted background light to enter the detector without any gain in the label emission intensity (e.g., long-decaying luminescence from the glass components). In time-resolved measurements, the intensity of the luminescence emission is usually significantly lower than the intensity of the luminescence from the conventional non-lanthanide labels excited with the same intensity. Thus, even low background light can cause problems, and the spectral blocking of the filters is even more critical.

5 Gated Detection

Effective time-resolved measurement of luminescence emission would need a way to detect the emission only after the excitation. One way is to keep the light path to the detector closed during the excitation and then open it after the excitation. Another way is to inactivate the detector during the excitation and activate it after the excitation. Yet another method is to use a fast detector that can resolve the detection time for each detected photons and record the decay histogram of the luminescence emission intensity – thus allowing gating of the measurement.

5.1 Analog Measurement Versus Photon Counting

Continuous light intensity can be easily measured by monitoring photocurrent from the detector. However, in the time-resolved measurement instead of measuring continuously signal is measured intermittently, after the excitation pulse and a delay time. In the gated analog integration methods, the total integrated intensity is measured during the exact time window specified by the gate pulse. In addition, usually the signal is averaged over several excitation cycles, with the consequent enhancement of the signal-to-noise ratio. This method is widely called boxcar integration. Boxcar integration can offer very high dynamic range without complicated electronics and is especially well suited to measure moderate to high intensities of emission light. The gated analog integration circuit as any other electric circuit adds some noise to the signal limiting the sensitivity of light detection.

Another and more sensitive way of measuring total integrated light intensity is to count the number of single photons hitting the detector during the measurement window. Single photon counting is especially well suited to record low intensities since detector gain variance is well eliminated by the photon counting principle. A limiting disadvantage is that the detector has to be able to resolve the successive pulses caused by single photons, and thus the technique is not suited for high intensity measurements.

Single photon counting offers better sensitivity than gated analog integration, because analysis of the height of the detected pulses can be used to reject part of the statistical noise as well as noise created by the photon-to-photon varying gain of the detector. Highest and lowest pulses are not normally caused by a single photon and can be effectively discriminated from the counting [11]. Single photon counting is more sensitive but does not allow measurement of as high intensities as gated analog integration. In time-resolved measurement of lanthanides, the light intensity is normally low and the sensitivity is a more important factor than the dynamic range making single photon counting the most practical and common choice.

5.2 *Shutters and Choppers*

Mechanical shutter and chopper are the oldest and simplest way to gate the detection. Mechanical methods also provide nearly 100% contrast between dark and light – this is not usually the case even with the newest optoelectronic shutters. Effective time-resolved measurement would need opening and closing times that are short compared to the luminescence lifetime of the label. The closing and opening times of mechanical shutters can be made quite fast if only a single operation is needed – with repeated opening and closing as with time-resolved measurement, the mechanical shutter as such is usually too slow. Significantly faster repeated opening and closing times can be achieved using a chopper, i.e., a rotating disk with holes/blades. In addition, the beam can be made narrow by focusing and placing the chopper or shutter at the focal point. This way the opening and closing times can be as low as microseconds with reasonable effort. Electro-optical or acusto-optical modulators have usually much faster opening and closing times, but the contrast remains in the range of 1,000:1 or worse. In addition, these modulators are quite expensive. A shutter or a chopper across the light path is in any case an extra component and is not practical in fluorometers for gating purposes, because often it is easier, cheaper, and faster to activate or inactivate the detection instead.

5.3 *Triggered Detection*

A sensitive light detector that can rapidly and repeatedly be activated and inactivated is an ideal detector for time-resolved measurement of luminescence intensity. Photomultiplier tubes (PMTs) in its variants and microchannel plates (MCPs) are detectors that fill these requirements to a large extent – and in practice the only detector types combining fast gating and high sensitivity. Thus, they are very well suited for time-resolved measurement of luminescence intensity. MCPs become especially useful when extremely fast timing is required – with the slow decaying lanthanides MCP is rather an overkill and not practical as the cheaper and simpler PM tubes serve well for the purpose.

6 Detectors

Wide selection of different detectors can be used in fluorometers. Most fluorometers rely on PMTs, but modern detection technology offer other choices like avalanche photodiodes (APDs).

6.1 Photomultiplier Tube

The PMT was invented already in 1936 and several new features enhancing the performance have been developed since then. It has several positive features such as high gain in excess of 10^5, low noise – in the range of a few photoelectrons per second, fast response time, etc. Photomultipliers are used either in current mode with moderate light intensities or in photon counting mode with low light intensities. In photon counting mode, "dead time" of the PM tube and electronics limits the practical count rates to ~10^7 photons/s and below [12].

Building photomultipliers is today routine industry and the most common models are relatively inexpensive, but compared to semiconductor-based products, the building process is still complicated. Thus high-volume low-cost production of PM tubes has not yet been possible. Photomultipliers are also relatively large in size and fragile – with the smallest and sturdiest ones approximately the size of a fingertip (e.g., Hamamatsu, R7400U series). They also require a high-voltage power to function increasing the complexity of a setup.

Sensitivity, low background noise, large detection area, and fast response time are the most prominent features of PMTs. The rather moderate – even low – quantum efficiency has maybe been the biggest drawback. Quantum efficiency can be in special cases approach 50%, but is often less than 10% – especially in the red end of the spectrum for the most common low-cost PM tubes. New photocathode constructions and materials (e.g., GaAs, GaAsP) have increased the red sensitivity and quantum efficiency to close 40% at lanthanide emission range with moderate increase in the background signal making them the detector of choice for time-resolved fluorometers. The further developments such as the μPMT based on MEMS technology (micro-electro-mechanical-systems) [13] carry a bright promise for further success for this old but yet evolving detector type.

6.2 APD and Silicon Photomultiplier

The quantum efficiency of the simple semiconductor detectors, photodiodes, can be up to 95%. Unfortunately, photodiodes as such are not suitable for low-level light detection. APDs (avalanche photodiode) are designed to operate with reverse bias voltage near the breakdown voltage (e.g., voltage between the anode and cathode at which the diode becomes conductive both ways). In photon counting, avalanche diodes operate in Geiger mode: the bias voltage is set slightly above the breakdown threshold voltage. In this condition, a single photon can trigger a strong avalanche current "breakdown," which can easily be detected. Since in the avalanche the diode becomes "short-circuited," special "quenching" electronics are needed to end the avalanche before the diode breaks itself. As photomultipliers also APDs with their quenching electronics have a "dead time" after a "count". The maximum count rate of APDs remains lower than the maximum count rate of photomultipliers. Another disadvantage over PMTs is the internal noise of the diodes, which is orders of magnitude larger than that of the photomultipliers and is dependent on the size of the diode's photoactive area. To reduce the internal noise, APDs are manufactured typically with only few tens up to 200 μm diameter photoactive area, making the use of the APDs challenging in terms of instrument design and limiting the advantage of high quantum yield to microfluorometry. Also the complicated quenching electronics and low quantum yield of manufacturing processes have kept the price of APDs relatively high.

Silicon photomultiplier was developed in Russia in the Moscow Engineering Physics Institute. It consists of a large array of APDs, which function in Geiger mode reaching often gain of more than 10^6. This is achieved with bias voltages below 100 V. These APDs are all built on common substrate. Each APD functions digitally giving a strong fixed pulse of current when "triggered" by a photon or even by several simultaneous photons. The output signal of the whole array is the sum of the current pulses from the APDs hit by photons, and so the number of photons hit onto the silicon photomultiplier is revealed by the height of this output current pulse. The number of simultaneous incident photons can be counted as long as these photons hit separate APDs [14]. Use of APDs with small active area minimizes the risk that they are hit by several photons simultaneously, but the use of a large array of APDs still forms a relatively large sensitive area where the photons can be counted. This way the small active area of a single APD can be combined with the bigger light sensitive area of the whole array. Compact size, mechanical robustness, high photon detection efficiency (over 60%), and insensitivity to magnetic fields are some other advantages compared to photomultiplier [15].

Silicon photomultiplier has nearly the same specifications and sometimes even better than the photomultiplier. It is a good candidate to replace the photomultiplier in many applications.

Although high sensitivity time-resolved instruments have been built for several decades, now the new developments with regard to light sources and detectors carry the promise that the high sensitivity time-resolved measurements may be brought into areas where hand-held low-cost readers are a must – an intriguing concept that can yet broaden the use of lanthanide luminescence into areas that until now have been considered out-of-reach.

References

1. Wieder I (1982) Luminescent antibody composition for immunofluorometric assay. US Patent 4,341,957
2. Soini E, Hemmilä I (1983) Luminescence spectroscopy assay means with luminescent chelate of europium. US Patent 4,374,120A
3. Diamandis EP (1999) Europium and terbium chelators for time-resolved fluorometric assay. EP0635006B1
4. Meurman OH, Hemmilä IA, Lövgren TN-E et al (1982) Time-resolved fluoroimmunoassay: a new test for rubella antibodies. J Clin Microbiol 16:920–925
5. Hamamatsu (2005) Technical information, xenon flash lamps. http://sales.hamamatsu.com/assets/applications/ETD/Xe-F_TLSX9001E05.pdf. Accessed 14 June 2009
6. Jury P (2007) Microchip lasers: beyond the visible. Laser Focus World, 1 April 2007
7. Laser diodes stretch further into the UV. Optics.org News. http://optics.org/article/37228. Accessed 22 March 2011
8. Eom IY (2005) Application of light emitting diodes as solid state light sources in analytical chemistry. Doctoral Dissertation, Texas Tech University
9. Hamamatsu (2005) Photon counting. Using photomultiplier tubes. http://sales.hamamatsu.com/assets/applications/ETD/PhotonCounting_TPHO9001E04.pdf. Accessed 14 June 2009
10. Sun X, Krainak M, Gentry B (2007) New developments in detector technology at wavelengths >1 micron. http://space.hsv.usra.edu/LWG/Jul07/Papers.jul07/Sun.jul07.pdf. Accessed 14 June 2007
11. Image Sensors World (2010) Hamamatsu develops MEMS-based photomultiplier tube technology. http://image-sensors-world.blogspot.com/2010/10/hamamatsu-develops-mems-based.html. Accessed 22 March 2011
12. Buzhana P, Dolgoshein B, Ilyin A et al (2001) ICFA instrumentation bulletin. http://www.slac.stanford.edu/pubs/icfa/fall01/paper3/paper3.pdf. Accessed 24 June 2009
13. Golovina V, Savelievb V (2004) Novel type of avalanche photodetector with Geiger mode operation. Nucl Instrum Meth Phys Res A 518:560–564

Frequency-Domain Measurements

Jouko Kankare and Iko Hyppänen

Abstract In the frequency-domain measurements of luminescence, the excitation intensity is modulated sinusoidally and the emission detected using a phase-sensitive amplifier. The present availability of conveniently modulatable light sources, such as light-emitting diodes and diode lasers, and relatively inexpensive lock-in amplifiers makes this technique well suited for the determination of lanthanide luminescence. The mathematical theory of luminescent lanthanide systems involves the application of matrix and complex analysis to the set of linear differential equations of the rate processes. The general solution is derived for the temporal populations of the excited species in the presence of an arbitrary functional form of excitation. The sinusoidal excitation and dual-phase lock-in detection of the emission provide a signal which can be expressed as a complex quantity with real and imaginary parts. It is shown that the imaginary part of the signal, i.e., the out-of-phase signal of the lock-in amplifier, is less prone to the interference from organic prompt fluorophores and external sources. The Kramers–Kronig relation can be used for checking the mutual compatibility of the real and imaginary parts of the signal. Two examples are given for the instrumentation and data treatment. The first example is the resonance energy transfer from a europium chelate to an organic acceptor held at a constant distance from the donor by oligonucleotide hybridization. The second example deals with the upconversion material, a mixture of lanthanide compounds. In both cases, the signal-to-noise ratio is excellent, allowing even the estimation of the continuous lifetime distribution.

Keywords Kramers–Kronig relation · Lifetime distribution · Light-emitting diode · Lock-in amplifier · Luminescence lifetime · Periodic modulation · Phase-sensitive detection · Rate matrix

---

J. Kankare (✉) and I. Hyppänen
Department of Chemistry, Laboratory of Materials Chemistry and Chemical Analysis, University of Turku, 20014 Turku, Finland
e-mail: kankare@utu.fi

P. Hänninen and H. Härmä (eds.), *Lanthanide Luminescence: Photophysical, Analytical and Biological Aspects*, Springer Ser Fluoresc (2011) 7: 279–312, DOI 10.1007/4243_2010_7,

Contents

Abbreviations

AC	Alternating current
AO	Acousto-optic
DC	Direct current
EO	Electro-optic
FD	Frequency-domain
GPIB	General purpose interface bus
K–K	Kramers–Kronig (relation)
LED	Light-emitting diode
MEM	Maximum entropy method
PM	Photomultiplier
RET	Resonance energy transfer
R–S	Riemann–Stieltjes integral
TD	Time-domain

1 Introduction

The methods used for determining the lifetimes and emission intensities can be divided into time-domain and frequency-domain methods. In the time-domain methods, the excitation of the system is performed by a light pulse or pulse sequence and subsequent emission is followed as a function of time. In the frequency-domain methods, the intensity of the excitation light follows a time-periodic function, and the response of the periodic emission intensity is measured as a function of frequency using phase-sensitive detection. The important difference between the methods is that in the frequency-domain methods the signal is measured simultaneously with the excitation, whereas in the time-domain methods the sample receives no excitation light during the reading. As a consequence, photon shot noise, which is proportional to the square-root of intensity, may give some contribution to the background noise in the frequency-domain methods but not in the time-domain methods. This means that at least in the case of very weak analyte luminescence in the presence of the strong background the time-domain methods are better in the quantitative determination of the luminophore. However, the frequency-domain methods have proved to be better for evaluating the time constants of multilevel systems. The FD methods are also significantly faster than the TD methods when used in the luminescence imaging.

The determination of lifetime of luminescence is one of the fundamental measurements when characterizing a luminescent system. The changes in lifetimes when changing the conditions during the measurements may give information on the mechanism of luminescence. In the case of systems possessing only a single lifetime the measurements are straightforward and the same can be often said even on the subsequent inference. There are two kinds of luminescent systems where several lifetimes can be expected. One system, presumably more common, is a mixture of fluorescent compounds. The other system is a single compound or a strongly interacting pair of compounds with a number of electronic excited energy levels, capable of reasonably slow internal conversion, intersystem crossing or energy transfer. A good example, presently of increasing practical importance, is lanthanide compounds in various forms. For instance, nanoparticles containing europium chelates or purely inorganic lanthanide compounds are used as labeling agents in various bioassays. These materials are often used together with organic fluorophors, undergoing Förster resonance energy transfer. A novel class of luminescent nanoparticles of growing importance is the upconverting nanoparticles. These materials, when excited with near infrared radiation, accept two (or more) photons and emit in the visible range. The luminescence of all these materials exhibits two or more lifetimes which may be at different spectral regions, and for resolving the mechanism of luminescence and perhaps for trying to optimize the spectral output, it is essential to estimate the lifetimes.

Although the measurement of a single lifetime is trivial, severe problems appear if the system shows several lifetimes. In the worst case, the system does not even have discrete time constants but a continuous distribution of lifetimes. The difficulties do not emerge as the lack of a numerical solution but vice versa, as a continuum of solutions fitting well to the experimental results. This is due to the notoriously ill-posed character of the equations used for estimating the time constants from the experimental data. An obvious way to overcome these difficulties is to increase the precision of the measurements. Another, actually more successful, way to obtain plausible values for lifetimes and their distribution is to impose suitable boundary conditions to the numerical solution. These conditions may be purely mathematical, based on the assumed regularity of the distribution function, or alternatively based on the physical feasibility of the solution. The most common assumption is that the lifetimes of the system are discrete numbers, which means that the cumulative distribution function of lifetimes is a step function. In a condensed heterogeneous phase, this is a doubtful assumption but yet commonly adopted.

The conventional instrumentation for the luminescence measurements has been adequately described in numerous textbooks on the instrumental analysis and not discussed here. The special properties of lanthanide luminophores impose different requirements for the instrumentation and data treatment, and these issues are the subject of this section. The method is not yet in common use, and for this reason, the theoretical part and examples are mainly based on the unpublished, rather recent results from the authors' laboratory. Proper understanding of the frequency-domain methods requires mathematics and the reader interested only in the use of the methods may skip the mathematical derivations.

2 Theory

2.1 Rate Equations

Customarily the energy levels of molecules and the transitions between them are illustrated by a Jablonski diagram. However, the Jablonski diagram is not used here, but instead, the graph-theoretic notation *directed graph* or shortly *digraph* where the vertices correspond to the energy states and the edges correspond to the transitions between them is used. The reason is that the location of energy levels is not of interest nor spin states, and this is a more illustrative way to present the system without unnecessary decorations, displaying the transitions in an easily grasped pattern. The electronic energy levels are treated as separate species which have concentrations or population densities. This is related to the concepts of *compartmental analysis* or global analysis of fluorescence decay [1–3].

Our goal is to evaluate the pertinent equations for the general luminescence systems, but in order to clarify the treatment we first take two examples from the three-state systems. Actually the only difference in these systems is in the different radiative transitions (Fig. 1).

In these hypothetical systems, there is a ground state $A^{(0)}$ and two excited states $A^{(1)}$ and $A^{(2)}$. These are considered phenomenologically as separate entities or *compartments* without any spectroscopic identities as singlets or triplets. Each entity has its time-dependent *population density* or *concentration*, which is denoted by square brackets in (1). The system is supposed to be closed, i.e.,

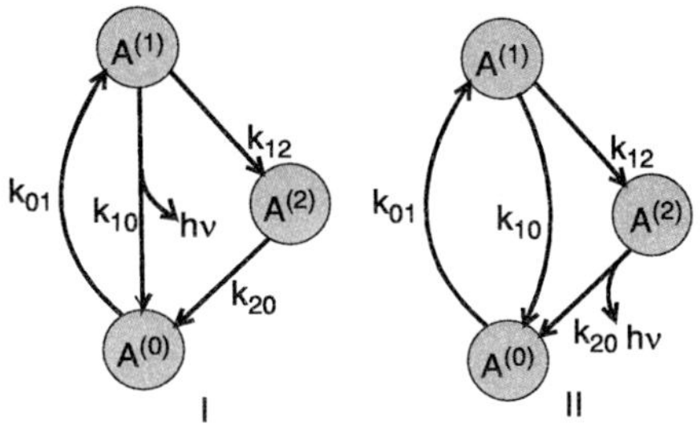

Fig. 1 Digraphs of hypothetical three-level systems with different radiative transitions

there is no exchange of entities with the environment and no complications due to diffusion. This means that the sum of concentrations is constant:

$$[\mathrm{A}^{(0)}] + [\mathrm{A}^{(1)}] + [\mathrm{A}^{(2)}] = c_0 = \text{constant} \tag{1}$$

Excitation $\mathrm{A}^{(0)} \rightarrow \mathrm{A}^{(1)}$ is assumed to occur either directly by absorbing a photon or indirectly by resonance energy transfer from a nearby excited atom or molecule.

The rate constants contain contributions from the radiative and nonradiative transitions. The stimulated emission is neglected. One example of systems I and II is a donor–acceptor complex where the donor is a lanthanide ion and the acceptor an organic fluorophore. An experimental example of this energy transfer system will be given at the end of this chapter.

The temporal behavior of systems I and II is governed by a set of linear differential equations:

$$\begin{aligned}
\frac{d\left[\mathrm{A}^{(0)}\right]}{dt} &= -k_{01}\left[\mathrm{A}^{(0)}\right] + k_{10}\left[\mathrm{A}^{(1)}\right] + k_{20}\left[\mathrm{A}^{(2)}\right];\\
\frac{d\left[\mathrm{A}^{(1)}\right]}{dt} &= k_{01}\left[\mathrm{A}^{(0)}\right] - (k_{10} + k_{12})\left[\mathrm{A}^{(1)}\right];\\
\frac{d\left[\mathrm{A}^{(2)}\right]}{dt} &= k_{12}\left[\mathrm{A}^{(1)}\right] - k_{20}\left[\mathrm{A}^{(2)}\right]
\end{aligned} \tag{2}$$

The emission signal I is proportional to the rate of radiative transitions and consequently to the concentration of the species from which the transition occurs. Systems I and II are in this sense different:

$$\begin{aligned}
&\text{System I}: I_{1\rightarrow 0} \propto k_{10}^{*}\left[\mathrm{A}^{(1)}\right];\\
&\text{System II}: I_{2\rightarrow 0} \propto k_{20}^{*}\left[\mathrm{A}^{(2)}\right]
\end{aligned} \tag{3}$$

The starred rate constants are constants for the radiative transitions and they may or may not be the same as the constants without a star. The main problem now is to derive the mathematical expressions for these concentrations when the time-dependence and intensity of the excitation are known.

When studying the general complicated systems it is convenient to use matrix formalism. The column matrices (vectors) will be denoted by bold lowercase letters and matrices by bold uppercase letters. The identity matrix is denoted by **I**. For example, the linear system (2) becomes then:

$$\frac{d\mathbf{a}}{dt} = \mathbf{A}\mathbf{a} \tag{4}$$

where **A** is the *rate matrix* of the system:

$$\mathbf{A} = \begin{pmatrix} -k_{01} & k_{10} & k_{20} \\ k_{01} & -(k_{10}+k_{12}) & 0 \\ 0 & k_{12} & -k_{20} \end{pmatrix} \tag{5}$$

and **a** is the vector of concentrations:

$$\mathbf{a} = \begin{pmatrix} [\mathrm{A}^{(0)}] \\ [\mathrm{A}^{(1)}] \\ [\mathrm{A}^{(2)}] \end{pmatrix} \tag{6}$$

The excitation is taken into account by assuming that the corresponding rate constant is a function of time. In the present example, the excitation occurs in the transition $\mathrm{A}^{(0)} \rightarrow \mathrm{A}^{(1)}$ and consequently k_{01} is time-dependent. The rate constant can be separated into an *amplitude factor* and a suitably normalized *modulation function* $f(t)$, for example:

$$k_{01} = \bar{k}_{01} f(t) \tag{7}$$

The amplitude factor is proportional to the excitation intensity and the absorption coefficient of the transition. Then the matrix **A** can be separated into time-independent and time-dependent parts, e.g.:

$$\mathbf{A} = \mathbf{A}_0 + f(t)\mathbf{A}_1 \tag{8}$$

In our example the *reduced rate matrix* $\mathbf{A}_0$ and *excitation matrix* $\mathbf{A}_1$ are as follows:

$$\mathbf{A}_0 = \begin{pmatrix} 0 & k_{10} & k_{20} \\ 0 & -(k_{10}+k_{12}) & 0 \\ 0 & k_{12} & -k_{20} \end{pmatrix}; \quad \mathbf{A}_1 = \bar{k}_{01} \begin{pmatrix} -1 & 0 & 0 \\ 1 & 0 & 0 \\ 0 & 0 & 0 \end{pmatrix} \tag{9}$$

Equation (2) becomes then:

$$\frac{d\mathbf{a}}{dt} = \mathbf{A}_0\mathbf{a} + f(t)\mathbf{A}_1\mathbf{a} \tag{10}$$

If $f(t)$ is constant when $t > 0$, (10) is easily solved. Then either $f(t) = u(t)$ or $f(t) = 1 - u(t)$, where $u(t)$ is the unit step function.[1] The former case corresponds to the constant excitation which is switched on at $t = 0$, whereas the latter case corresponds to the case where the excitation has been on until it is switched off at $t = 0$. The former case needs some further information on matrix **A**, but in the latter case (10) is easily solved:

$$\mathbf{a} = \exp(t\mathbf{A}_0)\mathbf{a}_0 \tag{11}$$

Here, $\mathbf{a}_0$ represents the initial values of concentrations and the matrix exponential is defined as a series:

$$\exp(t\mathbf{A}_0) = \mathbf{I} + t\mathbf{A}_0 + \frac{t^2}{2!}\mathbf{A}_0^2 + \frac{t^3}{3!}\mathbf{A}_0^3 + \dots \tag{12}$$

or by using the Sylvester formula and Frobenius covariants (vide infra).

The approximate solution of the full equation (10) can be shown to be a convolution integral:

$$\mathbf{a} - \mathbf{a}_{\mathrm{eq}} = \int_0^t f(\tau)\exp[(t-\tau)\mathbf{A}_0]d\tau\mathbf{A}_1\mathbf{a}_{\mathrm{eq}} \tag{13}$$

Here, $\mathbf{a}_{\mathrm{eq}}$ is the equilibrium state of the system. In our example, it is:

$$\mathbf{a}_{\mathrm{eq}} = \begin{pmatrix} c_0 \\ 0 \\ 0 \end{pmatrix} \tag{14}$$

In certain cases, it is also necessary to take into account the concentration of the lowest excited state.

Equation (13) is a general solution for the first-order perturbation in the case of a linearized reaction system. It includes a matrix exponential which is more conveniently computed by using a Sylvester formula. The Sylvester formula shows that an arbitrary analytical matrix function φ can be expanded to a sum:

$$\varphi(\mathbf{A}) = \sum_{i=1}^{N}\varphi(\lambda_i)\mathbf{Z}^{(i)} \tag{15}$$

where λ_is are the eigenvalues of an $N \times N$ matrix $\mathbf{A}_0$, and matrices $\mathbf{Z}^{(\mathrm{k})}$ are the Frobenius covariants for which distinct eigenvalues can be written as follows:

[1]The unit step function is defined as $u(t) = 0, t < 0$; $u(t) = 1, t > 0$; $u(0) = ½$.

$$\mathbf{Z}^{(k)} = \prod_{n \neq k}^{N} \frac{\mathbf{A}_0 - \lambda_n \mathbf{I}}{\lambda_k - \lambda_n} \tag{16}$$

It can be shown that the eigenvalues of matrix $\mathbf{A_0}$ are negative real numbers except the eigenvalue zero which corresponds to the equilibrium. The connection between the eigenvalues λ_i and lifetimes τ_i is as follows:

$$\tau_i = -\frac{1}{\lambda_i}\,;\ i = 1, 2, \ldots N-1;\ \lambda_N = 0 \tag{17}$$

When applying the Sylvester formula and the summation is extended over all non-zero eigenvalues, (13) becomes:

$$\mathbf{a} - \mathbf{a}_{\text{eq}} = \sum_{k=1}^{N-1} \int_0^t f(\tau) \exp[\lambda_k (t - \tau)] d\tau \, \mathbf{Z}^{(k)} \mathbf{A}_1 \mathbf{a}_{\text{eq}} \tag{18}$$

This can be considered as a general approximate solution to the set of differential equations (10). One should be aware of the approximations made in this expression when applied to the rate equations. First, the state of equilibrium $\mathbf{a}_{\text{eq}}$ is considered to be unperturbed, i.e., the saturation has not been taken into account. This is a good approximation as long as the modulation level is kept reasonably low. The second approximation is purely mathematical. The solution described by (18) is actually iterative, and we take only the first iteration step as the final solution. Again, this approximation is reasonably good if the perturbation induced by excitation is small enough. The most serious limitations in the use of (18) are based on the initial assumptions. The rate equation (4) is valid if the system is homogeneous at least in the luminescent region and the diffusion of luminescent species can be neglected. This is not always the case with solid samples, especially when the sample consists of nanosized particles where the surface/volume ratio is large.

2.2 *Periodic Excitation*

It is assumed now that the excitation is done using a sine-wave modulated light source. In this case, the DC *bias* excitation and the periodic signal are summed with the relative amplitude or modulation degree being μ. The emission from the luminescent sample has the same sinusoidal form as the excitation, but due to the rate processes the sine-wave has a time lag that appears as a phase shift θ (Fig. 2). In order to facilitate the phase calculations, the sine or cosine functions are customarily replaced by the complex exponential function.

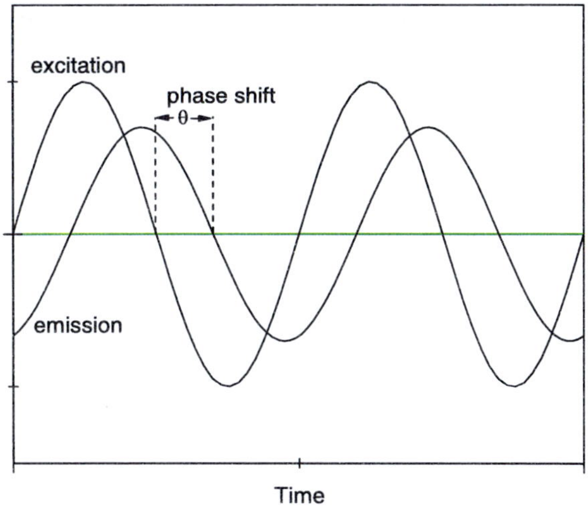

Fig. 2 Sinusoidal excitation and emission

Then the modulation function can be written as follows:

$$f(t) = 1 + \mu \exp(\mathrm{i}\omega t)\,;\; 0 < \mu < 1\,;\; \mathrm{i}^2 = -1 \tag{19}$$

It is assumed implicitly that the DC excitation is not strong enough to populate significantly the first excited state, i.e., $\mathrm{A}^{(1)}$.

The convolution integral then becomes:

$$\int_0^t f(\tau)\exp[\lambda_k(t-\tau)]d\tau = \lambda_k^{-1}\left(e^{\lambda_k t} - 1\right) + \mu\frac{e^{\mathrm{i}\omega t} - e^{\lambda_k t}}{\mathrm{i}\omega - \lambda_k} \tag{20}$$

With the progressing experiment, the exponential terms with $\lambda_k t$ in the exponent become negligible. Substitution into (18) gives:

$$\mathbf{a} = \mathbf{a}_{\mathrm{eq}} + \mu\exp(\mathrm{i}\omega t)\sum_{k=1}^{N-1}\frac{1}{\mathrm{i}\omega - \lambda_k}\mathbf{Z}^{(k)}\mathbf{A}_1\mathbf{a}_{\mathrm{eq}} - \sum_{k=1}^{N-1}\lambda_k^{-1}\mathbf{Z}^{(k)}\mathbf{A}_1\mathbf{a}_{\mathrm{eq}} \tag{21}$$

It can be shown that $\mathbf{Z}^{(N)}\mathbf{A}_1 = 0$ and using (15), (21) can be written as follows:

$$\mathbf{a} = \mathbf{a}_{\mathrm{eq}} - \sum_{k=1}^{N-1}\lambda_k^{-1}\mathbf{Z}^{(k)}\mathbf{A}_1\mathbf{a}_{\mathrm{eq}} + \mu\exp(\mathrm{i}\omega t)(\mathrm{i}\omega\mathbf{I} - \mathbf{A}_0)^{-1}\mathbf{A}_1\mathbf{a}_{\mathrm{eq}} \tag{22}$$

Actually, only the amplitudes $\mathbf{a}_{\text{per}}$ of the periodic components of the signal are measured:

$$\mathbf{a}_{\text{per}} = \mu \sum_{k=1}^{N-1} \frac{1}{\mathrm{i}\omega - \lambda_k} \mathbf{Z}^{(k)} \mathbf{A}_1 \mathbf{a}_{\text{eq}} = \mu (\mathrm{i}\omega \mathbf{I} - \mathbf{A}_0)^{-1} \mathbf{A}_1 \mathbf{a}_{\text{eq}} \tag{23}$$

A considerable advantage of (23) is that it does not include the Frobenius covariants. The derivation of expressions for the covariants requires eigenvalues and in the more complicated cases these are computed by numerical iterative methods. The inverse matrices (or solutions of linear equations) of (23) are calculated using finitely converging methods such as Gaussian elimination or LU decomposition, allowing symbolic solutions using proper computer programs for symbolic mathematics. The inherent assumption in the expansion (23) is the distinct eigenvalues of the rate matrix. In the case of multiple eigenvalues, the terms of the form $(\mathrm{i}\omega - \lambda)^{-\mathrm{n}}$ should be added to the expansion.

It is important to note that the functions (23) are complex having a discrete number of poles in the upper complex half-plane. This fact has implications in the form of the Kramers–Kronig relations between the real and imaginary parts as will be seen in Sect. 5.

3 Instrumentation

3.1 Excitation

In the frequency-domain methods, the excitation signal is a periodic function, commonly sine-wave, and the emission signal is followed by recording its periodic component. In some cases the excitation signal is a sum of sine-waves, and the emission signal is analyzed by the Fourier transform technique or its simplified analogs. The frequency-domain methods are very well known and commercial instrumentation has been available for prompt fluorescence of organic compounds such as fluorescein for a long time, but for the slowly decaying lanthanide luminescence the commercial instrumentation has only recently become available. This is in contrast to the time-domain methods where commercial instrumentation also has already been available for a long time for the long-lived luminescence. One reason for this is the thousand to million times longer time-scale which sets different requirements not only for the excitation and detection electronics but also for the data treatment. However, the most significant reason is probably the weak commercial demand for this kind of instrument, which does not show easily defined advantages over the pulse-excitation instruments commonly used, e.g., in the clinical chemistry. From the instrumental point of view, the situation is changing due to the inexpensive solid-state light sources which can be easily modulated. Moreover, the advantages of the frequency-domain methods have been gradually realized in settling the mechanisms of luminescence and in the applications to the luminescence imaging.

The basic requirement for the light source used in the frequency-domain instrument is the possibility to modulate the light intensity in the frequency range required by the luminophore. In the organic fluorophores, this range is from tens to hundreds of megahertz, whereas for the most interesting lanthanide luminophores it is from 10 to 100,000 Hz. In the case of the prompt organic fluorophores, the modulation is done by either electro-optic (EO) or acousto-optic (AO) modulators. The problems with these instruments has been adequately discussed in the textbooks, e.g., by Lakowicz [4]. Modulatable solid-state light sources, i.e., laser diodes and light-emitting diodes (LEDs), are being used more and more. The obvious reason is the relatively low cost of the sources and accompanying electronics. The limited spectral range has been a drawback, but presently LEDs are available in the UV range down to 255 nm and in the near infrared region at least up to 1,550 nm. More powerful but still reasonably priced laser diodes are available from the visible to NIR region.

Europium and terbium complexes are being used in many applications, such as cytometry, markers for bioassays, etc. In these cases the spectral range of excitation is in the near UV region. In the commercial time-domain instruments, the xenon flash lamps and nitrogen laser are commonly used. In the frequency-domain instruments, a He-Cd laser at 325 nm with an EO modulator can be used if a high intensity excitation is needed, but in practice for many applications a UV-LED is a good choice. The main drawback of the presently available UV-LEDs is their low output power. Another drawback is the spurious visible emission which, however, can be suppressed by a suitable shortpass filter, such as Schott DUG11 glass filter. The use and problems of UV-LEDs in connection with the time-resolved lanthanide luminescence has been reported by Jin et al. [5].

3.2 Detection

In the practical applications, the concentration of luminescent species is usually very low and consequently a photomultiplier is commonly used as a detector. If the emission can be efficiently focused into a small entrance, an avalanche diode can also be used. A reasonably fast detector is anyway needed also in the frequency-domain method due to the phase error induced by the detector. Still the requirements are not as strict as in the case of prompt organic fluorophores. In the case of the most commonly used lanthanides, the upper limit of modulation is around 100 kHz, but still the reasonably error-free phase detection requires ca. 1 MHz bandwidth from the detector and accompanying current-to-voltage preamplifier.

The accurate determination of the phase change between the excitation and emission waveforms is the central role in the frequency-domain measurements of luminescence. In the case of prompt fluorophores, the method of choice seems to be *heterodyne detection* – also called *cross-correlation* – and subsequent lock-in amplification. In the heterodyne detection, the gain of the photomultiplier is modulated by the frequency $\omega \pm \Delta\omega$ where ω is the frequency of the excitation modulation. As the consequence the PM signal contains a low-frequency component of $\Delta\omega$, which still has the same phase information which is now more easily and accurately retrieved. For the long-lived luminophores this is more challenging because the frequencies corresponding to the lifetimes are already in the low-frequency domain. Instead of the heterodyne detection, the signal from the PM, after passing through a current-to-voltage amplifier, can be directly taken into a lock-in amplifier, a method sometimes called *homodyne* detection.

A *lock-in amplifier* is a routine tool in the physicist's laboratory but not that common in the chemist's laboratory. The lock-in amplifier or sometimes called *phase-sensitive amplifier* is a device which with a very narrow bandwidth amplifies a signal that has the same frequency as the reference signal. The reference signal is generated by the lock-in amplifier or some external device. The input signal and reference signal must have the same frequency and constant phase shift θ. The main application of this instrument is the measurement of a weak AC signal buried in noise. The principles of this instrument have been described in numerous sources, and hence are not discussed in detail here. Only a short description is given in order to clarify the nomenclature and signal processing.

The basic parts of a lock-in amplifier are a multiplier and low-pass filter. The signal to be analyzed is multiplied with a reference signal, and the result of this electronic multiplication is subjected to low-pass filtering. Let the signal to be analyzed be:

$$S_{\mathrm{in}}(t) = A\sin(\omega t + \theta) \tag{24}$$

The reference signal is either:

$$S_{\mathrm{ref}}(t) = B\sin\omega t \quad \text{or} \quad C_{\mathrm{ref}}(t) = B\cos\omega t \tag{25}$$

and using trigonometric formulae, we get:

$$\begin{aligned} S_{\mathrm{in}}(t)S_{\mathrm{ref}}(t) &= AB\sin(\omega t + \theta)\sin\omega t \\ &= {}^1\!/_2 AB\cos(\theta) - {}^1\!/_2 AB\cos(2\omega t + \theta) \\ S_{\mathrm{in}}(t)C_{\mathrm{ref}}(t) &= AB\sin(\omega t + \theta)\cos\omega t \\ &= {}^1\!/_2 AB\sin(\theta) + {}^1\!/_2 AB\sin(2\omega t + \theta) \end{aligned} \tag{26}$$

Depending on the cutoff frequency, the low-pass filter more or less suppresses the time-periodic part of the signal. In a dual-phase lock-in amplifier, both $S_{\mathrm{ref}}(t)$ and $C_{\mathrm{ref}}(t)$ are used and the output signal is scaled to give the rms value of the amplitude, i.e., the two (digital) output signals of the lock-in amplifier are:

$$S_x = \frac{1}{\sqrt{2}}A\cos\theta; \quad S_y = \frac{1}{\sqrt{2}}A\sin\theta \tag{27}$$

Signal S_x is called *in-phase output* and S_y is *out-of-phase* or *quadrature output*. The phase shift between the input signal and reference signal is obtained from:

$$\tan\theta = \frac{S_y}{S_x} \tag{28}$$

In the mathematical data treatment, it is convenient to use a *complex amplitude* and Euler equations to represent the in-phase and out-of-phase signals as real and imaginary parts of a complex signal:

$$S = S_x + \mathrm{i}S_y = \frac{1}{\sqrt{2}}A(\cos\theta + \mathrm{i}\sin\theta) = \frac{1}{\sqrt{2}}Ae^{\mathrm{i}\theta} \tag{29}$$

The input signal is then written as:

$$S_{\mathrm{in}}(t) = Ae^{\mathrm{i}(\omega t+\theta)} \tag{30}$$

This notation simplifies considerably the mathematical expressions, and there is no need to use complicated trigonometric formulae when dealing with the changes in the phase and amplitude of AC signals.

In deriving (26), noise and possible DC bias are not included in the input signal. White noise contains all frequencies and a certain portion of noise seeps through the low-pass filter depending on the cutoff frequency. The lower the cutoff frequency (or equivalently higher time constant $\tau = 1/\omega_{\mathrm{cutoff}}$), the less noise there is in the output. The time constant cannot be increased too much because that might lengthen excessively the total measurement time.

Using a dual-phase lock-in amplifier, the in-phase and quadrature signals become:

$$\begin{aligned} S_x^{L\to M} &= Gk_{\mathrm{LM}}^{*}\mathrm{Re}\left[\mathbf{a}_{\mathrm{per}}\right]_{L+1} \\ S_y^{L\to M} &= Gk_{\mathrm{LM}}^{*}\mathrm{Im}\left[\mathbf{a}_{\mathrm{per}}\right]_{L+1} \end{aligned} \tag{31}$$

where G is the gain factor of the system.

As an example we take the systems I and II of Fig. 1. Direct substitution of $\mathbf{A}_0$, $\mathbf{A}_1$ and $\mathbf{a}_{\mathrm{eq}}$ into (23) gives:

$$\begin{aligned} \left(\mathbf{a}_{\mathrm{per}}\right)_2 &= \frac{\mu\bar{k}_{01}c_0}{k_{10}+k_{12}+\mathrm{i}\omega}; \\ \left(\mathbf{a}_{\mathrm{per}}\right)_3 &= \frac{\mu\bar{k}_{01}k_{12}c_0}{k_{10}+k_{12}-k_{20}}\left(\frac{1}{k_{20}+\mathrm{i}\omega} - \frac{1}{k_{10}+k_{12}+\mathrm{i}\omega}\right) \end{aligned} \tag{32}$$

The in-phase and out-of-phase signals of the lock-in amplifier become for the system I:

$$\begin{aligned} S_x^{1\to 0} &= G\mu\bar{k}_{01}k_{10}^{*}c_0\frac{k_{10}+k_{12}}{\left(k_{10}+k_{12}\right)^2+\omega^2}; \\ S_y^{1\to 0} &= -G\mu\bar{k}_{01}k_{10}^{*}c_0\frac{\omega}{\left(k_{10}+k_{12}\right)^2+\omega^2} \end{aligned} \tag{33}$$

and for the system II:

$$S_x^{2\to 0} = G\mu c_0 \bar{k}_{01} k_{12} k_{20}^* \frac{k_{20}(k_{10}+k_{12})-\omega^2}{(k_{20}^2+\omega^2)\left[(k_{10}+k_{12})^2+\omega^2\right]};$$

$$S_y^{2\to 0} = -G\mu c_0 \bar{k}_{01} k_{12} k_{20}^* \frac{\omega(k_{10}+k_{12}+k_{20})}{(k_{20}^2+\omega^2)\left[(k_{10}+k_{12})^2+\omega^2\right]} \quad (34)$$

4 Data Acquisition and Processing

Because the frequencies generally used for the measurement of lanthanide luminescence span several orders of magnitude, the natural choice of the frequencies is the geometric progression, i.e.:

$$f_k = d^k f_0 \text{ or } \log f_k = \log f_0 + k \log d \quad (35)$$

For instance, when measuring luminescence typically at 200 frequencies between 10 Hz and 100 kHz, the ratio d between consecutive frequencies is ca. 1.0471. The frequency generation is easily accomplished using a modern digital lock-in amplifier, which has a built-in programmable frequency synthesizer.

Many physical measurements yield data in the form of complex numbers. Examples are various impedance measurements in the fields of electronics and electrochemistry, and optical measurements of absorbance and refractive index. Many graphical representations have been developed for the preliminary inspection and elucidation of data. One of the simplest representations is the direct plotting of the lock-in output signals S_x and S_y vs logarithm of frequency. This representation is shown in Fig. 3 where our two example systems I and II (Fig. 1) are shown with the rate constants k_{10} = 1,000 s^{-1}, k_{12} = 1,500 s^{-1}, k_{20} = 2,000 s^{-1}. The right-hand side figure represents signals $S_y^{1\to 0}$ (system I) and $S_y^{2\to 0}$ (system II). The main differences are in the slight asymmetry of curve II as well as its narrower half-width – marked with dotted lines – compared with curve I. As a matter of fact, it is easy to show that if the system has a single relaxation time in the frequency range, the half-width of S_y in the logarithmic scale is always equal to $\log_{10}(7+4\sqrt{3}) \cong 1.144$. In this case also the time constant is easily obtained from the location of the maximum of the out-of-phase signal $\omega_{\max}$: $\tau = 1/\omega_{\max}$. The left-hand side figure represents signals $S_x^{1\to 0}$ and $S_x^{2\to 0}$. The outstanding feature is the zero intercept of $S_x^{2\to 0}$, marked by an arrow. It can be shown that the presence of zero intercept means that the transition is from the level that is not directly achieved from the ground state. However, its absence may show only that some transitions are too fast for the experimental frequency range. In our example we obtain from (34) that the frequency corresponding to the intercept is: $\omega = 2\pi f = \sqrt{k_{20}(k_{10}+k_{12})} \cong 2,236$ rad s^{-1}(356 Hz).

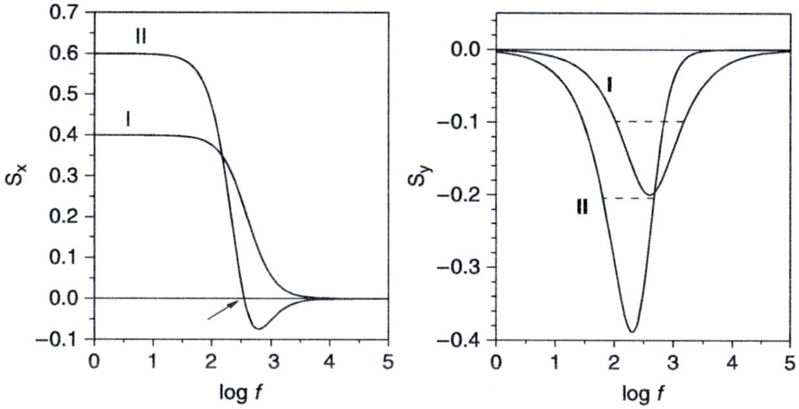

Fig. 3 In-phase (left) and out-of-phase (right) lock-in signals of systems I and II

In numerous methods in which the in-phase and out-of-phase signals are recorded using a lock-in amplifier or analogous instruments, it is customary to represent results in a graph where S_y is plotted vs S_x. Depending on the application, these plots are called parametric plots, complex plane plots, Argand diagrams, Nyquist plots, and Cole–Cole plots. In these plots frequency ω is an implicit parameter.

We can eliminate ω from (33) in the following way:

$$\left(S_x^{1\to 0}\right)^2 + \left(S_y^{1\to 0}\right)^2 = 2\rho S_x^{1\to 0} \tag{36}$$

where:

$$\rho = \frac{G\mu k_{10}^{*}\bar{k}_{01}c_0}{2(k_{10}+k_{12})} \tag{37}$$

This can be written as follows:

$$\left(S_y^{1\to 0}\right)^2 + \left(S_x^{1\to 0} - \rho\right)^2 = \rho^2 \tag{38}$$

If we now draw a graph in the S_x–S_y coordinates and take into account that the parameter ω obtains only positive values, we see that the graph is a semicircle with a center at:

$$S_x = \rho;\ S_y = 0 \tag{39}$$

and the radius ρ. Our example system I gives a semicircle in Fig. 4 using the same hypothetical parameters as in Fig. 2. System II does not give a semicircle and there is a zero intercept of $S_x^{2\to 0}$ at $\omega = \sqrt{k_{20}(k_{10}+k_{12})}$.

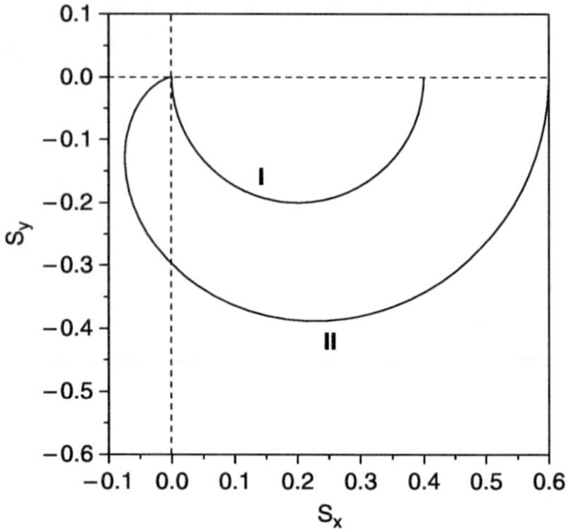

Fig. 4 Parametric plots of lock-in signals for the systems I and II

5 Integral Relations of the Lock-In Signals

There are several useful relations between S_x and S_y which can be applied even to the raw lock-in data before model fitting. The lock-in output signal in (23) and (31) can be written as follows:

$$\begin{aligned} S(\omega) = S_x(\omega) + \mathrm{i}S_y(\omega) &= \sum_{k=1}^{N} h_k \frac{1}{\mathrm{i}\omega - \lambda_k} \\ &= -\sum_{k=1}^{N} h_k \frac{\lambda_k}{\lambda_k^2 + \omega^2} - \mathrm{i}\sum_{k=1}^{N} h_k \frac{\omega}{\lambda_k^2 + \omega^2} \end{aligned} \tag{40}$$

where:

$$h_k = G k_{\mathrm{LM}}^{*} \mu \left(\mathbf{Z}^{(k)} \mathbf{A}_1 \mathbf{a}_{eq} \right)_{L+1}; \quad k = 1, 2, \ldots N-1; \; h_N = 0 \tag{41}$$

Integration of Re $S(\omega)$ in terms of ω gives then:

$$\begin{aligned} \int_0^\infty \mathrm{Re}\; S(\omega) d\omega &= -\int_0^\infty \sum_{k=1}^{N-1} h_k \frac{\lambda_k}{\lambda_k^2 + \omega^2} d\omega = {}^1\!/\!_2\pi \sum_{k=1}^{N-1} h_k \\ &= {}^1\!/\!_2\pi G k_{\mathrm{LM}}^{*} \mu \sum_{k=1}^{N-1} \left(\mathbf{Z}^{(k)} \mathbf{A}_1 \mathbf{a}_{\mathrm{eq}} \right)_{L+1} = {}^1\!/\!_2\pi G k_{\mathrm{LM}}^{*} \mu \left(\mathbf{A}_1 \mathbf{a}_{\mathrm{eq}} \right)_{L+1} \end{aligned} \tag{42}$$

The last expression resulted by applying the specific properties of the Frobenius covariants. Remembering how matrix $\mathbf{A}_1$ was defined, we see that term $\left(\mathbf{A}_1\mathbf{a}_{\mathrm{eq}}\right)_{L+1}$ is non-zero only when the level L is achieved by some time-dependent process, i.e., in most cases by direct excitation from the ground level. In other cases the integral is zero, meaning that Re $S(\omega)$ must have a zero point (cf. system II in our example). Conversely, if we do not observe any zero point, we cannot conclude anything because the decay process from level L may be so fast that it does not give any contribution to data within the experimental precision. For example, in the case of resonance energy transfer (RET) from lanthanide ion to an organic fluorophore, the natural decay process of the excited fluorophore is exceedingly fast compared with the timescale of lanthanide luminescence, and no zero point is observed. However, when dealing with the upconversion materials, the zero point is observed in most data. It is also very important to note that the amplitude factors h_k in (40) may attain both signs, simply because their sum in (42) may be zero. This fact has implications when choosing the algorithm for calculating the distribution of lifetime. For example, the methods based on the maximum entropy published so far are based on the positive amplitude factors and as such these methods cannot be used for the estimation of distributions in the general case.

Another interesting and useful integral is:

$$\int_0^\infty \frac{\mathrm{Im}S(\omega)}{\omega}d\omega = -\sum_{k=1}^{N-1} h_k \int_0^\infty \frac{d\omega}{\lambda_k^2+\omega^2} = \pm{}^1\!/_2\pi \sum_{k=1}^{N-1} h_k\left|\lambda_k^{-1}\right| \tag{43}$$

Comparing with (40) and taking into account the sign of λ, we see that:

$$\int_0^\infty \frac{\mathrm{Im}S(\omega)}{\omega}d\omega = -{}^1\!/_2\pi S(0) \tag{44}$$

This is an important relation which is used in connection with distributions.

The third important integral relation is the *Kramers–Kronig relation.* If complex-valued data from the experiments meet certain rather general regularity conditions, the real and imaginary parts are related to each other. This relation has several forms but for our purposes the most appropriate expression is:

$$\mathrm{Re}\ S(\omega) = -\frac{2}{\pi}P\int_0^\infty \frac{u\ \mathrm{Im}\ S(u)}{u^2-\omega^2}du \tag{45}$$

Letter P in front of the integral sign indicates the Cauchy principal value of the integral, which has a singularity at $u = \omega$. Note that (44) is a special case of (45). Due to the singularity, the numerical calculation of the K–K relation is not without problems. In our case, as data have been collected rather densely in the log-linear frequency scale (35), a fast approximate method for the K–K relation can be derived:

$$\mathrm{Re}\, S_n \approx -\frac{2\ln d}{\pi}\sum_{k=1}^{N_d}\frac{d^{-k}\mathrm{Im}\, S_{n-k} - d^{k}\mathrm{Im}\, S_{n+k}}{d^{-k} - d^{k}};\; S_m = S(\omega_m)$$
$$\mathrm{Im}\, S_m = 0,\; m > N_d\,;\; \mathrm{Im}\, S_{-m} = d^{-m}\mathrm{Im}\, S_0,\; m > 0 \qquad (46)$$

Here N_d is the number of frequencies taken to the experiment and d is the ratio of consecutive frequencies. The main use of Kramers–Kronig relation is in checking the compatibility of the imaginary and real parts of data. Various sources of interference have effect especially on the in-phase signal [6], and they are then revealed by the incompatibility of the measured and Kramers–Kronig transformed data. Because of the K–K relation, the real and imaginary parts are not independent and in principle, fitting only the imaginary part to the model is sufficient.

In all these integral relations, the upper limit of integration is infinity. In practice, *infinity* means the highest experimental frequency, and for instrumental reasons this is often in case of lanthanides about 100 kHz. Very often this is enough, but in those cases when we see that especially $S_y(\omega)$ is non-zero at the uppermost frequency, the results given by these relations are not reliable.

6 Distribution of Lifetimes

Although in many homogeneous systems fluorophores have distinct and discrete decay constants for their fluorescence, in heterogeneous systems the luminescent molecules have different environments and consequently different energy levels and also pathways for the energy dissipation. Moreover, in the RET processes the distance between the donor and acceptor is not constant but may vary slightly. Then it can be expected that the lifetimes are not sharply defined but they are actually continuously distributed. Mathematically this means that instead of the sum in (40), we have to use a Riemann–Stieltjes integral:

$$S(\omega) = \int_{-\infty}^{0} \frac{1}{\mathrm{i}\omega - \lambda}\, dH(\lambda) \qquad (47)$$

where $H(\lambda)$ is a cumulative distribution function of eigenvalues. Equation (47) is very general because it includes also the discrete case of (40) with a step or *staircase* function as $H(\lambda)$. If $H(\lambda)$ is continuously differentiable, the R–S integral (47) can be written as:

$$S(\omega) = \int_{-\infty}^{0} \frac{1}{\mathrm{i}\omega - \lambda}\frac{dH}{d\lambda}\, d\lambda \qquad (48)$$

The problem is to determine the function $H(\lambda)$ or its derivative from the experimental lock-in signals $S(\omega)$. Equation (48) is a Fredholm integral equation of the first kind which is usually strongly ill-posed. The solution that requires inverting the operator will be extremely sensitive to small changes or errors in the input. Various methods have been developed for this kind of problems, often called *inverse problems*. A common feature of these methods is that some prior knowledge of the nature of the solution is imbedded in the method. If this a priori knowledge concerns only some general properties of the solution, such as smoothness, we are talking about *regularization* method. One of these methods is the maximum entropy method (MEM) that has been used in the case frequency-domain fluorescence lifetime analysis [6–10]. A step further toward the Bayesian method is to give a certain mathematical form to the distribution. As a matter of fact, in many practical cases a discrete model with a finite number of lifetimes gives a rather good fit. A natural extension of the model is to retain the number of lifetimes and to allow the lifetimes to exhibit, e.g., a normal distribution in the logarithmic scale. This may not correspond accurately to the physical reality, but at least it gives an idea about the width of the lifetime distribution in those cases when the discrete solution is not too far.

It is assumed that the cumulative distribution can be written as (erfc = complementary error function):

$$H(s) = {}^1\!/_2 \sum_{k=1}^{N} h_k \operatorname{erfc} \frac{s_k - s}{\sqrt{2}\sigma_k}\,; \quad s_k = \ln\left(-\lambda_k^{-1}\right) = \ln \tau_k \tag{49}$$

This gives the normal distribution:

$$\frac{dH}{ds} = \frac{1}{\sqrt{2\pi}} \sum_{k=1}^{N} \frac{h_k}{\sigma_k} \exp\left[-\frac{(s - s_k)^2}{2\sigma_k^2}\right] \tag{50}$$

Equation (48) now becomes:

$$S(\omega) = \frac{1}{\sqrt{2\pi}} \sum_{k=1}^{N} \frac{h_k}{\sigma_k} \int_{-\infty}^{\infty} \frac{\exp\left[-{}^1\!/_2\sigma_k^{-2}(s - s_k)^2\right]}{i\omega + \exp(-s)}\, ds \tag{51}$$

If we allow $\sigma_k \to 0$, this expression tends to (40). Equation (51) looks formidable but with appropriate substitutions and approximations the integral can be tamed to a reasonably fast computable expression. The fractional steady-state intensities corresponding to each lifetime are then:

$$f_i = \frac{|h_i|\tau_i \exp\left({}^1\!/_2\sigma_i^2\right)}{\sum_{k=1}^{N} |h_k|\tau_k \exp\left({}^1\!/_2\sigma_k^2\right)} \tag{52}$$

A word of caution is in order for those pursuing the lifetime distributions. As mentioned, the Fredholm integral equation (48) is strongly ill-posed, and any result obtained without regularization on the basis of physical grounds should be regarded with suspicion. It seems that more robust solutions are obtained if one confines to discontinuous distributions, i.e., to discrete time constants. One possibility is to solve first the discrete time constants and then to keep these values constant when solving the widths σ_k.

7 Model Fitting

As mentioned before, the two output signals of a dual-phase lock-in amplifier are interpreted as the real and imaginary parts of the complex signal $S(\omega)$. Usually the fitting of a model to experimental data involves minimizing the sum of squares function. In the case of a complex-valued signal, there are two possibilities of representing data: either Cartesian or polar. Both representations have been used in fitting the frequency-domain data by the least squares method. In the Cartesian representation, the merit function *chi-square* is written as:

$$\chi^2 = \sum_{k=1}^{N_d} \left\{ \sigma_{x,k}^{-2} \left[S_x^{\text{obs}}(\omega_k) - S_x^{\text{calc}}\left(\omega_k; \hat{\beta}\right) \right]^2 + \sigma_{y,k}^{-2} \left[S_y^{\text{obs}}(\omega_k) - S_y^{\text{calc}}\left(\omega_k; \hat{\beta}\right) \right]^2 \right\} \tag{53}$$

where σ_xs and σ_ys are estimated standard deviations in the measurements and $\hat{\beta}$ is the vector of adjustable parameters. The standard deviations could be estimated to be equal within the frequency range used for the measurements or they could be estimated by some standard method (repeated experiments, moving average of deviations etc.). In practice, when comparing fittings, it is convenient to use the reduced chi-square:

$$\chi_R^2 = \frac{\chi^2}{\nu} \tag{54}$$

Here ν is the number of degrees of freedom. In the presence of only random errors, the value of χ_R^2 is expected to be close to unity.

Depending on the frequency range and on the quality of instrumentation, systematic deviations in the gain and phase shift within the range of measurement may be induced to the data. These instrumental changes can be compensated by measuring lock-in signals $S_x^{\text{comp}}(\omega_k)$ and $S_y^{\text{comp}}(\omega_k)$ for a sample which does not induce phase shift within the frequency range of measurement, e.g., an organic prompt fluorophore or even colloidal suspension (e.g., LUDOX®). Then the compensation can be done by complex division:

$$S_x^{\text{obs}}(\omega_k) + \mathrm{i}S_y^{\text{obs}}(\omega_k) = \frac{S_x^{\text{raw}}(\omega_k) + \mathrm{i}S_y^{\text{raw}}(\omega_k)}{S_x^{\text{comp}}(\omega_k) + \mathrm{i}S_y^{\text{comp}}(\omega_k)} \tag{55}$$

The merit function χ^2 is minimized by picking the best parameters β for $S^{\text{calc}}(\omega)$. Some adjustable parameters (h_k) are related linearly as seen in (40) and (51), but others (λ_k and σ_k) have a nonlinear relationship. A natural way to solve the problem is to use some nonlinear minimization program (e.g., Levenberg–Marquardt) for adjusting λ_k and σ_k and to solve h_k in the subprogram using standard linear techniques for minimizing χ^2.

Another way to write the merit function χ^2 is to use the polar representation of data, i.e., to write the sum of squares in terms of differences in the calculated and observed magnitude (modulus) and phase angle. In fact, this is more common than the Cartesian representation in the frequency-domain measurements of fluorophores. In principle, with proper weighting of data, this should give equivalent results. In practice, however, for the slowly decaying luminophores that are measured by the homodyne technique, the Cartesian representation is better. The reason is the unavoidable contamination of the in-phase signal $S_x(\omega)$ by periodic signals of the same frequency but with zero phase shift. These signals originate from scattering, from prompt fluorescence of impurities, or from crosstalk between the electrical circuits. These spurious signals give their contribution to the phase angle and magnitude but not to the out-of-phase signal S_y. Hence, if the Kramers–Kronig relation shows that the measured S_x is not in conformity with the K–K transformed S_y, it is worthwhile to try fitting on the basis of S_y only. As a matter of fact, the mere validity of the Kramers–Kronig relation shows that S_x and S_y do not carry independent information. If the fit is good for S_y, it should be good also with S_x unless there is in-phase disturbance. Nevertheless, it is obviously statistically advantageous if the simultaneous fit with both S_x and S_y is possible.

The merit function χ^2 is often rather shallow and contains a number of local minima. In order to find the global minimum, it is necessary to repeat the minimization many times by varying randomly the initial estimates of the adjustable nonlinearly related parameters. This makes the convergence of the minimization toward the global minimum rather slow, especially if also the distribution of lifetimes is sought for. The ill-conditioned system can be regularized by imposing further conditions such as (42) and (44) to the minimization of χ^2 using the method of the Lagrange multipliers.

As in all model fitting procedures, model validation is the most important part of the process. In the present case two different models can be devised. One is based on the expansion (40) or (51), and the other on the known form of matrix $\mathbf{A}_0$ and (23) and (31). In the latter case, the relationships between the linear parameters h_k are assumed to be known. Both models have their advantages. In the former model no assumptions are made on the mechanism and the number of lifetimes. This model should be fitted always first in order to get some knowledge on the statistically relevant number of lifetimes. A serious problem is the danger of overparametrization, i.e., using too many adjustable parameters for fitting.

Appropriate statistical tests such as *lag plot* or *autocorrelation plot* should be used, which effectively show the randomness of the data residuals [11, 12]. When the number of lifetimes is believed to be known, the mechanism and rate matrix of the process can be estimated and the latter model can be applied.

As cautioned before, it is advisable to calculate the distributions of lifetimes only when there is some a priori information on the maximum number of time constants and the accuracy of experimental data. There is no problem in obtaining nice distributions when fitting is started from good initial estimates, but there is no guarantee that completely different distributions with equally good statistical parameters are not obtained when starting from slightly different estimates.

8 Case Studies

Two simple examples will be given in order to illustrate the instrumentation and the primary data treatment. The discussion is focused on the technical problems and the interpretation of the results is left to a minimum. The first example deals with the resonance energy transfer (RET) from europium to an organic fluorophore. The distance between the donor and acceptor is kept constant by an oligonucleotide. RET is a very common technique in the case of lanthanides, and therefore the discussion is somewhat lengthy and detailed. The second example deals with a sample of upconversion phosphor containing yttrium, ytterbium and erbium.

8.1 Instrumentation

The instrumentation for both samples is essentially similar except for the excitation source and some optical filters. The optical part consists of the tubular excitation and emission chambers (Thorlabs) in the right angle configuration. The sample chamber is a cube containing either a 5-mm quartz NMR tube for the liquid samples or a 2-mm capillary tube for the solid materials. The photomultiplier is a head-on R1464 (Hamamatsu) equipped with suitable interference filters. The light source for the Eu measurements was a UV-LED T9H34C (Seoul Optodevice Co., Ltd) providing about 0.2 mW at 340 nm (half-width 15 nm). Light from the LED was focused to the sample by two fused silica plano-convex lenses with appropriate antireflection coating. For removing the spurious visible radiation a Schott DUG11 glass filter was used. The excitation source for the upconversion phosphors was a fiber-coupled laser diode G098PU11300M (Roithner, Austria) providing 1.3 W (max) at 970 nm. In this case a long-pass filter with the cutoff at 850 nm was used for suppressing possible visible radiation.

The signal from the photomultiplier was amplified in a high speed current amplifier DHPCA-100 (Femto Messtechnik GmbH, Germany) connected to a dual-phase lock-in amplifier SR830 (Stanford Research Systems, CA). The lock-in amplifier was connected to a personal computer using GPIB and data are collected using a program written in LabView (National Instruments).

The sine-wave modulation between 10 Hz and 100 kHz was generated by the internal frequency synthesizer of the lock-in amplifier. The number of data points collected equidistantly in the logarithmic scale between 10 Hz and 100 kHz was 200. The DC bias was added by a simple op amp circuit and the modulation degree was kept at 10%. The driver for the NIR laser diode was WLD3343 (Wavelength Electronics, Inc., Montana) with a 2 MHz bandwidth and 2 A maximum current.

8.2 Data Processing

A computer program was written in Visual Fortran 90 equipped with the DISLIN graphics package for the preliminary data processing and the final model fitting. Data are initially subjected to the optional compensation of instrumental gain and phase shifts using (55) and fluorescein as the compensation sample. Next a smoothed data set is created using the second order Savitzky–Golay method [13] with the 19-point window. The standard deviation for each data point is estimated from the deviations between the observed and smoothed data within the neighboring 19 points. Then the Kramers–Kronig transform is applied to S_y^{obs} using the approximate equation (46), and the resulting curve is compared with S_x^{obs} graphically on the screen. If there is an obvious discrepancy between the curve and experimental data, the model fitting is done solely on the basis of the out-of-phase signal S_y^{obs}. Now model fitting is done by starting from the discrete time constants (40) and the covariance matrix calculated using the bootstrap technique [14]. A slightly modified nonlinear direct search minimization program STEPIT is used [15]. First, the smallest possible number of time constants is chosen and one by one the number is increased until the statistical indicators show that further proceeding is not relevant. In addition to the chi-square function, lag plot and autocorrelation are used in assessing the randomness of residuals. The distributions of the lifetimes are now optionally computed by fitting (51) to the observed data. For the numerical computations, the integral in (51) was transformed to a form appropriate to the Chebyshev–Gauss quadrature.

8.3 RET with Labeled Oligonucleotides

The methods based on RET are commonly used for determining intramolecular donor–acceptor distances in macromolecules. The use of lanthanide chelates as donors and prompt fluorophores as acceptors is especially advantageous because of the large difference in their lifetimes. The present example shows the use of the frequency-domain method as an alternative in determining the relevant lifetimes in connection with a recently published work on RET [16].

Materials and labeling procedures were essentially the same as described by Kokko et al. [16]. The labeling reagent for the capture oligonucleotide was {2,2′,2″,2‴-{[2-(4-isothiocyanatophenyl)ethylimino]-bis(methylene)bis{4-{[4-(α-galactopyranoxy)phenyl]-ethynyl}pyridine-6,2-diyl}bis(methylenenitrilo)}-tetrakis(acetato)}europium(III) [17], which has the excitation maximum at 325 nm and the decay time of the luminescence ca. 1 ms at 615 nm. The target oligonucleotide was labeled with AlexaFluor 680 (Molecular Probes, Eugene, OR). The estimated distance between the Eu chelate and acceptor when the tracer and capture oligo are hybridized corresponds to 15 base pairs, i.e., 5.1 nm.

Measurements were done using a 615-nm filter for the Eu-donor and a 730-nm filter for the acceptor.

8.3.1 Kramers–Kronig Relations

In Fig. 5, we see the observed in-phase signal and K–K transformed out-of-phase signal at 615 and 730 nm. In neither case, the observed in-phase data and K–K transformed out-of-phase data are in conformity. This is actually expected because the excitation at 340 nm is to some extent capable of exciting the acceptor dye directly, and the time constant of the subsequent emission is in the nanosecond region, contributing only to the in-phase signal. Consequently, fitting is done using the out-of-phase signal only.

8.3.2 Model Fitting

Figures 6 shows the out-of-phase signals of the donor at 615 nm and acceptor at 730 nm. A crude estimation from the location of the minimum gives 0.9 ms for the lifetime of the donor but the asymmetry of the curve and its half-width of 1.50 suggest that the signal is composed of more than one lifetime. In the same way, the

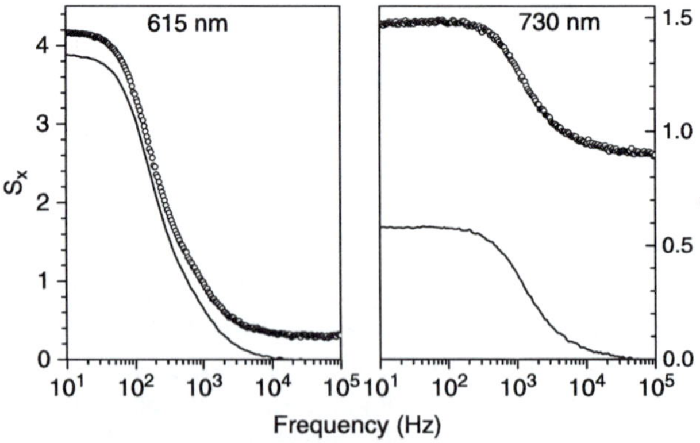

Fig. 5 The observed lock-in in-phase data (open circles) and the Kramers–Kronig transform of the corresponding out-of-phase data (solid lines) of the donor–acceptor pair

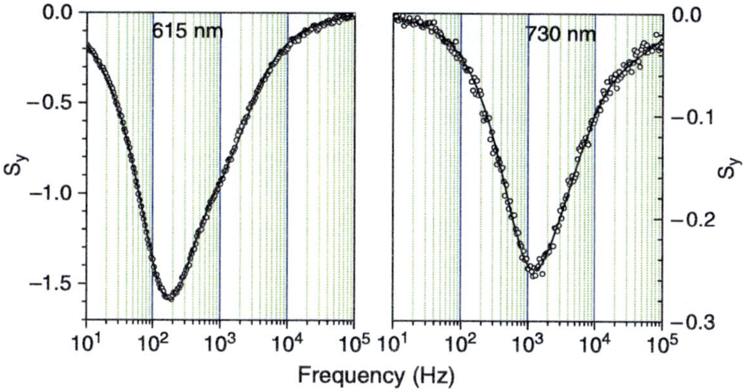

Fig. 6 The observed out-of-phase signals (open circles) and calculated values (solid lines) of the donor–acceptor system measured at the donor (615 nm) and acceptor (730 nm) emission wavelengths

signal corresponding to the acceptor gives 0.14 ms as the crude approximation but also in this case the half-width is 1.35, i.e., considerably greater than 1.144, the theoretical one-lifetime value.

The donor signal at 615 nm was fitted first with two time constants. The fit was reasonably good, giving $\tau_1 = 0.971$ ms and $\tau_2 = 0.128$ ms, $\chi^2_R = 2.74$ and the correlation coefficient of the lag plot $\rho = 0.671$. However, the autocorrelation diagram had a pattern clearly distinguished from noise. With three time constants, the results were $\tau_1 = 0.983$ ms, $\tau_2 = 0.148$ ms and $\tau_3 = 0.027$ ms, $\chi^2_R = 1.59$ and $\rho = 0.426$. The autocorrelation plot had no discernible pattern (Fig. 7). In the same way the acceptor signal at 730 nm was found to require three time constants. The results are shown in Table 1.

The standard deviations of lifetimes in Table 1 are rather small. They are actually underestimated because of the strong correlation. Based on the bootstrap method with 500 synthetic data sets the correlation coefficients in the case of donor emission are $r_{12} = 0.66$, $r_{13} = 0.36$, and $r_{23} = 0.76$.

RET of a donor–acceptor system resembles our examples I and II (Fig. 1). This involves the implicit assumption that the ligand-to-Eu transfer is rapid in the present time scale. Moreover, for simplicity it is assumed that the energy transfer occurs to the level $^5D_0$ of Eu^{III}. Here $A^{(0)}$ is the D–A pair in the ground state, $A^{(1)}$ is D–A where Eu^{III} is in the excited state, and $A^{(2)}$ is the D–A pair where RET has occurred and the acceptor is in the excited state. The donor emission at 615 nm corresponds to model I. We should observe only a single lifetime, the inverse of the sum of rate coefficients,$1/(k_{10} + k_{12})$. In fact, we observe three time constants. The presence of the first of them, 0.983 ms, corresponds to the donor alone without any RET and is expected due to incomplete labeling. Obviously $\tau_2 = 0.148$ ms corresponds to the radiative $^5D_0 \rightarrow {}^7F_2$ transition. This is within the error margins, the same as the first lifetime of the acceptor, 0.145 ms. According to (32) and (34), the

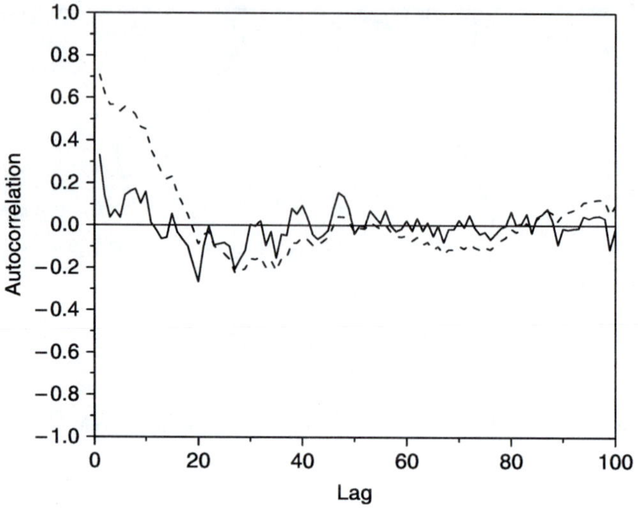

Fig. 7 Autocorrelation plot of the two-parameter (dotted line) and three-parameter (solid line) fits of the donor signal of the donor–acceptor pair

Table 1 Emission lifetimes of the donor–acceptor pair

	Donor 615 nm		Acceptor 730 nm	
	Lifetime (ms)[a]	Contribution[b]	Lifetime (ms)[a]	Contribution[b]
τ_1	0.983 (4)	74%	0.145 (5)	74%
τ_2	0.148 (3)	24%	0.036 (7)	20%
τ_3	0.027 (4)	2%	0.004 (1)	6%
χ_R^2	1.59	1.20		
ρ	0.426	0.097		

[a]The numbers in parentheses are $1 \times \sigma$ values
[b]Calculated from (52) by substituting $\sigma_k = 0$

acceptor emission should have two lifetimes, one the same as with the donor and the other corresponding to k_{20}. However, the latter process is a typical prompt transition in the nanosecond range and cannot be observed in the present time scale. Consequently the acceptor emission should possess a single lifetime, the same as with the donor emission. Still two additional lifetimes are observed. The donor lifetime τ_3 and acceptor lifetime τ_2 are the same within the error limits, about 30 μs. Heyduk et al. [18] have observed fast-decaying components in several cases of RET with lanthanide chelates. According to their interpretation, this may be due to a nonspecific aggregation whereby acceptor is held close to the donor. There are two possibilities: either a mixture of specific and nonspecific hybrids or an aggregate containing two acceptor moieties per one donor. The first one is a trivial case but the second possibility needs to be modeled by the graph of Fig. 8:

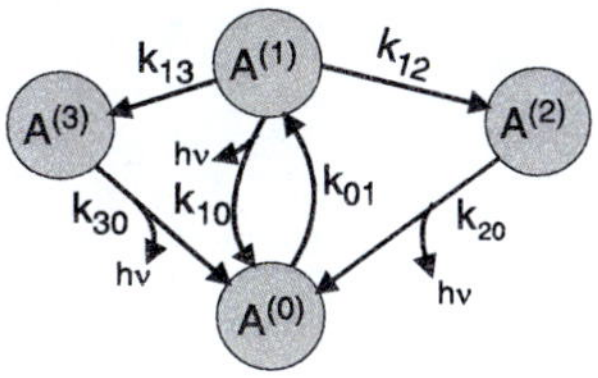

Fig. 8 RET system with two acceptors

Here the symbols $A^{(n)}$ denote the A′DA triplets at different excitation states:

$$\begin{aligned} A^{(0)} &\leftrightarrow \{A' - Eu^{III}(^7F_J) - A\}, \\ A^{(1)} &\leftrightarrow \{A' - Eu^{III}(^5D_0) - A\}, \\ A^{(2)} &\leftrightarrow \{A' - Eu^{III}(^7F_J) - A^*\}, \text{and} \\ A^{(3)} &\leftrightarrow \{A'^* - Eu^{III}(^7F_J) - A\} \end{aligned}$$

Using the same technique as for the simple models I and II, we obtain for the out-of-phase signal corresponding to the radiative transition $A^{(1)} \rightarrow A^{(0)}$:

$$S_y^{1\rightarrow 0} = -G\mu\bar{k}_{01}k_{10}^*c_0\frac{\omega}{(k_{10}+k_{12}+k_{13})^2+\omega^2} \tag{56}$$

By allowing the rates of the prompt fluorescence k_{20} and k_{30} to approach infinity, we have for the transitions $A^{(2)} \rightarrow A^{(0)}$ and $A^{(3)} \rightarrow A^{(0)}$:

$$\begin{aligned} S_y^{2\rightarrow 0} &= -G\mu\bar{k}_{01}k_{12}Q_{20}c_0\frac{\omega}{(k_{10}+k_{12}+k_{13})^2+\omega^2}; Q_{20} = \frac{k_{20}^*}{k_{20}} \\ S_y^{3\rightarrow 0} &= -G\mu\bar{k}_{01}k_{13}Q_{30}c_0\frac{\omega}{(k_{10}+k_{12}+k_{13})^2+\omega^2}; \ Q_{30} = \frac{k_{30}^*}{k_{30}} \end{aligned} \tag{57}$$

In addition to the scheme of Fig. 8, simultaneous excitation–emission processes are going on according to the systems I and II in Fig. 1. This process increases the signal with the lifetime $1/(k_{10} + k_{12})$, and consequently we can expect three lifetimes in the donor emission and two lifetimes in the acceptor emission.

According to the Förster equation, the donor–acceptor distance is obtained from:

$$R_{DA} = R_0\left(\frac{k_{10}}{k_{12}}\right)^{\frac{1}{6}} = R_0\left(\frac{\tau_{DA}}{\tau_D - \tau_{DA}}\right)^{\frac{1}{6}} \tag{58}$$

Here R_0 is the Förster radius, which depends on the spectral overlap of the donor emission and acceptor absorption, the quantum yield of the donor, the index of refraction, and the relative orientation factor of the donor and acceptor dipoles. We keep the other parameters as they are in the paper [16] but replace the total

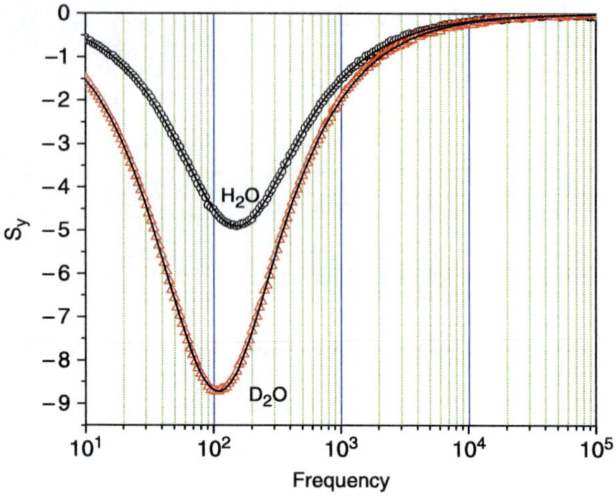

Fig. 9 The out-of-phase 615 nm emission of Eu^{III} complex with the nine-dentate ligand in water (circles) and in deuterium oxide (triangles)

quantum yield with the intrinsic quantum yield of europium Q_{Eu}. A crude approximation is obtained from the ratio of the lifetimes of the $^5D_0 \rightarrow {^7F_2}$ transition measured in water and deuterium oxide: $Q_{Eu} \approx \tau_D^{H_2O}/\tau_D^{D_2O}$. This is based on the assumption that the quantum yield is unity in deuterium oxide, shown to be a crude approximation by Xiao et al. [19] but still resulting in the intramolecular distances too long only by 4–7%. Figure 9 shows the experimental results and calculated fitting for our nine-dentate chelate measured in water and deuterium oxide. The numerical values are $\tau_D^{H_2O} = 1.060(6)$ ms, $\tau_D^{D_2O} = 1.466(6)$ ms, which give $Q_{Eu} \cong 0.723$. As an internal check, we may use these values for calculating the number of water molecules in the first coordination sphere of Eu^{III} using the corrected Horrocks formula [20]:

$$q_{Eu}(H_2O) = 1.11\left[\left(\tau_D^{H_2O}\right)^{-1} - \left(\tau_D^{D_2O}\right)^{-1} - 0.31\right] = -0.05 \tag{59}$$

Taking into account the estimated standard error ± 0.1, this is a proper value for a nine-dentate ligand. The intrinsic quantum yield and the other parameters give $R_0 = 7.35$ nm, which together with $\tau_{DA} = 1/(k_{10} + k_{12}) = 0.145$ ms and $\tau_D^{H_2O} = 1.060$ ms gives $R_{DA} = 5.4$ nm. This is in satisfactory agreement with 5.1 nm calculated from the number of intervening base pairs (15) and the increment caused by each base pair (0.34 nm). We can even estimate the donor–acceptor distance to the assumed second acceptor A′ by applying (58):

$$R_{DA'} = R_0\left(\frac{k_{10}}{k_{13}}\right)^{\frac{1}{6}} = R_0\left[\frac{\tau_{DA'}\tau_{DA}}{\tau_D(\tau_{DA} - \tau_{DA'})}\right]^{\frac{1}{6}} \tag{60}$$

With $\tau_{DA'} = 0.03$ ms, we obtain $R_{DA'} = 4.2$ nm. This is close to 4.1 nm obtained for the *trivial* case, i.e., the case where we assumed the presence of a nonspecific hybrid in the mixture instead of an aggregate with two acceptors.

8.3.3 Lifetime Distribution

In the present example, we assume that the values of the discrete lifetimes are at the maxima of the normal distributions in the logarithmic scale. The widths of the distributions are now the only adjustable parameters in the nonlinear fitting. Figure 10 shows the distributions for the donor and acceptor lifetimes. In both cases fitting was very good already when using three discrete lifetimes, but a slight decrease in the lag plot correlation coefficients and reduced chi-square could be still obtained by adjusting the distribution widths. One should understand that the half-widths do not reflect only the physical properties but also partly the insensitivity of the model fitting to the parameter variation. Consequently no in-depth conclusions should be drawn when comparing distributions obtained in different experiments. However, within the same experiment the relative half-widths of the lifetime distributions may be informative.

8.4 Upconversion Particles

The luminescence of upconversion materials has a rather complicated mechanism. There are several materials even commercially available, but one of the most

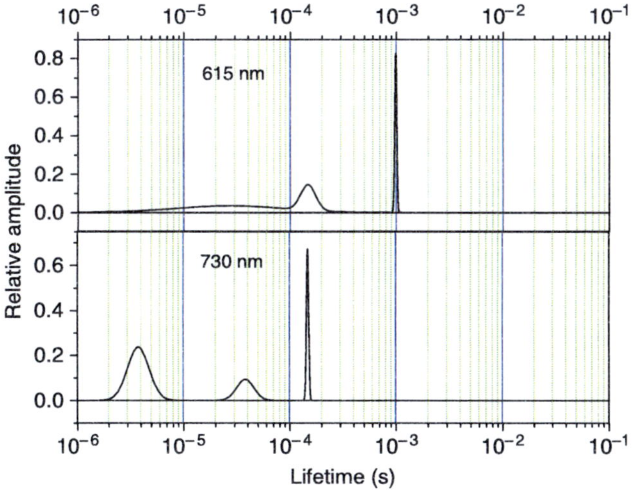

Fig. 10 Distribution of lifetimes of the donor emission (615 nm) and acceptor emission (730 nm) from the donor–acceptor pair

common material types which is also our example is based on a mixture of ytterbium and erbium compounds in a suitable inert supporting material, in our case an yttrium compound. The actual luminescent component is erbium(III), which can be excited to a higher energy level by two photons in the energy range of 10,000 cm^{-1}. These near infrared photons are received by ytterbium(III) which transfers energy to Er^{III}. The emission from Er^{III} consists of two sets of lines in the visible range, *green lines* at ca. 550 nm ($^2H_{11/2}$, $^4S_{3/2} \rightarrow {}^4I_{15/2}$) and *red lines* at ca. 650 nm ($^4F_{9/2} \rightarrow {}^4I_{15/2}$). The intensity ratio of red and green emission depends among other things on the size of particles [21].

The present example is a commercial upconversion material which is not of nanoparticle size. This material is an example of complicated luminescence systems consisting of many energy levels and transfer processes.

8.4.1 Materials

The upconversion phosphor in this example is PTIR550/F ($NaYF_4$: Yb^{3+}, Er^{3+}) from Phosphor Technology Ltd. (Stevenage, England). It consists of powder with ca. 5 μm particle size. The powder sample was packed into a 2-mm glass capillary tube. In this example, a 544/10 nm interference filter was used for recording the green emission and 650 nm filter for the red emission.

8.4.2 Kramers–Kronig Relations

In this example, the K–K transformed out-of-phase signal and the observed in-phase signal fit nearly completely (Fig. 11). This means that the in-phase and out-of-phase signals can be fitted together and also their Argand diagram (Fig. 12) is meaningful.

8.4.3 Model Fitting

Despite the seemingly good fit shown in Fig. 12, the simple linear model of (2) is not a satisfactory model for the upconversion materials. The excitation process is complicated by the two-photon absorption, energy transfer, and partial population of higher levels due to the strong laser excitation. The Argand diagrams of the lock-in signals of the green and red emission are very different as seen in Fig. 12. The green emission could be fitted with four lifetimes although the reduced chi-square remained rather high and further additions of adjustable parameters decreased it only slightly. The signal from the red emission proved to be very interesting. Fitting with an increasing number of adjustable lifetimes yielded nearly equal values for the lifetimes but with the amplitudes of opposite sign. Finally it was found that only a single lifetime was enough to give a very good fit but with the eigenvalue

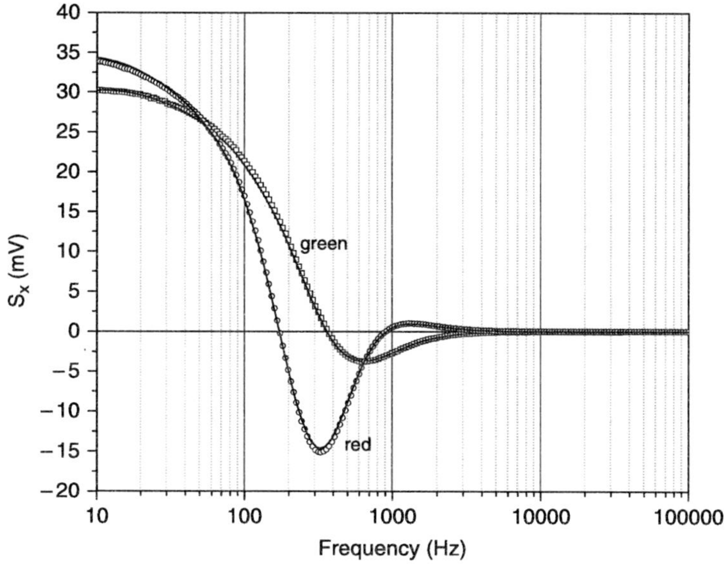

Fig. 11 The observed in-phase data (open circles and squares) and the Kramers–Kronig transform of the corresponding out-of-phase data (solid lines) on the green (squares) and red (circles) emission of upconversion material PTIR550/F. Emission at 544 and 650 nm, excitation at 970 nm

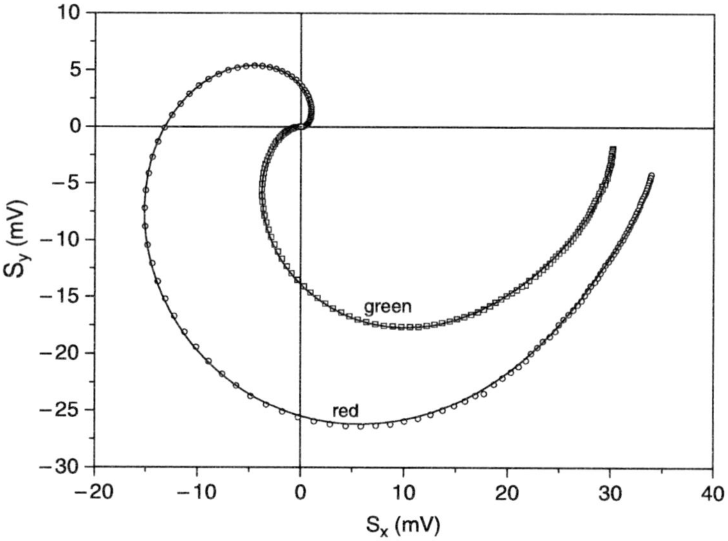

Fig. 12 Argand diagrams of lock-in signals from the 544 nm (green) and 650 nm (red) emission of the upconversion material PTIR550/F. Open circles and squares: experimental points; solid lines: calculated on the basis of four lifetimes (green) and one lifetime with multiplicity of 9 (red)

multiplicity of 9. This means that the lock-in signal can be described by the expansion (m = 9):

$$S(\omega) = \sum_{n=1}^{m} \frac{A_n}{(1 + \mathrm{i}\omega\tau)^n} \tag{61}$$

The signal from the green emission could not be fitted this way. The reason for this anomaly is still an open problem and is subject to further studies.

9 Summary

The frequency-domain methods are very suitable for measuring the lifetimes of most luminescent lanthanide compounds. The reason is the range of rate constants of decay processes, which fits well in the frequency range of the conventional inexpensive lock-in amplifiers. Together with the easily modulated light-emitting diodes or laser diodes as the excitation sources, frequency-domain instruments for the lifetime measurements can be constructed in any workshop with rather minimal resources and expertise.

The mathematical theory of the frequency-domain methods consists of beautiful applications of matrix and complex analysis. The general matrix rate equations have been derived for the monomolecular photochemical processes, and matrix analysis is used in deriving the general solution for the temporal concentrations of the excited species in the presence of an arbitrary functional form of excitation. The sinusoidal excitation and dual-phase lock-in detection of the emission lead to a signal which can be effectively treated as a complex number. For instance, the Kramers–Kronig relation, better known from the solid-state physics, can be used for checking the internal consistency of data.

The instrumentation and model fitting are exemplified by two examples. The first one is the resonance energy transfer from an europium chelate to an organic acceptor, held at a constant distance from the donor by oligonucleotide hybridization. The emission both from the donor and acceptor can be easily measured with an excellent signal-to-noise ratio. Here, the linear differential equations and their solution are the proper ways to treat the data. The method is easily extended to more complicated systems. The second example deals with the upconversion material which is a mixture of lanthanide compounds. In contrast to the previous example, the luminescence is generated in solid-state and complicated further by being a two-photon process. The Kramers–Kronig relation shows the good internal consistency of data, but the statistical tests show that the model fitting with the simple linear model is not satisfactory.

The power of the frequency-domain measurement of luminescence lifetimes of lanthanide derivatives is the excellent quality of data compared with the time-domain methods. This allows to use more sophisticated methods of data treatment to tackle even the mathematically ill-posed problem of lifetime distribution. Still the FD methods do not supersede the TD methods. Both methods have their range of application because of the different weighting of experimental data.

References

1. Beechem JM, Ameloot M, Brand L (1985) Global analysis of fluorescence decay surfaces: excited state reactions. Chem Phys Lett 120:466–472
2. Ameloot M, Beechem JM, Brand L (1986) Compartmental modeling of excited-state reactions: identifiability of the rate constants from fluorescence decay curves. Chem Phys Lett 129:211–219
3. Beechem JM, Gratton E, Ameloot M, Knutson JR, Brand L (1991) The global analysis of fluorescence intensity and anisotropy decay data: second-generation theory and programs. In: Lakowicz JR (ed) Topics in fluorescence spectroscopy, vol 2: principles. Plenum Press, New York
4. Lakowicz JR (2006) Principles of fluorescence spectroscopy, 3rd edn. Springer, New York
5. Jin D, Connally R, Piper J (2006) Long-lived visible luminescence of UV LEDs and impact on LED excited time-resolved fluorescence applications. J Phys D Appl Phys 39:461–465
6. Vinogradov SA, Fernandez-Searra MA, Dugan BW, Wilson DF (2001) Frequency domain instrument for measuring phosphorescence lifetime distributions in heterogeneous samples. Rev Sci Instrum 72:3396–3406
7. Brochon JC, Livesey AK (1990) Data analysis in frequency-domain fluorometry by the maximum entropy method – recovery of fluorescence lifetime distributions. Chem Phys Lett 174:517–522
8. McGown LB, Hemmingsen SL, Shaver JM, Geng L (1995) Total lifetime distribution analysis of fluorescence fingerprinting and characterization. Appl Spectrosc 49:60–66
9. Shaver JM, McGown LB (1996) Maximum entropy method for frequency domain fluorescence lifetime analysis. 1. Effects of frequency range and random noise. Anal Chem 68:9–17
10. Shaver JM, McGown LB (1996) Maximum entropy method for frequency-domain fluorescence lifetime analysis. 2. Timing, mismatched intensity, and reference lifetime errors. Anal Chem 68:611–620
11. NIST/SEMATECH e-Handbook of Statistical Methods, http://www.itl.nist.gov/div898/handbook/eda/section3/eda33.htm
12. Straume M, Frasier-Cadoret SG, Johnson ML (1991) Least-squares analysis of fluorescence data. In: Lakowicz JR (ed) Topics in fluorescence spectroscopy, vol 2: principles. Plenum Press, New York
13. Press WH, Teukolsky SA, Vetterling WT, Flannery BP (1994) Numerical recipes in FORTRAN, 2nd edn. Cambridge University Press, Cambridge, p 644
14. Press WH, Teukolsky SA, Vetterling WT, Flannery BP (1994) Numerical recipes in FORTRAN, 2nd edn. Cambridge University Press, Cambridge, p 686
15. Chandler J (1964) Quantum Chemistry Program Exchange, 11: 307
16. Kokko L, Kokko T, Lövgren T, Soukka T (2008) Particulate and soluble Eu(III)-chelates as donor labels in homogeneous fluorescence resonance energy transfer based immunoassay. Anal Chim Acta 606:72–79
17. von Lode P, Rosenberg J, Pettersson K, Takalo H (2003) A europium chelate for quantitative point-of-care immunoassays using direct surface measurement. Anal Chem 75:3193–3201
18. Heyduk T, Heyduk E (2001) Luminescence energy transfer with lanthanide chelates: interpretation of sensitized acceptor decay amplitudes. Anal Biochem 289:60–67
19. Xiao M, Selvin PR (2001) Quantum yields of luminescent lanthanide chelates and far-red dyes measured by resonance energy transfer. J Am Chem Soc 123:7067–7073

20. Supkowski RM, WDeW H (2002) On the determination of the number of water molecules, *q*, coordinated to europium(III) ions in solution from luminescence decay lifetimes. Inorg Chim Acta 340:44–48
21. Vetrone F, Boyer J-C, Capobianco JA, Speghini A, Bettinelli M (2004) Significance of Yb3+ concentration on the upconversion mechanisms in codoped Y2O3:Er3+, Yb3+ nanocrystals. J Appl Phys 96:661–667

Imaging of Lanthanide Luminescence by Time-Resolved Microscopy

H.J. Tanke

Abstract Time-resolved fluorescence microscopy (TRFM) of objects stained with luminescent lanthanides chelate-based reagents requires adaptation of microscope equipment to distinguish the long-lived fluorescence of the lanthanide from scattered excitation light and relatively short-lived autofluorescence of biological objects and optical components. How this is accomplished depends on the choice of image scanning, e.g., wide-field microscopy (image plane scanning) or object plane scanning [(confocal) laser microscopy]. Typically, excitation light is presented as pulses by applying mechanically chopping of a continuous light beam, or by using flash lamps or pulse lasers. The detector (CCD camera, photomultiplier tube) is time-gated to record the delayed signal in a defined time interval, either electronically or using phase-locked choppers in the emission pathway. This chapter discusses strategies for optimal TRFM and reviews the main existing configurations used for lanthanide imaging. In addition, special techniques such are Fluorescence Lifetime Imaging Microscopy (FLIM) and Time-Correlated Single Photon Counting (TCSPC) are presented and discussed as well.

Keywords Autofluorescence · Decay time · Delayed fluorescence · Fluorescence lifetime imaging microscopy · Lanthanides · Phosphorescence · Time-correlated single photon counting · Time-resolved fluorescence microscopy

---

H.J. Tanke
Department of Molecular Cell Biology, Leiden University Medical Centre, Einthovenweg 20, Bldg 2, PO Box 9600, 2300 RC Leiden, The Netherlands
e-mail: H.J.Tanke@lumc.nl

P. Hänninen and H. Härmä (eds.), *Lanthanide Luminescence: Photophysical, Analytical and Biological Aspects*, Springer Ser Fluoresc (2011) 7: 313–328, DOI 10.1007/4243_2010_2,

Contents

Abbreviations

CCD	Charge-coupled device
DELFIA	Dissociation enhanced lanthanide fluoroimmunoassay
FLIM	Fluorescence lifetime imaging microscopy
FRET	Fluorescence resonance energy transfer
LED	Light emitting diode
MRI	Magnetic resonance imaging
PMT	Photomultiplier tube
TCSPC	Time-correlated single photon counting
TR-FIA	Time-resolved fluoroimmuno assay
TRFM	Time-resolved fluorescence microscopy

1 Introduction

Time-resolved fluoroimmunoassays (TR-FIA) such as the Dissociation enhanced lanthanide fluoroimmunoassay (DELFIA) system have been successfully introduced in clinical chemistry in the 1970s. The principle of the assay is based on the difference between the fluorescence lifetime of the specific signal and the nonspecific background (scattered and reflected excitation light, autofluorescence of the object itself). When the fluorescence decay time of the label is sufficiently longer than the average background decay, the specific signal can be recorded after the background signal has decayed. This principle was used to suppress background signals [1, 2] using nanosecond-pulsed tunable lasers and for immunofluorescence applications using pyrene derivates as markers [3]. Whereas unwanted scattered excitation light could be suppressed in the early days of fluorescence microscopy by applying special optical configurations and by high quality multilayer optical filters, autofluorescence caused by intracellular components such as flavins, vitamins and co-enzymes was hard to suppress in this way, as it had broad emission characteristics ranging from the blue to the red part of the spectrum, at levels often equivalent to thousands of fluorescein molecules per cell [4]. Notably, for the majority of applications of fluorescence microscopy fluorescein and rhodamine derivatives were the fluorophores of choice. Modern labels with specific photophysical properties (photostability, far red emission), such as the Cyanin and Alexa dye series, were not available at that time. Autofluorescence (and also fixative induced fluorescence) is caused by many fluorescent substances, each with their own life time. Generally, the decay time is relatively fast (less than 100 ns) and therefore interferes fully with fluorophores such as fluorescein and rhodamine dyes with a decay time in the nanosecond range. However, disturbing autofluorescence could be easily suppressed by time-gated detection in case of emission of a dye with microsecond or millisecond decay.

The DELFIA system was introduced in clinical chemistry to successfully compete with radio-isotope based assays at that time, and therefore draw the attention of anyone interested in improving the sensitivity of fluorescence microscopy. However, before lanthanides could be successfully applied for TRFM a few principle hurdles and some inconveniences had to be overcome. Lanthanides based assays such as DELFIA consisted of two steps. The labeled antibody reacted with a target in an Elisa type format, after which the lanthanide (virtually non-fluorescent in aqueous solution) was separated from the site of binding and was encapsulated in the supernatant in a micelle type structure of beta-diketone to produce its fluorescent form: not applicable to microscopy, as the topological information was thereby lost. Erkki Soini and his colleagues in Turku recognized this problem, and developed a series of chemical compounds that strongly encapsulated the lanthanide (so-called chelates) thereby preventing the quenching effect of water molecules. These reagents were applied for immunocytochemical localization of smooth muscle myosin in histological sections. A standard epi-fluorescence microscope was used and the red lanthanide chelate was recorded (not time-resolved yet) on photographic film [5]. Furthermore, direct visualization of the lanthanide chelates by eye was cumbersome: the quantum efficiency of the lanthanide chelates is lower than of for instance that of fluorescein, the absorptivity was initially low (later on improved considerably by using chromophores – "antenna molecules" – in the chelate compounds) and their photobleaching was significant. But above all, prompt fluorescent dyes with nanosecond decay time have a much shorter duty cycle than delayed fluorescent or phosphorescent dyes with microsecond or millisecond decay time. Thus, at saturation of the dye – although that rarely occurs for most microscopic imaging methods – the amount of photons produced is inversely proportional to the decay time: the faster the dye decays (short duty cycle) the more photons can be emitted per time unit. This prompted the use of integrating detection devices, with or without signal amplification.

The introduction of a slow-scan charge-coupled device (CCD) camera for fluorescence microscopy [6] presented a suitable alternative to recording by film and CCD cameras were rapidly introduced as part of TRFM hardware [7–10]. Presently, two decades later a series of lanthanide based reagents with relatively high quantum efficiency and acceptable photostability exist and there are many chelating agents with relatively high quantum yield under aqueous conditions, and with improved absorptivity due to the chemical incorporation of chromophores in the chelate structure [13]. In addition, nanosized particulate labels in which lanthanides are encapsulated in polymers (polystyrene beads) or are part of an inorganic crystal structure (for instance yttriumoxysulfide in down- and upconverting phosphors) have been introduced for microscopic imaging. TRFM is obviously not restricted to imaging of specimens stained with lanthanide based reagents. Porphines and Pt and Pd co-proporphyrins have been successfully applied as well [11, 12].

This review will focus on hardware configurations and imaging principles applied to lanthanide stained specimens, discuss the main strategies for TRFM, and discuss the advantages and disadvantages of each of them. In addition, fluorescence life time imaging microscopy (FLIM) and time-correlated single photon counting (TCSPC) will be shortly discussed, as they have become powerful and widely used techniques for TRFM, albeit not primarily for the visualization of lanthanides.

2 Strategies for TRFM

As for other forms of fluorescence microscopy the choice for a certain TRFM technique is determined by image quality, sensitivity of detection and speed of recording, apart from the obvious issue of costs of hardware. As one would expect there is no ideal system for all applications, as for instance high speed acquisition generally negatively impacts sensitivity and the use of highly sensitive detection devices (for instance, some types of image intensifiers) may compromise image quality.

First, the photophysical properties of the lanthanides have to be considered. Whereas the lanthanides Eu^{3+}, Tb^{3+}, Sm^{3+}, and Dy^{3+} all have been used in fluoroimmunoassays, the vast majority of time-resolved microscopy applications relates to chelates of Eu^{3+} and Tb^{3+}, mainly because of their favorable chemistries and relative brightness. Europium and terbium chelates have a maximum excitation wavelength at 292 and 335 nm and maximum emission at 615 and 545 nm, respectively, with some variation depending on the type of chelate used. Their decay times vary considerably depending on the type of chelate and preparation embedding conditions of the specimen and range from microsecond to millisecond [13]. Thus, one needs (1) a UV light source, (2) a detector systems sensitive to approximately 500–650 nm light, and (3) hardware and specialized electronics that allows time-gating in the microsecond to millisecond range.

For TRFM of lanthanides light sources providing UV or near UV have been applied, either in pulsed form (xenon flash lamps) or as continuous radiation [xenon and mercury lamps, argon lasers, or light emitting diodes (LEDs)] in combination with light chopping modules. Importantly, for maximum emission the excitation wavelength should not only match the absorption maximum of the lanthanide complex (also taking the properties of the chelate into account) but the duration time of the resulting pulse must also be sufficiently long to effectively populate the excited state of the lanthanide to produce many emitted photons. Typically, the time needed to fully populate the excited state is in the same order as needed for depopulation, e.g., the decay time. Subsequently, the excitation is ended, and after a given time window, the emitted photons are integrated. Setting the time window is critical: sufficiently long to allow the process of prompt (auto)fluorescence to end but also to avoid disturbing "after glow" of for instance a pulsed xenon lamp; however, not too long in order not to miss specific emission (Fig. 1). Hennink et al. [14] calculated (and experimentally verified) that the detection efficiency of any time-resolved microscope is maximally 50%, whereas 42% was achievable in practice.

In adapting (digital) fluorescence microscopy one basically has to distinguish two main categories, depending on how the image is recorded. (1) *Image plane scanners*: that is the object is entirely illuminated, an intermediate image is formed by the objective lens which is recorded (scanned) by a detector (typically a CCD camera) at the image plane; essential is that the recording takes place *simultaneously* for each pixel. This type of scanning is applied for wide-field microscopes. (2) *Object plane scanners*: a focused beam of light is scanned over the sample and the fluorescence of each pixel is measured *sequentially* by a detector (photomultiplier or photodiode). Note: in principle one could also move the object through a fixed focused beam of light but that would be considerably slower. Confocal or multi-photon scanning to achieve optical sectioning of the specimen is a widely used variation of object plane scanning.

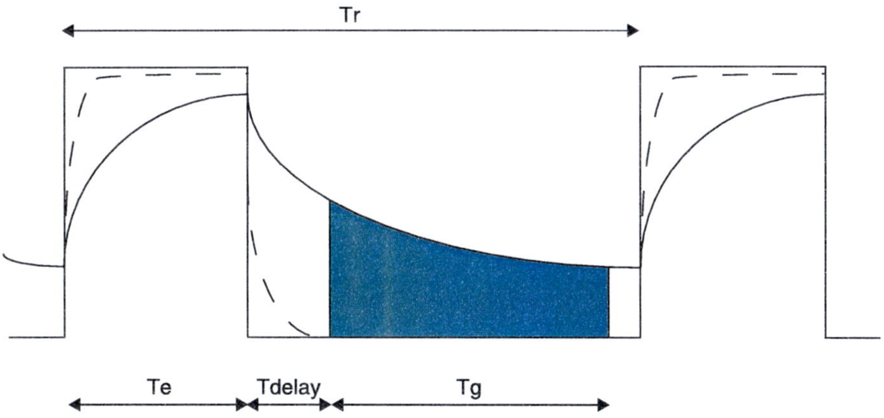

Fig. 1 The time-resolved detection scheme is based on a sequence of excitation pulses T_e, delay period T_{delay}, and detection window T_g, repeated with a repetition period T_r. The delay period allows the prompt fluorescence (*dashed line*) to decay before detection of the long lifetime luminescence (*solid lines*) starts. The *shaded area* indicates the detection window [14]

3 Configurations for Time-Resolved Fluorescence Microscopy

3.1 *Wide-Field Imaging Systems*

3.1.1 Systems Using Flash Lamps and Chopped Emission Light

The first wide-field TRFM systems applied xenon flash lamps for excitation, probably driven by the fact that this light source was originally used in the hardware of the DELFIA system. A rotating disc, synchronized with the xenon flash lamp was used at the emission side for the time-gated detection. Figure 2 shows two basic configurations. The system by Beverloo et al. [7] in Leiden was developed to visualize luminescent phosphor particles. As these particles were bright, chopping was performed in the primary image plane located below the prism by which the light was directed to the CCD camera and to the binoculars for direct visualization. The lamp was operated at 50 Hz, and the chopper (with transparent and non transparent segments) at 6,000 rpm. In Turku, Seveus et al. [9] describe a similar system to record the delayed luminescence of lanthanides, in which the disc was positioned in an image plane created above the prism, implying that delayed luminescence could not be directly visualized by the human eye.

The configurations shown in Fig. 2 used a relatively long T_{delay}, not only to suppress the autofluorescence but also to circumvent the tardy afterglow of the xenon lamp (about 6 μs), which is well suited to visualize lanthanides with decay times of several 100 μs. However, for shorter-lived dyes (10–50 μs) this setup is inefficient as a longer T_{delay} would lead to cut-off too much of the emitted lanthanide luminescence.

3.1.2 Systems Employing Choppers for Excitation and Emission

Marriott et al. [15] constructed a phosphorescence microscope, in which a laser or an arc lamp was used. The excitation light was focused onto a rotating excitation chopper blade so that the on–off transit time of the excitation pulse was kept at a minimum. A quartz optic fiber was used to lead the light into the microscope. The emission chopper and slit assembly were placed as close as possible to the CCD camera. The two choppers were rotated at the same speed and phase locked. Note that in all setups a small aperture plane of excitation light is created for fast and effective chopping (Fig. 3).

In 1994, ferroelectric shutters (FLC) were introduced as an alternative to mechanical shutters [11]. FLC devices are polarization based shutters that can be switched in about 50 μs. Using two shutters simultaneously prompt autofluorescence was effectively suppressed by a factor of 10^6. FLC shutters have poor transmission for UV light and were therefore only used at the emission side. Also, the switching time of 50 μs allows detection of relatively long-lived luminescence only.

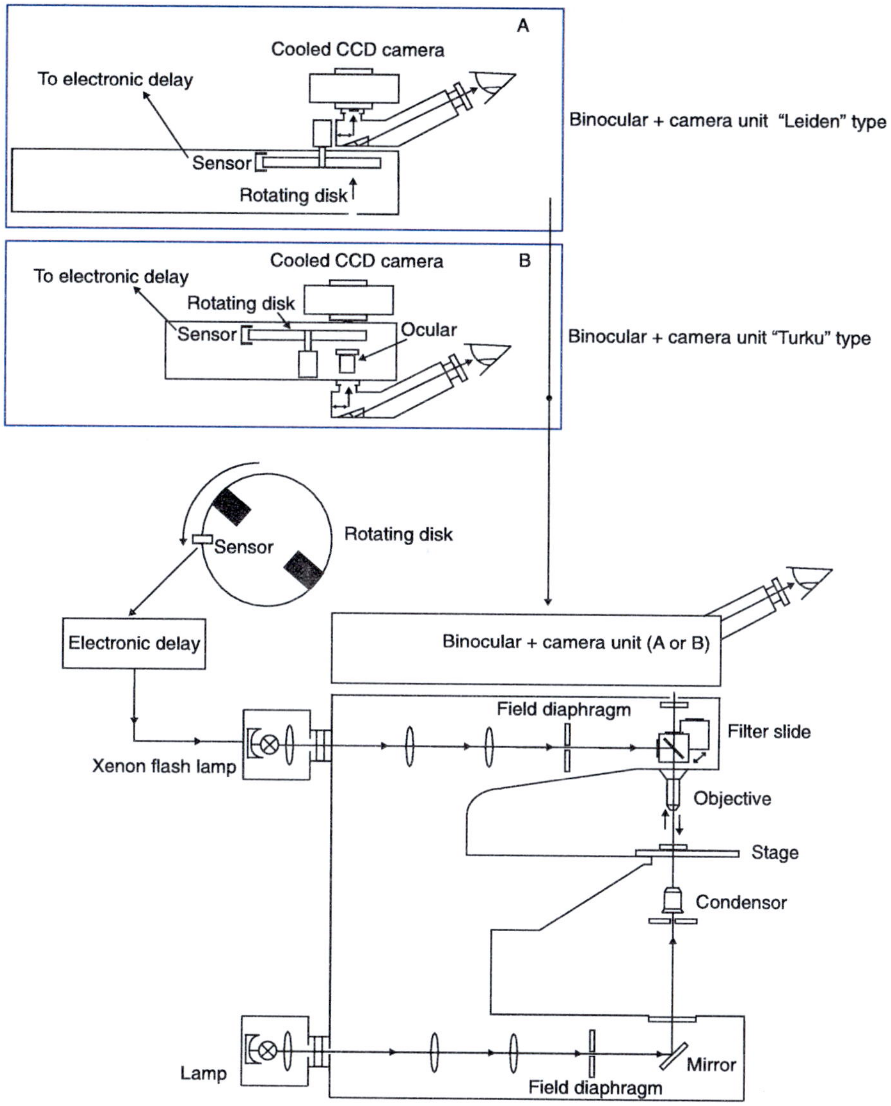

Fig. 2 Time-resolved microscope set up using a xenon flash lamp and the chopper at the emission side synchronized to the flash lamp to perform the time-gating of the luminescence. Depending on the position of the emission chopper, below (section A) or above (section B) the prism delayed luminescence can also be visualized by the human eye

3.1.3 Systems Using Gated Image-Intensified Cameras

A micro-channel plate (MCP) image intensifier can be considered as an array (typically 1,024 × 1,024) of small (about 10 μm in size) photomultiplier channels

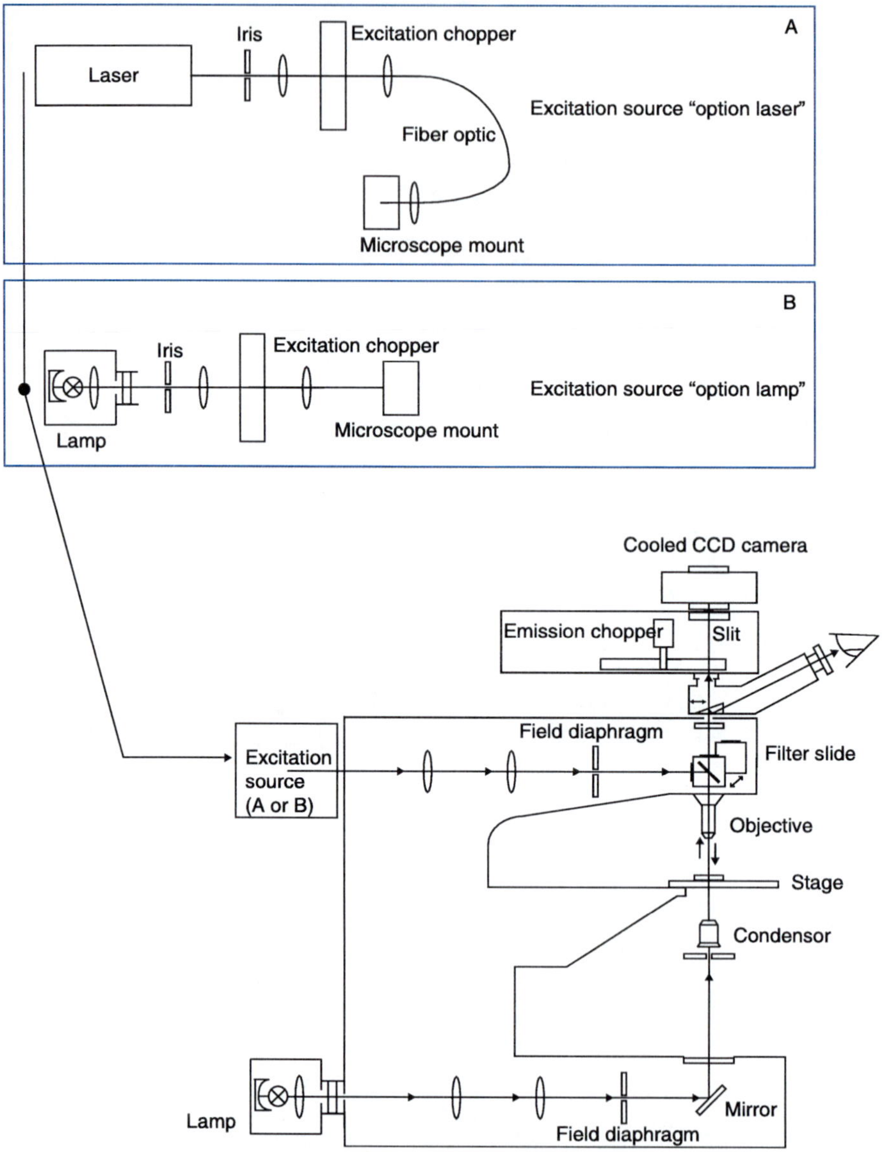

Fig. 3 Scheme of a phosphorescence microscope [10, 15]. Fluorescence excitation is performed by chopping a laser (section A) or an arc lamp (section B)

with a fluorescent end target, that produces an intensified image. The end target is coupled to a CCD detector. A gating electrode is used to rapidly switch the MCP in conduction, allowing amplification of the signal by a factor of 10,000 or more, and gating of the detector with nanosecond resolution. Gated image-intensified cameras

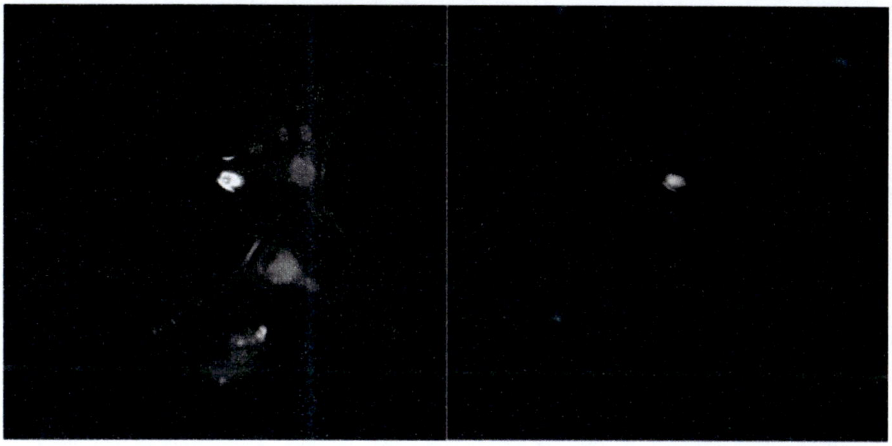

Fig. 4 A *Giardia lamblia* cyst labeled with monoclonal antibody conjugated to the europium chelate BHHST. The cysts were spiked into a matrix of algae, desmids and debris isolated from the municipal water supply and labeled in situ. Excitation was by means of a Nichia UV LED (NCSU033A) driven for 800 ms at 1 Amp and detection was delayed for 3.2 ms before the EM-CCD camera was triggered. *Left*: conventional fluorescence image (with much autofluorescent material present); *Right*: time-gated image, showing cyst only (kindly provided by Dr. Russell Connally; Sydney; see also [16])

were used by Hennink et al. [14] for time-resolved recording of phosphorescence of Pt-porphines (decay time 47 µs).

Recently an electron-multiplying CCD (EM-CCD) camera was introduced for time-gated imaging [16]. EM-CCD cameras are the solid-state equivalent of image-intensified CCD cameras. Their gain is lower which is compensated to some extend by a threefold better quantum sensitivity. For some applications a single exposure cycle is sufficient; however, if multiple delayed fluorescence cycles have to be integrated an external shutter mechanism is still needed [16]. The same group also introduced a relatively cheap but powerful UV LED as excitation light source [17]. Today's UV LEDs can deliver 270 mW of 365 nm light in CW mode and 500 mW in pulsed mode, which is more than sufficient for most lanthanides, despite the fact that excitation is not at the maximum absorption wavelength (Fig. 4).

Table 1 gives an overview of the various hardware configurations for wide-field time-resolved fluorescence microscopy (TRFM).

3.2 Fluorescence Lifetime Imaging Microscopy

Time-resolved image acquisition is also achievable by fluorescence lifetime imaging microscopy (FLIM). FLIM measures the fluorescence decay of each pixel of an image, on the basis of which time-resolved imaging can be performed. FLIM has two main variants: the time-resolved information is established either in the *time*

Table 1 History of time-gated luminescence (phosphorescence) microscope (modified after [16])

Year	Author[a]	Excitation	Detection	Luminophore
1988	Soini et al.	Hg lamp, CW	Eye	Eu^{3+} chelate (W1014)
1990	Beverloo et al.	Xe flashlamp	CCD + chopper	Y_2O_2 S:Eu
1992	Seveus et al.	Xe flashlamp	CCD + chopper	Eu^{3+} chelate (W1014)
1994	Marriott et al.	Hg lamp + chopper	CCD + chopper	Eu^{3+} BCPDA
1994	Verwoerd et al.	Xe-arc + chopper	CCD + ferro-electric shutter	Porphine labeled Sephadex
1995	Phimphivong et al.	He-Cd laser + chopper	CCD + ferro-electric shutter	Tb^{3+} chelate
1996	Hennink et al.	Argon laser + AOM	Gated MCP image int./CCD	Pt-porphine
1996	De Haas et al.	Xe flashlamp	CCD + ferro-electric shutter	W-8044 Eu^{3+}chelate
1997	Rulli	Xe flashlamp	Gated CCD	Eu^{3+} conjugated anti IgG
1997	De Haas et al.	Argon laser + AOM	Gated MCP image int./CCD	Pt porphyrin excited at 514 nm
1998	Vereb et al.	Xe flashlamp	Gated MCP image int./CCD	Eu-DTPA-CS124
1998	Phimphivong et al.	He-Cd laser + chopper	CCD + ferro-electric shutter	Tb^{3+}p-amino salicylaldehyde
1999	De Haas et al.	Hg-arc + chopper	CCD + ferro-electric shutter	Pt and Pd co-proporphyrins
2002	Connally et al.	Xe flashlamp	Gated MCP image int./CCD	Eu^{3+}-BHHCT
2003	Soini et al.	Xe flashlamp	CCD + chopper	Eu^{3+}, Tb^{3+}, Pt-Pd CPP
2004	Connally et al.	Xe flashlamp	Gated MCP image int./CCD	Eu^{3+}-BHHST
2006	Connally et al.	UV LED	Gated MCP image int./CCD	Eu^{3+}-BHHST
2008	Connally et al.	UV LED	EMCCD	Eu^{3+}-BHHST

[a]References [5, 7, 9–14, 17–22] can be found at the end of this chapter

domain or in the *frequency domain*. In the time domain, the specimen is excited by short pulses of light and the fluorescence decay curve is directly acquired after a given delay by sampling the fluorescence emission in time with a suitable detector; for instance, a high-speed gated image-intensified camera or a directly gated CCD camera. Frequency domain based FLIM uses intensity modulated excitation light and phase-sensitive detection, by which the fluorescence decay time for each image pixel can be calculated (see also Sect. 5b). In principle time and frequency domain FLIM can be applied for wide-field microscopy as well as for object plane scanning. However, as simultaneous acquisition of pixel information is attractive, most applications are in wide-field imaging systems. FLIM on the basis of TCSPC is an example of an (object plane) scanning technique. These three technologies are discussed in short below. Extensive reviews of FLIM methodology have been written by Herman et al. [23], Clegg et al. [24], Suhling et al. [25], and Suhling [26].

3.2.1 FLIM (Time Domain)

For a full description of the measuring principle, see Sect. 5a. Fluorescence lifetime imaging is easily implemented in (confocal) laser scanning systems [27, 28]. Light from an argon-ion laser is chopped with a fast electro-optical chopper to produce 2-ns pulses at 25 MHz. The fluorescence emission is detected sequentially in two time-gated windows, each delayed by a different time with respect to the excitation pulse. The system is quite flexible as both delay times and the time-width of the windows can be set independently. The fluorescence lifetime can be calculated from the accumulated photons in the time windows A and B. Assuming a mono-exponential decay of the fluorescence intensity and time windows of equal width, the fluorescence lifetime τ is given by

$$\tau = \Delta t / \ln(I_A / I_B),$$

where Δt is the time delay between the two windows and I_A and I_B are the corresponding fluorescence intensities. This process may be repeated as often as possible/necessary, after which the intensities accumulated in the two windows are read by a frame-grabber for all pixels in the image. This method was originally applied to image oxygen in living cells using RTDP, a fluorescent ruthenium derivative of which the lifetime linearly decreases with increasing oxygen concentration [27]. FLIM (time domain) is very well suited for imaging of long-lived lanthanide luminescence as the corresponding decay times are easily separated from fast decaying biological substances.

3.2.2 FLIM (Frequency Domain)

For a full description of the measuring principle, see Sect. 5b. A sinusoidally modulated excitation beam of light and detector are used to measure the phase shift and demodulation of the fluorescence signals with respect to their excitation signals. A fluorescence lifetime is then calculated from the phase shift and demodulation data. Particularly, wide-field frequency based FLIM is frequently used in cell biology and biophysics, as parallel acquisition of all image pixels is readily performed. The method is particularly suited to measure FRET (fluorescence resonance energy transfer) that occurs between donor and acceptor fluorophores at Forster distance (20–60 Å). FRET-FLIM allows the study of the interactions of proteins in living cells, which have been tagged with differently colored fluorescent proteins [29, 30] (Fig. 5).

3.2.3 Time Correlated Single Photon Counting

In TCSPC, the sample is excited with a picoseconds laser pulse and each single emitted photon is recorded with picoseconds accuracy. The sample is exited at MHz

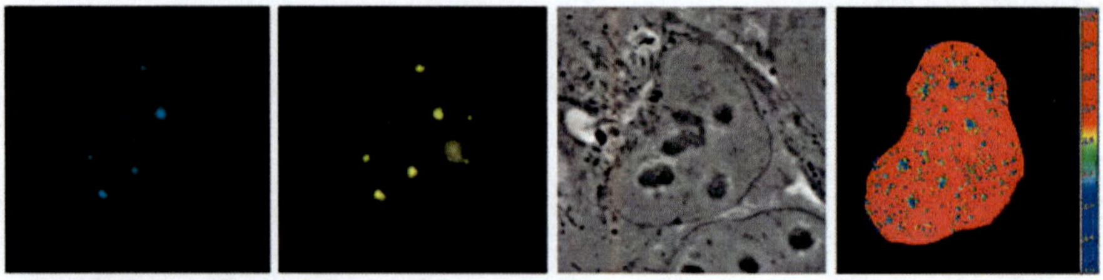

Fig. 5 Interaction of HP1β (a protein involved in chromatin organization) and SP 100 (a splicing factor) in PML nuclear bodies studied by FRET-FLIM . FRET between donor (CFP label) and acceptor (YFP label) leads to a significant reduction of the lifetime of the donor fluorescence. From *left* to *right*: CFP HP1β image, YFP-SP100 image, phase-contrast image, and lifetime image (scale: *red* = high; *blue* = low) (Images kindly provided by Dr. Roeland Dirks, Leiden, The Netherlands)

frequencies and dedicated electronics are used to build up a probability distribution histogram of fluorescence photons, which in fact represents the decay curve. TCSPC is a very powerful and accurate measuring technique. It provides high accuracy and reproducibility, single photon sensitivity, wide dynamic range and linear recording characteristics, not sensitive to fluctuations in excitation intensity and detector gain settings. Drawback of TCSPC is that for high accuracy the technique is time-consuming as the timing of the photons has to be performed on a pixel-by-pixel basis. The method is not very suitable for measuring long decay times, as for instance of the lanthanides that are discussed in this chapter. For further reading, see reviews on TCSPC by Becker et al. [31, 32] and Gratton et al. [33].

4 Discussion

Whereas time-resolved fluorescence based bioassays have achieved widespread use, till date TRFM for imaging of fluorescent lanthanide labels is not a frequently used technique compared to conventional wide-field or confocal fluorescence microscopy, despite its clear advantages. The main reason seems to be a "vicious circle situation." The first fluorescent lanthanide chelates by researchers at Wallac Oy and Turku University were not optimally suited for imaging applications (see introduction of this chapter): they had low quantum yields under aqueous conditions, some required secondary encapsulation/chelating steps, showed considerable photo-bleaching and relatively low absorption coefficients. Therefore, only a limited number of research groups developed imaging applications, and wide-scale introduction, for instance, for diagnostic applications (as was the case for DELFIA) never took place. Furthermore, other methods and fluorophores were introduced to circumvent the relatively high autofluorescence of biological material. Red and far-red sensitive CCD cameras allowed the use of fluorophores that emit in the red/far red region upon excitation with orange/red light, for instance the cyanine dyes [34]. At this high excitation wavelength autofluorescence of biological tissue and cells is less prominent compared to fluorophores that require UV, violet or blue excitation light. Secondly, using spectral unmixing, image processing and subtraction of autofluorescence the signal-to-noise of fluorescently stained specimens can be enhanced [35].

Consequently, the main suppliers of commercial fluorescence microscopes never seriously considered producing dedicated instrumentation for TRFM. In the past two decades, however, confocal microscopy and its multi-photon variants have

Table 2 Summary of advantages/strengths and disadvantages/weaknesses of various implementations of FLIM

	Advantages/strengths	Disadvantages/weaknesses
Time-gated Wide-field Time domain	• Fast • All pixels acquired in parallel • Suitable for long-lifetime probes such as lanthanides	• Low sensitivity; needs strong signal • Consecutive acquisition of time gates is sensitive to photobleaching and sample movement • Out-of-focus blur; lower resolution than provided by confocal FLIM • Pulsed laser needed; limited dynamic range
Wide-field Frequency Domain	• Fast • Parallel pixel acquisition • Easy to modulate continuous wave laser • No deconvolution (temporal) of system response and fluorescence decay necessary	• Sensitive to photobleaching and sample movement • Complex data and error analysis • Out-of-focus blur and lower spatial resolution than provided by confocal FLIM
Confocal/multi-photon scanning with TCSP	• Single photon sensitivity • Wide dynamic range • Linear recording characteristics independent of variations in excitation intensity or detector gain, and of photobleaching • Easy visualization of fluorescence decays and well-defined Poisson statistics • Inherent optical sectioning • Best signal-to-noise ratio • High temporal resolution	• Slow; each photon has to be timed individually • Pulsed laser needed • Unsuitable for very long lifetime probes, e.g., lanthanides • Saturates at high count rates
Confocal/multi-photon scanning with time-binning photon detection	• Fastest scanning technique • Inherent optical sectioning • Single-photon sensitivity	• Pulsed laser needed • Slower than wide-field imaging • Less accurate than TCSPC
Streak camera FLIM	• Potential for very high temporal resolution • Fast • Easy visualization of fluorescence decays	• Expensive • Limited dynamic range • Pulsed laser needed • Risk of non-linearities/ saturation

become a major tool for analysis of living cells, and variants of FLIM techniques are increasingly used to study protein–protein interactions (FLIM-FRET) or to measure the lifetime of a fluorophore to study its microenvironment. The advantages and disadvantages of the various FLIM approaches have been well reviewed [26] (see Table 2 for a summary).

As is clear from this Table not all FLIM approaches are suited for imaging of long-lived lanthanides and apart from this, some alternatives are much cheaper. The work by Connally and colleagues shows that relatively inexpensive equipment for wide-field TRFM can be built using UV LEDs as light source. Thus, wide-field (non-FLIM based) TRFM finally may become of age. Applications are to be expected in all areas where specimens notorious for autofluorescence have to be studied, such as forensic applications, plant research (chlorophyll fluorescence often dominates!) or microbiology of water and soil samples. Also for diagnostic pathology of paraffin embedded tissue sections applications are clearly there. There is a tendency in diagnostic pathology to look at more than two markers at a time, which is difficult to accomplish by bright field enzymatic staining. For multiple markers, fluorescence is the methods of choice, but up-till-now mainly restricted to frozen sections, as paraffin-embedded formalin-fixed specimens often suffer from too much autofluorescence. Notably, pathologists have become acquainted with digital microscopy, and increasingly look at digital images instead of seeing an analog image through a set of eyepieces.

Lastly, besides the fact that further improvement of hardware for TRFM is to be expected, new types of labels are being developed. Of particular interest is the new technology to produce "particulate" labels; that is, nanospheres consisting of polymers in which fluorescent molecules have been incorporated, surrounded by a nanosize layer of chemically modified silica to which bioreagents can be coupled. These labels can also be made electron dense (and therefore suitable for EM imaging) or filled with elements suited for magnetic resonance imaging (MRI), for instance gadolinium. The ideal nanosize particle has all three properties, and therefore is perfectly suited for correlative cell imaging. This technique provides a unique tool to visualize the affinity of new targeted contrast agents (and therapeutics) on a cellular level. Techniques combining EM and LM markers can be used for the *in vitro* validation of molecular imaging agents that target vulnerable plaques (e.g., in cardiovascular disease) and certain cancer markers (e.g., in breast and colorectal cancer). Labels combining LM and MRI are promising candidates for *in vivo* diagnostic imaging.

References

1. Thaer AA, Sernetz M (1973) Fluorescence techniques in cell biology. Spinger verlag, Berlin
2. Sacchi CA, Swelto O, Prenna G (1974) Pulsed tunable lasers in cytofluorometry. Histochem J 6:251–258
3. Mueller W, Hirschfeld T (1977) Background reduction in fluorochrome staining. Histochem J 9:121–123

4. Aubin JE (1979) Autofluorescence of viable cultured mammalian cells. J Histochem Cytochem 27:36–43
5. Soini EJ, Pelliniemi LJ, Hemmila IA et al (1988) Lanthanide chelates as new fluorochrome labels for cytochemistry. J Histochem Cytochem 36:1449–1451
6. Hiraoka Y, Sedat JW, Agard DA (1987) The use of a charge-coupled device for quantitative optical microscopy of biological structures. Science 238:36–414
7. Beverloo HB, van Schadewijk A, Gelderen-Boele S et al (1990) Inorganic phosphors as new labels for immunocytochemistry and time-resolved microscopy. Cytometry 11:784–792
8. Beverloo HB, van Schadewijk A, Bonnet J et al (1992) Preparation and microscopic visualization of multicolor luminescent immunophosphors. Cytometry 13:561–570
9. Seveus L, Vaisala M, Syrjanen S et al (1992) Time-resolved fluorescence imaging of europium chelate label in immunohistochemistry and in situ hybridization. Cytometry 13:329–338
10. Marriott G, Heidecker M, Diamandis EP (1994) Time-resolved delayed luminescence image microscopy using an europium chelate complex. Biophys J 67:957–965
11. Verwoerd NP, Hennink EJ, Bonnet J et al (1994) Use of ferroelectric liquid crystal shutters for time-resolved fluorescence microscopy. Cytometry 16:113–117
12. de Haas RR, van Gijlswijk RPM, van der Tol EB et al (1997) Platinum porphyrins as phosphorescent label for time-resolved microscopy. J Histochem Cytochem 45:1279–1292
13. Soini AE, Kuusisto A, Meltola N et al (2003) A new technique for multiparameter imaging microscopy: use of long decay time photoluminescent labels enables multiple color immunocytochemistry with low channel-to-channel crosstalk. Microsc Res Tech 62:396–407
14. Hennink EJ, de Haas R, Verwoerd NP et al (1996) Evaluation of a time-resolved fluorescence microscope using a phosphorescent Pt-Porphine model system. Cytometry 24:312–320
15. Marriott G, Clegg RM, Arndt-Jovin DJ et al (1991) Time-resolved imaging microscopy. Biophys J 60:1374–1387
16. Connally RE, Piper JA (2008) Time-gated luminescence microscopy. Ann NY Acad Sci 1130:106–116
17. Connally RE, Veal DA, Piper J (2006) High intensity solid-state UV source for gated time-gated luminescence microscopy. Cytometry 69A:1020–1027
18. Phimphivong S, Kolchens S, Edmiston PL et al (1995) Time-resolved, total internal reflection fluorescence microscopy of cultured cells using a Tb chelate label. Anal Chim Acta 307:403–417
19. de Haas RR, Verwoerd NP, van der Corput MPC et al (1996) The use of peroxidase-mediated deposition of biotin-tyramide in combination with time-resolved fluorescence imaging of europium chelate label in immunocytchemistry and in situ hybridization. J Histochem Cytochem 44:1091–1099
20. Rulli M, Kuusisto A, Salo J et al (1997) Time-resolved fluorescence imaging in islet cell autoantibody quantitation. J Immunol Methods 208:169–179
21. Vereb G, Jares-Erijman E, Selvin PR et al (1998) Temporally and spectrally resolved imaging microscopy of lanthanide chelates. Biophys J 74:2210–2222
22. Connally R, Veal D, Piper J (2004) Flash lamp excited time-resolved fluorescence microscope suppresses autofluorescence in water concentrates to deliver 11-fold increase signal to noise ratio. J Biomed Opt 9:725–734
23. Herman P, Lin HJ, Lakowicz JR (2003) Lifetime-based imaging. In: Biomedical photonics handbook. CRC, Boca Raton, pp 9.1–9.30
24. Clegg RM, Holub O, Gohlke C (2003) Fluorescence lifetime-resolved imaging: measuring lifetimes in an image. In: Methods in enzymology. Academic, New York, pp 509–542
25. Suhling K, French PMW, Phillips D (2005) Time-resolved fluorescence microscopy. Photochem Photobiol Sci 4:13–22
26. Suhling K (2006) Fluorescence lifetime imaging. In: Stephens S (ed) Cell imaging. Scion Publishing Ltd, Bloxham, UK
27. Gerritsen HC, Sanders R, Draaijer A et al (1996) Fluorescence imaging of oxygen in living cells. J Fluoresc 7:11–15

28. Gerritsen HC, Asselbergs NAH, Agronska V et al (2002) Fluorescence lifetime imaging in scanning microscopes: acquisition, speed, photon economy and lifetime resolution. J Microsc 206:218–224
29. Jovin TM, Marriott G, Clegg RM et al (1989) Photophysical processes exploited in digital microscopy: fluorescence resonance energy transfer and delayed luminescence. Ber Bunsenges Phys Chem 93:387–391
30. Lakowicz JR, Szmacinski H, Nowaczyk K et al (1992) Fluorescence lifetime imaging. Anal Biochem 202:316–330
31. Becker W, Bergmann A, Hink MA et al (2004) Fluorescence lifetime imaging by time-correlated single-photon counting. Microsc Res Tech 63:58–66
32. Becker W, Bergmann A, Haustein E et al (2006) Fluorescence lifetime images and correlation spectra obtained by multidimensional time-correlated single photon counting. Microsc Res Tech 69:186–195
33. Gratton E, Breusegem S, Sutin J et al (2003) Fluorescence lifetime imaging for the two-photon microscope: time-domain and frequency-domain methods. J Biomed Opt 8:381–390
34. Mujumdar RB, Ernst LA, Mujumdar SR et al (1989) Cyanine dye labeling reagents containing isothiocyanate groups. Cytometry 10:11–19
35. Mansfield JR, Hoyt C, Lebenson RM (2008) Visualization of microscopy-based spectral imaging data from multi-label tissue sections. Curr Protoc Mol Biol 12:14–19

Clinical Application of Time-Resolved Fluorometric Assays

Ulf-Håkan Stenman

Abstract The development of time-resolved fluoroimmunoassays (TRFIAs) was driven by the need to eliminate the use of radioactive isotopes in immunoassays and to develop automatic and ultrasensitive immunoassays. The use of time-resolved fluorescence became possible through the development of chelates that bound lanthanides with high affinity and could be coupled to proteins. This was combined with methods to dissociate the primary chelate together with the formation of a highly fluorescent new complex. The advantages of this technology are most efficiently used in sandwich assays using two antibodies, a solid phase antibody capturing the antigen and a detector antibody labeled with a lanthanide complex. This method facilitated a 10–100-fold lower detection limit in combination with an assay range covering 5–6 orders of magnitude. These advantages have facilitated elucidation of new biological phenomena and development of unique clinical applications. These have proved to be especially valuable for the determination of protein hormones and tumor markers. This review describes the clinical application of TRFIA.

Keywords Clinical applications · Hormone assays · Immunoassays · Time-resolved fluorescence

Contents

U.-H. Stenman
Department of Clinical Chemistry, Biomedicum, Helsinki University Central Hospital, Helsinki University, 00014 Helsinki, Finland
e-mail: ulf-hakan.stenman@helsinki.fi

P. Hänninen and H. Härmä (eds.), *Lanthanide Luminescence: Photophysical, Analytical and Biological Aspects*, Springer Ser Fluoresc (2011) 7: 329–342, DOI 10.1007/4243_2010_12,

Abbreviations

ACT	α1-Antichymotrypsin
Aio!	All-in-one
BPH	Benign prostatic hyperplasia
cpm	Counts per minute
DELFIAs	Dissociation-enhanced lanthanide fluoroimmunoassays
Dy	Dysprosium
Eu	Europium
fPSA	Free PSA
FSH	Follicle-stimulating hormone
GH	Growth hormone
hCG	Human chorionic gonadotropin
hCGβ	β Subunit of human chorionic gonadotropin
hK2	Human kallikrein 2
IEMA	Immuno-enzymometric assays
IFMA	Immunofluorometric assay
Ig	Immunoglobulin
IRMA	Immunoradiometric assay
KLK	Kallikrein-related peptidase
LH	Luteinizing hormone
MAb	Monoclonal antibody
PCR	Polymerase chain reaction
PSA	Prostate-specific antigen
PSA-ACT	Complex between PSA and ACT
RIA	Radioimmunoassay
Sm	Samarium
Tb	Terbium
tPSA	Total PSA (sum of free and complexed PSA)
TRFIA	Time-resolved fluoroimmunoassay, time-resolved fluorometric assay
TR-FRET	Time-resolved fluorescence resonance energy transfer

1 Introduction

The first quantitative immunoassays were based on the use of radioactive reporter molecules, mainly ^{125}I and ^{14}C. These radioimmunoassays (RIAs) have gradually been replaced by nonradioactive reporter molecules or "labels," e.g., enzymes, fluorescent, and luminescent molecules and combinations of these. The nonradioactive labels facilitated development of automated immunoassays, and they also contributed to the design of assays with substantially reduced detection limits. However, nonradioactive assay have an advantage only in immunometric sandwich assays; in inhibition assays, the use on nonradioactive labels has not provided any improvement in assay sensitivity as compared to RIA [1]. The development in immunoassay technology has been reviewed by Ekins [2].

2 Nonradioactive Reporter Molecules

The first automatic immunoanalyzers used fluorescent labels. These provided automation and short turn-around times, but due to a fairly high nonspecific background the sensitivity of assays was not improved over that obtained by RIA. Introduction of time-resolved fluorescence facilitated the development of assays with substantially reduced detection limit [1], here called improved sensitivity. This was based on reduction of the background, which facilitated the use of a large excess of label in sandwich-type immunometric assays and a much extended measuring range [3–5]. Other nonradioactive reporter molecules, e.g., luminescent labels or enzymes combined with luminescent or fluorescent substrates, can provide similar sensitivity of assays [2]. However, time-resolved fluorometry is unique in facilitating development of fully automated in-house assays, and it has therefore become widely used in research laboratories.

3 Time-Resolved Fluorometric Assays

The principle of time-resolved fluorometry has been described in detail elsewhere in this book (refer to the chapter by Bünzli [50]); this chapter only concerns clinical applications of time-resolved fluorometric assay (TRFIA) and comparison of this method with other assay techniques. The sensitivity of TRFIA is based on the use of chelates of rare earth metals (lanthanides), e.g., europium (Eu^{+3}), terbium (Tb^{+3}), samarium (Sm^{+3}), and dysprosium (Dy^{+3}). Of these, Eu usually provides the highest sensitivity and has therefore been most widely used [6]. The emission spectra and decay times of the lanthanides differ from each other, and it has therefore been possible to develop multilabel assays for simultaneous detection of up to four different antigens [7]. Simultaneous determination of several antigens is especially useful for screening purposes and for determination of two antigens [8], the ratio of which is clinically important [9, 10].

4 Early Developments

The first clinical application of time-resolved fluoroimmunoassay (TRFIA) was an ELISA-type assay for rubella antibodies [11], but the use of this technology for antibody determinations does not provide substantial advantages over ELISA assays. However, when applied to determinations of hormones and tumor markers occurring at low concentrations, the sensitivity of TRFIA can be fully used [3]. The high sensitivity is the result of several factors, i.e., the low fluorescence background in combination with the use of a large excess of tracer antibody that can be labeled with 5–10 Eu chelates. Furthermore, the large linear dynamic range of time-resolved fluorometry makes a measuring range of more than 4–5 orders of magnitude possible [3]. This is especially useful for assay on hormones and tumor markers that may occur at highly variable concentrations. The large linear dynamic range was also a result of the use of monoclonal antibodies (MAbs) for capture of the antigen to the solid phase. This is because more than 80% of MAb preparations are capable of binding antigen as compared to 1–3% of the antibodies in a polyclonal antiserum [12]. With polyclonal antibodies, the same binding capacity can only be achieved by affinity purification of the specific antibodies, which is tedious and requires large amounts of antiserum. Because monoclonal antibodies had become available when TRFIA was introduced, the advantages of this technology could be efficiently used.

5 TRFIA Methods

Application of time-resolved fluorescence in immunoassays became possible through the development of the dissociation-enhanced lanthanide fluoroimmunoassay (DELFIA) technology and time-resolved fluorometers for use with microtitration wells by Soini, Lövgren, Hemmilä, Pettersson, and Kojola [1, 3, 4]. The technology comprised methods for labeling of proteins with lanthanide chelates, dissociation of the lanthanide from its primary chelate, formation of a new fluorescent chelate in solution, and measurement of fluorescence in a time-resolved fluorometer (Fig. 1) [1, 3]. A large menu of assays for most clinically important hormones, tumor markers, and other disease markers are available for the Auto-DELFIA instrument (PerkinElmer, Turku, Finland), which is optimized for batch-wise assays. The Auto-DELFIA methods use relatively long incubation times to achieve high sensitivity in spite of small sample's volumes. A point-of-care instrument, DELFIA-Express, and a reader for manual assays, Victor, are also available. The Auto-DELFIA and Victor have been widely used for the development of in-house assays.

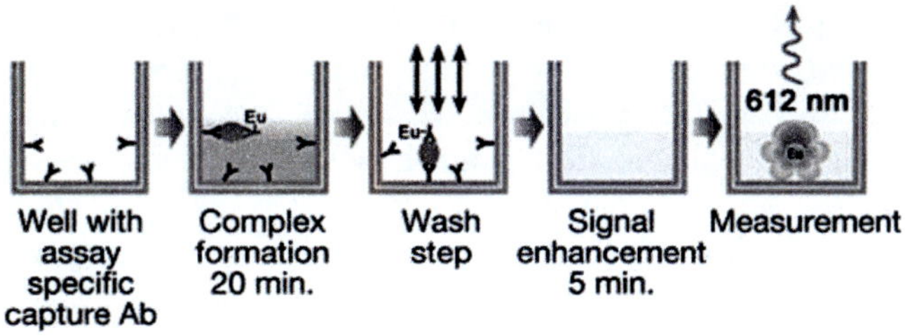

Fig. 1 Assay principle of DELFIA technology (courtesy of PerkinElmer)

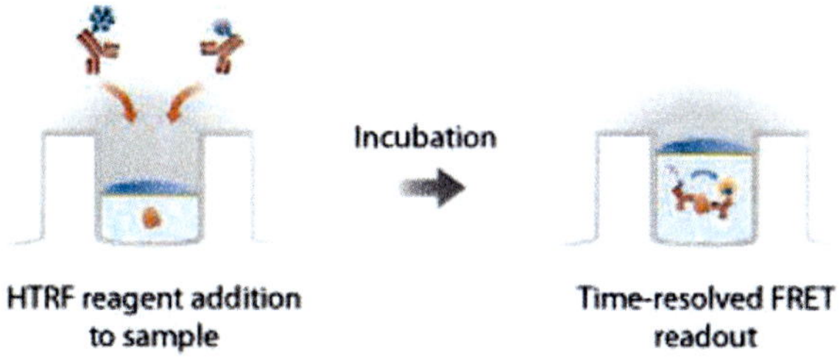

Fig. 2 Assay principle of TR-FRET (courtesy of Cisbio Bioassays)

The group of Lövgren has developed several novel variants of TRFIA. Using microparticles as the solid phase and an assay volume of only 330 nL, sensitive assays for CRP with 2–10 min assay times were developed [14]. They also developed an "all-in-one" (Aio!) analyzer in which all reagents are dried on the bottom of the assay wells. The system is characterized by rapid assay times and the possibility to analyze whole blood samples [15, 16]. Innotrac Diagnostics, Turku, Finland, introduced a point-of-care version of this analyzer.

Khoshravi et al. at Cybefluor used a different chelate, BCPDA, and measurement of fluorescence on the dried surface of microtitration wells [17]. They later enhanced the signal using alkaline phosphatase as a label and diflunisal phosphate as a substrate which, after cleavage by phosphatase, was detected by time-resolved fluorometry [18]. Commercial assays are not available, but a large number of sensitive research assays have been published by the group of Diamandis [19].

Mathis et al. combined time-resolved fluorometry with fluorescence energy transfer (TR-FRET) to develop homogeneous assays. Europium cryptate bound to one antibody is excited with a nitrogen laser and excites allophycocyanin coupled to another antibody (or antigen). Energy transfer occurs when the fluorophores are brought close to each other by binding to the same antigen (Fig. 2) [20, 21]. This technique was initially introduced by CiS-Bio under the brand name Kryptor, which was later acquired by Brahms Gmbh. Assays, for the most commonly determine tumor markers, many hormones and the acute phase reactant procalcitonin, and the screening markers PAPP-A, and hCGβ are available.

Lanthanide labels have also been used for the determination of nucleic acids. Eu chelates can be coupled to DNA probes that can be used to quantitate DNA amplification, e.g., in a polymerase chain reaction (PCR) [22, 23]. Linear DNA probes are labeled with a stable fluorescent terbium chelate, whose fluorescence is lower when coupled to the probe than when free in solution. When the probe anneals to a target amplicon, the exonucleolytic activity of DNA polymerase detaches the label increasing the terbium fluorescence [24]. An automatic instrument and assays for detection of pathogens in food products are available from Abacus Diagnostica (Turku, Finland).

6 Comparison of TRFIA with Other Techniques

The theoretical advantages of using TRFIA over RIA and conventional fluorescence methods have been discussed by Soini and Kojola. Their calculations indicate that Eu-labeled antigens would provide at least equal sensitivity as radiolabeled ones [1]. However, the full advantages of lanthanide labels are obtained in sandwich or immunometric assays using large amounts of labeled antibody [3]. The effect of antibody labeling has been studied in human chorionic gonadotropin (hCG) assays using the same antibodies and assay conditions and detector antibodies labeled with either a Eu chelate or ^{125}I. The results showed that the amount of tracer antibody was a key factor determining sensitivity [25]. With both labeling techniques, the highest signal-to-background ratio was obtained when using 5–50 ng of labeled antibody. Similar detection limits around 0.3 IU/L (corresponding to 1 pmol/L) was obtained for both assays when 5 ng of labeled antibody was used. The detection limit improved to 0.6 pmol/L with 50 ng of labeled antibody in the TRFIA but impaired slightly in the immunoradiometric assay (IRMA). The radioactivity of 5 ng of labeled antibody was 180,000 counts per minute (cpm) and that of 50 ng 1,800,000 cpm. In order to limit radiation hazards, the amount of radioactive tracer used in RIA methods is usually limited to 10,000–50,000 cpm. Thus, the advantages of using large amounts of monoclonal antibodies in immunometric assays cannot in practice be realized with isotope labeling [25].

The performance of TRFIA (IFMA), IRMA, RIA, and immuno-enzymometric assays (IEMA) for follicle-stimulating hormone (FSH) has been compared by Berger et al. [26]. When the same antibodies and assay condition were used, the IFMA had the lowest detection limit (2 ng/L) followed by the IEMA (8 ng/L), IRMA (100 ng/L), and RIA (10,000 ng/L). The assay range of IFMA (80,000-fold) was two orders of magnitude larger than that of any of the other assay [26].

Direct comparisons of TRFIA with other assays are rare, but based on information provided by various manufacturers indicate that similar sensitivity and a fairly wide assay range can be achieved with several automated immunoassay platforms, most of which use various forms of luminescence for detection [27]. However, in many cases a high sensitivity has been achieved by using a fairly large sample volume. In most DELFIA assays, the sample volume is 25 μL and total assay volume 225 μL. Improved sensitivity can be achieved using a larger sample volume and longer incubation times [28]. With increasing sample volumes, the effect of nonspecific interference increases, and this has to be compensated using calibrators containing antigen-depleted serum corresponding to the volume of the sample. This eliminates the average nonspecific effect [28], but because this effect may vary from one sample to another, the advantage of a large sample volume is often offset by variation in nonspecific interference [29]. The magnitude of this effect is seldom estimated.

Assay sensitivity is dependent on many aspects of assay design. In most automatic immunoanalyzers, the assay time is fairly short, i.e., in the range of 5–15 min. To reduce assay time, one-step incubation is mostly used. While this increases throughput, it adversely affects assay sensitivity and nonspecific interference. An advantage of the DELFIA technology is the possibility to develop in-house assays by optimizing incubation times and two-step incubation. Many research groups have used this to establish highly sensitive assays facilitating determination of very low concentrations of hormones and other biomarkers. Some TRFIA applications, in which high sensitivity or simultaneous determination of two or several antigens are important, are described below.

7 Clinical Applications of Time-Resolved Fluoroimmunoassays

7.1 Luteinizing Hormone and Follicle-Stimulating Hormone

An assay for luteinizing hormone (LH) was one of the first hormone assays based on the DELFIA technology [3]. The high sensitivity of this assay has facilitated studies on the role of LH during pubertal development. After the first years of life and until the onset of puberty, the serum concentrations of LH are low and undetectable by conventional immunoassays. At the onset of puberty, pulsatile secretion of LH can be detected by RIA methods, and this has been thought to initiate development of gonadal function causing increasing serum concentrations of estradiol in girls and testosterone in boys. When LH is measured by DELFIA methods, pulsatile secretion of LH can be detected already in prepubertal boys and girls [30]. Puberty is initiated by increasing concentrations of LH. An LH concentration in serum below 0.1 IU/L excludes puberty, while a concentration exceeding 0.2 IU/L reliably identifies the start of puberty. During pubertal development, the LH concentrations increase 100-fold [31]. When LH is measured in early morning urine, an increase in LH is detected up to 1 year before an increase in serum LH. This is apparently explained by an increase in nocturnal LH secretion before the increased serum levels are detected [32, 33]. The concentrations of FSH are higher than those of LH, and the increase during onset of puberty is less pronounced than that of LH [31].

Male contraception can be achieved by administration of androgens, e.g., testosterone, which suppresses pituitary gonadotropin secretion and thereby spermatogenesis. However, complete suppression of spermatogenesis is not always achieved and this can be expected to be related to the degree of gonadotropin suppression. However, in most cases, the concentrations of LH and FSH achieved by this treatment are below the detection limit of commercially available assays. To solve this problem, Robertson et al. developed DEFFIA-based LH and FSH assays with a sixfold improvement in sensitivity by increasing incubation times and sample volume from 25 to 100 μL. With these methods, the levels of LH and FSH could be measured in nearly all men undergoing testosterone treatment [28].

7.2 hCG, its Subunits and Fragments

Determination of hCG for diagnosis and monitoring of pregnancy is the most widely used immunoassay application [34]. High sensitivity is seldom needed for this purpose, but measurement of low concentrations of hCG and its beta subunit (hCGβ) is important in the diagnosis of cancer. hCG is produced by placental trophoblasts, trophoblastic tumors, and some germ cell tumors, especially testicular cancers. In addition, about 20–50% of virtually all nontrophoblastic tumors studied produce hCGβ [35, 36]. hCG has been considered very specific for pregnancy and trophoblastic tumors but use of highly sensitive immunofluorometric methods has shown that hCG occurs at low concentrations in men and nonpregnant women [37]. The upper reference range for hCG is age dependent increasing from 2 IU/L (6 pmol/L) in menstruating women to 6 IU/L (18 pmol/L) in postmenopausal women. In men under 60 years of age, the upper reference limit is 2 IU/L, increasing to 4 IU/L in older men. The concentrations of hCGβ are lower than those of hCG and the upper reference limit is 2 pmol in men and women irrespective of age [38]. Many hCG assays measure hCGβ together with hCG. Use of such assays for monitoring of tumors producing only hCGβ is of limited use because the upper reference for hCG + hCGβ (also called total hCG) is much higher (6–18 pmol/L) than for hCGβ alone [38]. Specific determination of hCGβ is especially useful for monitoring of patients with seminoma and nontrophoblastic cancers [35]. About half of the seminoma patients have elevated levels of hCGβ, while only 15% have elevated serum concentrations of hCG [39]. In half of those with elevated concentrations of hCGβ, the concentrations of total hCG would have been within the reference range [39].

7.3 Growth Hormone

Accurate determination of growth hormone (GH) requires highly sensitive assays. The use of monoclonal antibodies in combination with DELFIA technology has facilitated development of such assays [40, 41]. Using antibodies that differentiate between recombinant and pituitary GH, Bidlingmaier et al. developed assays for detection of recombinant GH for doping. They furthermore showed that methods based on DELFIA and chemiluminescence had similar sensitivity [42].

7.4 *Prostate-Specific Antigen and the Kallikrein-Related Peptidases*

Prostate-specific antigen (PSA) is a chymotrypsin-like enzyme belonging to the kallikrein-related peptidase (KLK) family, in which it is called KLK3. It is produced at high concentrations by the prostate and by prostatic cancer, for which PSA is a sensitive marker. However, because PSA is prostate specific rather than cancer specific, elevated serum levels are also caused by benign conditions of the prostate, of which benign prostatic hyperplasia (BPH) is very common in elderly men. In circulation, most of the PSA is bound to α1-antichymotrypsin (ACT), and the proportion of PSA-ACT is higher, while that of free PSA (fPSA) is lower, in patients with prostate cancer than in men with BPH [43]. Measurement of the proportions of either PSA-ACT or fPSA improves the specificity of PSA for cancer and, because pipeting errors are eliminated, simultaneous determination of the proportion of two forms improve measurement of the ratios of various forms [9, 10]. Total PSA, i.e., tPSA comprising fPSA and complexed PSA (mainly PSA-ACT) are captured by a monoclonal antibody; an Eu-labeled antibody detects fPSA, while total PSA is detected by a Sm-labeled antibody. Each form is detected by measuring the fluorescence of Eu and Sm using different wave lengths and time windows [9]. An analogous approach can be used to measure total PSA and PSA-ACT at the same time, but this approach is hampered by a fairly high nonspecific background caused by adsorption of ACT to the solid phase [10]. In women, PSA can be detected in breast fluid and in breast cancer tissue. A TRFIA method based on the Cyberfluor technology with a detection limit of 1 ng/L was used to demonstrate elevated serum levels in serum from breast cancer patients [18, 27].

Another member of the KLK family, KLK2 or human kallikrein 2 (hK2), is also specifically expressed in the prostate. Elevated serum concentrations of KLK2 occur in patients with prostate cancer and, contrary to PSA, the expression in cancer tissue is higher than that in the normal prostate. However, the serum concentrations of KLK2 are only about 1% of those of PSA. Thus, assays of very high sensitivity are needed to measure KLK2 in men with early prostate cancer. Using TRFIA, assays with a detection limit of 100 ng/L [44] and later 3 ng/L were developed [45]. Interestingly, Saedi et al. at Hybritech Inc. used the DELFIA technology to develop an assay for hK2 with a detection limit of <20 ng/L [46].

7.5 *Simultaneous Assay of Several Antigens*

The possibility to simultaneously determine several analytes is an advantage when both determinations are routinely performed for screening purposes. In addition to the simultaneous determinations of various forms of PSA, a dual label assay of alpha fetoprotein and hCGβ in maternal serum based on TRFIA has become widely used for screening of trisomy 21 in the fetus [47, 48].

7.6 Antibody Determinations

Determination of antibody concentrations in serum is important for diagnosis of microbial and autoimmune diseases. ELISA methods have become routine methods for these purposes. The enzyme labels used in ELISA methods can be replaced with Eu [11], but because the samples usually have to be diluted before assay, no advantage in sensitivity is achieved. However, the large measuring range of TRFIA can be used to eliminate the need for additional dilution of samples with high antibody titers, but this advantage has not been widely utilized. In conventional ELISAs, the antigen is coated onto microtitration wells, immunoglobulins (Ig) in the sample are allowed to bind to the antigen, and, after washing, the amount of bound Ig is quantitated with an enzyme-labeled antihuman Ig antibody. A high background caused by nonspecific adsorption of Ig in the sample to the solid phase often limits the sensitivity of the assay. A novel approach based on TRFIA eliminates the background. The serum sample is incubated in microtitration wells coated with streptavidin to which biotinylated antigen has been bound. After washing, antibodies captured to antigen on the solid phase are detected by incubation with the same Eu-labeled antigen, which binds to the other binding site of the captured antibody. The sensitivity and specificity of this novel approach, which uses the bivalent nature of antibodies, are similar to those of the much more tedious radio-binding assays. Other advantages are the wide, 1,000-fold, measuring range and the possibility of dual assay for two types of antibodies [49].

8 Conclusions

Time-resolved fluorescence was the first highly sensitive nonisotopic technique that was introduced to general use and, after nearly three decades, it still ranks among the most sensitive detection principles used for immunoassays. It has also found applications in nucleotide quantitation. Using nanoparticles doped with Eu as labels, sensitivity can be further improved. Even more sensitive detection methods have recently been described, e.g., methods using DNA amplification for detection. While these may provide a further reduction in detection limit, it remains to be shown whether they can be applied in clinical settings with biological samples outside research laboratories. One of the strengths of TRFIA is the possibility to establish fully automated methods for clinical and research purposes.

References

1. Soini E, Kojola H (1983) Time-resolved fluorometer for lanthanide chelates – a new generation of nonisotopic immunoassays. Clin Chem 29:65–68
2. Ekins RP (1998) Ligand assays: from electrophoresis to miniaturized microarrays. Clin Chem 44:2015–2030

3. Lovgren T, Hemmila I, Pettersson K, Eskola JU, Bertoft E (1984) Determination of hormones by time-resolved fluoroimmunoassay. Talanta 31:909–916
4. Hemmila I (1985) Fluoroimmunoassays and immunofluorometric assays. Clin Chem 31:359–370
5. Soini E, Hemmila I, Dahlen P (1990) Time-resolved fluorescence in biospecific assays. Ann Biol Clin (Paris) 48:567–571
6. Hemmila I, Dakubu S, Mukkala VM, Siitari H, Lovgren T (1984) Europium as a label in time-resolved immunofluorometric assays. Anal Biochem 137:335–343
7. Xu Y, Li Q (2007) Multiple fluorescent labeling of silica nanoparticles with lanthanide chelates for highly sensitive time-resolved immunofluorometric assays. Clin Chem 53:1503–1510
8. Pettersson K, Alfthan H, Stenman UH, Turpeinen U, Suonpaa M, Soderholm J et al (1993) Simultaneous assay of alpha-fetoprotein and free beta subunit of human chorionic gonadotropin by dual-label time-resolved immunofluorometric assay. Clin Chem 39:2084–2089
9. Mitrunen K, Pettersson K, Piironen T, Bjork T, Lilja H, Lovgren T (1995) Dual-label one-step immunoassay for simultaneous measurement of free and total prostate-specific antigen concentrations and ratios in serum. Clin Chem 41:1115–1120
10. Zhu L, Leinonen J, Zhang WM, Finne P, Stenman UH (2003) Dual-label immunoassay for simultaneous measurement of prostate-specific antigen (PSA)-alpha1-antichymotrypsin complex together with free or total PSA. Clin Chem 49:97–103
11. Meurman OH, Hemmila IA, Lovgren TN, Halonen PE (1982) Time-resolved fluoroimmunoassay: a new test for rubella antibodies. J Clin Microbiol 16:920–925
12. Tanner P, Stenman U-H, Seppälä M, Schröder J (1982) Sensitive and specific RIA for human chorionic gonadotropin (hCG) using monoclonal antibodies. Protides Biol Fluids 29:843–846
13. Dechaud H, Bador R, Claustrat F, Desuzinges C (1986) Laser-excited immunofluorometric assay of prolactin, with use of antibodies coupled to lanthanide-labeled diethylenetriamine-pentaacetic acid. Clin Chem 32:1323–1327
14. Tarkkinen P, Palenius T, Lovgren T (2002) Ultrarapid, ultrasensitive one-step kinetic immunoassay for C-reactive protein (CRP) in whole blood samples: measurement of the entire CRP concentration range with a single sample dilution. Clin Chem 48:269–277
15. von Lode P, Rainaho J, Laiho MK, Punnonen K, Peltola O, Harjola VP, Pettersson K (2006) Sensitive and quantitative, 10-min immunofluorometric assay for D-dimer in whole blood. Thromb Res 118:573–585
16. von Lode P, Rainaho J, Pettersson K (2004) Quantitative, wide-range, 5-minute point-of-care immunoassay for total human chorionic gonadotropin in whole blood. Clin Chem 50:1026–1035
17. Khosravi MJ, Chan MA, Bellem AC, Diamandis EP (1988) A sensitive time-resolved immunofluorometric assay of ferritin in serum with monoclonal antibodies. Clin Chim Acta 175:267–275
18. Black MH, Grass CL, Leinonen J, Stenman UH, Diamandis EP (1999) Characterization of monoclonal antibodies for prostate-specific antigen and development of highly sensitive free prostate-specific antigen assays. Clin Chem 45:347–354
19. Diamandis EP (1988) Immunoassays with time-resolved fluorescence spectroscopy: principles and applications. Clin Biochem 21:139–150
20. Mathis G (1993) Rare earth cryptates and homogeneous fluoroimmunoassays with human sera. Clin Chem 39:1953–1959
21. Mathis G, Socquet F, Viguier M, Darbouret B (1997) Homogeneous immunoassays using rare earth cryptates and time resolved fluorescence: principles and specific advantages for tumor markers. Anticancer Res 17:3011–3014
22. Dahlen P, Hurskainen P, Lovgren T, Hyypia T (1988) Time-resolved fluorometry for the identification of viral DNA in clinical specimens. J Clin Microbiol 26:2434–2436
23. Dahlen P, Syvanen AC, Hurskainen P, Kwiatkowski M, Sund C, Ylikoski J et al (1987) Sensitive detection of genes by sandwich hybridization and time-resolved fluorometry. Mol Cell Probes 1:159–168

24. Nurmi J, Ylikoski A, Soukka T, Karp M, Lovgren T (2000) A new label technology for the detection of specific polymerase chain reaction products in a closed tube. Nucleic Acids Res 28:E28
25. Alfthan H (1986) Comparison of immunoradiometric and immunofluorometric assays for serum hCG. J Immunol Methods 88:239–244
26. Madersbacher S, Shu-Chen T, Schwarz S, Dirnhofer S, Wick G, Berger P (1993) Time-resolved immunofluorometry and other frequently used immunoassay types for follicle-stimulating hormone compared by using identical monoclonal antibodies. Clin Chem 39:1435–1439
27. Ferguson RA, Yu H, Kalyvas M, Zammit S, Diamandis EP (1996) Ultrasensitive detection of prostate-specific antigen by a time-resolved immunofluorometric assay and the immulite immunochemiluminescent third-generation assay: potential applications in prostate and breast cancers. Clin Chem 42:675–684
28. Robertson DM, Pruysers E, Stephenson T, Pettersson K, Morton S, McLachlan RI (2001) Sensitive LH and FSH assays for monitoring low serum levels in men undergoing steroidal contraception. Clin Endocrinol (Oxf) 55:331–339
29. Stenman UH (1997) Immunoassay standardisation. In: Price C, Newman D (eds) Principles and practice of immunoassay, 2nd edn. Macmillan, London, pp 245–268
30. Dunkel L, Alfthan H, Stenman UH, Tapanainen P, Perheentupa J (1990) Pulsatile secretion of LH and FSH in prepubertal and early pubertal boys revealed by ultrasensitive time-resolved immunofluorometric assays. Pediatr Res 27:215–219
31. Apter D, Cacciatore B, Alfthan H, Stenman UH (1989) Serum luteinizing hormone concentrations increase 100-fold in females from 7 years to adulthood, as measured by time-resolved immunofluorometric assay. J Clin Endocrinol Metab 68:53–57
32. Demir A, Voutilainen R, Juul A, Dunkel L, Alfthan H, Skakkebaek NE, Stenman UH (1996) Increase in first morning voided urinary luteinizing hormone levels precedes the physical onset of puberty. J Clin Endocrinol Metab 81:2963–2967
33. Demir A, Dunkel L, Stenman UH, Voutilainen R (1995) Age-related course of urinary gonadotropins in children. J Clin Endocrinol Metab 80:1457–1460
34. Stenman UH, Unkila KL, Korhonen J, Alfthan H (1997) Immunoprocedures for detecting human chorionic gonadotropin: clinical aspects and doping control. Clin Chem 43:1293–1298
35. Alfthan H, Haglund C, Roberts P, Stenman UH (1992) Elevation of free beta subunit of human choriogonadotropin and core beta fragment of human choriogonadotropin in the serum and urine of patients with malignant pancreatic and biliary disease. Cancer Res 52:4628–4633
36. Stenman UH, Alfthan H, Hotakainen K (2004) Human chorionic gonadotropin in cancer. Clin Biochem 37:549–561
37. Stenman UH, Alfthan H, Ranta T, Vartiainen E, Jalkanen J, Seppala M (1987) Serum levels of human chorionic gonadotropin in nonpregnant women and men are modulated by gonadotropin-releasing hormone and sex steroids. J Clin Endocrinol Metab 64:730–736
38. Alfthan H, Haglund C, Dabek J, Stenman UH (1992) Concentrations of human choriogonadotropin, its beta-subunit, and the core fragment of the beta-subunit in serum and urine of men and nonpregnant women. Clin Chem 38:1981–1987
39. Lempiainen A, Hotakainen K, Blomqvist C, Alfthan H, Stenman UH (2007) Increased human chorionic gonadotropin due to hypogonadism after treatment of a testicular seminoma. Clin Chem 53:1560–1561
40. Strasburger C, Barnard G, Toldo L, Zarmi B, Zadik Z, Kowarski A, Kohen F (1989) Somatotropin as measured by a two-site time-resolved immunofluorometric assay. Clin Chem 35:913–917
41. Albertsson-Wikland K, Jansson C, Rosberg S, Novamo A (1993) Time-resolved immunofluorometric assay of human growth hormone. Clin Chem 39:1620–1625
42. Bidlingmaier M, Wu Z, Strasburger CJ (2003) Problems with GH doping in sports. J Endocrinol Invest 26:924–931

43. Stenman UH, Leinonen J, Alfthan H, Rannikko S, Tuhkanen K, Alfthan O (1991) A complex between prostate-specific antigen and alpha 1-antichymotrypsin is the major form of prostate-specific antigen in serum of patients with prostatic cancer: assay of the complex improves clinical sensitivity for cancer. Cancer Res 51:222–226
44. Piironen T, Lovgren J, Karp M, Eerola R, Lundwall A, Dowell B et al (1996) Immunofluorometric assay for sensitive and specific measurement of human prostatic glandular kallikrein (hK2) in serum. Clin Chem 42:1034–1041
45. Vaisanen V, Peltola MT, Lilja H, Nurmi M, Pettersson K (2006) Intact free prostate-specific antigen and free and total human glandular kallikrein 2. Elimination of assay interference by enzymatic digestion of antibodies to F(ab′)2 fragments. Anal Chem 78:7809–7815
46. Saedi MS, Hill TM, Kuus-Reichel K, Kumar A, Payne J, Mikolajczyk SD et al (1998) The precursor form of the human kallikrein 2, a kallikrein homologous to prostate-specific antigen, is present in human sera and is increased in prostate cancer and benign prostatic hyperplasia. Clin Chem 44:2115–2119
47. Qin Q, Christiansen M, Lovgren T, Norgaard-Pedersen B, Pettersson K (1997) Dual-label time-resolved immunofluorometric assay for simultaneous determination of pregnancy-associated plasma protein A and free beta-subunit of human chorionic gonadotropin. J Immunol Methods 205:169–175
48. Norgaard-Pedersen B, Alfthan H, Arends J, Hogdall CK, Larsen SO, Pettersson K et al (1994) A new simple and rapid dual assay for AFP and free beta hCG in screening for Down syndrome. Clin Genet 45:1–4
49. Ankelo M, Westerlund A, Blomberg K, Knip M, Ilonen J, Hinkkanen AE (2007) Time-resolved immunofluorometric dual-label assay for simultaneous detection of autoantibodies to GAD65 and IA-2 in children with type 1 diabetes. Clin Chem 53:472–479
50. Bünzli J-CG, Eliseeva SV (2010) Basics of lanthanide photophysics. Springer Ser Fluoresc doi: 10.1007/4243_2010_3

Electrochemiluminescence of Lanthanides

S. Kulmala, T. Ala-Kleme, and J. Suomi

Abstract Electrogenerated luminescence of lanthanides is reviewed with emphasis on the electrochemiluminescence (ECL) of lanthanide chelates. Main application area of lanthanide chelates in this field is their use as electrochemiluminescent labels in bioaffinity assays such as in immunoassays or DNA probe assays. With lanthanide chelates as labels, hot electron-induced ECL at thin insulating film-coated cathodes outperforms ECL based on traditional electrochemistry at active metal electrodes. ECL excitation of lanthanide(III) chelates occurs normally by ligand-sensitized mechanisms wherein the ligand is first excited by redox reactions followed by energy transfer from ligand to the central ion, which finally emits by f–f transitions. Luminescent lanthanide ions can be excited at oxide-coated metal electrodes when these ions act as luminescence centers in the oxide film and/or at the oxide/electrolyte interface or as solvated in the close vicinity of the electrode surface. These ions typically show high field-induced solid-state electroluminescence when embedded inside of the oxide films and ECL at the surface of the electrode or in solution close to the electrode surface. These lanthanide-doped oxide films can be fabricated either by anodization of certain lanthanide-doped valve metals or from pure valve metals by anodization in lanthanide ion-containing anodization bath preferably with AC voltages. Some lanthanide ions can be electrically excited also in strongly acidic sulfuric acid solutions at platinum electrodes with mechanisms not known with certainty.

Keywords Electrogenerated chemiluminescence · Galvanoluminescence · Hot electron · Hydrated electron · Hydroxyl radical · Labels · Lanthanoids · Rare earths · Sulfate radical · Valve metals

S. Kulmala (✉) and J. Suomi
Laboratory of Analytical Chemistry, Aalto University, P.O. Box 16100, 00076 Aalto, Finland
e-mail: sakari.kulmala@aalto.fi

T. Ala-Kleme
Department of Chemistry, University of Turku, 20014 Turku, Finland

P. Hänninen and H. Härmä (eds.), *Lanthanide Luminescence: Photophysical, Analytical and Biological Aspects*, Springer Ser Fluoresc (2011) 7: 343–360, DOI 10.1007/4243_2011_17,

Contents

Abbreviations

C/I/E	Conductor/insulator/electrolyte solution
CL	Chemiluminescence
ECL	Electrochemiluminescence, electrogenerated CL
ECLIA	Electrochemiluminoimmunoassay
EL	Electrogenerated luminescence
FITC	Fluorescein isothiocyanate
HECL	Hot electron-induced electrogenerated chemiluminescence
HOMO	Highest occupied molecular orbital
LUMO	Lowest unoccupied molecular orbital

1 Introduction to Different Forms of Electrogenerated Luminescence

Luminescent lanthanide ions have numerous practical applications in chemistry and technical devices, some of them involving electrical excitation of the lanthanide ions. Photophysical properties of lanthanide ions as well as perhaps the most important application field, photoluminescence of lanthanide chelates, and lanthanide-containing nanoparticles are treated elsewhere in this volume.

Electrical excitation of lanthanide ions has been commonly used in luminous discharge lamps and thin-film electroluminescent displays for a long time [1]. Presently, the utilization of lanthanides in different variants of light emitting diodes has attracted plenty of interest [2, 3]. In these fields, mainly Eu^{3+}, Tb^{3+}, Sm^{3+}, Dy^{3+}, and Tm^{3+} with 4f → 4f transitions are utilized, but some applications have also used Eu^{2+}, Ce^{3+}, and Yb^{2+} displaying 5d → 4f transitions.

Elbanowski has reviewed general chemiluminescence (CL) of lanthanides repeatedly [4, 5]. In this chapter, only electrochemical excitation of lanthanide ions and chelates is considered.

It would be practical if the term electroluminescence in science would be a general name for all the luminescence phenomena generated by electrical means, but it has been established in the scientific community to cover only the solid-state electroluminescence. Therefore, electrogenerated luminescence (EL) is selected here as a general name to cover all the forms of luminescence induced by electrical excitation. Thus, if one suspects that a novel, poorly known process may be composed of several luminescent processes, which are all induced by electrical excitation but have different mechanisms, it is considered safe to call it EL.

Electrochemiluminescence (ECL) is treated here as CL produced directly or indirectly as a result of electron transfer between an electrode and some solution species or species bound to the electrode surface. ECL as a term is considered here to have no difference with the term electrogenerated CL. ECL can have as a final luminophore-exciting step (1) a heterogeneous electron transfer between the luminescent species and an electrode or (2) a homogeneous electron transfer between a solvated species and the luminescent species. Thus, in ECL at least one of the reagents is generated *in situ* in an electrolytic process. ECL shares many analytical advantages with CL, the most important one being the low detection limit owing to the low background emission.

In analytical chemistry, ECL can be used for determination of electrochemiluminescent compounds, or other more interesting compounds using electrochemiluminescent labels in techniques and protocols analogous to fluoroimmunoassays or chemiluminoimmunoassays. Electrochemiluminoimmunoassay (ECLIA) methods are not very mature yet, and their value for analytical science will certainly grow in the future. General reviews on ECL have been published quite frequently [6–11].

EL produced by electrolysis of aqueous solution was found for the first time already in the end of nineteenth century [12]. In Braun's experiment, AC current was driven through a cell containing dilute sulfuric acid solution and the working electrode was composed of aluminum with unknown purity and platinum counter electrode was used. In this case, Braun actually observed electroluminescence from the forming of aluminum oxide film during anodic half cycles and, probably, electroluminescence and ECL of the impurities and defects of the oxide film during cathodic half cycles. This kind of EL observed at metal electrodes forming anodic oxide films has traditionally been called galvanoluminescence [13], but actually there is no need for a separate term because the observed total luminescence in all of these experiments is always composed of electroluminescence and ECL with varied importance of these phenomena in different experiments.

2 Excitation of Ions and Molecules by Redox Reactions in ECL

When light is generated by chemical reactions, there are two main pathway types providing the necessary excitation energy to the finally emitting species. One type is based on the disintegration of the luminophore itself when the energy is mainly available from the breakage of its bonds. The other type is based on a sequence of electron transfer reactions. One-electron redox reactions can lead to excited states

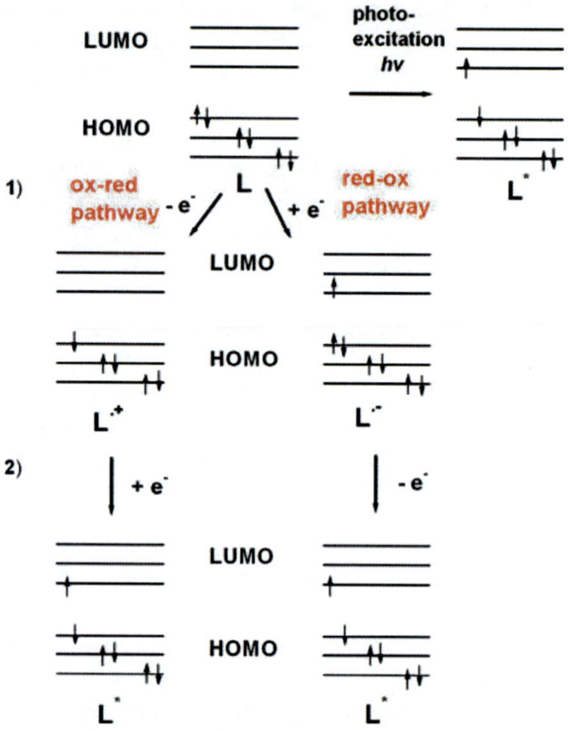

Fig. 1 The analogy between photoexcitation and redox-excitation of an uncharged compound L is presented. In the step (2), the product is formed either in the ground state or in an excited state. The generation of an excited end product depends on the characteristics of the reductant/oxidant of this step

similar to the photo-excitation of a molecule (or metal ion or metal chelate), which can be described by the schematic molecular orbital energy diagram presented in Fig. 1. There are three possible excitation pathways, namely (1) oxidation-initiated and (2) reduction-initiated pathways, and (3) comproportionation pathway. Pathways (1) and (2) can be shortened to ox–red and red–ox excitation routes. If the luminophore radical intermediates are produced in aqueous solution, there can be considerable differences in their stability, which alone may make either the ox–red or the red–ox excitation route very unfavorable, in which case the other one becomes predominant.

In the oxidation-initiated (ox–red) pathway, an electron is first removed from the highest occupied molecule orbital (HOMO) of the compound (or from an occupied level of a metal ion). Then, a reducing agent may donate an electron to fill the hole in the HOMO levels of the compound (or low energy metal ion levels) and a ground state product is formed. However, if the energy levels of the lowest unoccupied orbitals of the oxidized form of the molecule (or high energy levels of an ion) and the reductant are suitably matched, an electron can be donated preferably to the lowest unoccupied level (LUMO), which results in the formation of a product in its excited state.

Analogously, in the reduction-initiated (red–ox) pathway, the properties of the oxidant of the second step determine whether the end product is formed in its ground or excited state. The prerequisite for comproportionation pathway is that the oxidized and reduced forms of the molecule are produced simultaneously or successively within the lifetime of the radical first formed. Comproportionation pathway results in the formation of one excited and one ground-state molecule from the reacting radicals.

For simplicity, Fig. 1 was drawn as if only singlet state excitation would be possible (electrons remain all the time "net-spin paired"). Actually electron transfer can result in either excited triplet or singlet state end-products. In organic, non-aqueous solvents, some molecules produce ECL involving singlet state emission both by so-called energy-insufficient T-route via triplet–triplet annihilation, and by the energy-sufficient S-route [14–20].

In aqueous media, the necessary highly cathodic and/or highly anodic working electrode potentials can be reached only in very special cases. In addition, if the luminophore anion and/or cation radicals can be formed at all, they are typically so reactive with water that most of them are consumed by unwanted side reactions. ECL methods requiring strictly nonaqueous conditions can have much value only in very limited areas of analytical chemistry.

3 EL of Solvated Lanthanide Ions in Solution and at Electrode/Electrolyte Interface

In the 1960s Ganley and coworkers observed that Nd-doped, Tb-doped and Dy-doped aluminum electrodes produced both anodic and cathodic lanthanide ion-specific EL during AC-excitation, but the luminescence generation mechanisms remained quite obscure [21, 22]. Much later, Ohtake et al. have shown that also anodic polarization of ZnO semiconductor electrodes doped with either Sm^{3+}, Eu^{3+}, Dy^{3+}, Ho^{3+} or Er^{3+} induces lanthanide ion-specific EL in strongly acidic aqueous solution [23].

Bulgakov [24] has shown that during electrolysis at a platinum electrode in highly acidic sulfuric acid solution, Tb^{3+} and Dy^{3+} ions show an ECL spectrum similar to their photoluminescence spectra. He assumed that the emission was somehow induced by generation of ozone. Later it was established that the generation of sulfate radicals by dissolution of UV-irradiated peroxodisulfate in aqueous solution in the presence of Tb^{3+} ions produces Tb^{3+}-specific CL without the presence of any other stronger obvious reductants than water. The exact mechanism of the CL and EL generation under these conditions is not resolved yet [25]. In the light of these discoveries, it seems possible that Bulgakov was oxidizing sulfate to peroxodisulfate at the anode, and part of the forming sulfate radicals at the anode surface could react with these lanthanide ions resulting in CL.

We observed much later that lanthanide ion-specific cathodic EL could be generated and quantitatively measured also when some of the lanthanide ions were added in solution and working electrodes made of pure aluminum were used [26–28]; however, the mechanisms again remained obscure. The mechanism of Eu^{3+}-specific cathodic EL in the presence of nitrate ions, which was observable only at a rotating disc electrode and not at stationary electrode, remains unresolved. This system was later reevaluated by another group without any significant new results [29].

It was also noticed that certain anodically oxidizable metals could produce Tb^{3+}-doped oxide films. Tb^{3+} ions are embedded in the oxide during its formation, especially during AC anodization in an anodization bath containing Tb^{3+} ions. These doped oxide films can be used for determination of hydrogen peroxide and bioanalytes, such as glucose and cholesterol, that can be enzymatically coupled with production of hydrogen peroxide from dissolved oxygen [30, 31]. The mechanism in these applications has been proposed to be hole injection into the oxide by cathodically produced hydroxyl radical from hydrogen peroxide. It results in one electron oxidation of Tb^{3+} centers in the oxide, and is rapidly followed by Tb^{4+} reduction back to the original oxidation state by cathodically injected electrons from the conduction band of the oxide in question. As a result of the energy levels, the Tb ion returns to its original oxidation state in its excited $^5D_4$ state [30, 31]. Later, Tb (III)-specific EL at oxide-covered aluminum electrode was also studied. It was found out that Tb(III)-specific EL is enhanced by the presence of azide ions or coreactants such as hydrogen peroxide and peroxodisulfate ions, which produce highly oxidizing radicals by one-electron reduction. This EL is produced via several parallel mechanisms usually involving oxidation state changes of Tb(III), which are normally difficult to produce in an aqueous medium. The cathodically produced primary species are tunnel-emitted hot electrons and F^+ centers [32] of the oxide film. These primary species are capable of hard reductions and oxidations not normally encountered in the electrochemistry of aqueous solutions. They can probably produce hydrated electrons and hydroxyl radicals, which are able to mediate reductions and oxidations in aqueous solution beyond the tunneling distance from the oxide/electrolyte interface [33]. This EL seems to be formed as a sum of the luminescence emitted by Tb (III) ions adsorbed at the oxide/electrolyte interface, partially incorporated in the oxide, and located in the solution at the close vicinity of the oxide/electrolyte interface [34]. Tunnel emission of hot electrons from thin insulating film-coated cathodes and injection of these hot electrons to aqueous electrolyte solutions were discussed for the first time in 1995 [35] but in more detailed way a bit later [36–38]. They are shortly discussed below in connection of lanthanide chelates.

Figure 2 displays the energetics of Al/Al_2O_3/Electrolyte interface together with formal potentials of various species in the system at about pH 5. It shows e.g., that charge transfer via F and F^+ center band makes it energetically possible to oxidize Tb(III) to Tb(IV) between cathodic pulses and different radical species create several energy-sufficient routes to excited Tb^{3+} ions by redox reactions.

It was concluded that in this system [34] the most important routes for the generation of the EL are the energy transfer excitation pathways from intrinsic emission centers of the oxide film to Tb(III) ion, which do not need the redox reactions of Tb(III), as well as the different variants of the red–ox and ox–red excitation pathways. The oxidation of Tb(II) ($4f^8 5d^1$) by a sufficiently strong oxidant (an oxidant capable of abstracting an electron from the 4f orbitals) as well as the reduction of Tb(IV) ($4f^7$) by a sufficiently hot electron, results in the formation of Tb(III) in its $4f^7 5d$ excited state. This excited state is known to feed

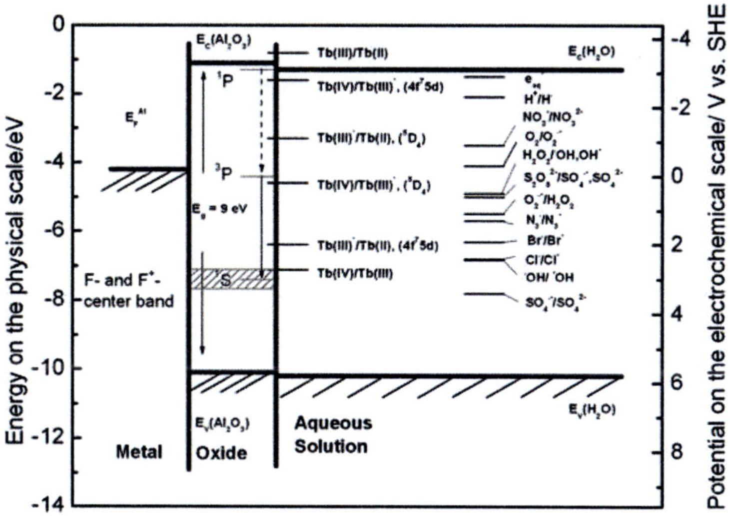

Fig. 2 Energy diagram of Tb(III) ion EL at oxide-covered aluminum electrode. Formal potentials of oxygen, hydrogen peroxide and oxyradicals at pH 5–6 are used in the diagram [34]

directly the $^5D_4$ excited state. The EL-quenching effect of the free radical scavengers used in the experiments can be based on the reactions with the intrinsic reactive species produced by cathodic pulse-polarization of oxide-covered aluminum and with the Tb(II) or Tb(IV) ions. As a result, it could only be stated that all the above-mentioned excitation pathways are energetically possible. The predominant excitation mechanism of Tb(III) ions at a greater distance from the electrode surface and of Tb(III) ions partially embedded in the oxide film is probably different. In the case of highly cathodic pulse potential excitation of hydrated Tb(III) ions, the predominant excitation mechanism, at least in air-saturated solutions in the presence of hydrogen peroxide or high concentrations of azide or bromide ions, is the reduction initiated excitation pathway. This pathway is initiated by tunnel-emitted hot electrons, which reduce Tb(III) to Tb(II). Tb(II) is then oxidized by F^+ centers, dissolved oxygen, hydrogen peroxide, reductively produced superoxide or hydroxyl radicals, halide or pseudo-halide radicals. The oxidation to Tb(III) occurs to either its $^5D_4$ state or to a higher excited state, which feeds the $^5D_4$ state. Finally, the Tb(III) emits exhibiting the typical $^5D_4 \rightarrow {}^7F_J$ transitions [36–38].

4 ECL of Lanthanide Chelates

We observed already in the mid 1980s that some Tb^{3+} complexes could produce much stronger cathodic EL than hydrated Tb^{3+} ions [39]. We also experimentally found out that some more stable multidentate ligands could produce strong cathodic EL, especially when Tb^{3+} was as a central ion, but also weaker EL when Dy^{3+} and

Sm^{3+} were the central ions [40]. At the same time, we experimentally developed the first ECLIA utilizing Tb(III) chelate label [41]. At that time, the excitation and luminescence mechanisms were still waiting to be resolved, although it already seemed possible that a solvated electron might be a mediating species in the cathodic excitation steps at oxide-covered aluminum electrodes.

Redox excitation mechanisms of Tb^{3+}-chelates were studied by generating hydrated electrons in aqueous Tb^{3+} chelate solutions with known methods [25, 42–44] and comparing the results with those observed at cathodically pulse polarized oxide-covered aluminum electrodes [35]. It was proposed that in all of these cases the excitation mechanism is a ligand-sensitized mechanism in which the aromatic multidentate ligand is first excited by electrochemical steps. Then the excited ligand transfers the energy to the central ion, which finally emits with $^5D_4 \rightarrow {}^7F_J$ transitions. These results were later supported by Richter and Bard [45], and their ECL measurements using Eu^{3+}-chelates in nonaqueous solutions. In addition to Richter and Bard, also Yu et al. [46] have studied ECL of europium(III) chelates in nonaqueous solutions.

The primary cathodic step of hot electron-induced electrochemiluminescence (HECL) was postulated to be an injection of hot electrons into aqueous electrolyte solution by tunnel emission through an insulating barrier, followed by reduction reactions induced either by presolvated hot electrons or by fully hydrated electrons [37, 38]. The introduction and details of this kind of hot electron electrochemistry and HECL has been reviewed very recently [33] and only short descriptions of the basic features are given here.

Electrodes coated with a thin layer of insulating oxide (ca. 4–6 nm thick film) are usable for generation of HECL. For materials forming insulating oxides by anodic oxidation, in situ generation of suitable oxide film thickness is possible by anodic pulse preceding the cathodic excitation pulse. With other types of base conductor materials in Conductor/Insulator/Electrolyte solution (C/I/E) structures, the insulating film must be prepared in advance upon the conductor. Atomic layer deposition (ALD) is an example of an excellent method of depositioning ultra-thin insulating oxide films upon suitable sufficiently conducting materials [47].

The working electrode applied in HECL can have various shapes and structures. The conductor in the C/I/E junctions can either be a metal with a rather low work function [37, 48] or a strongly doped degenerate semiconductor [36, 47, 49]. Earlier we have assumed that only insulating oxides with a very wide band gap (of the order of 8–9 eV) such as alumina, magnesia, and silica could be used as insulators in C/I/E junctions [36, 38]. However, it seems that much smaller band gaps are sufficient, whenever the oxide is still a true insulator (i.e., the oxide has Fermi level in the mid gap region) and is not behaving as an intrinsic semiconductor, which is common to many metal oxides [50]. It is rather interesting that also double barrier tunnel emission electrodes of type M/I/M/I work nicely [51], and thus possibly layers of different materials could be applied as well.

For the actual HECL measurements, borate buffer seems to be the best buffer whenever its buffer capacity coincides in the right pH region. This is due to the very low reactivity of boric acid and borate ions with hydrated electron and many other radicals [52, 53].

It is likely that the cathodic reductions at these electrodes can be produced simultaneously by different types of electrons: (1) presolvated hot electrons, (2) hydrated electrons, (3) heterogeneously transferred electrons from the conduction band of the insulating film at insulating film/electrolyte interface, and (4) probably also from the surface states of the insulating film [35, 36, 38, 50, 54, 55]. Normally, only presolvated hot electrons or hydrated electrons are sufficiently energetic to participate in reaction pathways leading to ECL in aqueous solutions.

In addition, presolvated hot or hydrated electrons can efficiently produce strongly oxidizing radicals from added coreactants such as dissolved oxygen and peroxo compounds [32, 35, 37]. This means that highly reducing and highly oxidizing conditions exist simultaneously in the vicinity of the electrode, and as a result it is possible to carry out chemiluminescent reactions, which are not possible in aqueous solutions at active metal electrodes. For example, sulfate and hydroxyl radicals cannot be produced electrochemically in fully aqueous solutions because peroxodisulfate and hydrogen peroxide are two-electron reduced at active metal electrodes [55–57]. However, hydrated electrons are able to produce hydroxyl radicals in reaction with hydrogen peroxide and sulfate radical in reaction with peroxodisulfate ion at near diffusion controlled rates [52].

Hydrated electron is a very strong reducing species [E° (e_{aq}^-) $= -2.9$ V versus SHE] [52], having only slightly lower energy than that of the conduction band edge of liquid water. This explains why they are able to induce CL and ECL from a large variety of luminophores [58, 59]. In addition, hydrated electron is not following the Marcus electron transfer theory, and therefore it is reducing fast those species with only moderately lower reduction potential than that of the hydrated electron but also extremely strong oxidants as well. Thus, Marcus inverted region is not observed in its reactivity and the hydrated electron can easily carry out the last step of the above-mentioned ox–red excitation pathway of luminophores as well as very often also to initiate the red–ox excitation pathway. Sulfate radical is one of the strongest one-electron oxidants known with standard reduction potential of about 3.4 V versus SHE [60]. In the simultaneous presence of hydrated electrons and sulfate radicals, there is more than 6 eV of energy available in excitation reaction pathways.

Under these harsh conditions, it has been possible to simultaneously excite luminophores with very different redox properties and with emission wavelengths either in the UV, VIS or NIR range [54, 61, 62]. The possibility of simultaneous excitation allows the development of multiparametric assays using wavelength discrimination to separate the signals emerging from different labels. Hot electron electrochemistry allows excitation of a variety of organic luminophores such as derivatives of fluorescein and coumarines, Rhodamine B, 9-fluorenylmethyl chloroformate or water soluble naphthalene derivatives that emit in the range from UV to NIR and have very different redox properties [63–65]. In principle, many of the common labels in fluoroimmunoassay methodology are anticipated to be applicable as ECL labels excitable by the hot electron electrochemistry.

We have shown that even simple derivatives of benzene can be excited by hot electron electrochemistry at thin insulating film-coated cathodes which induces the forming of both singlet and triplet excited states by electron transfer reactions [66].

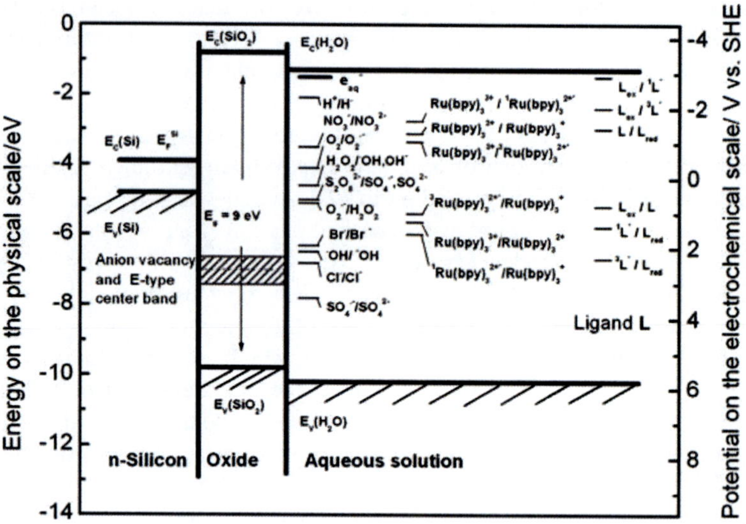

Fig. 3 Energy diagram of strongly doped *n*-Si/SiO$_2$ electrode in contact with aqueous electrolyte solution. The diagram contains the formal redox potentials of the primary radicals and combined redox and luminescence properties of Ru(bpy)$_3^{2+}$, and 2,6-bis[N,N-bis(carboxymethyl)amino-methyl]-4-benzoylphenol (ligand L). The diagram demonstrates clearly that on thermodynamical basis, both of these luminophores could be excited even to their singlet excited states in the simultaneous presence of hydrated electrons and sulfate radicals via both ox–red and red–ox pathways. However, in practice the ox–red pathway seems to be dominant in both cases. E-type electron centers in silica fully correspond to F and F$^+$ electron centers in alumina [36]

Thus, direct observation of the forming of the triplet excited states in aqueous solution at room temperature is possible. This is usually possible by photoexcitation only in the solid (frozen) state of the same compounds. Figure 3 shows the energetics of an aromatic 7-dentate ligand, which forms strongly electrochemiluminescent chelates with some lanthanide(III) ions, together with those of the most intensively studied electrochemiluminescent compound in the world, Ru(bpy)$_3^{2+}$. The energetics is shown at an oxide-coated silicon electrode. Ru(bpy)$_3^{2+}$ can be excited both using traditional electrochemistry at active metal electrodes and hot electron electrochemistry, but ligand L and Tb(III)-L chelate can only be excited by hot electron electrochemistry in fully aqueous solution.

It has been shown that aromatic Gd(III) and Y(III) chelates produce ligand-centered emissions during cathodic pulse-polarization of thin insulating film-coated electrodes, while the corresponding Tb(III) chelates produce metal-centered $^5D_4 \rightarrow {}^7F_J$ emissions [67]. It was observed that a redox-inert paramagnetic heavy lanthanoid ion, Gd(III), seems to enhance strongly intersystem crossing in the excited ligand and to direct the deexcitation toward a triplet-state emission, while a lighter diamagnetic Y(III) ion directs the photophysical processes toward a singlet-state emission of the ligand. From these central ions, Tb(III) ion also enhances intersystem crossing but it can also accept energy from the ligand triplet state to its resonance level. Thus, the following emission routes can be presented in the above-mentioned cases of strongly redox inert central ions [Ln(III) marks any lanthanide ion] and ox–red excitation route:

$$\mathrm{Ln(III)\text{-}L} + \mathrm{Ox.}^{\cdot} \rightarrow \mathrm{Ln(III)\text{-}L_{ox}} + \mathrm{Ox}^{-} \tag{1a}$$

$$\mathrm{Ln(III)\text{-}L_{ox}} + \mathrm{e}^{-}{}_{\mathrm{hot}}\left(\mathrm{or\,e}^{-}{}_{\mathrm{aq}}\right) \rightarrow \mathrm{Ln(III)\text{-}{}^{1}L^{*}} \tag{1b}$$

$$\mathrm{Ln(III)\text{-}{}^{1}L^{*}} \rightarrow \mathrm{Ln(III)\text{-}L} + h\nu \tag{1c}$$

$$\mathrm{Ln(III)\text{-}{}^{1}L^{*}} \rightarrow \mathrm{Ln(III)\text{-}{}^{3}L^{*}} \tag{2a}$$

$$\mathrm{Ln(III)\text{-}{}^{3}L^{*}} \rightarrow \mathrm{Ln(III)\text{-}L} + h\nu \tag{2b}$$

$$\mathrm{Ln(III)\text{-}{}^{3}L^{*}} \rightarrow {}^{*}\mathrm{Ln(III)\text{-}L} \tag{3a}$$

$${}^{*}\mathrm{Ln(III)\text{-}L} \rightarrow \mathrm{Ln(III)\text{-}L} + h\nu_{\mathrm{J}} \tag{3b}$$

Here $\mathrm{Ox.}^{\cdot}$ is an oxidizing radical such as sulfate or hydroxyl radical (or an empty electron center, i.e., an anion vacancy of the oxide) and L is the chelating aromatic ligand. Thus, Y(III)-L reacts according to reactions (1a)-(1b), Gd(III)-L according to the steps (1a)-(1b)-(2a)-(2b) and Tb(III)-L via the whole reaction series (1a)-(1b)-(2a)-(3a)-(3b). Eu(III) is easily reduced, which makes its use to differ from that of Tb(III) as a central ion. Generally, its use needs the conditions to be changed to more oxidizing ones by addition of a relatively high concentration of peroxodisulfate ions. Quite complicated details of the mechanism variants are presented in the original publication [68].

Ala-Kleme et al. [62] have shown that Yb^{3+} chelates display near-infrared ECL in aqueous electrolytes. They have also shown that the HECL emission of Dy(III) and Sm(III) can be strongly enhanced by making heterodinuclear chelates containing Y(III) [69].

Lis et al. [70] have shown recently by using Al/Al_2O_3 electrodes that a 1:2 chelate of Eu(III) and coumarin-3-carboxylate is electrochemiluminescent in aqueous solution. The same group has also studied some other weak complexes that show ECL but cannot be used as labels due to their too small stability constants [71].

5 Future Prospects of Analytical Applicability of HECL and Lanthanide Labels

The HECL intensity of lanthanide chelates is sufficiently strong for them to be used as labels in bioaffinity assays, especially when time-resolved detection is utilized. The organic luminophores showing their singlet state emission suffer from the fact that during tunnel emission the electric field in the insulating film is high enough to generate weak solid-state electroluminescence during the cathodic excitation pulses. Therefore, labels showing long-lived luminescence give much lower blank signals as the short-lived solid-state electroluminescence is not a problem. The lanthanide chelates we have used are not quenched by the presence of dissolved oxygen and are therefore much better suited to the analysis than metalloporphyrins, which otherwise would be a quite attractive second choice. However, some of the metalloporphyrins are also quite strongly electrochemiluminescent and their luminescence lifetime allows time-resolved detection [72].

Tb(III) chelates can be detected down to picomolar levels by ECL (Fig. 4), but the best Tb(III) labels presently known still cannot quite compete with the best photoluminescent lanthanide(III) labels. The synthesis routes of several phenolic Tb(III) chelates displaying strong HECL are available in the literature [74], and any skilled organic chemist can synthesize these ligands and chelates. Some of the slightly less efficient HECL-emitting Tb(III) labels are commercially available from PerkinElmer Life Sciences, Wallac, Finland [58, 73].

Tb(III) chelate labels have so far been applied in heterogeneous [37, 41, 51, 62, 66, 75, 76] as well as homogeneous [41, 73, 76] immunoassays and recently also in DNA probe assays [77].

Both immunometric (noncompetitive) [37, 41, 51, 62, 66, 75, 76] and competitive [73] immunoassay formats can be applied. The reason why homogeneous,

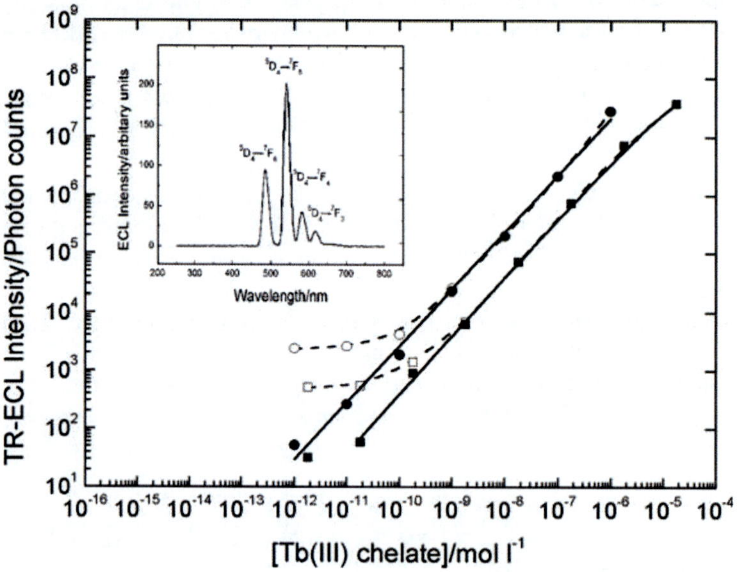

Fig. 4 Calibration curves of Tb(III) label chelates. (*Filled circle*) Tb^{3+}-2,6-bis[N,N-bis(carboxymethyl)aminomethyl]-4-methylphenol, (*filled square*) Tb^{3+}-N^1-(*p*-isothiocyanatobenzyl)-diethylenetriamine-N^1, N^2, N^3, tetraacetate. Conditions: 0.2 M borate buffer, pH 7.75, 0.1%NaN_3, pulse amplitude −10 V. *Open circles* and *open squares* denote signal before subtracting the blank. Measured with semiautomatic electrochemiluminometer [73]

i.e., separation free immunoassays [41, 73, 76] are possible is that the presolvated hot and hydrated electrons are able to carry out the reductions only up to a distance of about 200 nm from the oxide film surface and, hence, only the labels within this range can be excited. In practice, this means that if the amount of the electrochemiluminescently labeled antibody is appropriately optimized for the immunometric assay, only the labels involved in the surface antibody$_1$–antigen-labeled antibody$_2$ immunocomplexes are cathodically excited. Most of the unreacted labeled antibodies are too far from the electrode surface to be excited [58]. However, heterogeneous assays based on HECL detection have always lower detection limits than homogeneous assays. In the heterogeneous assays, an optimal HECL measuring buffer can be used without the radical scavenging additives useful for the incubation buffer of the bioaffinity reaction stages [51, 62, 76, 78]. Presently, very few DNA-probing assays have yet been carried out [77].

An advantage of HECL over the other ECL systems utilizing mainly $Ru(bpy)_3^{2+}$-based labels is that HECL excitation easily allows the use of both wavelength and time discrimination as well as simultaneous excitation of labels having very different redox and optical properties. Figure 5 displays as an example the

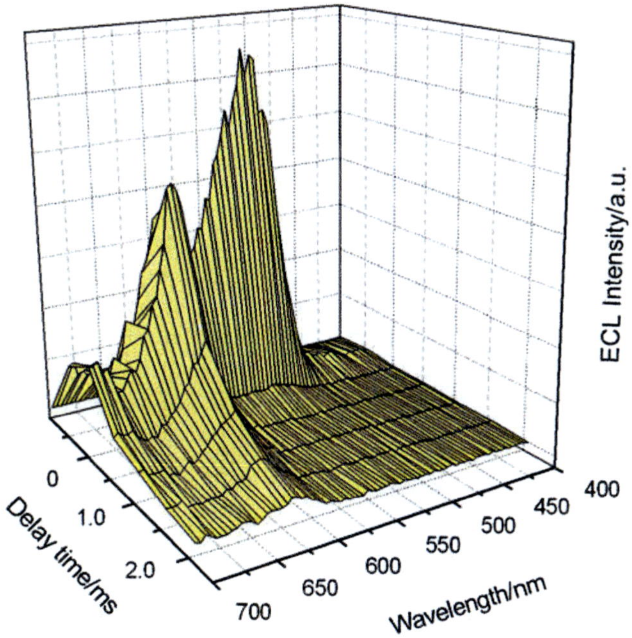

Fig. 5 Simultaneous excitation of FITC and Eu(III)-1-NCS at oxide-coated aluminum electrode [68]. Conditions: Oxide-covered Al-strip working electrode, Pt-wire counter electrode, coulostatic pulse generator, applied pulse voltage −40 V, pulse frequency 20 Hz, pulse charge 120 μC, 0.2 M boric acid buffer at pH 9.2, 1×10^{-3} M $K_2S_2O_8$. Time-resolved spectra were measured with an instrument having relatively good sensitivity but poor resolution [79]

time-resolved HECL spectrum of simultaneously excited aromatic Eu(III) chelate and fluorescein isothiocyanate (FITC).

Most important recent development of HECL and lanthanide labels for the point-of-need analysis using low-cost disposable cell chips is the finding that suitable HECL working electrodes can be fabricated not only by various traditional metallization and deposition techniques [80] but also by printing techniques [81]. It is believed that for some applications printing techniques are the best suited already now, and it is quite possible that later the same accuracy and reproducibility will be achieved by printing that is presently obtained by conventional manufacturing techniques of electronics [82] when fabricating disposable HECL chips.

Among the aromatic lanthanide chelates, Tb(III) chelates are the best labels for utilization in bioanalytical HECL methods due to their long luminescence lifetime and high HECL intensity [37, 67]. Some Eu(III) chelates seem to be usable, although their HECL intensity is lower and luminescence lifetime is shorter in comparison with those of the corresponding Tb(III) chelates [27, 68]. Unfortunately, other lanthanide (III) chelates have much poorer performance in applications in which high sensitivity is required.

Phenolic ligands with DTPA or EDTA-type chelating side arms and sufficient stability constants have so far been the best ligands for efficient HECL production [37, 38, 74]. Nitrogen in the ring of an aromatic ligand usually means poor HECL production efficiency [37–39, 66, 67], although sufficient stability constants can be achieved also by using pyridine-based aromatic moieties in the ligands.

6 Conclusions

ECL of lanthanides is reviewed with emphasis on the HECL of lanthanide chelates.

Some lanthanide chelates produce strong luminescence under the HECL conditions, and therefore these compounds are well suited to be used as labeling compounds in e.g., bioassays and other analyses requiring sensitive detection. Time-resolved detection, which decreases the background noise, i.e., the blank signal, increases the sensitivity yet further. Many of the lanthanide chelates have long enough luminescence lifetimes for time-resolved detection. In addition, wavelength discrimination can be used to simultaneously detect several labels in one sample.

The application of lanthanide chelates as HECL labels in immunoassays and DNA probe assays shows great promise. The labels are suitable both for homogeneous assays and for heterogeneous assays; heterogeneous assays are at a disadvantage for needing more steps than homogeneous assays, but on the contrary the measuring buffer can be chosen more freely in the heterogeneous assays and therefore it can be more readily optimized than the one in homogeneous assays.

Printed electrodes in disposable HECL chips and cartridges seem most promising when the low cost is the most important issue and the accuracy demands are not very strict.

References

1. Godlewski M, Leskelä M (1994) Excitation and recombination processes during electroluminescence of rare earth-activated materials. Crit Rev Solid State Mater Sci 19:199–239
2. He X-H, Lian N, Sun J-H, Guan M-Y (2009) Dependence of luminescence properties on composition of rare-earth activated (oxy)nitrides phosphors for white-LEDs applications. J Mater Sci 44:4763–4775
3. Evans RC, Douglas P, Winscom CJ (2006) Coordination complexes exhibiting room-temperature phosphorescence: evaluation of their suitability as triplet emitters in organic light emitting diodes. Coord Chem Rev 250:2093–2126
4. Elbanowski M, Makowska B, Staninski K, Kaczmarek M (2000) Chemiluminescence of systems containing lanthanide ions. J Photochem Photobiol A Chem 130:75–81
5. Elbanowski M, Staninski K, Kaczmarek M (1993) Chemiluminescence used in biochemical investigations. An application of the lanthanide ions as chemiluminescent probe. Acta Phys Pol A 84:993–1002
6. Russ Algar W, Massey M, Krull UJ (2009) The application of quantum dots, gold nanoparticles and molecular switches to optical nucleic-acid diagnostics. Trends Anal Chem 28:292–306
7. Richter MM (2004) Electrochemiluminescence (ECL). Chem Rev 104:3003–3036
8. Richter MM (2008) Electrochemiluminescence. In: Ligler FS, Taitt CR (eds) Optical biosensors, 2nd edn. New York, Elsevier
9. Miao W (2008) Electrogenerated chemiluminescence and its biorelated applications. Chem Rev 108:2506–2553
10. Wang Z, Li J (2008) Nanostructure presented chemiluminescence and electrochemiluminescence. Annu Rev Nano Res 2:63–101
11. Marquette CA, Blum LJ (2007) Applications of the electrogenerated luminescent reactions in biosensor and biochip developments. In: Marks RS (ed) Handbook of biosensors and biochips. Wiley, UK
12. Braun F (1898) On evolution of light at some electrodes in electrolyte solutions. Ann Physik Chem WF 65:361
13. Ikonopisov S (1975) Problems and contradictions in galvanoluminescence. Crit Rev Electrochim Acta 20:783–793
14. Armstrong NR, Wightman RM, Gross EM (2001) Light-emitting electrochemical processes. Ann Rev Phys Chem 52:391–422
15. Fahnrich KA, Pravda M, Guilbault GG (2001) Recent applications of electrogenerated chemiluminescence in chemical analysis. Talanta 54:531–559
16. Andersson A-M, Schmehl RH (2001) Radiative charge recombination and electrochemiluminescence. Electron Transf Chem 1:312–341
17. Bard A, Debad J, Leland J, Sigal G, Wilbur J, Wohlstadter J (2000) In: Meyers RA (ed) Encyclopedia of Analytical Chemistry. Applications, Theory and Instrumentation. Wiley, New York
18. Gerardi RD, Barnett NW, Lewis SW (1999) Analytical applications of tris(2,2'-bipyridyl) ruthenium(III) as a chemiluminescent reagent. Anal Chim Acta 378:1–41
19. Knight AW (1999) A review of recent trends in analytical applications of electrogenerated chemiluminescence. Trends Anal Chem 18:47–62
20. Lee WY (1997) Tris(2,2'-bipyridyl)ruthenium(II) electrogenerated chemiluminescence in analytical science. Microchim Acta 127:19–39
21. Ganley WP (1972) Galvanoluminescence of rare earth doped aluminum. Thin Solid Films 11:91–97
22. Ganley WP, Mooney PM, Huminik D (1969) Effect of impurities on the anode, cathode and a.c. galvanoluminescence spectra of aluminum. Thin Solid Films 3:377–385
23. Ohtake T, Hijii S, Sonoyama N, Sakata T (2006) Electrochemical luminescence of n-type ZnO semiconductor electrodes doped with rare earth metals under the anodic polarization. Appl Surface Sci 253:1753–1757

24. Bulgakov RG (1973) Phototransfer and photoluminescence in europium(II) solutions as competing processes. Khim Vysokomol Soedin Neftekhim 8–9
25. Kulmala S, Haapakka K (1994) Terbium(III) lyoluminescence induced by the dissolution of UV-irradiated potassium peroxodisulfate in aqueous solutions. Anal Chim Acta 294:13–25
26. Haapakka K, Kankare J, Kulmala S (1985) Feasibility of low-voltage cathodic electroluminescence at oxide-covered aluminum electrodes for trace metal determinations in aqueous solutions. Anal Chim Acta 171:259–267
27. Haapakka K, Kankare J, Kulmala S (1988) Cathodic europium(III) electroluminescence at an oxide-coated aluminum electrode. Anal Chim Acta 209:165–174
28. Kulmala S, Kankare J, Haapakka K (1991) Electrogenerated luminescence of terbium(III) in aqueous solutions. Anal Chim Acta 252:65–76
29. Meulenkamp EA, Kelly JJ, Blasse G (1993) Electrochemically induced characteristic luminescence of metal ions at anodic valve metal oxides. J Electrochem Soc 140:84–91
30. Haapakka K, Kulmala S (1988) Cathodic electroluminescence at the terbium(III)-doped oxide-covered tantalum electrode and its feasibility for the determination of hydrogen peroxide. Anal Chim Acta 208:69–79
31. Lipiainen K, Kulmala S, Haapakka K (1992) Hydrogen peroxide-induced cathodic terbium (III) electroluminescence at a terbium-doped oxide-covered zirconium electrode. Anal Chim Acta 266:51–66
32. Kulmala S, Ala-Kleme T, Hakanen A, Haapakka K (1997) F-centre luminescence from oxide-covered aluminium cathode induced by two step reduction of peroxydisulfate anions. J Chem Soc Faraday Trans 93:165–168
33. Suomi J, Kulmala S (2009) Hot electron-induced electrochemiluminescence. Reviews in Fluorescence, in press (Chapter 3)
34. Kulmala S, Kulmala A, Helin M, Hyppanen I (1998) Hot electron-induced time-resolved electrogenerated luminescence of Tb(III) ions at stationary oxide-covered aluminium electrodes. Anal Chim Acta 359:71–86
35. Kulmala S, Haapakka K (1995) Mechanism of electrogenerated luminescence of terbium(III)-{2, 6-bis[N, N-bis(carboxymethyl)aminomethyl]-4-benzoylphenol} chelate at an oxide-covered aluminum electrode. J Alloys Compd 225:502–506
36. Ala-Kleme T, Kulmala S, Latva M (1997) Generation of free radicals and electrochemiluminescence at pulse-polarized oxide-covered silicon electrodes in aqueous solutions. Acta Chem Scand 51:541–546
37. Kulmala S, Kulmala A, Ala-Kleme T, Pihlaja J (1998) Primary cathodic steps of electrogenerated chemiluminescence of lanthanide(III) chelates at oxide-covered aluminum electrodes in aqueous solution. Anal Chim Acta 367:17–31
38. Kulmala S, Ala-Kleme T, Joela H, Kulmala A (1998) Hot electron injection into aqueous electrolyte solution from thin insulating film-coated electrodes. J Radioanal Nucl Chem 232:91–95
39. Kankare J, Haapakka K, Puhakka O, Kulmala S (1985) Electrogenerated luminescence of terbium complexes on an aluminium electrode. 36th meeting of ISE, Salamanca, September 1985
40. Kankare J, Fälden K, Kulmala S, Haapakka K (1992) Cathodically induced time-resolved lanthanide(III) electroluminescence at stationary aluminum disk electrodes. Anal Chim Acta 256:17–28
41. Kankare J, Haapakka K, Kulmala S, Näntö V, Eskola J, Takalo H (1992) Immunoassay by time-resolved electrogenerated luminescence. Anal Chim Acta 266:205–212
42. Kulmala S, Raerinne P, Takalo H, Haapakka K (1995) Tb(III) chelate chemiluminescence induced by the dissolution of additively colored potassium chloride in aqueous solutions of peroxydisulfate ions. J Alloys Compds 225:492–496
43. Kulmala S, Kulmala A, Laine E, Takalo H, Haapakka K (1995) Terbium(III)-{2,6-bis[N, N-bis(carboxymethyl)aminomethyl]-4-benzoylphenol} chelate enhanced lyoluminescence of x-ray irradiated sodium chloride. J Alloys Compds 225:497–501

44. Kulmala S, Kulmala A, Latva M, Haapakka K (1997) X-ray irradiated sodium chloride as an excitation source for the sensitized terbium(III)-specific chemiluminescence of aromatic Tb(III) chelates in aqueous solutions. Anal Chim Acta 347:333–350
45. Richter MM, Bard AJ (1996) Electrogenerated chemiluminescence. 58. Ligand-sensitized electrogenerated chemiluminescence in europium labels. Anal Chem 68:2641–2650
46. Yu H-X, Cui H, Guan J-B (2006) Cathodic electrochemiluminescence of acetonitrile, acetonitrile-1,10-phenanthroline and acetonitrile-acetonitrile-ternary Eu(III) complex systems at a gold electrode. Luminescence 21:81–89
47. Håkansson M, Helin M, Putkonen M, Jiang Q, Kotiranta M, Suomi J, Ala-Kleme T, Kulmala S (2005) Electrochemiluminescence of Tb(III) chelates at optically transparent tunnel emission electrodes fabricated by atomic layer deposition. Anal Chim Acta 541:137–143
48. Håkansson M, Jiang Q, Helin M, Putkonen M, Niskanen AJ, Pahlberg S, Ala-Kleme T, Heikkilä L, Suomi J, Kulmala S (2005) Cathodic Tb(III) chelate electrochemiluminescence at oxide-covered magnesium and n-ZnO:Al/MgO composite electrodes. Electrochim Acta 51:289–296
49. Jiang Q, Suomi J, Håkansson M, Niskanen AJ, Kotiranta M, Kulmala S (2005) Cathodic electrogenerated chemiluminescence of $Ru(bpy)_3^{2+}$ chelate at oxide-coated heavily doped silicon electrodes. Anal Chim Acta 541:159–165
50. Morrison SR (1980) Electrochemistry at semiconductor and oxidized metal electrodes. Plenum Press, New York
51. Håkansson M, Jiang Q, Suomi J, Loikas K, Nauma M, Ala-Kleme T, Kankare J, Juhala P, Eskola JU, Kulmala S (2006) Cathodic electrochemiluminescence at double barrier Al/Al2O3/Al/Al2O3 tunnel emission electrodes. Anal Chim Acta 556:450–454
52. Buxton GV, Greenstock CL, Helman WP, Ross AB (1988) Critical review of rate constants for reactions of hydrated electrons, hydrogen atoms and hydroxyl radicals (•OH/•O-) in aqueous solution. J Phys Chem Ref Data 17:513–886, and references therein
53. Neta P, Huie RE, Ross AB (1988) Rate constants for reactions of inorganic radicals in aqueous solution. J Phys Chem Ref Data 17:1027–1284
54. Kulmala S, Ala-Kleme T, Heikkilä L, Väre L (1997) Energetic electrochemiluminescence of (9-fluorenyl)methanol induced by injection of hot electrons into aqueous electrolyte solution. J Chem Soc Faraday Trans 93:3107–3113
55. Helin M, Jiang Q, Ketamo H, Håkansson M, Spehar A-M, Kulmala S, Ala-Kleme T (2005) Electrochemiluminescence of coumarin derivatives induced by injection of hot electrons into aqueous electrolyte solution. Electrochim Acta 51:725–730
56. White HS, Bard AJ (1982) Electrogenerated chemiluminescence. 41. Electrogenerated chemiluminescence and chemiluminescence of the tris(2,2'-bipyridine)ruthenium(2^+)-peroxydisulfate(2-) system in acetonitrile-water solutions. J Am Chem Soc 104:6891–6895
57. Ege D, Becker WG, Bard AJ (1984) Electrogenerated chemiluminescent determination of Ru $(bpy)_3^{(2+)}$ at low levels. Anal Chem 56:2413–2417
58. Kulmala S, Suomi J (2003) Current status of modern analytical luminescence methods. Anal Chim Acta 500:21–69, and references cited therein
59. Ala-Kleme T, Kulmala S, Väre L, Juhala P, Helin M (1999) Hot electron-induced electrogenerated chemiluminescence of $Ru(bpy)_3^{2+}$ chelate at oxide-covered aluminum electrodes. Anal Chem 71:5538–5543
60. Memming R (1969) Mechanism of the electrochemical reduction of persulfates and hydrogen peroxide. J Electrochem Soc 116:785–790
61. Kulmala S, Helin M, Ala-Kleme T, Väre L, Papkovsky D, Korpela T, Kulmala A (1999) Electrochemiluminescent labels for applications in fully aqueous solutions at oxide-covered aluminium electrodes. Anal Chim Acta 386:1–6
62. Ala-Kleme T, Haapakka K, Latva M (1999) Near-infrared electrogenerated chemiluminescence of ytterbium(III) chelates in aqueous electrolytes. Anal Chim Acta 395:205–211
63. Kulmala S, Ala-Kleme T, Kulmala A, Helin M, Väre L (1999) Electrochemiluminescent labels for applications in fully aqueous solutions at oxide-covered aluminium electrodes. Anal Chim Acta 386:1–6

64. Helin M, Håkansson M, Canty P, Spehar A-M, Kulmala S (2002) Hot electron–induced electrogenerated chemiluminescence of 1-aminonaphtalene-4-sulphonate at oxide-covered aluminium electrodes in aqueous solution. Anal Chim Acta 454:193–201
65. Jiang Q, Spehar A-M, Håkansson M, Suomi J, Ala-Kleme T, Kulmala S (2006) Hot electron-induced cathodic chemiluminescence of rhodamine B at disposable oxide-coated aluminum electrodes. Electrochim Acta 51:2706–2714
66. Ala-Kleme T, Kulmala S, Jiang Q (2006) Generation of free radicals and electrochemiluminescence from simple aromatic molecules in aqueous solutions. Luminescence 21:118–125
67. Kulmala S, Ala-Kleme T, Latva M, Loikas K, Takalo H (1998) Hot electron-induced electrogenerated chemiluminescence of rare earth(III) chelates at oxide-covered aluminum electrodes. J Fluoresc 8:59–65
68. Jiang Q, Håkansson M, Spehar A-M, Ahonen J, Ala-Kleme T, Kulmala S (2006) Hot electron-induced time-resolved electrogenerated chemiluminescence of a europium(III) label in fully aqueous solutions. Anal Chim Acta 558:302–309
69. Ala-Kleme T, Haapakka K, Latva M (1998) Y(III)-enhanced Dy(III) and Sm(III)-specific electrogenerated luminescence of heterodinuclear 1-Y(III)-Dy(III)-1 and 1-Y(III)-Sm(III)-1 chelates. J Alloys Compds 275–277:911–914
70. Lis S, Staninski K, Grzyb T (2008) Electrochemiluminescence study of europium(III) complex with coumarin3-carboxylic acid. Int J Photoenergy. doi:10.1155/2008/131702
71. Staninski K, Lis S (2008) Electrogenerated luminescence of chosen lanthanide complexes at stationary oxide-covered aluminum electrode. J Alloys Compds 451:81–83
72. Väre L, Canty P, Håkansson M, Spehar A-M, Papkovsky D, Ala-Kleme T, Kankare J, Kulmala S (2002) Time-resolved electrochemiluminescence of metalloporphyrins. Anal Chim Acta 453:269–279
73. Eskola J, Mäkinen P, Oksa L, Loikas K, Nauma M, Jiang Q, Håkansson M, Suomi J, Kulmala S (2006) Competitive immunoassay by hot electron-induced electrochemiluminescence detection and using a semiautomatic electrochemiluminometer. J Lumin 118:238–244
74. Kankare J, Karppi A, Takalo H (1994) Novel labeling agents for immunoassay by time-resolved electrogenerated chemiluminescence. Anal Chim Acta 295:27–35
75. Helin M, Väre L, Håkansson M, Canty P, Hedman H-P, Heikkilä L, Ala-Kleme T, Kankare J, Kulmala S (2002) Electrochemiluminoimmunoassay of hTSH at disposable oxide-coated n-silicon electrodes. J Electroanal Chem 524–525:176–183
76. Kulmala S, Håkansson M, Spehar A-M, Nyman A, Kankare J, Loikas K, Ala-Kleme T, Eskola J (2002) Heterogeneous and homogeneous electrochemiluminoimmunoassays of hTSH at disposable oxide-covered aluminum electrodes. Anal Chim Acta 458:271–280
77. Spehar-Délèze A-M, Suomi J, Jiang Q, de Rooij N, Koudelka-Hep M, Kulmala S (2006) Heterogeneous oligonucleotide-hybridization assay based on hot electron-induced electrochemiluminescence of a rhodamine label at oxide-coated aluminum and silicon electrodes. Electrochim Acta 51:5438–5444
78. Ala-Kleme T, Mäkinen P, Ylinen T, Väre L, Kulmala S, Ihalainen P, Peltonen J (2006) Rapid electrochemiluminoimmunoassay of human C-reactive protein at planar disposable oxide-coated silicon electrodes. Anal Chem 78:82–88
79. Pihlajamäki SV, Kankare JJ (1986) Photon counter system with a rotating interference filter for measuring time-resolved electroluminescence spectra. Anal Instrum 15:171–192
80. Kulmala S, Niskanen A J, Ylinen-Hinkka T, Franssila S, (2010) Accurate integrated low-cost electrode chips for point-of-need analysis and methods of utilization. Patent application FI 20100251
81. Kulmala S, Eskola J, Laaksonen T, Korpela T, Håkansson M, Suomi J, (2010) Integrated carbon electrode chips for electrical excitation of lanthanide chelates and analytical methods utilizing these chips. Patent application FI 20100253
82. Niskanen AJ, Ylinen-Hinkka T, Kulmala S, Franssila S (2009) Ultrathin tunnel insulator films on silicon for electrochemiluminescence studies. Thin Solid Films 517:5779–5782

Sensitized Bioassays

Ilkka Hemmilä and Ville Laitala

Abstract Luminescent lanthanides are well-suited and widely applied markers for bioaffinity assays, both separation-based heterogeneous assays, such as immunoassays, as well as for homogeneous assays, such as assays based on energy transfer. Their properties allow efficient combination of spectral and temporal resolution to gain good sensitivity, perfect discrimination of produced sensitized signal, and robust assays. Both europium and terbium, as their stable and highly luminescent chelates, have found wide applications in research aiming for new drug discovery. The complicated but efficient energy transfer process in the chelates, and between chelate and acceptor, provides versatile tools to stipulate good assays for various applications and also to gain limited multiplexing to allow measurement of several biomarkers simultaneously. The chapter reviews the recent development in the area and gives overview of interesting applications.

Keywords Europium · FRET · HTS (high-throughput screening) · Multiplexing · Terbium

Contents

I. Hemmilä (✉) and V. Laitala
PerkinElmer, Wallac Oy, PO Box 10, 20101 Turku 10, Finland
e-mail: Ilkka.Hemmila@perkinelmer.com

P. Hänninen and H. Härmä (eds.), *Lanthanide Luminescence: Photophysical, Analytical and Biological Aspects*, Springer Ser Fluoresc (2011) 7: 361–380, DOI 10.1007/4243_2010_10,

1 Lanthanides as Labels

Lanthanides, particularly when complexed with suitable strong chelating and light-absorbing ligands, provide quite unique luminescent (in the context of applications referred to as fluorescence) properties, which has made them attractive and increasingly popular labels in in vitro immunoassays and research applications.

To be a good alternative label for a practical bioaffinity binding assay, lanthanide chelate has to fulfill a set of requirements – making the choice of organic ligand a challenge. Table 1 lists the main requirements for a chelate label suitable for bioaffinity assays and sensitized assays in particular. Solubility and minimal hydrophobic interactions are a necessity in bioassays applying delicate reagents with specific recognition. High thermodynamic and kinetic chelate stabilities are also needed in assays where the chelate has to withstand high ionic strength, competing complexing agents, variable pH (e.g., lysosome), and high concentrations of other ions. Stability is very much application-dependent, and, e.g., for in vivo imaging, this is encountered very strongly with gadolinium chelates used for MRI. Intravenous applications set particularly high demand for not only to ascertain good signal and contrast in imaging but also to avoid toxicity with high doses of free lanthanides. In addition, to be practical in a luminescence-based assay technology, the lanthanide ion

Table 1 Main issues upon optimization of lanthanide donor chelates for bioaffinity assays

Feature	To be optimized	Effect
Size	Small preferred	Less steric effect on binding
Charge	May vary depending on assay	Has major effect on nonspecific interactions and ion-pair quenching
Solubility	As high as possible	Minimal hydrophobic interactions and easy labeling
Stability	As high as possible	Tolerance against in vivo condition, cellular compartment high metallic ions and chelating compounds
Good protection	Confirmed nona-dentate	Avoid water and minimize other external quenching
Intensity	As high as possible	To provide maximal ET potency and signal
Decay	0.1–1 ms	Current instruments in the field not able to measure below 10 μs. Over 1 ms makes measurement slow
Emission profile	Most emission concentrated into donating peak	High ET and low background at spectral minimum area used for measurement
ET rate constant	As high as possible	

has to produce high and intense emission with desired properties in terms of excitation wavelength, emission distribution, and decay time. The complexing agent needs to provide also a good protection against aqueous and other ambient quenching reactions (metallic ions, other ion-pair interactions, and diffusion related reactions). These requirement are fulfilled, e.g., with stable complexones containing suitable antenna as a part of the chelating group providing high light collection efficiency – preferably at as long wavelength as practical to ascertain excitation at over 300 nm – a well situated triplet state to ascertain efficient energy transfer, suitable ligand field symmetry to create desirable emission profile, and a good protection (nona-dentate) to avoid inner-sphere OH and other ambient surrounding-related excited state deactivating processes (Table 1).

The long excited state lifetimes of lanthanide chelates are, perhaps, the most important and differentiating features amongst fluorescent labels. The transitions in the 4*f* orbital are electric forced and have long lifetimes [1, 107]. The lifetimes depend on the "natural" lifetime of the central ion in the complex in addition to the sum of all deactivation processes, such as inner- and outer-sphere OH vibrations [2, 3], all the proximal C–H and N–H vibrations [4, 5], and eventual energy back-flow and energy charge transfer quenching [6] the chelate may express.

Lanthanide chelates are most commonly deactivated through the manifold of O–H vibration [5]. In practical assay conditions, there are, however, variable other quenching routes, which potentially diminish the desired signals and interferes with the assay. Quenching depends to a great extend on the ligand field, its protective ability, and on the overall charge of the complex. In addition to colored compounds able to take energy either through collisions [7] or through resonance energy transfer, species such as free radicals [8], paramagnetic metallic ions [9], metallic surfaces [10], or organic compounds having suitable absorbance create interference. For example, free manganese ions (Mn^{2+}), generally applied in cellular signaling assays, have to be blocked with EDTA to avoid complete quenching of negatively charged complexes.

1.1 Ligand-Enhanced Lanthanide Emission

Ligand-enhanced excitation and efficient intrachelate energy transfer is necessary to obtain highly fluorescent lanthanide chelates and also to gain practical excitation wavelength (Fig. 1). The generally applied chelates have excitation maxima in the range 300 to 350 nm, even though chelates with excitation up to around 400 nm are reported. The practical excitation maximum is limited by the required gap between the singlet state and its triplet state as the triplet state energy has to be above the energy accepting level of the ion [11].

By theoretical model [12] and by analyzing experimentally with a set of chelates and their triplet states [13], it seems likely that lanthanides accept excited energy from the ligand through ligand triplet state [14], and in the case of Eu, the actual accepting level seems to be any of the several ^5D_j levels, primarily $^5D_1$ or $^5D_2$ levels (Fig. 1).

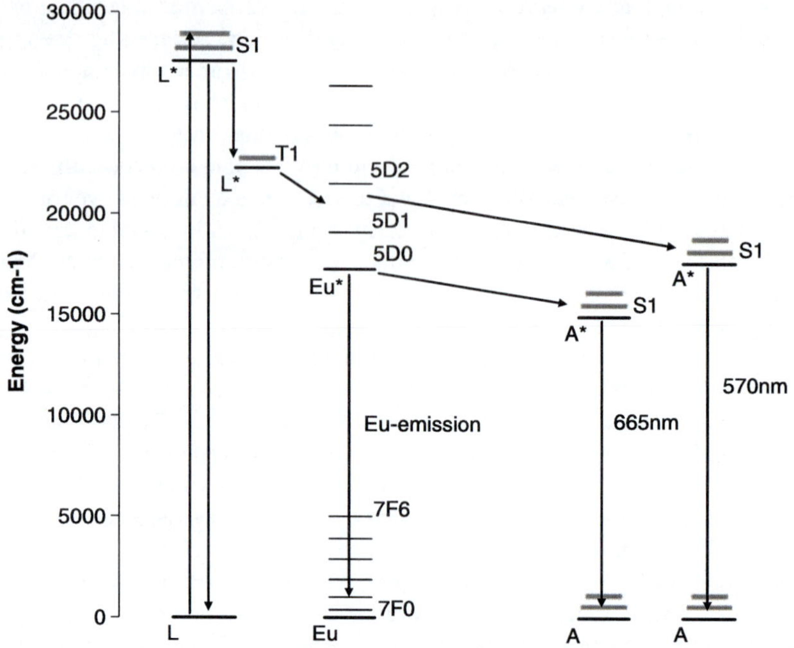

Fig. 1 Energy flow diagram within a europium chelate and an energy transfer assay. *L* ligand, *A* acceptor, *asterisk* shows excited state level

In those chelates, lanthanide ions function as triplet sensitizers routing the energy with high efficiency through the singlet state, triplet state, to the final emittive level of the ion. Due to the long lifetime of the lowest excited state of the lanthanides, the overall decay time is primarily determined by its natural lifetime.

2 Resonance Energy Transfer

The Förster-type resonance energy transfer (FRET) has become a common practice in the study of macromolecular structures, conformational changes, in vitro and cellular complex formation, and so on [15]. Ullman introduced its use for clinical diagnostics as a way to make homogeneous immunoassays [16].

The major limitations of a conventional FRET technology relate, firstly, to the rather short distance required for efficient energy transfer to take place even with perfectly matching donor–acceptor pairs (generally below 10 nm), and secondly, to the difficulty in discriminating the energy-transfer-sensitized acceptor emission from the emission emanating from directly excited acceptor emission, and also the ambient background interference. While the distance limitation is valid for all FRET assays, the interference of directly excited acceptor can be avoided by using temporal discrimination.

In addition to the use of emittive acceptors to measure sensitized emission, various nonemitting compounds have also been used in assays where the proximity creates signal quenching. Those assays are mainly applied in enzymatic assays where hydrolysis interrupts the proximity and reaction is followed by recovery of the signal.

2.1 Lanthanides as Energy Donors

The luminescent properties of lanthanide chelates make them ideal energy donors for FRET applications. Firstly, the application of background insensitive time-gated (TR) detection [17] gives the advantage of discrimination of ambient background from desired signal. The long excited state lifetime provides a tool to avoid the interference derived from direct acceptor excitation by using temporal discrimination, first reported by Morrison [18]. In addition, the sharp, line-type emissions provide a good and background free area for energy transfer measurement to take place by a combination of a good spectral and temporal resolution. Further, each of the emission lines of a lanthanide donor can be separately utilized with different acceptors providing an advantage in multianalyte applications. An additional advantage having widely different decays is the fact that in TR-FRET, a single short decay acceptor can take energy from a large number of donors during their excited state.

An example of energy-transfer scheme in a time-resolved mode (TR-FRET) is shown in Fig. 2. Regardless of the complicated energy flow starting from intrachelate

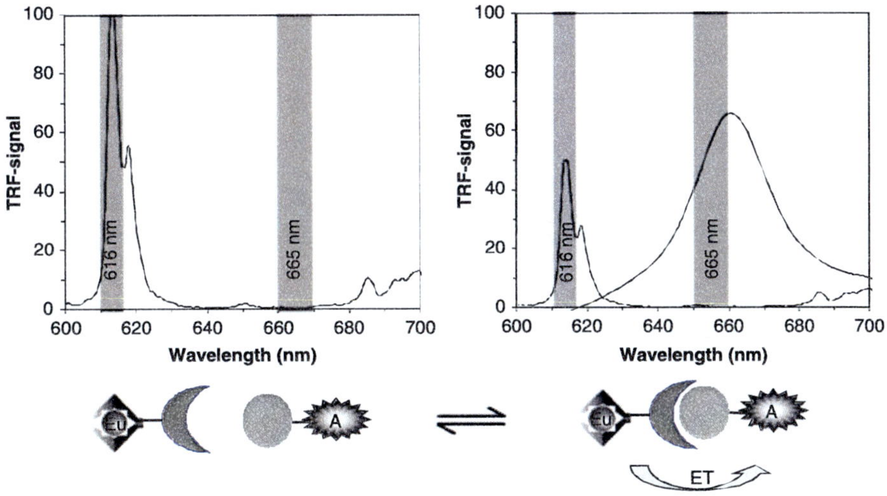

Fig. 2 Time-resolved FRET assay in a protein–protein interaction with europium-chelate and acceptor-labeled proteins as partners. Respective gated emissions are shown before and after the binding reaction showing the emerged sensitized long decay acceptor emission. Typical wavelength windows are shown at 613 nm for europium donor and 665 nm for sensitized emission

energy transfer and ending into intermolecular energy transfer in a biochemical complex, the overall efficiency can be reasonable high and the good signal discrimination enables relatively robust and sensitive detection in in vitro assays. As a result, the lanthanide-based resonance energy transfer assays have become standard tools in the area of drug discovery and high-throughput screening [19, 20].

2.1.1 Europium Chelates as the Energy Donors

Europium is the most commonly used energy donor ion in practical TR-FRET applications, in diagnostics and in high throughput screening. A wide variety of different structures have been synthesized and optimized for europium [20]. By far, the most applications rely on energy transfer from the main emittive ionic level, $^5D_0$, to a suitable acceptor having good spectral overlapping integral. This FRET obeys well the regular Förster equations in respect to spectral overlapping and Förster radius. Regular organic fluorochromes or fluorescent proteins with highly overlapping absorbance at 613 nm area are used as energy acceptors. The local emission minimum, at 665 nm, provides a low background wavelength window for energy transfer measurement. The near infrared area, over 700 nm, can also be used for sensitized acceptor measurement, provided that red-shifted acceptors are applied. Europium chelates have very practical decay times ranging from about 350 μs to above 1 ms, making the measurement of TR-FRET signal easy, robust and rapid.

Europium has several emittive transitions, all of which with the exception of magnetic dipole transition [7] can participate into energy transfer at about the same potency, with slight emphasis at the longer wavelength transitions ($^5D_0$–$^7F_4$). On the other hand, optimization of the ligand field and its symmetry can be exploited to stipulate emission profile and to concentrate most of the energy to the single hypersensitive transition ($^5D_0$–$^7F_2$), allowing efficient ET on that region and low background on the areas of measurement of the sensitized acceptor emission [21].

In most of the fluorescent Eu chelates, the energy flow goes actually through the higher excited states of the ion, via $^5D_2$ and $^5D_1$ (Fig. 1). Those levels are not generally regarded emittive but can be recorded by measuring fluorescent chelates with a short delay time. Figure 3 shows emission spectra of a typical fluorescent europium chelate after two different delays expressing both $^5D_0$ and $^5D_1$ emission. It has also be reported that $^5D_1$, and possibly also $^5D_2$, can be used directly as energy-donating levels when using acceptors energetically overlapping with these levels, resulting in relatively high energy transfer efficiency taking into account the minimal overlapping integral and exhibiting a curious decay time most likely reflecting the decay of the higher ionic excited levels [21, 22]. Energy transfer directly from $^5D_1$ seems, however, to require narrower proximity, and its use may be limited to specific applications, such as DNA hybridization assays [23].

Europium chelates are most often used donors in the commercialized TR-FRET product lines. Europium cryptates are applied in the diagnostic system of Brahms, Kryptor® [24], and in the HTS system of Cis-Bio International (HTRF®). Various

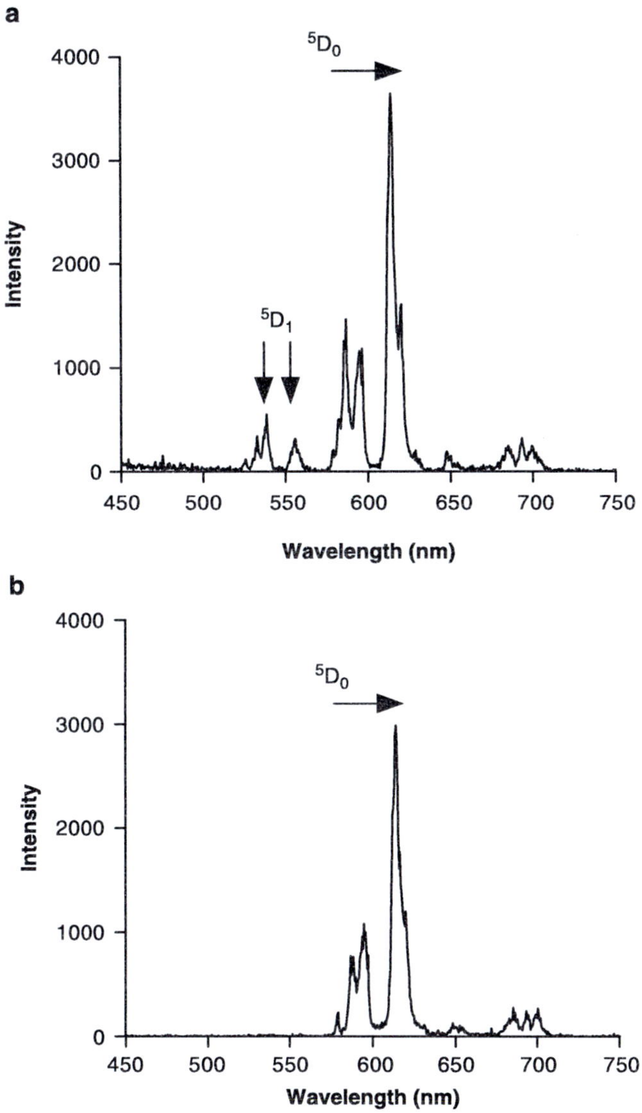

Fig. 3 Emission spectra of a fluorescent europium chelate with two different delay times. (**a**) using a delay of 2 μs showing also $^5D_1$ emission, (**b**) with a delay of 7 μs whereafter only the long decay $^5D_0$ emissions are visible

polycarboxylate-based complexes are applied in the LANCE® product line of PerkinElmer, as well as in LFRET® of GE Healthcare. Other product offering based on europium as energy donors are from Hidex Proxyscreen (beads) and Upstate KinEASE [19].

2.1.2 Terbium Chelates as the Energy Donors

Compared to europium, the excited energy level on terbium ion is at a higher energy level (above 20,000 cm^{-1}) and constitutes of the single excited state, $^5D_4$, which is functioning both as the energy-acceptive and -emittive level (Fig. 4). The higher excited state naturally requires higher triplet state for the ligand reflecting to higher singlet state and, consequently, shorter wavelength excitation. Those features seem to entail into higher ambient background and also into relatively more pronounced energy backflow probability and thus higher sensitivity to environmental changes. On the other hand, the energy gap between the lowest excited state and the highest ground state level is relatively large, and terbium is not very sensitive to aqueous quenching and it seems feasible to find very bright terbium chelates.

Due to energy backflow problem, the decays of even the stable fluorescent Tb chelates vary rather widely from below 100 μs to as long as 2.6 ms [21, 23, 25]. Fluorescent terbium chelates have also found commercial applications in HTS in Lanthascreen product line of Invitrogen and later on by CisBio with the Lumi4 chelates of Lumiphore company.

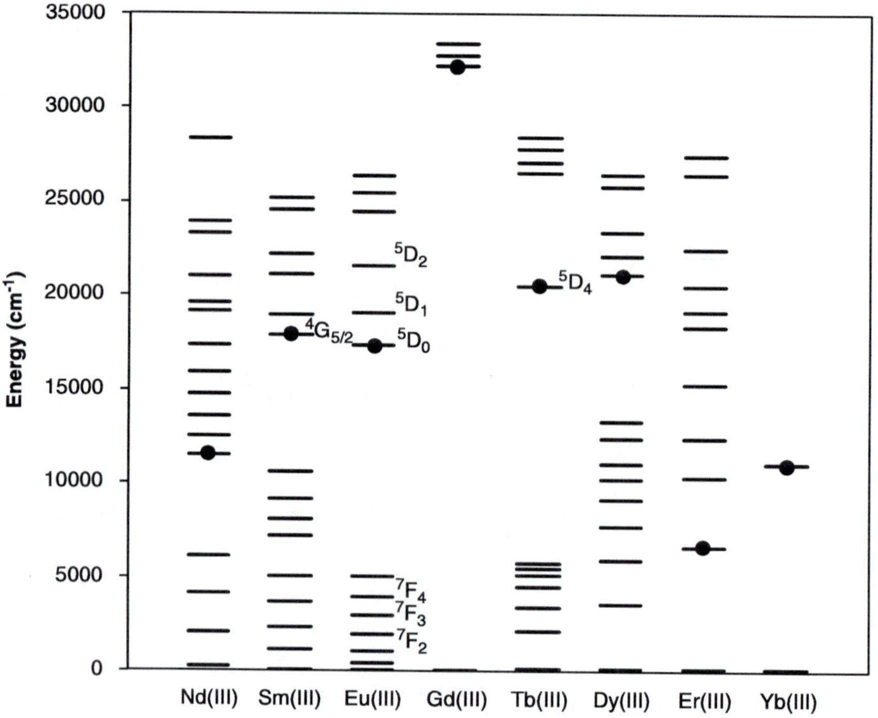

Fig. 4 Some excited and ground state energy levels with luminescent lanthanides. Lowest excited state is indicated by a *spot*. The levels referred to at the text are indicated

Selvin et al. [26, 27] developed a series of chelates based on a carbostyril moiety as the energy harvester attached to complexones including an amino-reactive CS_{124}-DTPA-phenyl-ITC and thiol-reactive CS_{124}-DTPA-EMCH-Tb. Hydroxyisophtalamide derivatives is a new group of structures examined with different lanthanides [25, 28] and have found applications in TR-FRET applications through Lumiphore company and their licenses, e.g., with Echelon Biosciences in their True-FRET® assays and in the HTRF products of CisBio International.

With fluorescent terbium chelates, both the transitions $^5D_4$–$^7F_6$ and $^5D_4$–$^7F_4$ have been used as the energy donating lines, the former together with a short wavelength acceptor, such as fluorescein (FITC), and the latter with red shifted acceptors, such as Cy-3 or rhodamines [26, 29, 30]. A clear advantage with terbium relates to its suitability as a donor for green fluorescent protein (GFP) generally used as a reporter protein allowing in vivo applications, such as cellular MAP kinase monitoring [31].

Lanthanide chelates with long decay times can also be used to make efficient tandem probes where both emission wavelength and decay times can be stipulated by selecting the spacer arm and distances between donor moiety and acceptor. Selvin reported a series of tunable probes composed of fluorescent terbium- and europium-chelates coupled to fluorescein or Cy-5, respectively [32].

2.1.3 Other Lanthanides

Besides europium and terbium, the other lanthanides have gained relatively little attention and practical applications in TR-FRET. In a study of DNA hybridization with labeled oligonucleotides, we have found that samarium shows unexpectedly high energy transfer kinetics and good signals applicable in proximity assay at a short distance, such as oligonucleotide hybridization [33]. Regardless of a very low overlapping integral between samarium and Alexa-Fluor 532, the sensitivity of homogeneous delta-F508 assay was very high, allowing 12 pM of the amplified sequence to be detected, demonstrating a very efficient energy transfer emanating directly from the $^4F_{3/2}$ or $^4G_{5/2}$ levels (Fig. 4). In a study using different Alexa dyes in hybridization assay with samarium-labeled probe, the energy transfer from the higher excited states produced longer and stable decay time with nonoverlapping ET pair (samarium and Alexa Fluors −488, −514 and −532) as compared to traditional overlapping pair (Sm and Alexa Fluor 647) (6.6–8.5 μs as compared with 1.3 μs with Alexa-647) [33].

The long wavelength-emitting lanthanides, neodymium, ytterbium, and erbium, are interesting alternatives and potentially applicable at near infrared area, above 1,000 nm. More wide practical applications of the other lanthanides are hampered by a general lack of good and efficient excitation-gating cycle in submicrosecond time domain because the instruments in the field are not generally able to measure short-lived emissions.

2.2 Nanobeads as Donors

Europium-chelate impregnated nanobeads, e.g., those commercialized by Seradyn (Fluoro-Max, Seradyn Inc), are tested as potential energy donors in homogeneous immunoassay of estradiol [34, 35]. Relatively large beads embedded with fluorescent Eu chelates tend to entail large Eu related background and compromised signal due to large amounts of chelates inside the beads. High background at 665 nm area can be partly overcome using $^5D_0$–$^7F_3$/$^7F_4$ as the nominal energy donating transition and NIR area overlapping acceptor, such as Alexa 680, because the europium cross-talk interference is considerable lower at NIR area. An assay of prostate-specific antigen, PSA, was constructed using two set of beads, europium-doped donor beads (Seradyn) and larger sized NIR-emittive acceptor beads (TransFluorSphere® from Invitrogen), reportedly with high sensitivity down to 0.1 ng/ml [36].

Surface-modified nanobeads, beads where the chelates have shorter distance to the respective acceptor-labeled reactant, are also used to create proximity in TR-FRET assays such as cAMP immunoassay and assays of kinase activities by the Molecular Devices IMAP® system. The kinase assays apply fluorescent peptides as substrates, which after phosphorylation are collected to beads by complex formation between phosphor and the trivalent metallic ions on the beads. Fluorescent terbium chelates, substituted with a phosphor group, provide thereafter the energy-donating groups to create sensitized signal on bead surfaces [37].

A quite different type of sensitized emission is applied in oxygen channeling assays, where excitation light-produced free oxygen radicals sensitizes the second bead embedded europium chelates through a chemiluminescence reaction [38]. This is applied in diagnostics as Loci assay and in research as AlphaLisa® assays (PerkinElmer). Upconverting particles, reviewed by Nann in this book, are also used as energy donors in assays.

2.3 Suitable Energy Acceptors for Lanthanides

The choice of acceptor follows simple Förster rules. Firstly, there has to be a good spectral overlap integral between the main emission lines of the donor and, secondly, the acceptor should have relatively high emission intensity at the local donor emission minimum. For europium chelates, 665 nm area is well-suited for acceptor measurement due to low donor background. Another suitable low background area is at 730 nm, requiring, however, a suitable NIR emitting acceptor. Terbium chelates allow sensitive FRET measurement at 520, 570, or at over 650 nm areas. Naturally, high quantum efficiencies and molar absorptivities are advantageous for all acceptors.

A wide range of fluorescent compounds is today available and well-suited for energy acceptors. In addition to allophycocyanin (APC), which used to be the most common acceptor for europium in all applications, commercially applied small molecular acceptors include undisclosed D2 of CisBio (see e.g., cAMP HTRF assay; www.htrf.com), ULight of PerkinElmer, in addition to Alexa Fluors and carbocyanines. GFP is a generally used reporter group for protein expression and can also be used as energy acceptor for terbium chelates allowing in cellulo screening [31]. In quenching-based assays, various colored quenchers have been applied, including black hole quencher (Biosearch Technologies) and QSY-7 (Molecular Probes), as well as more simple absorbing compounds, such as dabcyl and even Cy-5, have been used.

Quantum dots have created interest as highly fluorescent particles having extremely high absorptivity and quantum yield. Their use in energy transfer assays has so far found less applications. Hildebrand et al. reported, in a very simple assay system based on biotin–streptavidin binding, an efficient energy transfer (TR-FRET) between a terbium chelate (open chain bis-bipyridine scaffolded complexone) and europium cryptate to quantum dots [39, 40]. They reported R_o of 81–104 Å for terbium and 96 Å for europium chelate. In a respective study of Härmä et al, using different fluorescent europium and terbium chelates in a competitive binding assay of biotin and estradiol, it was calculated that, however, the quantum dots are not particularly efficient energy acceptors for lanthanide chelates [41].

3 Multiplexing

Lanthanides are particularly well-suited for multilabel assays due to their narrow line type emissions and varying decay times. Combination of different lanthanide donors where each of them may apply a set of different acceptors and create ET signal at different wavelength and decay times is also applied in bioanalytical assays, particularly in DNA hybridization assays in measuring single nucleotide polymorphic (SNP) alleles related to disease risk factors.

Combination of europium- and terbium-labeled oligonucleotide probes with complementary probes labeled with quenchers has been used in competitive hybridization assay as to screen celiac disease-related alleles [23] and cystic fibrosis risk alleles [42]. Doubly labeled beacon-type probes containing europium chelate as donor and either Alexa 647 or Alexa Fluor 700 as acceptors has been utilized by measuring both the wavelength and decay change upon hybridization [43].

Dual-label TR-FRET substrates have also been used for simultaneous monitoring of the enzymes related to apoptosis; caspases 1, 3, and 6 [44]. Kokko et al. utilized fluorescent terbium chelates as the donors for two different acceptors, Alexa Fluor 488 and 680, in a dual-label immunoassay of free and total PSA [45]. Free, uncomplexed PSA were measured with TR-FRET from terbium labeled (universal) antibody to free-specific Alexa 488 labeled antibody, and the total PSA was quantified with TR-FRET from terbium to another generic antibody labeled with Alexa 680. Kupcho et al. followed nuclear receptor coactivator recruitment using a TR-FRET combination of a terbium donor and tetramethylrhodamine and Alexa Fluor as the acceptors [46].

Jeyakumar and Katzenellenbogen developed a dual-acceptor assay to monitor a nuclear receptor (thyroid hormone – retinoid X receptor heterodimer) regulation by measuring both coactivator and corepressor binding by using fluorescein and Cy-5 as acceptors for a terbium chelate-labeled response element-bound receptor complex [47].

4 Gated Detection and Signal Processing

TR-FRET assays are, by definition, assays applying gated detection. Due to their relatively wide use in drug discovery, practically all plate reader vendors have adapted their instruments to read TR-FRET assays too. Both xenon flash lamps and nitrogen lasers are applied in the plate readers to create short-lived and intense excitation pulses at the area of 330–340 nm.

Due to a tiny, but visible, signal crossover, detection of the ET-sensitized acceptor signal is generally done through a relatively short time window (from 50 to 150 μs or 100 to 500 μs) in order to avoid excessive background from the long-lasting donor emission. In most applications, the europium-sensitized acceptor emission is integrated through a 665 nm narrow band-pass interference filters.

The advent of other lanthanides and applications allowing shorter proximity may bring along improvements in the current instrumentation in terms of excitation pulse width and purity. LED and solid state lasers might be the sources of choice in the next generation of detection. Especially the recent development in UV-LEDs has raised interest in using them as a low price light source. Still at the present time, the low pulse power of LEDs at UV has prevented their wide use. Sufficient pulse powers can also be achieved with solid state lasers, but for practical applications, their prices are higher than those of traditional light sources.

Gated detection is generally accomplished with photomultiplier gating, but mechanical choppers, in combination with cooled CCD (ViewLux instrument of PerkinElmer), can be applied as well. Modern high performance CCD or CMOS arrays or CCDs equipped with individual amplifiers (EMCCD) have found less applications. Multi-pixel photon counter (MPPC, Hamamatsu) may provide a next-generation gated detector. MPPC uses multiple avalanche photodiode pixels in Geiger mode, enabling good quantum efficiency and short response time. Dark current remains to be an issue with MPPC as compared to traditional photomultipliers.

Delay time and measuring window optimization is naturally available in all TR-FRET applications to distinguish specific signal from all the noise sources. Delay optimization is particularly important in applications where the TR-FRET signal has a short decay time or the donor quenching is to be measured. Optimized delay time prevents losing the acceptor signal in case of efficient FRET and, correspondingly, more efficient donor quenching can be measured using adequate delay times [33, 48]. Short measurement windows can be utilized to further suppress

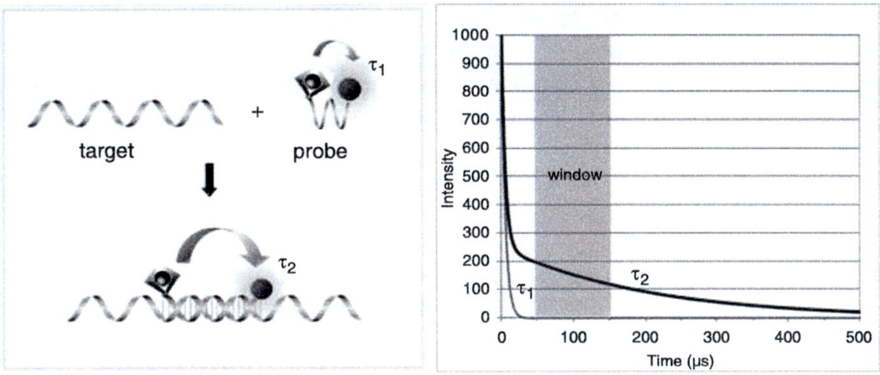

Fig. 5 An example of TR-FRET decay change upon binding. Hybridization assay of target (related to celiac disease) is accomplished with dual-label flexible probe, which upon hybridization is extended resulting in prolonged decay time as shown in the measured values. Redrawn from [43]

background coming from the long-decay cross-talk of donor signal through acceptor filter. This is applicable in TR-FRET assays where the energy transfer efficiency is high and the FRET signal has a short decay time so that most of the ET-sensitized acceptor signal fits into a narrow measuring window.

Window-optimized measurement is accomplished in a dual-label hybridization with dual-label (donor–quencher) probes composed of either europium or terbium donors and fluorescent or nonfluorescent acceptors. In unhybridized free form, flexibility of the probe allows very efficient energy transfer resulting in very short-lived sensitized signal. Upon hybridization, the energy transfer is decreased and decay is longer [43]. Figure 5 shows the assay principle and the respective decays in an assay.

5 Quench Interference and Correction in Assays

In the assays of biological samples, and in HTS with micromolar concentrations of library compounds of variable source, the signal interference is frequently encountered regardless of the time-resolved detection. Two of the main interferences taking place in TR-FRET applications are related to the direct compound-related quenching of donor (here referred as chemical quenching) and the compound absorbance-related signal attenuation. In the chemical quenching, interfering compounds annul the excited state by diffusion, collision, or ion-pair mechanisms. The quenching results in shortening of the decay times, and the sensitized emission is decreased and shortened at the same rate.

The second relatively common interference is related to colored compounds having interfering absorbance (color quenching). This is a problem mainly at higher volumes with relatively long optical path length (96-well plates) and at higher concentrations. The most common attenuation takes place at UV light through decreased chelate excitation. In case the absorbance extends to visible range, it potentially diminishes donor emission measurement at a different rate than acceptor emission causing differential quenching. On the contrary to chemical quenching, color quenching does not affect the decay times of the donor nor of the acceptor.

To avoid excessive false positives in a screening, a suitable quench correction is generally needed. The simplest model is to use a ratio-fluorometry, traditionally applied in fluorescence assays [18, 49]. The actual ratio, the pure ET-sensitized acceptor emission divided by the bulk of donor emission, is a good measure of biocomplex formation, unless the free ligand and the complexes are differentially quenched. Differential quenching is encountered in cases where a small molecular ligand with high diffusion rate constant binds to a large membrane fragment or a whole cell having much slower diffusion rate. In such a condition, the free ligand is quenched at a higher rate than the complex, and the ratio tends to overcompensate the correction.

Due to frequent compound interference in high-throughput screening, a recommendation has been made for TR-FRET to record donor emission, acceptor emission, choice of appropriate donor acceptor pair, and optimize the delay times [50]. In addition, collecting all the available data, including absorbances at excitation and emission wavelengths, the whole decay profiles of donor emission and acceptor emission in addition to their intensities could provide tools to improve the analytical quality of the screening.

6 TR-FRET Applications in Cellular Signaling

In drug discovery, all the biochemical reactions starting from receptor–ligand interaction to protein expression and posttranslational modifications are targets for potential medical intervention. TR-FRET has become one of the major assay technologies in high-throughput screening [19]. As a real homogeneous assay, TR-FRET fits very well to assay miniaturization and automation [51] used, e.g., in kinase screening in 1,536-well plates, and even 8,640 separate assays have been screened within the same footprint using TR-FRET in a gel-based reagent spot array plate (named μArc system) at Abbott [52, 53].

For screening of receptor–ligand interactions, TR-FRET has mainly be used by applying lanthanide chelate-labeled ligands, agonists, or antagonists, and binding is followed and quantitated by suitable secondary carrier coupled with acceptor, such as antireceptor antibody, or antitag antibody with recombinant receptors or lectins binding to extracellular surfaces. For example, interleukin-2 receptor–ligand binding assay has been reported with europium-labeled IL-2 and Cy-5-labeled antibody [54] and lipoprotein LDL receptor with europium-labeled ligand (PSK9) and Alexa 647-labeled antibody [55]. Hu et al. used hemagglutinin as a tag in the expression of human complement 5a receptor assayed with terbium-labeled ligand (C5a) and Alexa 488-labeled antihemagglutinin antibody [56].

In addition to phosphorylation, the activity of receptors is controlled by the formation of homo- or heteromers. TR-FRET is a perfect tool to follow receptor activation on the membranes exemplified by opioid receptor [57], dopamine receptor [58, 59], histamine receptor [60], adrenoceptors [60, 61], IL-13 [62, 63], chemokine receptors [64, 65], and receptors related to EBV infection [66]. Most commonly, the assays have been accomplished with receptors coexpressed with suitable tags, such as c-myc, polyhistidine, glutatione S transferase (GST) or FLAG, and using europium- and APC-labeled antibodies.

Activation of G-protein-coupled receptors is followed by a number of secondary signaling reactions. G_i- and G_s-coupled receptor activations can be monitored by increase or decrease, respectively, of cyclic AMP (cAMP) [67], and accordingly, TR-FRET immunoassay of cAMP has become a common tool for screening of GPCR targets [19, 68–70] (Fig. 6). G_q coupled receptor activation is generally followed by calcium-flux assays, but it can also be monitored by myo-inositol-1,4,5-triphosphate (IP3) formation. Trinquet et al. developed a surrogate assay for IP3 by measuring the more stable end product, IP1 by a TR-FRET immunoassay [71, 72].

Kinases form today one of the most used drug target group [73]. Both tyrosine kinases, as well as threonine/serine kinases, have been used as targets in search for activity modulators. Commonly potential kinase inhibitors are searched by

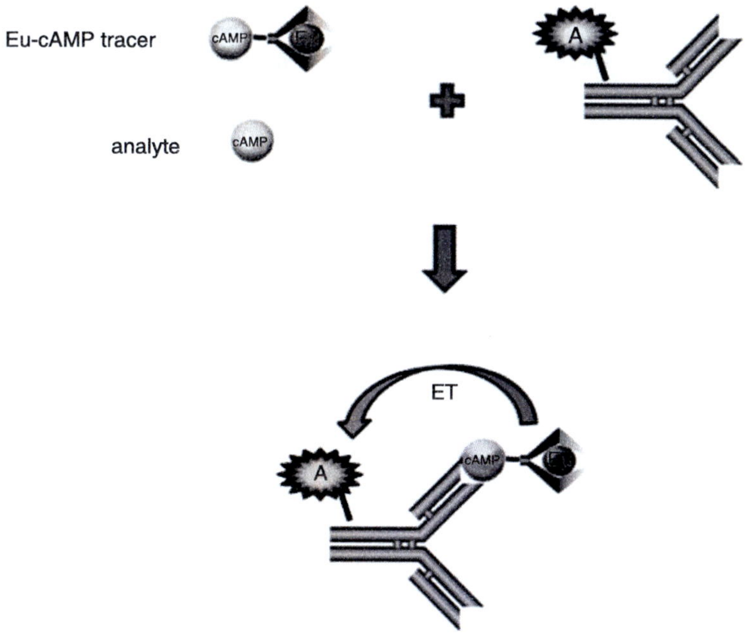

Fig. 6 A typical competitive immunoassay of a small antigen (cAMP) with TR-FRET. Sample cAMP competes in the assay with europium-labeled tracer for limited amount of anti-cMAP antibody. In the absence of sample cAMP, maximum TR-FRET signal is produced

using peptide substrates coupled with acceptor or biotin with combination of antiphosphor-specific antibodies labeled with lanthanide chelate donors [19, 73, 74] (Fig. 7). Hundreds of different kinases have been screened and reported by TR-FRET. Kinase assay applications vary from membrane-bound phosphoinositide-3-kinase (PI3K) done with europium [75] or terbium [76], to MAP kinase family [77–80], to Tie and AbI kinases [81–84] and VEGFR autophosphorylation [85–88], just to give some examples. Transcreener® assay of BellBrooks measures kinase activities by quantitation of created ADP [89], and the Invitrogens kinase-binding assays and enzymes are measured as ATP-binding proteins using ATP competing tracer and kinase proteins expressed with suitable tags [90].

Nuclear receptor activation in nucleus is the following step in signaling cascade before protein expression. TR-FRET is frequently applied in nuclear receptor ligand bindings and coactivator- or corepressor-recruitment assays, reported, for example, for peroxisome proliferator-activated receptors (PPAR) [91–94], estrogen receptors [29, 95], and others [96–101]. Jeyakumar et al. measured simultaneously [47] and sequentially [102] the activation of thyroid and steroid receptors using terbium-chelate-labeled response element sequence, with fluorescein labeled coactivator (Src3) and fluorescein or cyanine 5-labeled corepressor (NCoR). After

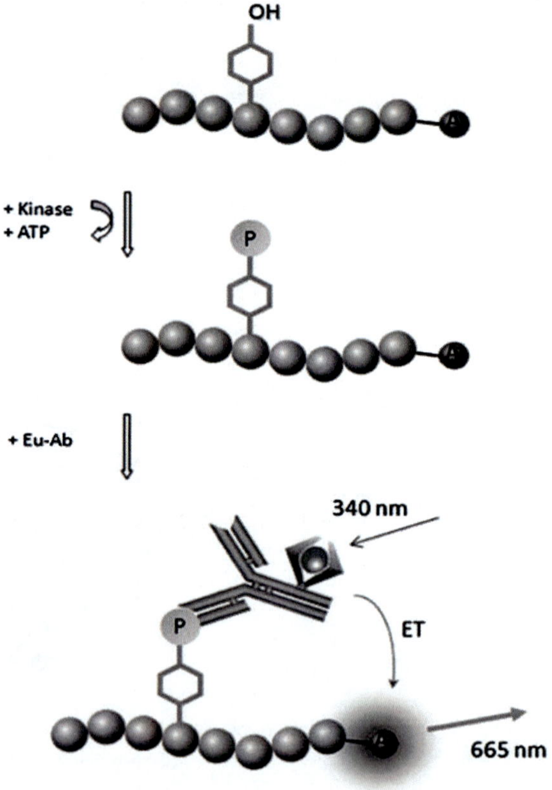

Fig. 7 A scheme of an assay of tyrosine kinase activity measurement as done in kinase inhibitor screening with TR-FRET. Acceptor-labeled peptide substrate is phosphorylated by the kinase under investigation, and the resulting phosphorylated peptide is quantitated with europium-labeled antiphospho-specific antibody

protein translation, there are several cellular machineries to modify the activities or to target the proteins. For example, ubiquitination of various proteins, such as NFκB, TRAF, and p53, have been followed by TR-FRET technologies [103–106].

7 Conclusions

Lanthanide chelates due to their exceptional excited state lifetime have become popular tools in developing sensitive assay technologies, for example, for diagnostics. In addition, the combination of long decay time to narrow banded emission lines provides a clear opportunity to make second generation of FRET assays where both robustness, through time-resolved detection and the sensitivity, through signal discrimination by temporal and spectral combinations, are applied. Today, lanthanide-based TR-FRET assays are one of the leading technologies in high-throughput screening, in search for novel drug targets.

An additional advantage, the feasibility to limited multiplexing, is not yet applied very much. Taking into account the spectral features (different emission lines), temporal features, and the varying energy levels that can be applied, there is a huge unused potential in lanthanide chelates to make robust and easy multiplexed assays, e.g., for single nucleotide polymorphic assays.

The current development is mostly concentrated in finding good, stable bright donor chelates with high energy transfer rate constant and low background. The alternative approach has been to find chelate clusters, or nanobeads, to further improve the signal strength. It remains to be seen whether those beads, latex beads, microcrystals, upconverted crystals, or quantum dots are able to provide clear advantages for sensitized assays in the future.

References

1. Bünzli J-CG, Chopin GR (eds) (1989) Lanthanide probes in life, chemical and earth sciences. Theory and practice. Elsevier, Amsterdam
2. Horrocks WDeW Jr, Sudnick DR (1979) J Am Chem Soc 334–340
3. Supkowski RM, De HW, Jr W (2002) Inorg Chim Acta 340:44–48
4. Buono-Cuore GE, Li H, Arciniak B (1990) Coord Chem Rev 99:55–87
5. Beeby A, Clarkson IM, Dickins RS, Faulkner S, Parker D, Royle L, de Sousa AS (1999) J Chem Soc Perkin Trans 2:493–502
6. Sabbatini N (1987) J Phys Chem 91:4681–4685
7. Dexter DL (1953) J Chem Phys 21:836–850
8. Matko J, Jenei A, Wei T, Edidin M (1992) Cytometry 19:191–200
9. Sabbatini N, Perathoner S, Lattanzi G, Dellonte S, Malzani V (1988) Inorg Chem 27:1628–1633
10. Pérez-Luna VH, Yang S, Rabinovich EM, Buranda T, Sklar LA, Hampton PD, Lopez GP (2002) Biosens Bioelectron 17:71–78
11. Dadabhoy A, Faulkner S, Sammes PG (2000) J Chem Soc Perkin Trans 2:2359–2360
12. Silva FRG, Malta OL (1997) J Alloys Compd 250:427–430

13. Latva M, Takalo H, Mukkala V-M, Matachescu C, Rodrígues-Ubis JC, Kankare J (1997) J Lumin 75:149–169
14. Weissman SI (1942) J Chem Phys 10:214–217
15. Stryer L, Thomas DD, Meares CF (1982) Annu Rev Biophys Bioeng 11:203–222 (tai: Stryer L (1978) Ann Rev Biochem 47: 819; as a spectroscopic ruler)
16. Ullman EF, Schwarzberg M, Rubenstein KE (1976) J Biol Chem 251:4172
17. Soini E, Hemmilä I (1979) Clin Chem 25:353
18. Morrison LE (1988) Anal Biochem 174:101
19. Comley J (2006) Drug Discov World Spring:22
20. Hemmilä I, Mukkala V-M (2001) Crit Rev Clin Lab Sci 38:441
21. Hemmilä I, Laitala V (2005) J Fluoresc 15:529
22. Laitala V, Hemmilä I (2005) Anal Chem 77:1483
23. Ollikka P, Raussi H-M, Laitala V, Jaakkola L, Hovinen J, Hemmilä I, Ylikoski A (2009) Anal Biochem 386:20
24. Mathis G (1993) Clin Chem 39:1953
25. Samuel APS, Moore EG, Melchior M, Xu J, Raymond KN (2008) Inorg Chem 47:7535
26. Li M, Selvin PR (1995) J Am Chem Soc 117:8132
27. Selvin PR, Hearst JE (1994) Proc Natl Acad Sci USA 91:10024
28. Petoud S, Cohen SM, Bünzli J-CG, Raymond KN (2003) J Am Chem Soc 125:13324
29. Gowda K, Marks BD, Zielinski TK, Ozers MS (2006) Anal Biochem 357:105
30. Wang G, Yuan J, Matsumoto K, Hu Z (2001) Talanta 55:1119
31. Riddle SM, Vedvik KL, Hanson GT, Vogel KW (2006) Anal Biochem 356:108
32. Chen JC, Selvin PR (2000) J Am Chem Soc 122:657
33. Laitala V, Hemmilä I (2007) Anal Chim Acta 551:73
34. Kokko L, Sandberg K, Lövgren T, Soukka T (2004) Anal Chim Acta 503:155
35. Kokko T, Kokko L, Lövgren T, Soukka T (2007) Anal Chem 79:5935
36. Valanne A, Lindroos H, Lövgren T, Soukka T (2005) Anal Chim Acta 539:251
37. Sharlow ER, Leimgruber S, Yellow-Duke A, Barrett R, Wang QJ, Lazo JS (2008) Nat Protoc 3:1350
38. Ullman EF, Kirakossian H, Switchenko AC, Ishkanian J, Ericson M, Wartchow CA, Pirio M, Peace J, Irvin BR, Singh S, Singh R, Patel R, Dafforn A, Davalian D, Skold C, Kurn N, Wagner DB (1996) Clin Chem 42:1518
39. Hildebrandt N, Charbonniére LJ, Beck M, Ziessel RF, Löhmannsröben H-g (2005) Angew Chem Int Ed 44:7612
40. Hildebrandt N, Charbonniére LJ, Ziessel RF, Löhmannsröben H-G (2007) Proc SPIE 6448:K1
41. Härmä H, Soukka T, Shavel A, Gaponik N, Weller H (2007) Anal Chim Acta 604:177
42. Ylikoski A, Elomaa A, Ollikka P, Hakala H, Mukkala V-M, Hovinen J, Hemmilä I (2004) Clin Chem 50:1943
43. Laitala V, Ylikoski A, Raussi H-M, Ollikka P, Hemmilä I (2007) Anal Biochem 361:126
44. Karvinen J, Elomaa A, Mäkinen M-L, Hakala H, Mukkala VM, Peuralahti J, Hurskainen P, Hovinen J, Hemmilä I (2004) Anal Biochem 325:317
45. Kokko T, Liljenback T, Peltola MT, Kokko L, Soukka T (2008) Anal Biochem 80:9763
46. Kupcho KR, Stafslien DK, DeRosier T, Hallis TM, Ozers MS, Vogel KW (2007) J Am Chem Soc 129:13372
47. Jeyakumar M, Katzenellenbogen JA (2009) Anal Biochem 386:73
48. Johansson MK, Cook RM, Xu J, Raymond KN (2004) J Am Chem Soc 126:16451
49. Lim CS, Miller JN, Bridges JW (1980) Anal Biochem 108:176
50. Imbert P-E, Unterreiner V, Siebert D, Gubler H, Parker C, Gabriel D (2007) Assay Drug Discov Technol 5:363
51. Klumpp M, Boethcer A, Becker D, Meder G, Blank J, Leder L, Forstner M, Ottl J, Mayr LM (2006) J Biomol Screen 11:617
52. Freiberg G, Wilkins J, David C, Kofron J, Hirst GC, Burns DJ, Warrior U (2004) J Biomol Screen 9:12

53. Gobalakrishnan SM, Karvinen J, Kofron JL, Burns DJ, Warrior U (2002) J Biomol Screen 7:317
54. Stenroos K, Hurskainen P, Eriksson S, Hemmilä I, Blomberg K, Lindqvist C (1998) Cytokine 10:495
55. Pandit S, Wisniewski D, Santoro JC, Ha S, Ramakrishnan V, Cubbon RM, Cummings RT, Wright SD, Sparrow CP, Sitlani A, Fisher TS (2008) J Lipid Res 49:1333
56. Hu LA, Zhou T, Hamman BD, Liu Q (2008) Assay Drug Dev Technol 6:543
57. McVey M, Ramsay D, Kellett E, Rees S, Wilson S, Pope AJ, Milligan G (2001) J Biol Chem 276:14092
58. Gazi L, Lopez-Gimenez JF, Rudiger MP, Strange PG (2003) Eur J Biochem 270:3928
59. So CH, Varghese G, Curley KJ, Kong MMC, Alijaniaram M, Ji X, Nguyen T, O'Dowd BF, George SR (2005) Mol Pharmacol 68:568
60. Backer RA, Dees G, Carillo JJ, Booth RG, López-Gimenez JF, Milligan G, Leurs R (2004) J Pharmocol Exp Ther 311:131
61. Ramsay D, Carr IC, Pediani J, López-Gimenez JF, Thurlow R, Fidock M, Milligan G (2004) Mol Pharmacol 66:228
62. Yang X, Capotosto R, DiBlasio E, Hu Z, Kriz R, Lorenzo M, Malakian K, Wolfrom S, Wilhelm J, Wolf SF (2006) Anal Biochem 351:158
63. Yang X, Lee J, Brooks J, Wilhelm J, Myszka D, Kasaian MT, Goldman S, Wolf S, Fitz LJ (2008) Anal Biochem 367:206
64. Wilson S, Wilkinson G, Milligan G (2005) J Biol Chem 280:28663
65. Parenty G, Appelbe S, Milligan G (2008) Biochem J 412:245
66. Mehle A, Wildon H, Zhang C, Brazier AJ, McPike M, Pery E, Gabuzda D (2007) J Virol 81:13235
67. Williams C (2004) Nat Rev Drug Discov 3:125
68. Hamilton BS, Doods HN (2008) Eur J Pharmacol 580:55
69. Wang L, Henrich M, Buckler KJ, McMenamin M, Mee CJ, Sattelle DB, Paterson DJ (2007) J Mol Cell Cardiol 43:717
70. Von Bülow V, Dubben E, Engelhardt G, Hebel S, Plümäkers B, Heine J, Rink L, Haase H (2007) J Immunol 179:4180
71. Trinquet E, Fink M, Bazin H, Grillet F, Maurin F, Bourrier E, Ansanay H, Leroy C, Michaud A, Durroux T, Maurel D, Malhaire F, Goudet C, Pin JP, Naval M, Hernout O, Chrétien F, Chapleur Y, Mathis G (2006) Anal Biochem 358:126
72. Bergsdorf C, Kropp-Goerkis C, Kaehler I, Ketscher L, Boemer U, Parczyk K, Bader B (2008) Assay Drug Dev Technol 6:39
73. Eglen RM (2008) Drug Discov World Summer:21
74. Ma H, Deacon S, Horiuchi K (2008) Expert Opin Drug Discov 3:607
75. Gray A, Olsson H, Batty IH, Priganica L, Downes CP (2003) Anal Biochem 313:234
76. Neilsen P, Ferguson C, Thompson J, Branch AM, Miller A (2007) Gen Eng News Aug 30
77. Han S, Zhou V, Pan S, Liu Y, Hornsby M, McMullan D, Klcok HE, Haugen J, Lesley SA, Gray N, Galdwell J, Gu X-J (2005) Bioorg Med Chem Lett 15:5467
78. Ross S, Chen T, Yu V, Tudor Y, Zhang D, Liu L, Tamayo N, Domiguez C, Powers D (2006) Assay Drug Devl Technol 4:397–409
79. Yeh TC, Marsh V, Bernat BA, Ballard J, Colwell H, Evans RJ, Parry J, Smith D, Brandhuber BJ, Gross S, Marlow A, Hurley B, Lyssikatos J, Lee PA, Winkler JD, Koch K, Wallace E (2007) Clin Cancer Res 13:1576
80. Hall JP, Kurdu Y, Hsu S, Cuozzo J, Liu J, Telliez J-B, Seidl KJ, Winkler A, Hu Y, Green N, Akew GR, Tam S, Clark JD, Lin L-L (2007) J Biol Chem 282:33295
81. Deng S-J, Liu W, Simmons CA, Moore JT, Tian G (2001) Cob Chem High Throughput Screen 1:525
82. Lin TH, Metzger A, Diller DJ, Desai M, Henderson I, Ahmed G, Kimble EF, Quadros E, Webb ML (2006) J Pharmacol Exp Ther 318:495

83. Zhou T, Parillon L, Li F, Wang Y, Keats J, Lamore S, Xu Q, Shakespeare W, Dalgarno D, Zhu X (2007) Chem Biol Drug Discov 70:171
84. Miyazaki Y, Nakano M, Sato H, Truesdale AT, Stuart JD, Nartey EN, Hightower KE, Kane-Carlson L (2007) Bioorg Med Chem Lett 17:250
85. Patel N, Sun L, Moshinsky D, Chen H, Leahy KM, Le P, Moss KG, Wang X, Rice A, Tam D, Laird D, Yu X, Zhang Q, Tang C, McMahon G, Howlett A (2003) J Pharm Exp Ther 306:838
86. Conway JG, McDonald B, Parham J, Keith B, Rusnak DW, Shaw E, Jansen M, Lin P, Payne A, Crosby RM, Johnson JH, Frick L, Lin M-HJ, Sepee S, Tadepalli S, Votta B, James I, Fuller K, Chambers TJ, Kull FC, Chamberlain SD, Hutchings JT (2005) Proc Natl Acad Sci USA 102:16078
87. Alberti MJ, Auten EP, Lackey KE, McDonald OB, Wood ER, Preugschat F, Cutler GJ, Kane-Carlson L, Liu W, Jung DK (2005) Bioorg Med Chem Lett 15:3778
88. Kumar R, Knick VB, Rudolph SK, Johnson JH, Crosby RM, Crouthamel M-C, Hopper TM, Miller CG, Harrington LE, Onori JA, Mullin JA, Gilmer TM, Truesdale AT, Epperly AH, Boloor A, Stafford JA, Luttrell DK, Cheung M (2007) Mol Cancer Ther 6:2012
89. Klink TA, Kleman-Leyer KM, Kopp A, Westermeyer TA, Lowery RG (2008) J Biomol Screen 13:476
90. Riddle SM (2009) Gen Eng News Feb: 28
91. Schearer BG, Steger DJ, Way JM, Stanley TB, Lobe DC, Grillot DA, Iannoone MA, Lazar MA, Wilson TM, Billin AN (2008) Mol Endocrinol 22:523
92. Pochetti G, Godio C, Mitro N, Caruso D, Galmozzi A, Scurati S, Loiodice F, Fracchiolla G, Tortorella P, Laghezza A, Lacecchia A, Novellino E, Mazza F, Crestani M (2007) J Biol Chem 282:17314
93. Burgermeister E, Schnoebelen A, Flament A, Benz J, Stihle M, Gsell B, Rufer A, Ruf A, Kuhn B, Märki HP, Mizrahi J, Sebokova E, Niesor E, Meyer M (2006) Mol Endocrinol 20:809
94. Fujimore T, Sakuma H, Ohkubo-Suzuki A, Aramori I, Mutoh S (2006) Biol Pharm Bull 29:423
95. Liu K, Knappenberger KS, Käck H, Andersson G, Nilsson E, Darthsch C, Scott C (2003) Mol Endocrinol 17:346
96. Abad MC, Askari H, O'Neill J, Klinger AL, Milligan C, Lewandowski F, Springer B, Spurlino J, Rentzeperis D (2008) J Steroid Biochem Mol Biol 108:44
97. Gunther JR, Du Y, Rhoden E, Lewis I, Revennaugh B, Moore TW, Kim SH, Dingledine R, Fu H, Katzenellenbogen JA (2009) J Biomol Screen 14:181
98. Kassam A, Miao B, Uoung PR, Mukherjee R (2003) J Biol Chem 278:10028
99. Hultman ML, Krsnoperova NV, Li S, Du S, Xia C, Dietz JD, Lala DS, Welsch DJ, Hu X (2005) Mol Endocrinol 19:1460
100. Albers M, Blume B, Schlueter T, Wright MB, Kober I, Kremoser C, Deushle U, Koegl M (2006) J Biol Chem 281:4920
101. Urizar NL, Liverman AB, Dodds DT, Silva FV, Ordentlich P, Yan Y, Gonzáles FJ, Heyman RA, Mangelsdorf DJ, Moore DD (2002) Science 296:1703
102. Jeyakumar M, Webb P, Baxter JD, Scanlan TS, Katzenellenbogen JÁ (2008) Biochemistry 47:7465
103. Boisclair MD, McClure C, Josiah S, Glass S, Bottomley S, Kamerkar S, Hemmilä I (2000) J Biomol Screen 5:319
104. Xu K, Belunis C, Chu W, Weber D, Podlaski F, Huang K-S, Reed SI, Vassilev LT (2003) Biochem J 371:957
105. Hong CA, Swearinger E, Mallari R, Gao X, Cao Z, North A, Young SW, Huang S-G (2003) Assay Drug Discov Techn 1:175
106. Murray MF, Jurewicz AJ, Martin JD, Ho TF, Zhang H, Johansson KO, Kirkpatrick RB, Ma J, Lor LA, Thrall SH, Schwartz B (2007) J Biomol Screen 12:1050
107. Bünzli G J-C (2010) Basics of lanthanide photophysics. Springer Ser Fluoresc doi: 10.1007/4243_2010_3

Index

P. Hänninen and H. Härmä (eds.), *Lanthanide Luminescence: Photophysical, Analytical and Biological Aspects*, Springer Ser Fluoresc (2011) 7: 381–386, DOI 10.1007/4243_2011,